GLENCOE

PHYSICS

PRINCIPLES & PROBLEMS

Mc
Graw
Hill
Education

COVER: Spacedreamer/Shutterstock.com

mheducation.com/prek-12

McGraw Hill Education

Send all inquiries to:
McGraw-Hill Education
STEM Science
8787 Orion Place
Columbus, OH 43240

ISBN: 978-0-07-677478-4
MHID: 0-07-677478-3

Printed in the United States of America.

2 3 4 5 6 7 8 9 RMN 20 19 18 17 16

TABLE OF CONTENTS

TABLE OF CONTENTS

Mechanics

Energy

Waves and Light

Electricity and Magnetism

TABLE OF CONTENTS

Electricity and Magnetism

Subatomic Physics

Quick Start Guide
Glencoe Physics | Teacher Center

Login information

1. Go to **connected.mcgraw-hill.com**
2. Enter your registered Username and Password.
3. For **new users** click here to create a new account
4. Get **ConnectED Help** for creating accounts, verifying master codes, set up classes and more.

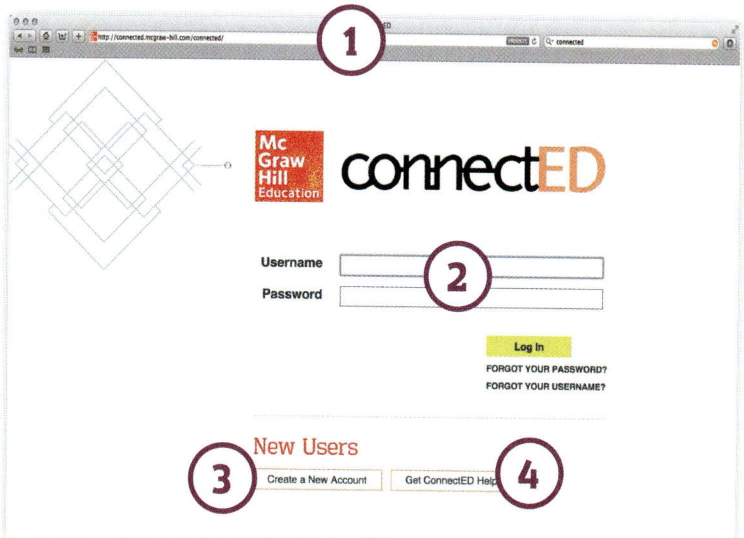

Your ConnectED Center

5. Scroll down to find the program from which you would like to work.

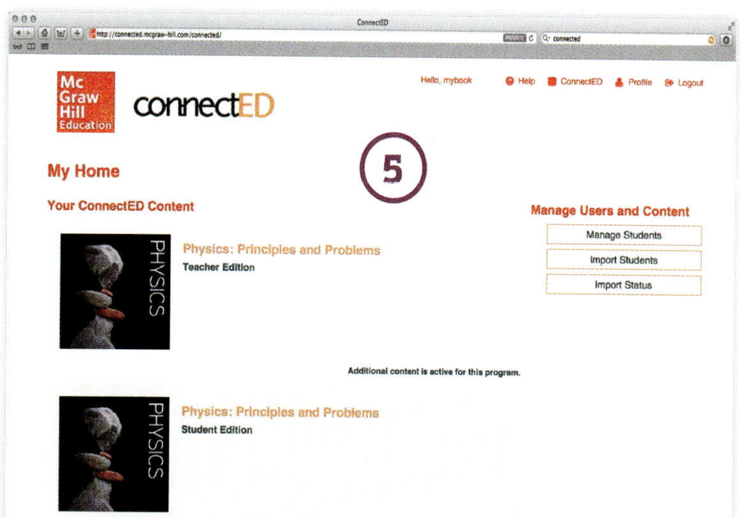

Quick Start Guide
Glencoe Physics | Teacher Center

1. The Menu allows you to easily jump to anywhere you need to be.

2. Click the **program icon** at the top left to **return to the main page** from any screen.

3. **Select a Chapter and Lesson** Use the drop down boxes to quickly jump to any lesson in any chapter.

4. Return to your **My Home** page for all your **ConnectED** content.

5. The **Help** icon will guide you to online help. It will also allow for a quick logout.

6. **Search Bar** allows you to search content by topic or standard.

7. **Access the eBook** **Use** the **Teacher Edition** to see content and suggested answers.

8 Quickly access helpful links to **multiple resources**, such as:

- LearnSmart
- Enrichment, Reinforcement, Transparency
- Animations, Videos
- Self-check, Quizzes, Tests
- Project-Based Learning Activities
- Lab Manuals, Safety Videos, Virtual Labs & other tools
- Vocabulary, Multilingual eGlossary, Vocab eGames, Vocab eFlashcard
- Personal Tutors
- Classroom Presentation Toolkit with Powerpoints
- Science Notebook
- Physics Math Handbook, Science and Engineering Practices Handbook
- Professional Development

9 **Daily Spotlight** is your preview of that day's lesson plans and assignments for each class.

connected.mcgraw-hill.com

Quick Start Guide
Glencoe Physics | Teacher Center

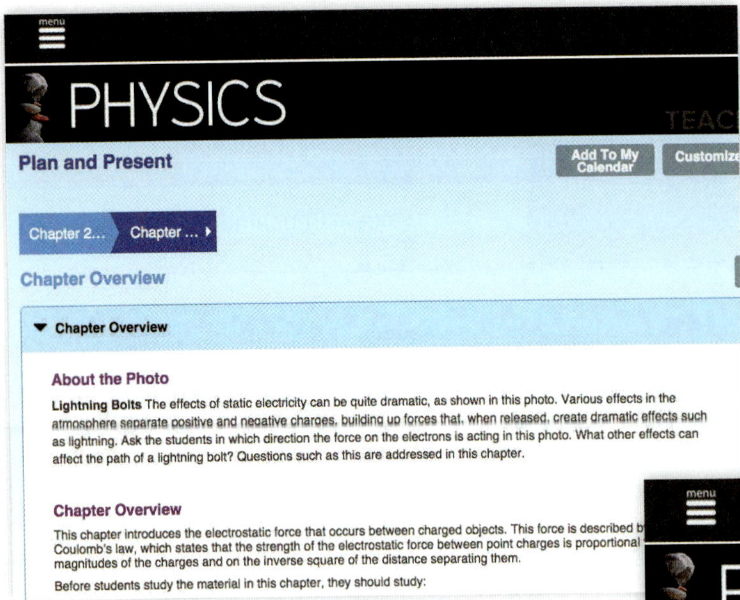

Use Plan and Present to:

- Review teacher instructional content
- Access ready-to-use lessons
- Customize pre-built lessons with your own resources

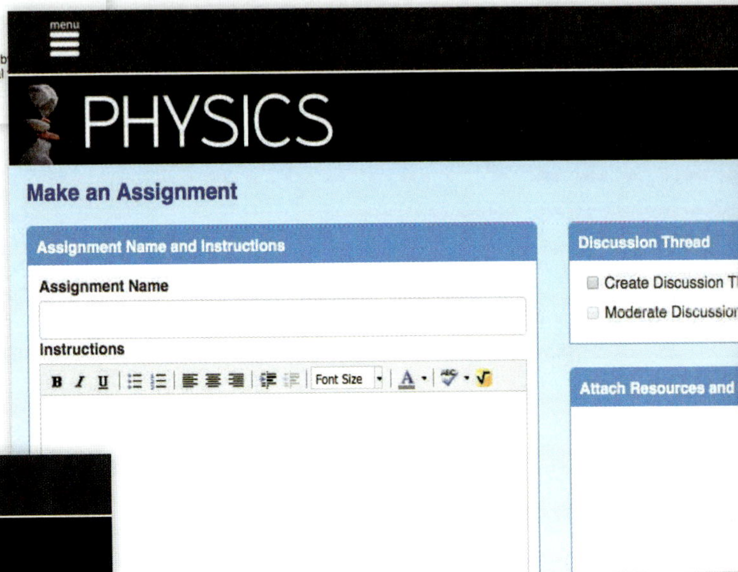

Use Assignment Tracker to:

- View recently created assignments
- Assign homework and activities
- Send and reply to student messages

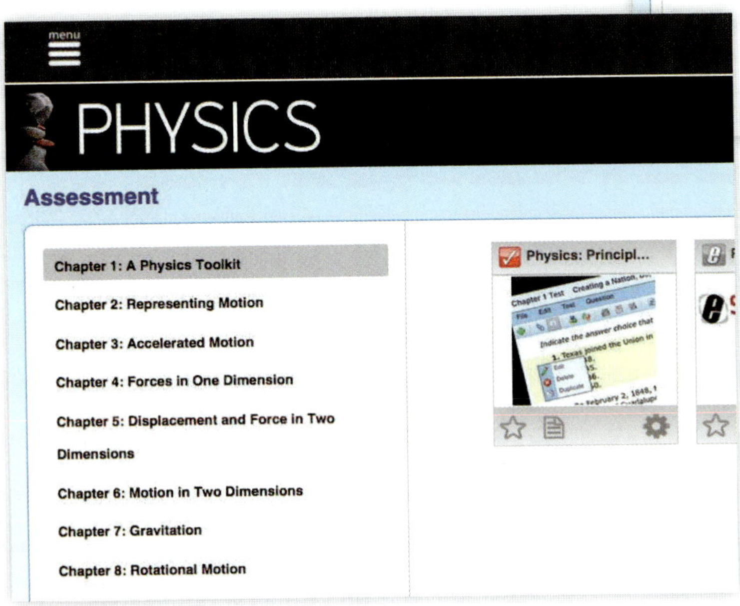

Use Assessment to:

- Access test banks to tailor assessments
- Create customized worksheets to assign online or print and distribute
- Prepare students for assessments with online testing

Mc Graw Hill Education

connected.mcgraw-hill.com

ABOUT THE AUTHORS

The authors of *Physics: Principles and Problems* used their physics content knowledge and teaching expertise to craft manuscript that is accessible, accurate, and geared toward student achievement.

Paul W. Zitzewitz, lead author

was an emeritus professor of physics and of science education at the University of Michigan–Dearborn. He was awarded posthumously a Carleton College 2014 Alumni Association Distinguished Achievement Award. He received his BA from Carleton College and his MA and PhD from Harvard University, all in physics. Dr. Zitzewitz taught physics to undergraduates at the University of Michigan-Dearborn for 36 years and, as an active experimenter in the field of atomic physics, published more than 50 research papers. He was named a Fellow of the American Physical Society for his contributions to physics and science education for high school and middle school teachers and students. He is now treasurer of the American Association of Physics Teachers and has been the president of the Michigan Section of the American Association of Physics Teachers and chair of the American Physical Society's Forum on Education.

David G. Haase

is an Alumni Distinguished Undergraduate Professor of Physics at North Carolina State University. He earned a BA in physics and mathematics at Rice University and an MA and a PhD in physics at Duke University where he was a J.B. Duke Fellow. He has been an active researcher in experimental low-temperature and nuclear physics. He teaches undergraduate and graduate physics courses and has worked many years in K-12 teacher training. He was the founding director of The Science House at NC State, a science-math learning center that leads teacher training and student programs across North Carolina. He has co-authored over 100 papers in experimental physics and in science education. He is a Fellow of the American Physical Society. He received the Alexander Holladay Medal for Excellence, NC State University; was awarded the Pegram Medal for Physics Teaching Excellence; and was chosen 1990 Professor of the Year in the state of North Carolina by the Council for the Advancement and Support of Education (CASE).

Kathleen A. Harper

is a senior lecturer in the Engineering Education Innovation Center at The Ohio State University. She received her MS in physics and BS in electrical engineering and applied physics from Case Western Reserve University, and her PhD in physics from The Ohio State University. She has taught introductory physics, astronomy, and engineering courses to undergraduates for over 20 years and has been instrumental in offering Modeling Instruction workshops to in-service high school teachers, both in Ohio and nationwide. Her research interests include the teaching and learning of problem-solving skills and the development of alternative problem formats. She is active in AAPT, on both the local and national levels, often presenting talks and workshops on teaching problem solving. Additionally, she is co-editor of a selection of articles available through AAPT's Compadre portal entitled, "Getting Started in Physics Education Research."

BE THE SCIENTIST! BE THE ENGINEER!

ConnectED is your one-stop online resource to give students real-world challenges that deepen understanding of core ideas and cross-cutting concepts!

- Physics ebook
- Science and Engineering Practices Handbook
- Applying Practices activities
- PBLs
- Design-your-own Labs
- Guided investigations

Get the LearnSmart© advantage! Improve performance by using this interactive and adaptive study tool. Personalized learning paths will help students practice and master key physics concepts.

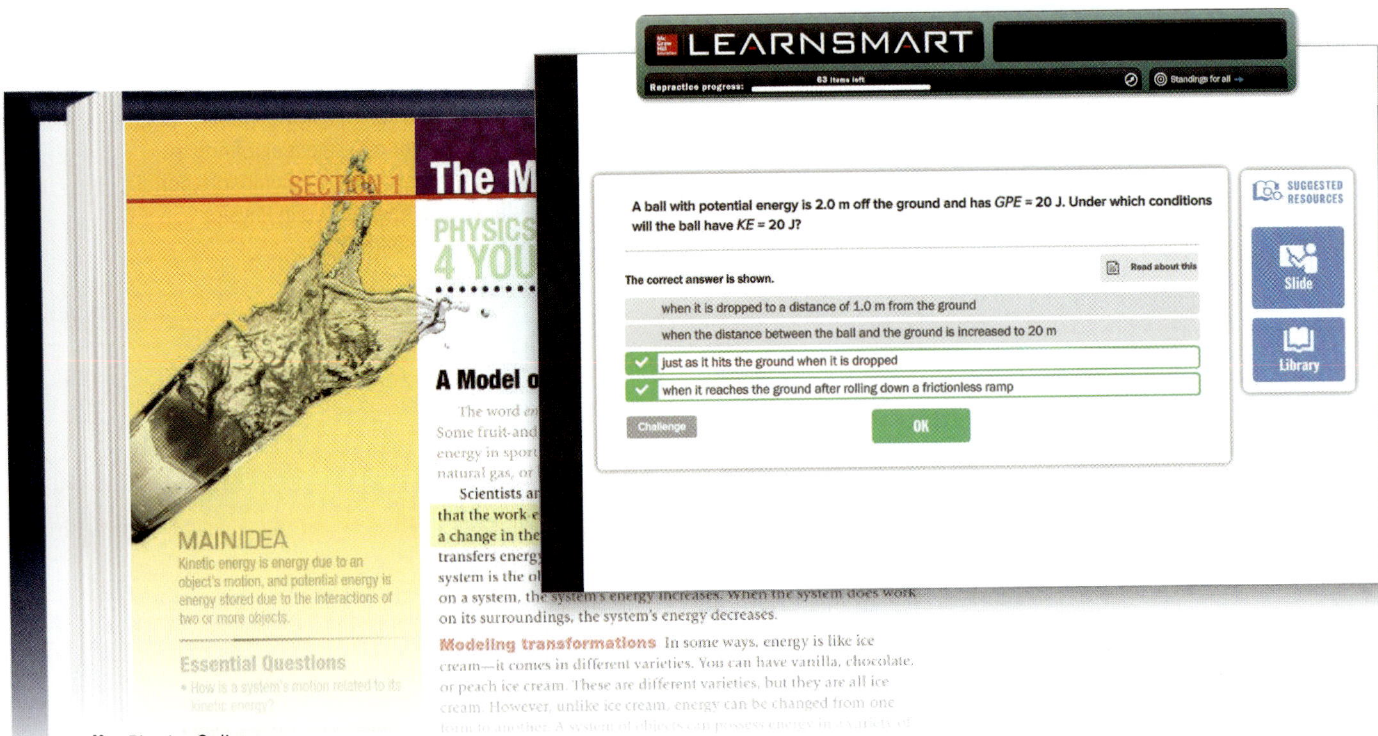

MORE for You Online!

All of the essential content at hand in *Physics: Principles and Problems Teacher Essentials* is also available online. Additionally, the ConnectED Teacher Center serves as a comprehensive resource for numerous detailed references. Online you'll find extended solutions, interactivities, videos, and more!

McGraw Hill Education connectED

Make content work for you—

- Customizable labs and activity worksheets
- Lesson presentations
- McGraw-Hill eAssessment
- Planner

Bring ideas to life—

- Animations
- Simulations
- Virtual Investigations
- Physics TV videos
- Personal Tutors
- Games

Find math support when you need it—

- Online Solutions Manual
- Math Handbook

Enrich your teaching strategies—

- Professional development videos
- Multilingual eGlossary
- And more!

MORE for Students Online!

The ConnectED Student Center let you provide further instruction and practice outside the classroom.

- **Online Problems, Help, and Investigations**
  Example Problems and Practice Problems throughout each chapter give students step-by-step instructions and hints on how to solve different types of physics problems.
- **Online Study Tools**
  LearnSmart, quizzes, assessments, and study tools provide students with opportunities for self-assessment, review, and additional practice.

Digital Vision/Getty Images

How to Use the *Physics: Principles and Problems Teacher Essentials*

- Everything you need in a logical, efficient design
- Easy-to-use organization
- Find what you need, when you need it

Within the chapter, each section is organized by the Student Edition topic.

Corresponding Student Edition page numbers for each topic are clear and easy to find.

Additional example problems at point of use allow you to provide timely in-class practice.

SECTION 1 Periodic Motion

student text pp. 382–387

1 Introduce

Bellringer

Periodic Motions Have students observe the periodic motions of various objects, such as swinging pendulums, a plucked string, as well as a metal ruler or meterstick with one end clamped to a tabletop and set in motion by releasing the opposite end from an up or down displacement. Help students generate and discuss questions such as: What is necessary to start the motion? What maintains the motion? What eventually happens to the motion? Is the motion regular in displacement and/or time? BL Visual-Spatial EL

Tie to Prior Knowledge

x, v, a,* and *F Have students review concepts of displacement, velocity, acceleration, and force introduced in previous chapters.

2 Teach

Mass on a Spring and Hooke's Law

student text pp. 382–385

Use Figure 2

Force Magnitude-Stretch Distance Graphs Have students look at **Figure 2.** Have students recall that the area under the curve of a force versus displacement (stretch distance on the *x*-axis) graph represents work. Review the connection for students:

The area of a triangle is $A = \frac{1}{2}bh$.

Thus, $W = \frac{1}{2}x_1F_1 = \frac{1}{2}x_1kx_1$

$W = \frac{1}{2}kx_1^2$

Use Figure 3

MAINIDEA Have students look at the bar graphs in **Figure 3.** Be sure students recognize that the force exerted on the object by the spring is not the same magnitude (bars) and direction (vectors) throughout the compression-stretching cycle. Point out that if this were so, the bar graphs in **Figure 3** would show *PE* and *KE* the same in each phase. For example, point out that the bar graphs for $t = 0.2$ s and $t = 0.4$ s show how *PE* and *KE* change as the spring is compressed from its equilibrium position (the vertical dotted line) to the $v = 0$ position. Also, show that the force acts in the opposite direction of the object's motion (as shown by the F_{sp} vector) as the coils of the spring are compressed.

ADDITIONAL IN-CLASS EXAMPLE

Use with Example Problem 1.

Problem A 560-N cyclist sits on a bicycle seat and compresses the two springs that support it. The spring constant equals 2.2×10^4 N/m for each spring.

a. How much is each spring compressed?

b. By how much does the compression increase each spring's elastic *PE*?

Response

a. $F = kx$

$x = F/k = \left(\frac{1}{2}F_w\right)/k$

$x = \frac{\frac{1}{2}(560\ \text{N})}{2.2\times10^4\ \text{N/m}} = 1.3\times10^{-2}$ m

b. $PE_{sp} = \frac{1}{2}kx^2 = \frac{1}{2}(2.2\times10^4\ \text{N/m})(1.3\times10^{-2}\ \text{m})^2$

$= 1.9$ J

Pendulums and Resonance

student text pp. 386–387

Concept Development

Force Direction For the simple pendulum shown in **Figure 4,** $F_g + F_T = F_{net}$. Model this by drawing a vector-addition diagram for when the pendulum bob is at the position on the right, as seen in **Figure 4.**

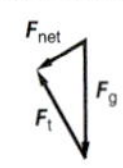

Ask students why F_T is greater than F_g. The bob travels an arc of a circle, so the net force at the bottom of the path must be centripetal (toward the center of the circle). OL Visual-Spatial EL

224 Chapter 14 • Vibrations and Waves

Finding answers to Student Edition exercises is quick and easy.

- Section answers are grouped together at the end of each section.
- Chapter Assessment answers are grouped together at the end of each chapter.

SECTION 1 ANSWERS

student text pp. 382–387

SECTION 1 • ANSWERS

Reading Checks and Caption Checks

Reading Check, *student text p.382*

The requirement for an object to be in simple harmonic motion is that the magnitude of the restoring force on the object is proportional to the magnitude of the displacement of the object from equilibrium. (Remember that both force and displacement are vector quantities, which have both magnitude and direction.)

Caption Check, *student text p. 382*

The displacement would be 0.5*x* m.

Reading Check, *student text p. 383*

The importance of the negative sign in Hooke's law is that it indicates that the restoring force on the object (the restoring force being a vector pointing from the object toward the equilibrium point) is opposite in direction to that of the displacement of the object from equilibrium (the displacement being a vector pointing from the equilibrium point to the object).

Caption Check, *student text p. 384*

Physics Challenge

student text p. 385

1. Conservation of energy implies that the car's gravitational potential energy at the top of the hill will be equal to the spring's elastic potential energy when it has brought the car to rest. The equations for these energies can be set equal and solved for *x*.
   $PE_g = PE_{sp}$, so $mgh = \frac{1}{2}kx^2$
   $x = \sqrt{\frac{2mgh}{k}}$
2. The height is doubled and *x* is proportional to the square root of the height, so *x* will increase by $\sqrt{2}$.
3. In the case of an ideal spring, the spring will propel the car back to the top of the hill.

Section 1 Review

student text p. 387

8. The pendulum swings back and forth, following the same path each cycle and requiring the same amount of time to complete each cycle.

All of your teaching essentials are here!

- **Identify Misconceptions** to catch and correct students' ideas
- Quick **DEMOs** to illuminate ideas and excite students about physics
- **Activities** to reinforce concepts with hands-on experience

SECTION 2 *student text pp. 388–393*

Wave Properties

student text pp. 389–393

Identify Misconceptions

Amplitude and Pulse Speed Some students might think that waves with greater amplitude have higher speed. Ask students whether a wave's amplitude affects its speed, and have them write a paragraph supporting their statement. Then, do the following Quick Demo. OL Linguistic

Quick DEMO

Amplitude and Pulse Speed

Estimated Time 10 minutes

Materials spring, stopwatches

Procedure Use a long spring as described in this section's Bellringer or conduct the demonstration in a long hallway by extending the spring and having a student hold one end. Have several students use stopwatches to measure the time for a small-amplitude pulse, and then for a large-amplitude pulse, to travel the length of the spring. Because the time measurements and distances are similar, students might infer that the pulse speeds are similar.

PHYSICS PROJECT ACTIVITY

The Perfect Wave Have students research information on the characteristics of ocean waves with regard to wave speed, amplitude, wavelengths, and other wave characteristics. Research might include descriptions of wave trains, fully-developed seas, tidal waves, wave interference, why and how waves break, the different types of breakers, and how a surfer identifies the "perfect" wave. Information on the instruments used to measure waves, how someone on a ship measures a wave, and the biggest wave ever recorded could also be of interest to students. Encourage students to develop comparative graphics to depict these concepts. OL Linguistic-Visual

Content Background

Dispersive Media In some materials, the velocity of a wave depends not only on the medium but also on the frequency of the wave. Such a medium is called a dispersive medium. Shallow water is a dispersive medium for surface waves. Similarly, light waves undergo dispersion as they travel though glass and diamond. Even though little dispersion takes place as the sound waves of audible sounds travel through air, the speed of higher frequency ultrasonic sound waves in air is measurably greater than the speed of sound waves in the audible range.

Content Development

Amplitude, Energy, and Medium Ask students if a wave of small amplitude in a heavy spring carries more energy than a wave with a larger amplitude in a lighter spring. Have them explain their answers. It is possible, but not necessarily so. The factors that influence the energy of a wave and thus its amplitude include such characteristics as the force exerted on the spring, the resistance of the spring, and the elasticity or springiness of the material. Exerting the same force on springs of different materials would result in different amplitudes, with the more massive, heavier spring exhibiting a smaller amplitude. If the two springs exhibited equal amplitudes, you could conclude that the force exerted on the larger spring was greater than the force exerted on the smaller spring because more work was needed to overcome the resistive forces of the larger spring. AL

Use Figure 11

Wave Properties Have students look at **Figure 11.** List wave properties on the board, and have students identify which ones are shown in, or can be determined from, each graph. **Figure 11** left side: *amplitude, crest, trough, and wavelength;* **Figure 11** right side: *amplitude, crest, frequency, period, trough* OL Visual-Spatial EL

Reinforcement

Simple Harmonic Motion Have students who performed the Extension activity in Section 1 compare the graphs with **Figure 11**'s right side. Point out that the displacement of most mechanical waves can be modeled as simple harmonic motion plotted over time.

Section 2 • Wave Properties **229**

Several end-of-chapter features further connect physics to the real world:

- Physics: That's Entertainment
- A Closer Look
- On the Job
- Frontiers in Physics
- How It Works

CHAPTER 14 *student text p. 400*

HOW IT **WORKS**

Events of Magnitude

Background

Major earthquakes transfer energy to buildings and other structures through surface waves. Without careful design, buildings can be significantly damaged or even collapse due to oscillations from these waves. In Japan, which is prone to frequent earthquakes, builders are especially concerned with how earthquakes might damage structures. This concern helped drastically reduce damage and loss of life during the March 2011 Tohoku earthquake in Japan.

Teaching Strategies

- Have students discuss how they think a computer-controlled pendulum might control the sway of buildings. Inform students that inertia causes the large masses of the pendulum to move more slowly than the building around them. In effect, this means that the motion of the pendulum is sometimes in the opposite direction of the swaying building, preventing the building from swaying too far and exceeding the flexibility of the materials from which it is built.

TEACHING PHYSICS

Help Your Students Understand Physics

This program is organized based on Big Ideas, Main Ideas, and Essential Questions.

- Each chapter begins with the **BIGIDEA**—a succinct statement of the key take-away concept of the chapter.
- Each section begins with the **MAINIDEA**, which shifts the focus to the core idea of the section.
- Essential Questions reflect learning objectives for the section. Each Section Review assesses the Essential Questions.

Introduce the BIGIDEA

At the beginning of each chapter, Teacher Essentials includes an interesting way to grab students' attention and introduce the **BIGIDEA** for the first time. This might include a quick demonstration or activity, or questions designed to get students thinking and talking about the **BIGIDEA**.

> **Introduce the BIGIDEA**
> **Thermal Energy** Hold an inflated balloon in front of students, and ask them if anything is moving inside the balloon. Students might answer that molecules are moving inside the balloon. Then ask students if these molecules have energy. Students should recognize that moving particles have kinetic energy. Inform students that the energy associated with an object's atoms and molecules is called thermal energy.

(t)Jetta Productions/Iliana Neely/Getty Images

Teach the MAINIDEA

Near the beginning of each section, Teacher Essentials explicitly addresses the **MAINIDEA**. This teaching element aligns directly with the **MAINIDEA** and leads students to a better understanding of it.

> **2 Teach**
> **Changes of State**
> *student text pp. 330–333*
> **Content Development**
> **MAINIDEA** One way to demonstrate the energy required to change a solid to a liquid and then to a gas would be the melting of an ice sculpture. Begin with small ice sculptures (colored water frozen in a candy mold). Place two in dishes: one at room temperature a...

Assess the MAINIDEA

Near the end of each section, Teacher Essentials provides strategies for assessing student understanding of the **MAINIDEA**.

> **3 Assess**
> **Assess the MAINIDEA**
> **Efficiency and Entropy** Ask students what would happen to the temperature in a room if an air conditioner were left running on a table in the middle of the room. The temperature of the room would increase. The thermal energy removed from the room by the cool a... reintroduced to the room through the heat exhaust co... on the other side of the air conditioner. Even if the sa... amount of thermal energy was removed by the cool ai... that was reintroduced to the room by the exhaust coi... the heat produced by the compressor motor would st...

Various elements support and enhance each chapter's BIGIDEA and MAINIDEAs.

- Tie to Prior Knowledge
- Reinforcement
- Identify Misconceptions
- Visual Literacy
- Additional In-Class Examples
- Concept Development
- Critical Thinking
- Content Background
- Check for Understanding
- Extension
- And more!

Differentiated Instruction

Students' abilities vary widely. *Physics: Principles and Problems Teacher Essentials* contains strategies for reaching all students.

Differentiated Instruction tags follow each activity throughout the chapter. See the key below for the meaning of each Differentiated Instruction tag.

BL Below Level	Below-level activities should be appropriate for students working below grade level.
OL On Level	On-level activities should be appropriate for students working on grade level.
AL Above Level	Above-level activities are designed for students working above grade level.
EL English Language Learner	EL activities are appropriate for students learning English.
COOP LEARN Cooperative Learning	Activities designed for small cooperative group work

Learning Styles appear after each **BL**, **OL**, **AL**, **EL**, or **COOP LEARN** when appropriate.

Kinesthetic learners learn through touch, movement, and manipulating objects.

Visual-Spatial learners think in terms of images, illustrations, and models.

Logical-Mathematical learners understand numbers easily and have highly developed reasoning skills.

Linguistic learners write clearly and easily understand the written word.

Auditory-Musical learners remember spoken words and can create rhythms and melodies.

Interpersonal learners understand and work well with other people.

Intrapersonal learners can analyze their own strengths and weaknesses and may prefer to work on their own.

The **DIFFERENTIATED INSTRUCTION** activities are not only meant for below-level students. They offer additional support for any student struggling with a concept.

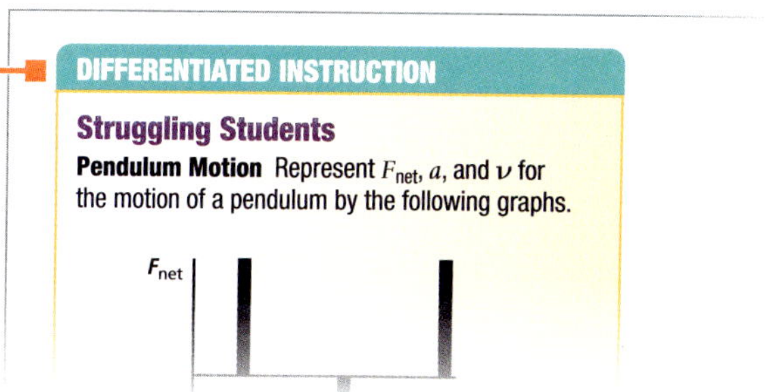
DIFFERENTIATED INSTRUCTION

Struggling Students

Pendulum Motion Represent F_{net}, a, and v for the motion of a pendulum by the following graphs.

Math Support for Physics

Understanding math can greatly enrich the physics learning experience. This program provides many tools to help you strengthen and enrich students' math skills. From remediation to challenge problems, you can find math support for all of your students.

Math Guidance

- Example Problems
- Problem-Solving Strategies
- Customizable Connecting Math to Physics Online Workbook

Extra Practice

- Practice Problems
- Additional Practice Problems Online
- Physics Challenge Problems
- Additional customizable problem workbooks online

Useful Resources

- Personal Tutors
- Math Handbook

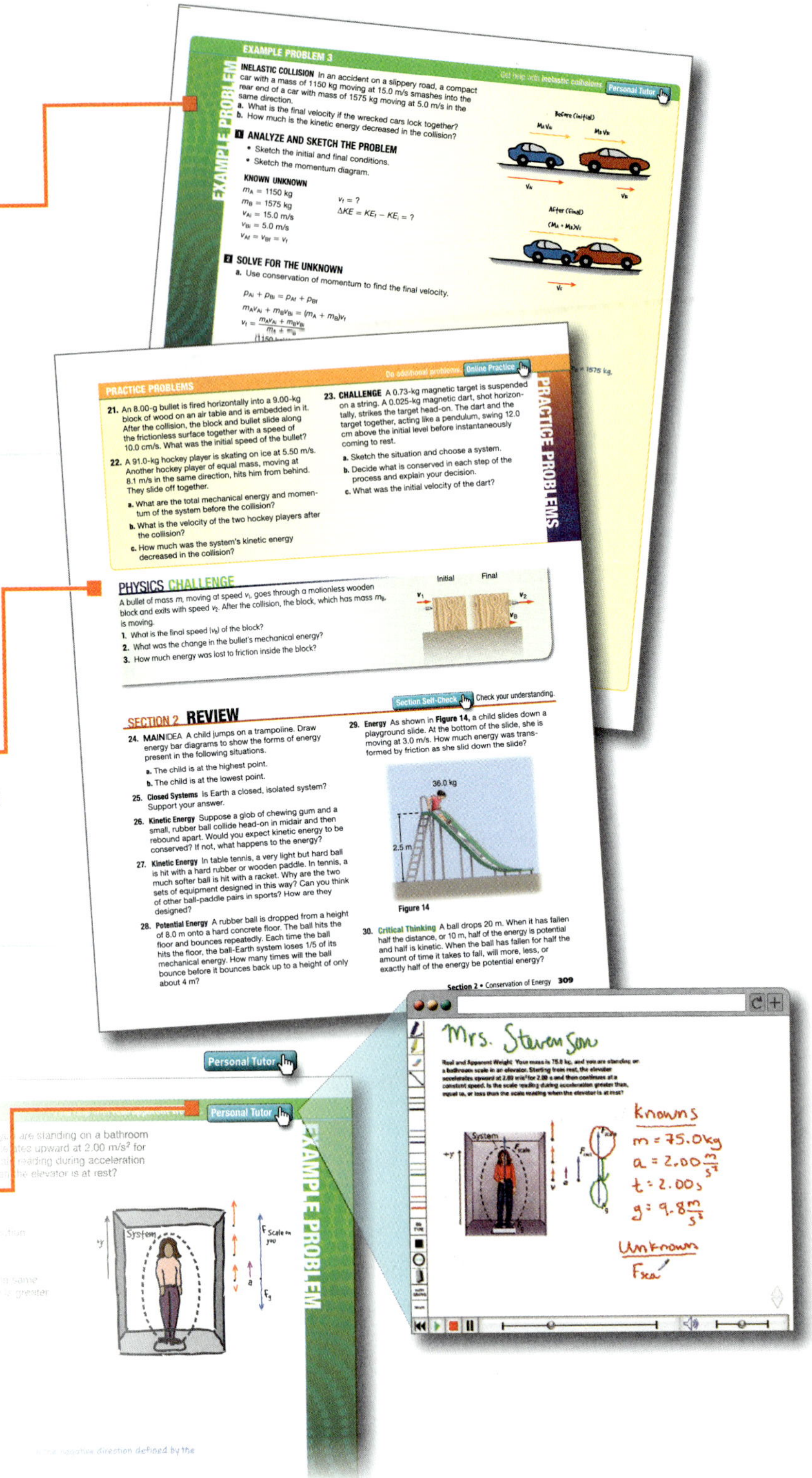

Reading Support for Physics

While most all students know how to read by the time they graduate from high school, many do not have the skills necessary to read and use a nonfiction, technical text effectively. *Physics: Principles and Problems* includes many strategies to help students further transition to independent reading and learning.

- **PHYSICS 4 YOU** creatively links the lesson to students' lives.
- **READING CHECKS** help students self-check their understanding of what they have just read by summarizing, explaining, describing, and applying.
- **CAPTION CHECKS** prompt students to thoroughly examine photos, graphs, and diagrams and apply what they have learned.
- **SECTION REVIEWS** revisit the Essential Questions at the end of each section.
- **VOCABULARY** features provide definitions and samples of both science usage and common usage for the same word.
- **REAL-WORLD PHYSICS** connects students' reading to real-world applications.

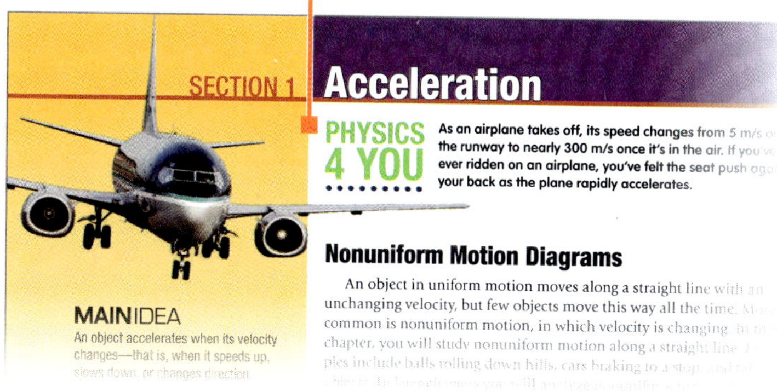

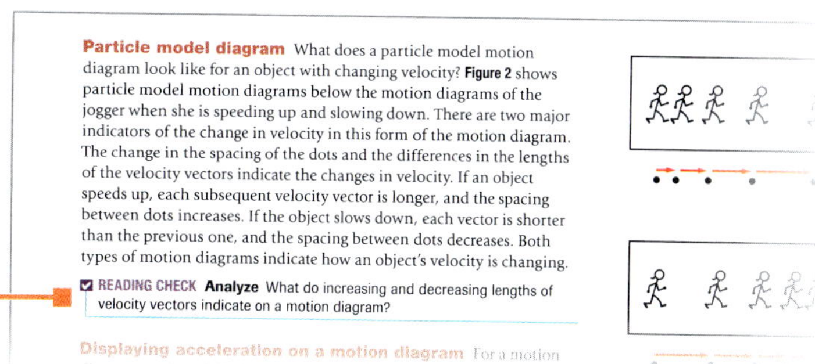

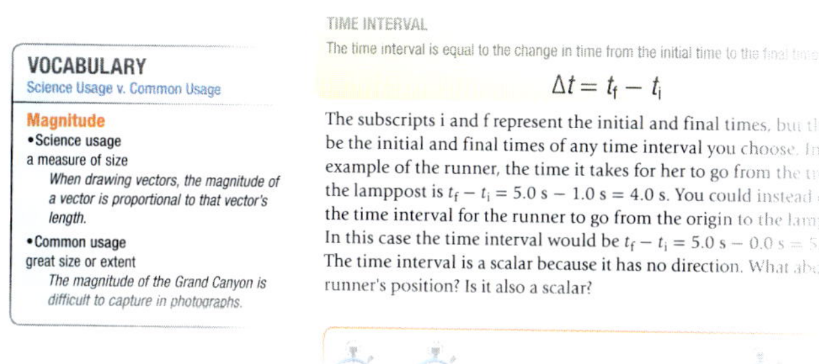

MODELING PHYSICS

Physics: Principles and Problems and Modeling Instruction

An increasing number of science instructors throughout the United States are utilizing elements of Modeling Instruction in their courses. Below are highlighted several current features of that fit particularly well with this approach.

What is Modeling Instruction?

Modeling Instruction is an instructional approach that developed in the late 1980s through the collaboration of an award-winning high school physics teacher and a physics professor in the Phoenix area. In 2001, Modeling was the first program designated by the U. S. Department of Education as an exemplary program in high school math and science instruction.

One of Modeling's strengths is that it is a pedagogical framework and not a tightly prescribed curriculum. The backbone of this framework is what is known as the Modeling Cycle. In a typical Modeling Cycle, the students are not executing a lab to verify an equation or to test an explicitly articulated prediction; they are engaging in guided scientific inquiry.

A Modeling Cycle investigating constant velocity might follow the progression shown below.

1. The instructor shows students a toy car moving across the room and asks the students to share their observations. All observations are recorded on the board. The teacher guides the students to focus on observations that can be quantified.

Hutchings Photography/Digital Light Source

2. The teacher asks students to describe a way in which they might determine whether there is a relationship between these quantities. In this case the resulting class discussion culminates in small groups of students designing laboratory investigations to find the relationship between the distance traveled by the car and the elapsed time.

3. Each group shares their results on student-sized whiteboards, presenting their results graphically. The teacher engages them in a discussion of the graphs's implications. New representational tools may be introduced, such as the motion diagram. In the constant velocity example, the graphs of student data lead to the familiar kinematic equation:

$$x = x_0 + vt$$

4. Students apply the common understanding developed in the lab in a variety of situations, which may include problem solving, discussions, projects, and lab practica.

Hutchings Photography/Digital Light Source

MODELING PHYSICS

How does this program support Modeling?

Physics: Principles and Problems **contains many elements recommended by physics education research that are incorporated in most Modeling classrooms.**

The empirical nature of science: This program engages students in making observations of their surroundings (in the real-world text examples and also in the chapter opening photos, Physics 4 You, and Physics TV) and in reasoning about how these observations lead to accepted mathematical relationships.

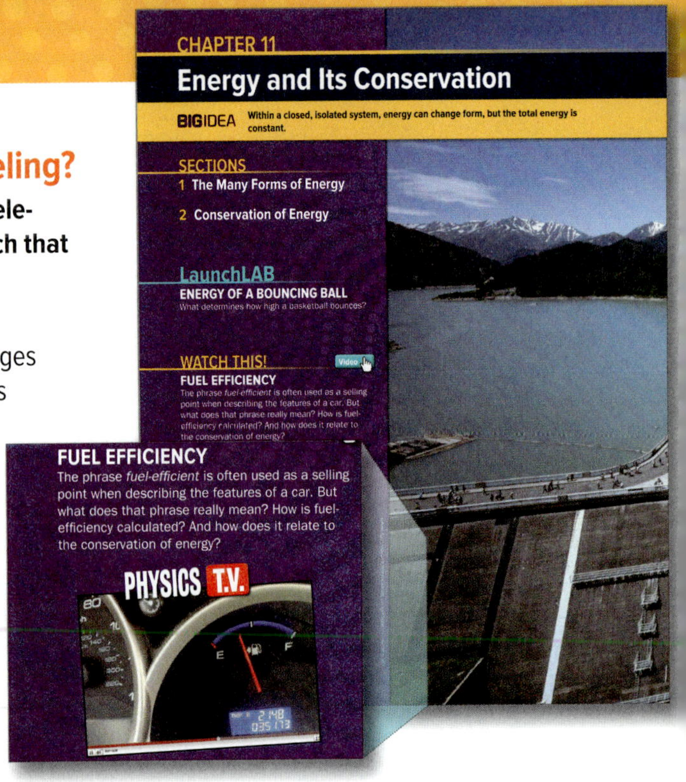

Multiple representations: Students more readily grasp topics when they have several representational tools available. Examples include the use of motion diagrams for kinematics and force problems and the use of work-energy bar charts to explore energy conservation.

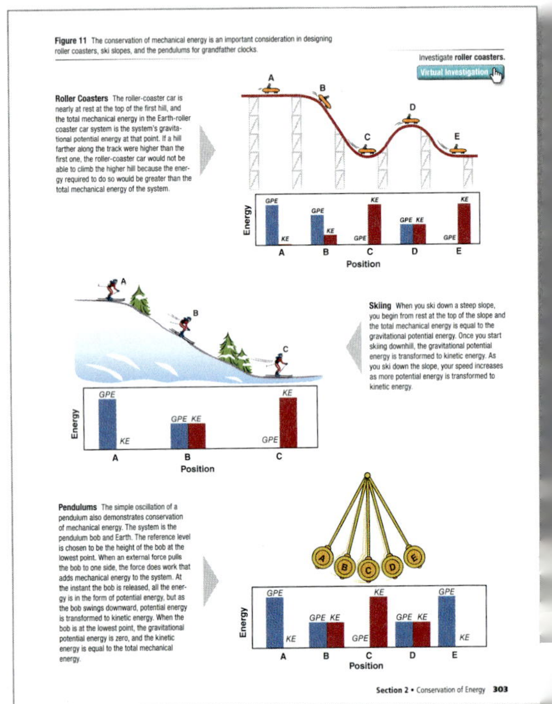

A rich array of application activities: The end-of-chapter problems, as well as the teacher materials, contain many activities that resonate with Modeling instructors, including lab practica, ranking tasks, reverse problems, and problem posing.

Increasing the Impact of Modeling

If you are interested in learning more about Modeling, many professional societies offer introductory workshops at regional and national meetings. In the summer, universities around the country host intensive immersion workshops. Visit connected.mcgraw-hill.com for professional development videos and an introduction to Modeling Instruction.

Professional Development

Learn more about Modeling Instruction.

CHAPTER 1

A Physics Toolkit

About the Photo

Have students examine the photo of the first bionic hand with bending fingers. Tell the students that the i-LIMB is capable of peeling a banana, typing on a keyboard, and picking up a paper clip. Ask students what kind of considerations must be taken into account to design such a device like i-LIMB. Relate to the students that these design requirements all need a firm grasp of many physics concepts to develop such a complex bionic device.

Chapter Overview

Math is the language of physics, and students need to know how to use math as a tool to make the most of their studies. The first section familiarizes students with scientific methods. It also distinguishes scientific law from scientific theory. In the second section, students will read about units and significant figures. In the third section, students learn about precision, accuracy, and uncertainty in measurement. Finally, students learn about graphing data and how equations and graphs reveal the relationship between variables.

Introduce the BIG IDEA

Separate students into groups of three or four, and give each group a portable white board and dry erase marker. Ask groups to complete the following statements: "Biologists study...", "Chemists study...", "Geologists study...", and "Physicists study...". Select some groups to present their ideas to the class while holding up their white board. Pull together their lists for physics onto the board, and categorize items as either energy or matter.

Next, ask students to brainstorm on the steps of scientific investigation. After the brainstorm session, select groups to present their white boards, and cull together their ideas (pose questions, make hypotheses, experiment, acquire data/results, and analyze/make conclusions) that relate to the scientific method.

Go online!
connected.mcgraw-hill.com

Your online experience includes:

- **Video**
- **Concepts In Motion**
- **Virtual Investigations**
- **All Lab, Applying Practices activities, and PBL worksheets**
- **Online Practice**
- **Personal Tutor**
- **Multilingual eGlossary**
- **Presentation**

Use the LaunchLAB

In *Mass and Falling Objects,* students can investigate whether mass affects the rate at which an object falls.

Use Physics T.V.

Video

ROCK STACKING How do you get a stack of rocks to remain upright and balanced? It's all about understanding the physics! Explore the science behind the art of rock stacking.

1 Introduce

Bellringer

Importance of Science Hold up several devices, such as a cell phone, a compact fluorescent bulb, and a calculator. Point out that almost every device students' use each day was originally developed by basic scientific research and scientific methods. An idea led to research, and after a great deal of testing, a device was developed. **BL** **Visual-Spatial**

Tie to Prior Knowledge

Methods of Investigation Students have performed investigations in prior science classes. Ask them to describe the procedures they followed. Lead them in understanding that although the procedures might vary, all investigations have some things in common. Students had to make measurements, record data and analyze data, and draw conclusions.

2 Teach

What is physics?

student text p. 4

Concept Development

MAINIDEA Give groups of students small sealed, shoe-sized black boxes with different unknown items in them, e.g. rubber balls, slotted masses, breakfast cereal pieces, acorns, corks, crumpled up paper or tissue, foam package peanuts, etc. The boxes should have very small holes in the side in which chopsticks can be inserted to poke around at the object. The students need to figure out what is in their box based on their observations—sounds, feeling the item move inside, rolling, slipping, poking it through the hole, etc. Ask students to make several hypotheses of what the item could be using observations to support their reasoning. Inform students that while each of their steps may not have been done identically, all groups followed similar methods to determine the hidden item inside. Once groups have made their guesses, reveal what items are inside. Inform students that physics uses a similar method of experiment and observation called the scientific method to study energy and matter. **COOP LEARN**

Identify Misconceptions

Science and Scientists Find out what sort of preconceived notions your students might have about what science is and what sort of people do science. It is helpful to stress throughout the school year that science is much more than looking up facts in a book. Scientists constantly test their ideas against new situations and adjust their ideas based on the results.

Scientific Methods

student text pp. 5–7

DIFFERENTIATED INSTRUCTION

Struggling Students Many students carry with them the notion that scientists rigidly adhere to a common series of steps. Inform students that scientists approach and solve problems with imagination, creativity, prior knowledge, and perseverance. These, of course, are the same methods that all effective problem solvers use. What distinguishes science from other endeavors is scientists' focus on testing ideas against observations. **BL**

Use the MiniLAB

In *Measuring Change,* students learn how mass affects the length of a spring.

Quick DEMO

Light and Solar Cells

Estimated Time 15 minutes

Materials solar cell, multimeter, sunny day

Procedure Ask students how they think the electrical energy from a solar cell relates to the amount of sunlight striking the surface of the cell. Once students have generated some theories, ask how these theories might be tested. Set up the multimeter to measure the current that the solar cell generates. Place the solar cell in direct sunlight, facing the Sun. Record the current. Cover various fractions of the cell's surface, and record the corresponding currents. With students, do a brief data analysis, discuss the process used, and decide what conclusions they could reasonably draw from the data. Students will discover that different methods of investigation are possible, but the results are the same: increased sunlight results in increased current.

Extension

Measure from Afar Galileo, with the aid of his telescope, estimated the heights of mountains on the Moon by estimating the lengths of the shadows. Ask a student to hold a shoebox and a 30-cm ruler upright about a meter in front of a white projection screen. Use a flashlight to simultaneously make shadows of the two objects on the screen. Ask students to point out which object makes the longer shadow. the taller object Have students use ratios to compare measured dimensions of the objects and the shadows cast. Students should find that each shadow's length is proportional to the height of the object. OL Visual-Spatial

Content Background

Galileo and Scientific Methods In 1609 Galileo Galilei (1564–1642) built a telescope and used it to study the sky. He found that the Moon was neither a perfect sphere nor a smooth sphere. In fact, the Moon had mountains, whose heights he could estimate from the shadows they cast. Through his telescope, Galileo also discovered four moons that orbit Jupiter, that the Milky Way is made up of more stars than anyone had imagined previously, and that Venus has phases similar to our Moon. Based on this new perspective, Galileo argued that Earth and the other planets circle the Sun.

Models

student text pp. 7–8

Use Models

Equations Provide students with an example of how they can use equations to model a phenomenon. Explain that the equation $v = \frac{d}{t}$ describes how velocity (v), distance (d), and time (t) relate to each other. If you know any two of these quantities, you can easily solve for the third. Ask students to provide you with an equation that solves for d and another equation that solves for t. $d = vt$ and $t = \frac{d}{v}$

OL Logical-Mathematical

3 Assess

Assess the MAIN IDEA

Outline an Experiment Ask students to write an outline for an experiment to learn which substances break down faster in a landfill: a fast food hamburger, a plastic grocery store bag, an orange, a piece of paper, an old leather shoe, and a polystyrene cup. Have several students present their proposed experimental outline. Stress that though student's steps will differ, the basic steps of the scientific method are similar.

Extension

Applications of Science Show students a laser pointer. Explain that scientists originally developed lasers and masers to demonstrate a particularly interesting property of matter—stimulated emission. Although today lasers have many useful applications, for many years lasers were called a solution in search of a problem. Divide the class into teams and ask each to come up with a list of applications of lasers today. Have students compare and validate their lists.

OL Interpersonal

Reading Checks and Caption Checks

Caption Check, *student text p. 7*
Models have changed because scientists have continued to make new discoveries about the structure of the atom.

Caption Check, *student text p. 8*
Answers will depend on the models that are present in the classroom. Possible responses: a globe, a model of the solar system, a toy car, a model skeleton.

Reading Check, *student text p. 8*
Computers allow scientists to model systems that are too big or to test a proposed explanation of how a process occurred. Computer simulations allow pilots to practice under simulated harsh or dangerous conditions without putting themselves in harm's way.

Section 1 Review

student text p. 9

1. Possible answer: I would make observations and ask a question based on the observations. I would research what is already known about the problem and form a hypothesis. I would design and carry out an experiment to test my hypothesis and analyze the results. I would check to see whether the results supported my hypothesis. I might ask another question based on the results or observations I made during the experiment.
2. A hypothesis is a possible explanation for a problem using what you know and what you observe. A hypothesis can be tested by making observations, by building a model, or by performing an experiment.
3. Bias can affect the results or conclusion of an investigation, making them invalid.
4. Scientists use models to help explain or learn more about things that are too large, too small, or too far to visualize or observe easily. Examples may include the solar system, a cell, DNA model, or the aerodynamics of an aircraft.
5. A scientific theory is an explanation of an event based on knowledge from observations and investigations. A scientific law is a statement about what happens in nature and that seems to be true all the time. Since a theory explains why something happens and a law does not, a theory cannot change into a law.
6. Testing opinions is not scientific. It is impossible to prove that an opinion is true for everyone. In addition, the survey was based on a small part of the population, and it only included students at one school. The results cannot be extended to the entire population.
7. No; the value of 9.8 m/s^2 has been established by many other experiments, and to discard the finding you would have to explain why it is wrong. There are probably some factors affecting your calculation, such as friction or how precisely you can measure the different variables.

1 Introduce

Bellringer

Unit Systems Ask students to measure something—the length or width of the room, the width of a table—without using any formal measuring device. They will need to use their hand widths, foot lengths, or some similar unit to make this measurement. Then ask everyone in the class to report what they find. Discuss how difficult it is to compare results with all of these different unit systems. BL Kinesthetic

Tie to Prior Knowledge

Units Students who have already taken chemistry should be somewhat familiar with SI and scientific notation, but this textbook does not presume any such prior knowledge.

2 Teach

Mathematics in Physics

student text p. 10

Content Background

Solving for the Unknown Measuring and computing physical quantities requires the use of mathematics. Our modern system of mathematics owes a great deal to Hindu and Islamic mathematicians—and it's not just for the concept of zero and the Arabic numerals we use. Classical algebra developed over a period of 4000 years. The word *algebra* comes from the Arabic word *aljabr,* meaning "the (science of) reuniting." The word *algorithm* is a corruption of the name of the mathematician Mohammed ibn Musa al-Khwarizmi, who wrote a definitive text on algebra in 830 A.D.

Concept Development

MAIN IDEA Tell students that they may often hear that math is the language of physics. Have students work in pairs to develop expressions for the following scenarios. Car B drove three times as far as car A. Together they both drove 120 miles. Develop the expressions that allow for distance of the cars to be solved for. $d_B = 3d_A$, and $3d_A + d_A = 120$. Issac Newton discovered that the force of gravity between two masses was directly proportional to the product of the two masses divided by the distance of separation squared. $F \propto \frac{m_1 m_2}{r^2}$. In a circuit lab, a student measures that the current running through a resistor is simply the ratio of voltage across the resistor to resistor value. $I = \frac{V}{R}$

SI Units

student text pp. 10–11

Discussion

Problem Why do scientists use the metric system, instead of using English units or some other system?

Response In the metric system, with its base of ten, it is easy to convert from one scale of measurements to another. For instance, it's much easier to convert from centimeters to meters than from inches to yards. OL

Concept Development

Unit Systems Ask students why it is important to have an agreed-upon system of units. For one thing, this makes comparisons between different groups easier. It also helps everyone get an idea of the magnitudes of various measurements. For instance, most students will have some idea of how fast 25 mph is. But do students know how fast 10 furlongs per fortnight is? As the course progresses, students will gain an idea of how fast 25 m/s is. OL

PHYSICS PROJECT ACTIVITY

Ancient Standards All civilizations have had to develop standards for measuring. For example, in Mesopotamia (3500–1800 B.C.), workers built the first cities using cubits, measuring roughly the length of one's forearm from elbow to wrist bone (one cubit could range from 43–56 cm, or 17–22 in.). Have students research the measurement systems of various civilizations. Their reports should include the advantages and disadvantages of the systems, the units of measure, the rationale or origin of the unit, and their equivalent measure in SI. They also should convert some commonly measured items (for example, a football field) to provide additional perspective on the system. OL Linguistic

Reinforcement

Prefix Game Activity Write the metric prefixes that you intend to use frequently in your class on index cards. Make several sets. Divide the students into teams, and give each team a set of the cards. Have each student on the team randomly select a card, then have a contest to see which team can line up its members most quickly according to the magnitude of the prefix on the card each member has drawn. BL Interpersonal

Reinforcement

Conversion Factors and Cubic Units Students can have trouble with conversion factors as they relate to cubic units. For example, students know that 100 cm equals 1 m. They may conclude that 100 cm^3 is equal to 1 m^3, rather than 1,000,000 cm^3 (1×10^6 cm^3) is equal to 1 m^3. Provide the class a model of a 1 m^3-cube. Have students determine the volume of the cube in cm^3 and mm^3. From their calculations, have them create conversion factors.

Dimensional Analysis, Significant Figures, and Solving Problems

student text pp. 11–12

Concept Development

Rounding Students might have trouble rounding to the correct number of significant figures when a number is halfway between two numbers. Students should follow the following rules. (1) When the leftmost digit to be dropped is 5 followed by a nonzero number, that digit and any digits that follow are dropped. The last digit in the rounded number increases by one. For example, 8.7519 rounded to two significant figures equals 8.8. (2) If the digit to the right of the last significant figure is equal to 5 and 5 is not followed by a nonzero digit, look at the last significant figure. If it is odd, increase it by one; if even, do not round up. For example, 92.350 rounded to three significant figures equals 92.4 and 92.25 equals 92.2.

Concept Development

Significance In common English *significant* means "important," while in science it means "measured." Nonsignificant figures are important as placeholders. A measurement of 8000 m has four important digits, but only one of them is significant.

Reinforcement

Significant Figures Ask students to write a four-digit number that has one zero that is not significant and one zero that is significant. Sample answer: 1020; the first zero is significant, but the second zero is not significant. After about 30 s, have each of them exchange papers with another student and evaluate. OL

3 Assess

Assess the MAIN IDEA

Conservation Equation Physics uses mathematics to represent nature because nature can be described in logical and quantitative terms. Bring in several beakers, cups or other vessels that can hold water. Label each beaker using masking tape with energy terms such as *potential energy to thermal energy, potential energy to kinetic energy, potential energy to sound energy, and potential energy to radiant energy.* Take a large pitcher of water and label it *potential energy.*

Have a student distribute the water between the beakers. Next, ask the students to develop a mathematical expression relating total energy before the distribution and after. Inform students that conservation laws in physics (energy and momentum) are related to this activity where water was conserved, not lost.

Check for Understanding

Scientific Notation Give students a list of numbers in scientific notation. Ask students to rank the numbers from least to greatest. Make sure to include some negative quantities and some quantities with negative exponents. OL Logical-Mathematical

Reading Checks and Caption Checks

Caption Check, *student text p. 11*
It is important to have standards so that measurements can be determined to be accurate and can be compared worldwide.

Reading Check, *student text p. 11*
gigabytes

Caption Check, *student text p. 12*
It is uncertain because the ruler measures to the millimeter. The last digit is based on an estimated distance between two tic marks on the ruler.

Section 2 Review

student text p. 13

8. The formulas are concise and can be used to predict new data.

9. Answers might include that SI units help people communicate their results, that SI units are the units of measurement in most countries around the world, or that SI units are easy to manipulate because they are based on multiples of ten.

10. 750,000 kHz

11. 31,622,400 s

12. **a.** 2.5 g after rounding

b. 4.33 m after rounding

c. 3.2×10^2 cm^2

d. 1.22 g/mL

e. 93.6 cm after rounding

f. 1600 m after rounding

13. $v = \frac{F}{Bq}$

14. Sample answer: For most cars, the answer is unreasonable because 290 km/h is equivalent to 81 m/s or 180 mph. The speed might be reasonable for a race car.

1 Introduce

Bellringer

Measurement Technique Ask two students to act out taking a measurement. One should use good technique, and one should make some fairly obvious errors. Ask the class which student's results would be more believable and why. OL Kinesthetic

Tie to Prior Knowledge

Accuracy and Precision These are concepts students should be familiar with, even if they don't use the terms in a scientific way. Have students consider aspects of everyday experiences they might be able to measure, such as waiting for the bus, competing on the track, and building bookshelves. BL

2 Teach

What is measurement? and Comparing Results

student text pp. 14–15

Concept Development

MAIN IDEA Ask students to bring in their favorite cookie or cake recipe. Ask each student to describe the physical properties of the food: taste, texture, consistency, etc. Ask students to make predictions of how their favorite recipe would turn out if they strayed from the recipe by doing such things as changing the baking temperature or time, by estimating their measurements rather than using measuring cups or by eliminating ingredients? Alter taste, may burn, change texture, may not rise, etc. Inform students that in order for an experiment to be repeated, the measurements must be made carefully and a detailed recipe or notes must be recorded. Only careful measurements and detailed "recipes" will produce results that can be compared, and reproduced in the future.

Discussion

Question Draw a cartoon of a beach. A sign says, "Pond Shallow—average depth 3 feet." A swimmer is standing on the shore. Ask students whether the swimmer could wade into water over his head. yes Ask students whether the pond could be 30 ft deep in spots. yes Ask whether the sign is useful. An average may not accurately convey the highest and lowest measurements. OL Logical-Mathematical

DIFFERENTIATED INSTRUCTION

Visually Impaired Ask students to put the index finger of one hand on their noses and extend the other arm as far as possible to the side. Tell them that the distance between their two index fingers is approximately 1 m in length. Then have students feel the thickness of their little fingers, and tell them this width is about 1 cm; the width of the fist is about 10 cm. Remind students that they will always have these distances handy to help them make estimates. OL Kinesthetic

Precision Versus Accuracy

student text pp. 15–16

Identify Misconceptions

Unit Systems and Accuracy Some students might think the metric system is more accurate than the English system because it is the one scientists choose to use. In truth, neither system is inherently more accurate than the other.

Use Figure 10

Uncertainty Have three to six students measure identical items, such as the width of a textbook or desk. Ask them to write down their measurements without announcing or comparing results. Their measurements will most likely vary slightly. Make a graph with uncertainty bars. Repeat with three groups, or three classes, to make a graph similar to the one in **Figure 10.** OL Kinesthetic

Discussion

Problem Show the results of a calculation on two different calculators where one displays more decimal places on its readout than the other. Which is more accurate?

Response The accuracy of the calculation has little (usually nothing) to do with the calculator. Instead, the accuracy relates more to how careful the calculator's user is when observing the readings. Just because a calculator gives more decimal places does not mean that the person using it will obtain results that are more accurate. OL Logical-Mathematical

Critical Thinking

Average Accuracy Refer back to the cartoon that you drew earlier of the person at the shore. Ask students to discuss whether increasing the number of significant figures would have helped the situation. Even if the measurement were given as 3.0000 ft, the swimmer could still be in over his head because the number was an average. OL

Techniques of Good Measurement

student text pp. 16–17

Use Figure 13

Parallax in Measurement Ask students to look at an object in the distance with one eye shut. Have them position their thumbs at arm's length in front of the object so as to block it from view. Then have them view the object with the other eye shut, then the first. Ask them to tell you what happens to the object. It appears to shift. Explain that the farther the apparent shift, the closer the object is to the observer. OL Kinesthetic

REAL-WORLD PHYSICS

Global Positioning Systems These systems allow sailors, hikers, and drivers to determine their locations on the surface of Earth to within a few meters. Remind students that the coordinates are three-dimensional. Ask students whether anyone has a GPS device. If so, ask the student to demonstrate it to the class.

Use the PhysicsLAB

In *Mass and Volume,* students will determine the relationship between mass and volume for various substances.

3 Assess

Assess the MAIN IDEA

Drops on a Penny Assign students to small groups, and give each group a penny and an eye dropper. Assign one group to be the "sloppy" or uncareful group which does not take their time or do all five trials. Next, have the remaining students take great care in their procedures. Wash the pennies, dry them completely, and count the number of water drops that fit onto the penny. Repeat the procedure for five trials, and calculate an average for each team. Report the results on the front board. Ask students what similarities and differences they see in the results. Discuss reasons for these. Ask students why the experiment was done with multiple trials. Ask students if they would expect to get similar results if they were to perform the water drop experiment again.

Check for Understanding

Accuracy Ask students to generate a list of real-world examples where it is important that measurements be done accurately and precisely. OL

Reteach

Simple Math Give the students the numbers 5.87 km and 1.2×10^{-2} km. Ask them to add, subtract, multiply, and divide these numbers. add, 5.88 km; subtract, 5.86 km; multiply, 7.0×10^{-2} km^2; divide, 4.9×10^2 OL

Reading Checks and Caption Checks

Caption Check, *student text p. 15*
The results of Student 1 and Student 2 overlap, so they are in agreement. The results of student 3 do not overlap with the other two measurements, so they are not in agreement. Student 3's result may not be reproducible. The measurements are likely to be inaccurate.

Caption Check, *student text p. 16*
More information is needed to determine if the balance is accurate. It is properly zeroed, but whether it gives a correct reading when measuring an accepted standard is unknown.

Reading Check, *student text p. 16*
Both are important when making measurements. Precision is the degree of exactness of a measurement. Accuracy is the how well a measurement agrees with the accepted value.

Caption Check, *student text p. 16*
Parallax shifted the measure by about 0.1 N, or about 10 g.

Section 3 Review

student text p. 17

15. It would be more precise but less accurate.

16. As the edge of the ruler gets worn away over time, the first millimeter or two of the scale would also be worn away if the scale started at the edge.

17. No; it doesn't change the fineness of the divisions on its scale.

18. His height would be between 181.5 and 182.5 cm. Precision of a measurement is one-half the smallest division on the instrument. The height 182 cm would range ± 0.5 cm.

19. a. 7.05×10^3 cm^3

b. nearest tenth of a cm; nearest 10 cm^3

c. 243.6 cm

d. nearest tenth of a cm; nearest tenth of a cm

20. You should not have much confidence in the precision of the report. A result can never be more precise than the least precise measurement. The calculated average lap time exceeds the precision possible with the clock.

1 Introduce

Bellringer

Graphs First, show students a data table and a graph of the same data. Ask them to tell you which one they can understand more quickly. Second, show students a graph from some nonphysics context—something like number of sales versus hour of the day or number of cars through an intersection versus day of the week. Ask students what the graph represents and who might be interested in the information it conveys. Finally, show students a graph with no labels on the axes. Ask them what it represents. When students respond that they don't know, ask them why they can't tell; emphasize the importance of labeling. This activity can be expanded to show other graphs missing important elements. **OL** **Visual-Spatial**

Tie to Prior Knowledge

Graphs Students have graphed equations in math classes before. They should be somewhat familiar with exponents and with equations for lines and parabolas. Students who have had chemistry should know about the importance of units.

2 Teach

Identifying Variables

student text pp. 18–19

Concept Development

MAINIDEA Collect the following data from students about the shoes they are wearing: color, material, brand, size, style, and the gender of the owner. Assign small groups to organize the data into different kinds of bar graphs that examine the relationships between these variables. Have students sketch their bar graphs on portable white boards. Choose several successful bar graphs that show clear trends and relationships among the data as examples in class. Also, show some bar graphs that clearly do not identify a trend among the data. Reinforce to the students that visual interpretations of data are helpful tools to see trends and to take in a lot of data at once.

PHYSICS PROJECT ACTIVITY

Graphing Applications Ask students to go through newspapers or magazines to find examples of graphs that are trying to promote a product or viewpoint. Have them alter the graph somehow, such as modifying the scaling or the numbers on the axes to make a different visual impression. Ask students to write a short paragraph on how people can design graphs with the purpose of misleading readers. **OL** **Visual-Spatial**

Linear Relationships

student text p. 20

Use the MiniLAB

In *How Far Around?,* students determine the relationship between circumference and diameter.

Identify Misconceptions

Slope Show students an *x-y* graph of two parallel lines, with one line shorter than the other. Ask them which has a greater slope. Some may respond that the longer line has a greater slope, but inspection shows that the slopes in both graphs are the same. Mathematically, $\frac{\Delta y_1}{\Delta x_1} = \frac{\Delta y_2}{\Delta x_2}$ **OL** **Logical-Mathematical**

Use an Analogy

Steps and Slopes Point out to students how climbing steps is analogous to finding the slope. Use math to explain that rise over run is the same as $\frac{\Delta y}{\Delta x}$, where Δy is equal to the number of steps they climb up times the height of each step and Δx is the number of steps times the depth of each step. That is, Δx measures distance as the length along the *x*-axis.

Use Figure 16

Unit Swap Ask students how the line graph in **Figure 16** would change if the data had been measured and recorded in English units instead of metric units. The graph using English units would be the same shape, but the numbers on the axes would be different. **OL**

Nonlinear Relationships
student text pp. 21–22

Quick DEMO
Types of Graphs

Estimated Time 15 minutes

Materials windup car or truck, meterstick, graph paper

Procedure Show students some sort of windup car or truck. Plot the distance that it travels as a function of the number of turns it is wound. From looking at the graph, have the students give their best guess as to what kind of relationship there is between the two variables (turns and distance). Is it linear, parabolic, inverse, or something else?

Reinforcement
Linear and Inverse Relationships Ask students to make a list of linear relationships and inverse relationships. After several minutes, ask a few students to write their ideas on the board and then lead a discussion for review. Make sure students state why they think a particular relationship is one or the other. OL Interpersonal

PHYSICS CHALLENGE ACTIVITY
Filling a Beaker Give students a large, graduated metric beaker, and ask them to fill it using a slow trickle of water from a faucet. Repeat with a fast flow. Tell students that the volume of the beaker is measured in milliliters and the flow rate of the water can be measured in terms of milliliters per second. Ask students to graph an inverse relationship associated with filling the beaker of water. Students should identify the constant, the dependent variable, the independent variable, and the inverse relationship equation. The elapsed time needed to fill the beaker is inversely related to the flow rate. The constant is the capacity of the beaker (V_f) in mL; the independent variable is the flow rate (q) the dependent variable is elapsed time (t_f). The inverse relationship is $t_f = \frac{V_f}{q}$. AL Visual-Spatial

DIFFERENTIATED INSTRUCTION
Struggling Students Give students several different sizes of cardboard squares. Have them measure the sides of each and compute the corresponding areas, then plot the area (A) as a function of the side length (x). Their line graphs will be parabolic curves. Have them compare their graphs to **Figure 17** on page 21. Point out that the line graph shows the quadratic relationship because one variable depends on the square of another. BL Kinesthetic

Use the PhysicsLAB
In *It's in the Blood,* students use blood splatter patterns to provide clues.

Predicting Values
student text p. 23

Use the PhysicsLAB
In *Exploring Objects in Motion,* students will explore how to determine the speed of a vehicle.

3 Assess

Assess the MAINIDEA
Interpreting Graphical Information Go online to find a few "famous" graphs, e.g. Napoleon's 1812–1813 Russian invasion, Hubble's law, atmospheric carbon dioxide concentration per year plot, etc. Ask students to interpret the graphs and draw conclusions. What type of measurements did the scientists need to make each graph?

Check for Understanding
Linear Graphs Ask each student to make a graph that shows a linear relationship. Have them put numerical values on the graph and calculate the slope of the graph. After a few minutes, instruct students to trade with a neighbor to check their work. OL

Extension
Radius and Circumference Ask students to imagine a rope tied around Earth's equator and assume that the surface is completely smooth all the way around ($C = 2\pi r$, let $r = 6400$ km). Then ask how far above the surface the rope would be if the length of rope were increased by approximately 200 km. approximately 32 km (mathematical shortcut: divide the length added to the circumference, 200 km, by 2π) AL

Reading Checks and Caption Checks

Caption Check, *student text p. 20*
As mass decreases, the length of the spring decreases.

Reading Check, *student text p. 21*
In a quadratic relationship, one variable depends on the square of the another.

Caption Check, *student text p. 22*
As speed increases, time decreases.

Reading Check, *student text p. 22*
One variable depends on the inverse of the other variable.

Practice Problems

student text p. 22

21. a.

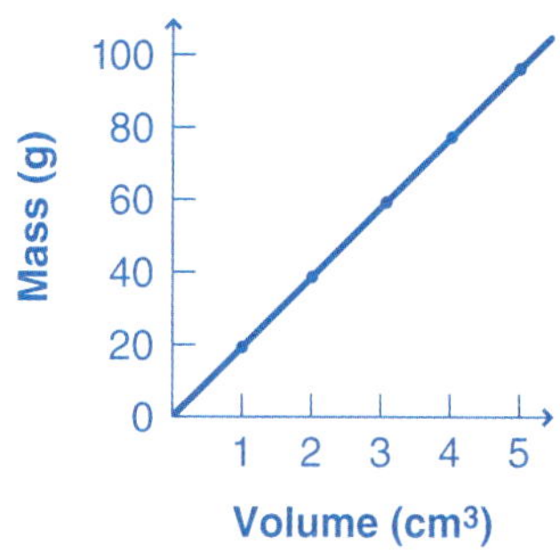

b. a straight line

c. The relationship is linear.

d. 19 g/cm^3

e. $m = (19\ \text{g/cm}^3)V$

f. The mass for each cubic centimeter of pure gold is 19 g.

Physics Challenge

student text p. 21

1.

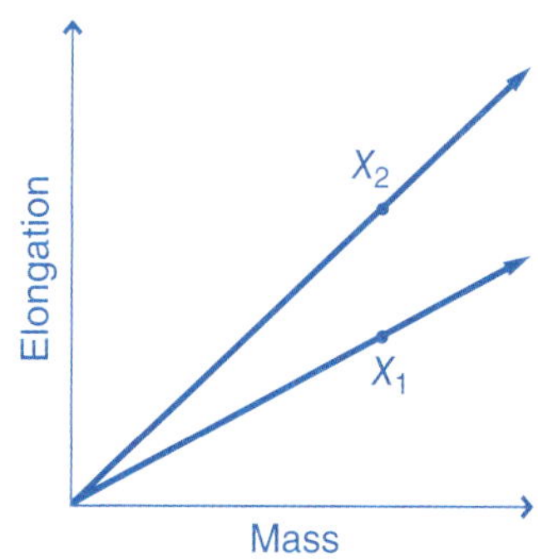

2. Yes; the origin corresponds to 0 elongation when the mass is 0.

3. The slope for the second spring is steeper.

4. $x_2 = 1.6x_1$, $5.3\ \text{cm} = 1.6x_1$, $3.3\ \text{cm} = x_1$

Section 4 Review

student text p. 23

22.

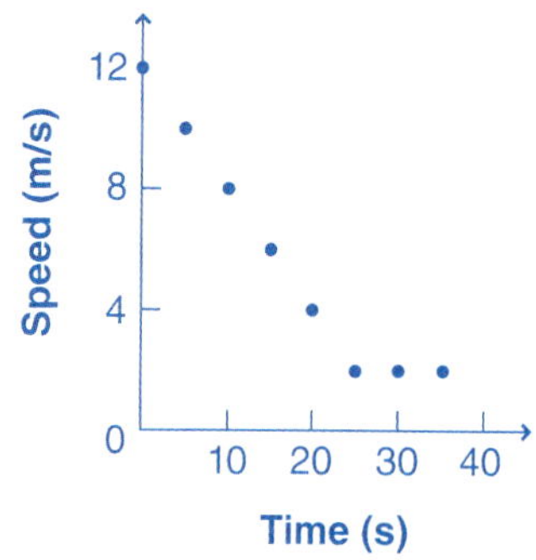

23. There is a nonzero total mass when the volume of the material is zero. This could happen if the mass value includes the material's container.

24. 16 g

25. About 2.6 h

26. The spring whose line has a smaller slope is stiffer and, therefore, requires more mass to stretch it 1 cm.

PHYSICS THAT'S ENTERTAINMENT!

Coming to Life

The Physics Behind Animation

Background

The first all-CGI (Computer Generated Image) feature-length animation was Pixar Studio's *Toy Story,* released in 1995. Pixar utilizes math-based animation techniques, as opposed to competing, big feature animation studios that primarily use motion-capture technology.

Teaching Strategies

- Show a clip from a Pixar film such as *The Incredibles* and as contrast, Dreamworks' *The Polar Express,* which filmed its human characters using motion capture. Ask students which appears to be more realistic.
- Ask students if they've seen an animated feature recently and if so, what aspects seemed physically accurate and whether they thought this was done by design or a result of limited technology.
- Ask students which topics in physics they think would be most helpful for animators. Student answers will vary. Possible answers include: particle dynamics (the interactive relationship between atoms and molecules), the representation of motion, the wave nature of light, forces within liquids and other mediums.
- Have students research the Special Interest Group on Computer Graphics and Interactive Techniques (SIGGRAPH), an annual conference that showcases developments in CGI technologies.

GOINGFURTHER >>>

Expected Outcomes Student answers will vary. Benefits include the money and time saved by not employing actors, not requiring the large area and special equipment needed for motion capture, and the ability to make changes in the computer rather than reshooting problematic scenes. Negatives include the substantial computing power needed to render complex CGI and the fact that motion capture can recreate more physically accurate movement in certain situations.

SECTION 1

Mastering Concepts

27. Possible answer: Identify a problem, gather information about it by observing and experimenting, create a model or theory to explain the results, analyze the information to test the model, and use the model to predict new results.

28. a. The solar system is very large.

b. Airplane aerodynamics are complex and dynamic.

c. The mathematical model can quantify the force each object is exerting.

SECTION 2

Mastering Concepts

29. Mathematics allows you to be quantitative, to say "how fast," not just "fast."

30. The International System of Units, or SI, is a base 10 system of measurement that is the standard in science. The base units are the meter, kilogram, second, kelvin, mole, ampere, and candela.

31. The derived units are combinations of the base units.

32. a. Zeros are necessary to indicate the magnitude of the value, but there is no way of knowing whether the instrument used to measure the values actually measured the zeros. The zeros may serve only to locate the 1.

b. Write the number in scientific notation, including only the significant figures.

33. a. centimeter

b. millimeter

c. kilometer

34. $\frac{60 \text{ min}}{1 \text{ h}}$

35. a. 3.49×10^{5} g

b. 2.87×10^{5} J/cm^3

Mastering Problems

36. a. 0.423 m

b. 6.2×10^{-12} m

c. 2.1×10^{4} m

d. 2.3×10^{-5} m

e. 2.14×10^{-4} m

f. 5.7×10^{-8} m

37. a. 6.12×10^{9} s

b. 2.94×10^{-6} m

c. 1.250×10^{-4} kg

d. 7.50×10^{7} g

38. 1.234 and 7.603 are tied with 4, 0.250 has 3, 0.13 has 2, and 0.08 has 1

39. a. 1

b. 4

c. 5

d. 1

e. 3

40. a. 34.7 m

b. 25.022 m

c. 46.00 cm^2

d. 3.1 kg

41. a. 2.9×10^{9} m^2

b. 2.0×10^{5} m/s

c. 1.3×10^{-6} km^2

d. 1.9×10^{2} kg/m^3

42. a. 408 N

b. 64.5 kg

43. No; it is in kg•s.

SECTION 3

Mastering Concepts

44. the precision of a measuring device, which is limited by the finest division on its scale

45. The final digit is estimated.

Mastering Problems

46. 48.2 kg

47. 2.4×10^{2} m^3

48. 362.1 m

49. ±0.05 g

50. 3.6 ± 0.1 A

51. A standard residential doorframe height is about 80 inches, which is about 200 cm. The precision depends on the measurement instrument used.

52. a. 1.2°C/h

b. about 8°C

c. No; temperature is unlikely to continue falling sharply and steadily that long.

SECTION 4

Mastering Concepts

53. The slope of a linear graph is the ratio of the vertical change to the horizontal change, or rise over run.

54. a. Positive. As speed increases, reaction distance increases.

b. Larger. The driver who was distracted would have a longer reaction time and thus a greater reaction distance at a given speed.

55. Temperature is the independent variable; volume is the dependent variable.

56. quadratic; $y = ax^2 + bx + c$

57. a. inverse relationship

b. linear relationship

c. quadratic relationship

Mastering Problems

58. a. (A) 80 g, (B) 260 g, (C) 400 g

b. (A) 36 cm^3, (B) 12 cm^3, (C) 7 cm^3

c. The slope represents the increased mass of each additional cubic centimeter of the substance.

d. The y-intercept is (0, 0). It means that when $V = 0\ cm^3$, there is none of the substance present ($m = 0$ g).

59. a.

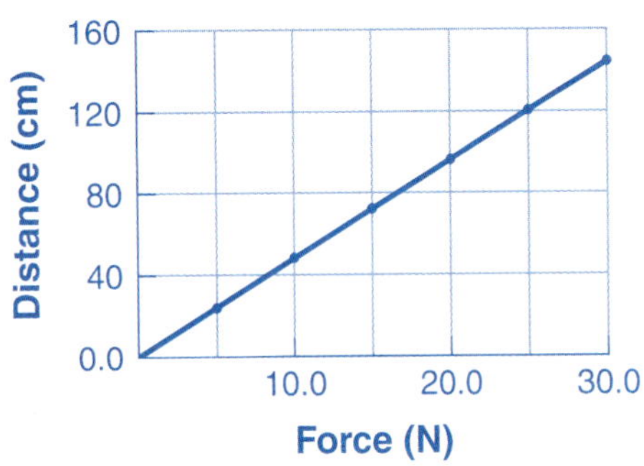

b. a straight line

c. $d = 4.9F$

d. The constant is 4.9 and has units cm/N.

e. 108 cm or 110 cm using 2 significant figures

60. a.

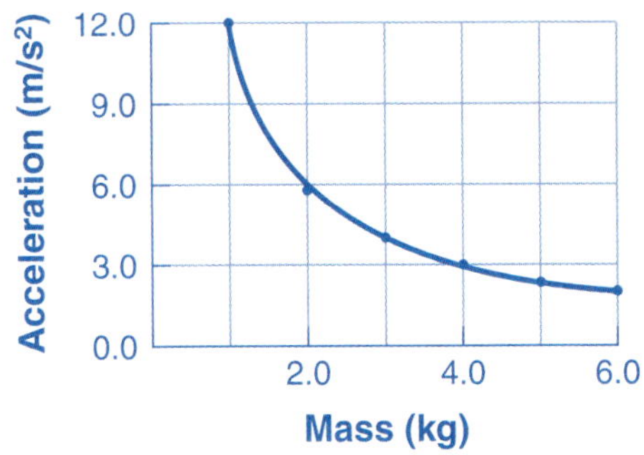

b. a hyperbola

c. $a = \dfrac{12}{m}$

d. $kg \cdot m/s^2$

e. 1.5 m/s^2

61. a.

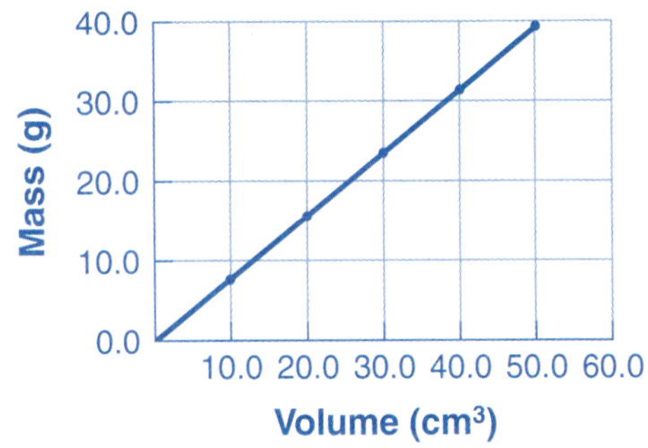

b. a straight line

c. $m = 0.79V$

d. g/cm^3; density

e. 25.7 g

Applying Concepts

62. There is no definite order of specific steps. However, whatever approach is used, it always includes close observation, controlled experimentation, summarizing, checking, and rechecking.

63. A scientific law is a rule of nature, where a scientific theory is an explanation of the scientific law based on observation. A theory explains why something happens; a law describes what happens.

64. When $t = 0$ and $t = 2$, the ball's height will be about 20 m. When $t = 5$, the ball will have landed on the ground, so the height will be 0 m.

65. a. possible answers include g/cm^3, kg/m^3

b. derived unit

66. **a.** cm

b. mm

c. m

d. km

67. The chart might include: radius of an atom, 5×10^{-11} m; virus, 10^{-7} m; thickness of paper, 0.1 mm; width of paperback book, 10.7 cm; height of a door, 1.8 m; width of town, 7.8 km; radius of Earth, 6×10^{6} m; distance to the Moon, 4×10^{8} m.

68. The chart might include: half-life of polonium-194, 0.7 s; time between heartbeats, 0.8 s; time to walk between physics class and math class, 2.4 min; length of school year, 180 days; time between elections for the U.S. House of Representatives, 2 years; time between U.S. presidential elections, 4 years; age of the United States, (about) 235 years

69. **a.** $(3.001 \pm 0.001)\times10^{8}$ m/s

b. $(2.999 \pm 0.006)\times10^{8}$ m/s

70. In addition and subtraction, you ask what place the least precise measure is known to: in this case, to the nearest centimeter. So the answer is rounded to the nearest centimeter. In multiplication and division, you look at the number of significant figures in the least precise answer: in this case, 2. So the answer is rounded to 2 significant figures.

71. The slope would be negative, because the change in vertical distance is negative for a positive change in horizontal distance

72. The slope is zero. The change in vertical distance is zero. *y* does not depend on *x*.

73. The units in each term of the equation must be in meters because distance (d) is measured in meters. $av^2 = a(\text{m/s})^2$, so a is in s^2/m; $bv = b(\text{m/s})$, so b is in s^{-1}.

74. 8.3 cm $\pm$ 0.05 cm or 83 mm $\pm$ 0.5 mm

75. A scientific theory has been tested and supported many times before it becomes accepted. A hypothesis is an idea about how things might work; it has much less support than a theory.

76. Possible answers include Newton's laws of motion, law of conservation of energy, law of conservation of charge, law of reflection

77. Air resistance affects many light objects. Without controlled experiments, everyday observations might have suggested to the ancient Greeks that heavier objects did fall faster.

78. $\pm$0.5 mL

79. Answers will vary, but a correct form of the answer is, "Every minute, three more people enter a room. If the room was initially empty at time = 0, how many people will be there after 8 minutes?"

Mixed Review

80. 0.0034 m, 45.6 m, 1234 m

81. 80 m is equivalent to about 260 feet, which would be very large. 5 meters would be a more reasonable value.

82. 162 shorts

83. volume = 1.87×10^{-4} m^3, density = 8.87 g/cm^3

84. 5.4×10^{7} y

85. 8.00 g/cm^3

Thinking Critically

86. The "right" question is one that points to fruitful research and to other questions that can be answered.

87. 286 kg

88. 0.0494 g/cm^3

89. mass of ball, footing, practice, and conditioning

90. Answers will vary. A possible form of the correct answer would be, ".... You then add 46.3 mL of rubbing alcohol to it. How much total liquid do you have?"

Writing in Physics

91. Answers will vary. For example, students might describe how scientists' views of the basic forces of nature have changed over time or how scientists' views of radiation have changed.

92. For example, students might suggest that improved precision can lead to better observations.

Standardized Test Practice

student text p. 31

Multiple Choice

1. C

2. C

3. B

4. A

5. A

6. B

Free Response

7. a. $a = \frac{F}{m}$

b. $\frac{1\ \text{kg}}{1000\ \text{g}}$

c. $a = \left(\frac{2.7\ \text{kg}\cdot\text{m/s}^2}{350\ \text{g}}\right)\left(\frac{1000\ \text{g}}{1\ \text{kg}}\right) = 7.7\ \text{m/s}^2$

8. $d = -\left(\frac{6}{7}\right)t + 11$

Rubric

The following rubric is a sample scoring device for free response questions.

Points	Description
4	The student demonstrates a thorough understanding of the physics involved. The response may contain minor flaws that do not detract from the demonstration of a thorough understanding.
3	The student demonstrates an understanding of the physics involved. The response is essentially correct and demonstrates an essential but less-than-thorough understanding of the physics.
2	The student demonstrates only a partial understanding of the physics involved. Although the student may have used the correct approach to a solution or may have provided a correct solution, the work lacks an essential understanding of the underlying physical concepts.
1	The student demonstrates a very limited understanding of the physics involved. The response is incomplete and exhibits many flaws.
0	The student provides a completely incorrect solution or no response at all.

CHAPTER 2

Representing Motion

About the Photo

Constant Speed Have students examine the photograph. Ask students how they know the horses are moving. Possible answers: the horses' manes are flying; the blur in the photograph. Then have students describe the horses' motion in more detail. Students might describe the positions of the horses' hooves, the angles of the legs, or how the horses' tails are shown.

Chapter Overview

This chapter introduces students to the idea of formally describing and analyzing motion. Students learn how to construct and analyze motion diagrams. Students then work with measurements of position, displacement, and time interval. Position-time graphs are introduced, along with some qualitative analysis of these graphs. Finally, students are introduced to velocity as the slope of the position-time graph and learn to distinguish between speed and velocity.

Before students study the material in this chapter, they should study:

- Graphing data
- Measurement in science
- Scientific method

To solve the problems in this chapter, students will need a strong grasp of:

- Significant figures
- Solving linear equations

Introduce the BIG IDEA

Motion of Objects Show students a street map with 2 points 10 blocks apart along a straight street–the points marked A and B. Have students imagine someone walking from A to B. Have students describe the person's motion in getting from A to B. The person would start walking from A, walk from block to block, and stop at B. Ask students to describe the person's speed as they walked. The person's speed was constant. How could the person know his or her speed? Time the walk from A to B with a watch, and divide the time by

(l)John Giustina/Photodisc/Getty Images; (r)Steve Allen/Brand X Pictures

Go online!
connected.mcgraw-hill.com

Your online experience includes:

- **Video**
- **Concepts In Motion**
- **Virtual Investigations**
- **All Lab, Applying Practices activities, and PBL worksheets**
- **Online Practice**
- **Personal Tutor**
- **Multilingual eGlossary**
- **Presentation**

Use the LaunchLAB

In *Toy Car Race,* students investigate the type of data needed to describe and compare speeds.

Use Physics T.V.

Video

MEASURING SPEED

Have you ever been passed by another car on the freeway? If you know a few important details, it's possible to determine how fast that car is going. It's physics in action on the freeway.

1 Introduce

Bellringer

Motion Categorization Show students objects that exhibit different motion, such as toys that accelerate (faster or slower) or move with constant velocity, swing back and forth, vibrate, or move in circles. Have students categorize them by how they move.
BL Kinesthetic EL

Tie to Prior Knowledge

Motion Students will have experienced motion and should be able to describe it. Ask them how they know something is moving, or what evidence it would take to convince them that something is moving.
BL Linguistic

2 Teach

All Kinds of Motion and Motion Diagrams

student text pp. 34–36

Concept Development

Motion Diagram Demo Help students connect motion diagrams with strobe photos by demonstrating with a strobe and a constant-velocity toy. Focus on the equal time intervals between flashes and the images produced.

Particle Models

student text p. 36

Use Figure 3

Motion Diagrams Go over particle model motion diagrams in **Figure 3** with students to make sure that they understand the major features. Remind students that the time intervals between any two adjacent dots in **Figure 3** are the same. OL

Critical Thinking

MAINIDEA Ask students why it is important that the time intervals used to show the motion in the motion diagram part of **Figure 3** are the same. It is important to change only one variable—the distance the runner moves. If time intervals vary, it would be harder to find how distance changes with time. OL

Use the PhysicsLAB

In *Motion Diagrams,* students will compare and contrast motion diagrams of toy cars.

Quick DEMO

Generating Motion Diagrams

Estimated Time 10 minutes

Materials video camera, television, plastic wrap or acetate sheet, transparency pen, overhead projector

Procedure Using the video camera, record a few seconds of a moving object. For a good connection to the content, choose an object with a relatively constant velocity. Play the recording for students and then return it to the beginning. Place the plastic wrap or acetate on the video display so that students see the object through the plastic or acetate. Choose a central spot on the object on which to apply the particle model first, then use the transparency pen to indicate that point on the plastic or acetate. Advance the recording by a few frames. Indicate the object's position again. Repeat this procedure until you have gone through the object's entire motion as recorded in the video. By placing the plastic or acetate on the overhead, you can show the motion diagram to the class easily.

Reinforcement

Particle Model Activity Ask students to explain the particle model and to give an example of when it does not apply. A simplified model is not useful when the motions of irregularly shaped objects are measured over short distances, especially when objects are being compared. One such example is a racehorse winning by a nose. BL Interpersonal

3 Assess

Assess the MAINIDEA

Motion Have students explain why it is important to use the same spot on an object each time its motion is marked. so you can use the dots to measure the motion of the object

Check for Understanding

Motion Diagram Show a constant-velocity motion diagram that has seven evenly spaced dots and tell students that it represents 12 s. Ask how much time passed between adjacent dots. 2 s BL

Reteach

Motion Diagram Tell students to imagine running with a wet paintbrush. Every 10 s they touch the brush to the ground. The series of paint marks is the motion diagram.

Reading Checks and Caption Checks

Caption Check, *student text p. 34*
The train would not appear blurry.

Reading Check, *student text p. 35*
You must know where an object is located at different times and when it is at each position.

Reading Check, *student text p. 35*
No; the spaces are only equal if the object is traveling at a constant speed.

Caption Check, *student text p. 36*
The dots are evenly spaced.

Section 1 Review

student text p. 36

1. A motion diagram shows the positions of a moving object at equal time intervals.
2. 
3. See Online Solutions Manual. The point should be close to the center of the car.
4. See Online Solutions Manual. The point should be close to the center of the bird.
5. 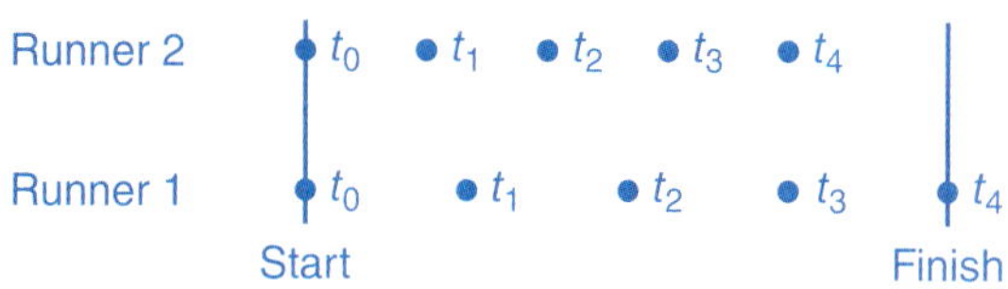

SECTION 2 Where and When?

student text pp. 37–40

1 Introduce

Bellringer

Where? Ask students how they know where something is. Then ask them about the location of a specific place, such as the cafeteria. In order for students to describe the location of the cafeteria accurately, they will have to give a reference point. This is a useful springboard from which to begin talking about coordinate systems and origins. **BL** **Visual-Spatial**

Tie to Prior Knowledge

Distance and Time Intervals Students are familiar with the concepts of distance and time interval, but their knowledge is probably not scientifically precise. The introduction of coordinate systems will help to formalize their knowledge. Students should be familiar with origins and axes from their math courses.

2 Teach

Coordinate Systems

student text pp. 37–38

Discussion

Problem Show students a constant-velocity particle model motion diagram without the beginning or end point indicated and no dot spacing information. Ask students what information might be useful to know that is not given by that representation.

Response It does not reveal in which direction the object was moving, where it started, where it ended, how much time had elapsed between the dots, or the distance between the dots. **OL** **Visual-Spatial**

Concept Development

Coordinate Systems Help students recognize why coordinate systems are important. Ask students to explain how to find their homes to someone from out of town. When students finish, ask them what their reference point or points were. **BL**

Critical Thinking

Position and Distance Ask students the difference between the position of an object and the distance of an object from the origin. Have them relate their answers to the position and distance of cities. Sample answer: The position of an object indicates exactly where the object is located. For example, when we say New York City is about 3950 km east of Los Angeles, we are giving New York City's position. When we say New York City is about 3950 km from Los Angeles, we are giving the distance from Los Angeles to New York City. **OL**

Vectors and Scalars

student text pp. 38–40

Reinforcement

Everyday Scalars and Vectors To emphasize the distinction between scalar and vector quantities, list one-by-one everyday words or examples that are scalars or vectors. Ask individual students to describe the scalar or vector nature of quantities such as 98.6°F, a ten-yard penalty, a kilogram of flour, and winds east at 10 to 25 km/h. **BL** **Linguistic**

Use the MiniLAB

In *Vector Models,* students will use construction toys to model vector addition.

DIFFERENTIATED INSTRUCTION

Physically Challenged If students have physical impairments that make it difficult to use a ruler and pencil to draw vectors, have students work in small cooperative groups with straws cut to different lengths. Cut each straw in half lengthwise to prevent it from rolling, cut a point in one end, and mark the length on the straw. Set this up as a challenge by mixing together sets of straws that add in one dimension and others that add at right angles (lengths in a ratio 3:4:5 or 5:12:13). This activity will help all students get a feel for vector addition and can easily be adapted to vector subtraction. **BL** **Kinesthetic**

Identify Misconceptions

Distance and Displacement Students might think that distance and displacement are interchangeable. The distance an object moves is a scalar quantity because it has no direction. The displacement of an object is the change of its position. The displacement has magnitude and direction. If an object moves 5 m to the right, it has a different displacement than if it moves 5 m to the left.

PHYSICS CHALLENGE ACTIVITY

MAINIDEA Refer students to **Figure 9,** and ask students what the coordinate system is. the straight line on which the person is running Ask students where the coordinate system places the biggest tree. The coordinate system places the biggest tree at 5 m. Have students describe what would happen to the positions and displacements of the runner if the coordinate system were changed. The positions would change, but the displacements would remain the same. AL

Use an Analogy

Vector Subtraction Activity Ask each student to express, in writing, step-by-step instructions for vector subtraction. Then ask them to illustrate their written instructions with an example. Alternatively, ask students to trade instructions with another student and to follow those instructions to see if they are accurate. For example, a hiker walks 5 km in a straight line away from camp and rests. The hiker then walks an additional 2 km away from camp along the same direction and rests again. The hiker's displacement between the two resting places is a vector pointed away from the camp with a magnitude of 2 km. The hiker's displacement from the beginning of the trip to the second rest is a vector pointing away from the camp with a magnitude of 7 km. BL Linguistic

REAL-WORLD PHYSICS

Olympic Timing Timing the races at the Olympic Games and at other major sporting contests is a very important aspect of the games. The times runners take to finish the race are recorded to the hundredth of a second and are used to determine world and Olympic records.

Use Figure 10

Vector Subtraction Ask students to use a ruler to prove the vector subtraction shown in **Figure 10.** Point out that although the mathematics is simple, they should be careful to watch the directions to avoid errors. OL

3 Assess

Assess the MAINIDEA

Position and Displacement Tell students that a runner starts in the middle of a straight block, and runs to one end. Have students describe the coordinate system of the runner. The coordinate system would be along the straight line from the middle of the block (the zero) to whichever end of the block the runner went. Have students explain under what condition might the runner's position be negative and displacement positive. if the runner ran back through the origin to the opposite end of the block

Check for Understanding

Vectors and Scalars Ask students to give examples of vector and scalar quantities, along with explanations of why those examples are appropriate. Sample answer: The mass of an object is a scalar quantity—it makes no sense to ask what direction the mass is in grams. Velocity, on the other hand, is a vector— when an object is moving, it makes sense to ask what direction the object is moving. BL

Extension

Time Intervals Ask students to think of three real-world instances in which there is a need to measure precise time intervals. For example, a competitive runner will want to know how long it takes her to run 400 m. OL Logical-Mathematical

Reading Checks and Caption Checks

Caption Check, *student text p. 38*

The arrow would indicate a position of +5.

Caption Check, *student text p. 39*

20 m to the left

Reading Check, *student text p. 39*

Distance is the length of the entire trip, which might include back and forth motion. Displacement is the distance and direction from the starting point.

Reading Check, *student text p. 39*

The direction of the arrow points in the direction from the origin to the object's position. The length represents the distance from the origin to the object's position.

Caption Check, *student text p. 40*

20 m north

Section 2 Review

student text p. 40

6. Sample answer: The position direction could be along the length of the pool. The origin could be at one of the short sides of the pool.

7. Start  End

8. A position vector goes from the origin to the object. When the origins are different, the position vectors are different. On the other hand, a displacement vector has nothing to do with the origin.

9. Home School

10. The two students should agree on the displacement, distance, and time interval for the trip because these three quantities are independent of where the origin of the coordinate system is placed. The two students would not agree on the car's position because the position is measured from the origin of the coordinate system to the location of the car.

1 Introduce

Bellringer

Position-Time Data Show students a toy car moving at a constant velocity. As a class, collect data about its position and time. Use these data to plot a position-time graph. You can then refer to this graph throughout the section. BL Visual-Spatial

Tie to Prior Knowledge

Graphs and Word Problems Students will be familiar with graphs from their math classes, but this graphing may not have had much context associated with it. Students should be familiar with independent and dependent variables, plotting points, line of best fit, and so on. Students also will be familiar with solving word problems from their math classes.

2 Teach

Finding Positions

student text pp. 41–43

Concept Development

MAIN IDEA To help students fully understand the graphing process and to connect it to data, take students through the process of constructing a position-time graph. Use the data from the Bellringer activity or data generated from equations of motion and prepared before class. BL Logical-Mathematical EL

Identify Misconceptions

Position-Time Graphs Some students might confuse a position-time graph with the particle model of a motion diagram. Ask students what information is in the position-time graph that is not in the motion diagram. The position-time graph is a pictorial representation of a table of data. A table of data contains more information than a motion diagram because the elapsed times and distances are actually recorded on the table. BL

Use Figure 11

Position In **Figure 11,** ask students to identify the position of the runner whose motion is represented after 2.0 s. 10.0 m

Critical Thinking

Position-Time Graph Ask students to describe what **Figure 11** would look like if the runner started from the same place but moved in the opposite direction. It would be a graph in the fourth quadrant that shows each subsequent point being farther away from the origin. AL

Content Background

Scientific Method Students can think of the graph of a runner's motion as the result of a scientific investigation. The hypothesis might be that the distance the runner covers increases as the time increases. The mass of the runner and the velocity of the runner are called the controlled variables because they do not change. If the measurements of position are made at equal intervals of time, time is called the independent variable since the experimenter determined the interval. The distance is called the dependent variable. It is customary to plot the independent variable on the horizontal axis and the dependent variable on the vertical axis.

Concept Development

Representations A student's house is on the same street as her school 10 blocks away. After 1 min, the student has walked 1 block; after 2 min, the student has walked 2 blocks; after 3 min, 4 blocks; after 4 min, 7 blocks; after 5 min, 9 blocks; and after 6 min, the student reaches the school. Ask students to represent the motion in three different ways. Students may draw motion diagrams, position-time graphs, and data tables to represent motion. OL

Use Figure 14

Show Motion Students should understand the different ways of representing motion. Have students look at each of the representations in **Figure 14** and discuss how each one shows how the object moves. The table gives exactly the same information as the graph, but the particle model gives much less information than the other two representations. You may want to revisit this again in Section 4, after the concept of velocity has been formalized. OL

Use Models

Position-Time Representations at Rest Data tables and position-time graphs are models that describe motion. Have students create these equivalent representations of motion for a migrating bird starting from rest and traveling at 10 km/h for 4 h, resting for 1 h, restarting at the same speed for another 2 h, and then resting again for 1 h. Explain that when an object is motionless, its slope on a position-time graph is a horizontal line.

Multiple Objects on a Position-Time Graph

student text pp. 44–45

Reinforcement

Objects Passing Ask students how you can use the same coordinate system to describe the motion of two different cars, A and B, on the same road. Each car has its own line on the graph. Ask students what the graph looks like if car A passes car B or if car B passes car A. The lines cross each other. **BL** **Linguistic**

Discussion

Problem In Example Problem 2, which runner started ahead of the other, runner A or runner B? What does ahead mean? If the slopes depicting runners A and B started at the same points and were rotated downward below the time axis, which runner would be considered starting ahead, and what would ahead mean in this case?

Response In the first case, runner A started ahead of runner B. Runner A started out at the origin, runner B started out from behind the origin, and both were running in the same direction. The downward graph would show that both runners were running away from the origin in the direction opposite the first situation. Runner B would have started ahead in this case because he started 50 m away from the origin in the direction of travel, while runner A started at the origin.
OL **Logical-Mathematical**

ADDITIONAL IN-CLASS EXAMPLE

Use with Example Problem 2.

Problem When and where does Tai pass Luis?

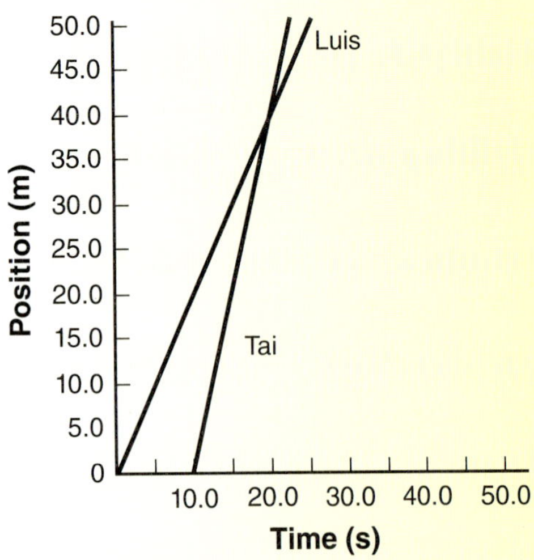

Response The point at which the two graphed lines intersect is at 20.0 s and 40.0 m.

3 Assess

Assess the MAIN IDEA

Motion Diagrams Have students imagine a bus that travels 2.0 km from rest, stops, backs up 0.5 km, then comes to rest again. Then it moves forward 1.0 km from rest and stops. Have students draw two motion diagrams, one for each of the directions of motion. Then have students assign one coordinate system for the complete motion. Ask students which displacements are positive and which are negative. If forward direction is positive, backing up is negative.

Check for Understanding

Graphs and Motion Diagrams Present students with a graph of an object moving away from the origin at a constant speed. Ask students to draw a motion diagram that corresponds to it. Make sure that they specifically indicate their coordinate system and the object's starting point. The motion diagram will be a series of dots evenly spaced. The first dot should be labeled *origin.*
OL **Logical-Mathematical**

Extension

Position-Time Graphs Have students look at **Figure 17.** Ask them to extrapolate the position of the hockey puck at 10 s. 200 m **BL** **Logical-Mathematical**

Reading Checks and Caption Checks

Caption Check, *student text p. 41*
The line corresponds to distance from the origin, which is increasing, not the direction of motion.

Caption Check, *student text p. 43*
Sample answer: The table provides exact values for the data at a glance.

Reading Check, *student text p. 43*
The instantaneous position of a runner is the runner's position at a specific moment.

Reading Check, *student text p. 44*
The intersection of two lines on a position-time graph tells you when objects have the same position.

Practice Problems

student text p. 42

11. The car begins at a position of 125.0 m and moves toward the origin, arriving at the origin 5.0 s after it begins moving. The car continues beyond the origin.

12.

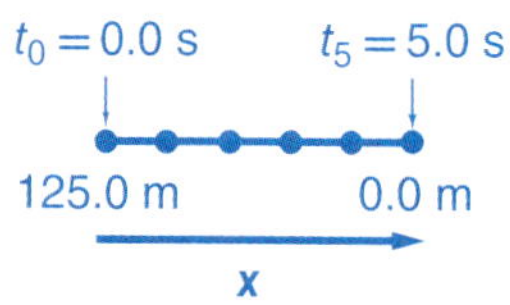

13. a. at 4.0 s

b. 100.0 m

c. 50.0 m

14. The pedestrians walk the same distance during each time interval, and they both walk east the entire time. Pedestrian A starts west of the origin, walks toward the origin, and continues walking east. Pedestrian B starts at the origin and walks east.

15. a. 19 s

b. 58 s

c.

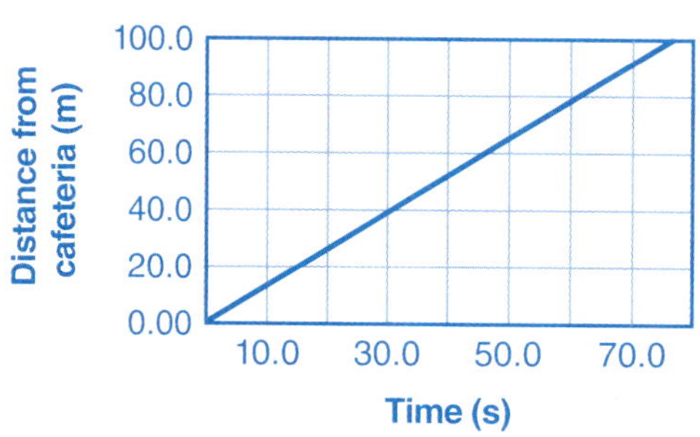

student text p. 45

16. Runner A passed the origin.

17. runner B

18. at –50.0 m

19. approximately 30 m

20. a. 6.0 min

b. No; the lines representing Juanita's and Heather's motions get farther apart as time increases. The lines will not intersect.

c. 1 km

d. $t = 1.8$ hr

Physics Challenge

student text p. 43

1.

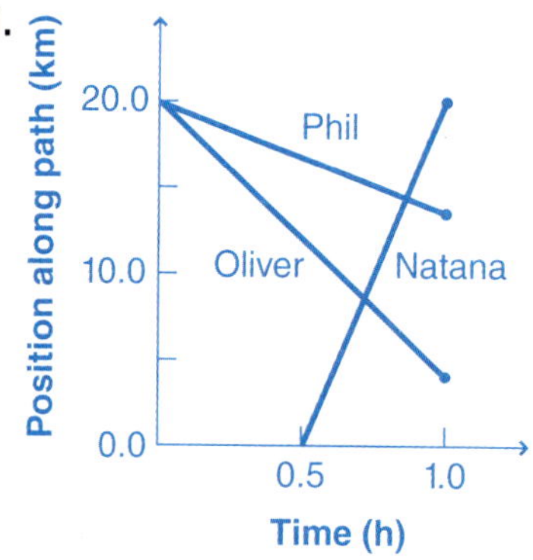

2. Natana passes Olivia at 12:13 P.M.

3. Phil is roughly 6.8 km north of Natana's and Olivia's positions.

Section 3 Review

21. See Online Solutions Manual.

22. See Online Solutions Manual.

23. 0.5 s

24. 100 m

25. 2.0 s

26. No, they don't describe the same motion. Although both objects are traveling in the positive direction, one is moving more quickly than the other. Students can cite a number of different specific examples from the graph and particle model to back this up. For example, the particle model shows a position of 2 m after 2 s, but the graph shows a position of 8 m after 2 s.

1 Introduce

Bellringer

Speed Ask a student to walk across the room two times—first slowly and then more quickly. Ask the other students if the first student did what you asked. Ask them how they know. Specifically, ask what evidence they used to make their decision. Have students generate a list of what physical quantities need to be known to determine how fast an object is going. The quantities are the initial position, the final position, and the time it took the student to go from the initial position to the final position. BL Visual-Spatial

Tie to Prior Knowledge

Speed Students will be familiar with the concept of speed. They might not know the difference between speed and velocity, however, and many will use these two terms interchangeably. If students introduce the term *velocity* into the discussion before you are ready to develop it, ask them what they mean.

2 Teach

Velocity and Speed

student text pp. 46–49

Concept Development

Average Velocity Demo You can model the example of the two joggers in class by two toys that move with different constant speeds. This could be done as an interactive demonstration or as a small group activity. You may even choose to have students take data and calculate the speeds of the toys. BL Kinesthetic

Critical Thinking

Interpreting Graphs Ask students if it is possible to have an accurate position-time graph that is a vertical line. no; because that would mean that the object would be in more than one place at a time or that it would have an infinite velocity AL

Applying Physics

Speed of Light According to the theory of relativity, elaborated in 1905 by Albert Einstein (1879–1955), the fastest possible speed anything can have is the speed of light. The first scientist to attempt to measure the speed of light was Galileo Galilei (1564–1642). His method was to place two men on mountaintops that were a known distance apart. Each man had a lantern with a shutter, and one of the men had an assistant with a device to measure time. However, over such a short distance Galileo was not able to make an accurate measurement. He could conclude only that light moves faster than sound. The speed of light was first successfully measured by Ole Roemer (1644–1710). Roemer's measurement was based on observations of the moonrise of one of Jupiter's moons.

Quick DEMO

Collision Experiment Problem

Estimated Time 10–15 minutes

Materials two slow-moving toy cars or toy tractors, metersticks, stopwatch or clock with a second hand

Procedure Have students determine the speed of each toy car by using the meterstick and the stopwatch. Have students write the speeds on the board. Set up the cars pointing toward each other, and ask students to predict where the two cars will collide. Run the demonstration to see whether the predictions are true. Lead students to understand that if the speeds of the toy cars are equal, the cars will collide in the middle, whereas if one car is faster than the other, the cars will meet at a point farther from the point where the faster car started.

DIFFERENTIATED INSTRUCTION

Struggling Students Students sometimes have trouble interpreting the units m/s properly, particularly the idea that an object's speed tells you how many meters that object travels in 1 s. To help students understand this, ask students to consider a car starting at the origin and traveling at 20 m/s. Ask them where the car is at 1 s, 2 s, 3 s, and so on; then ask them how far the car traveled during the first second, during the second second, and so on. Help students use this information to interpret speed properly. BL Logical-Mathematical

Use the PhysicsLAB

In *Constant Speed,* students will measure distance and time to find average speed.

Use the PhysicsLAB

In *Measure Velocity,* students will use a motion detector to measure velocity.

ADDITIONAL IN-CLASS EXAMPLE

Use with Example Problem 3.

Problem What is the average velocity of the object whose motion is represented in this graph? What is its average speed?

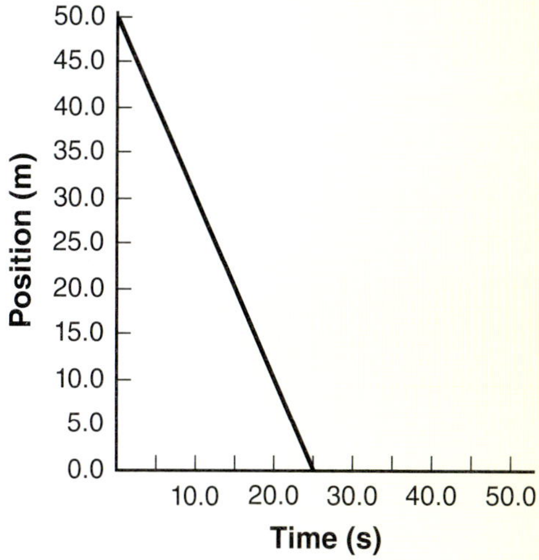

Response $\bar{v}$ = (50.0 m − 0.0 m)/(0.0 s − 25.0 s) − 2.00 m/s, 2.00 m/s

PHYSICS PROJECT ACTIVITY

Motion in a Straight Line Have students go around the school and record examples of different moving objects. Three examples should be of objects whose motion students can describe using the physics in this chapter. There also should be three examples of objects whose motion students cannot yet describe accurately. Students should specifically state why the current model of motion does not apply to these objects. For example, students should be able to represent the motion of a ball rolling on a horizontal floor from what they have learned in this chapter. They will not be able to model the motion of a ball bouncing down stairs, however. **OL** **Kinesthetic**

Reinforcement

MAIN IDEA Place a ball at the top of a long ramp and release it. Ask students to describe the instantaneous velocity of the ball as it rolls down the ramp. Explain that the velocity is zero at the top and increases until the ball reaches the bottom. Using a ruler and a stopwatch or clock, ask students to determine the average velocity of the ball as it changes position by moving from the top of the ramp to the bottom. **BL** **Visual-Spatial**

Identify Misconceptions

Instantaneous and Average Velocity In everyday language, the word *velocity* might be used to mean either instantaneous velocity or average velocity. As a result, students might confuse the two meanings. Clarify the difference by relating the terms to the motion of a car. Ask students how they know how fast they are going when they are traveling in a car. The speedometer indicates the instantaneous velocity. Ask students what the average speed of a car is in stop-and-go traffic if the car travels 40 km in 2 h. 20 km/h

Equation of Motion

student text pp. 49–51

Use Figure 23

Ask students how they know which car in **Figure 23** has the greater average velocity. The distance between intervals for the car traveling to the right is greater than those for the car traveling to the left. Also, the vector for the car traveling to the right is longer than the vector for the car traveling to the left. **BL** **Logical-Mathematical**

PHYSICS CHALLENGE ACTIVITY

Races Give students or have them find the speeds of several kinds of animals of different sizes. These speeds will be in different units. Ask students to predict the order of finish for a 100-m race among the animals. You can also ask them to figure out the time of each contestant. Then have them represent visually and explain mathematically how they arrived at their answers.
AL **Logical-Mathematical**

Reinforcement

Position Have students form pairs. Ask students to explain to each other what the four methods of representing the motion of an object moving at constant speed are. Each student should explain two of the methods. If a student does not understand the method explained, the student should ask questions. **BL** **Interpersonal**

3 Assess

Assess the MAIN IDEA

Velocity of an Object Have students draw a motion diagram for a truck moving at constant speed from point A to point B. Explain to students that the truck is traveling from city A to city B, a distance of 100 km. Tell students that the truck covers the first half of the trip at 50 km/h and the second half of the trip at 100 km/h. Have students find the average velocity of the truck.

$$\text{avg speed} = \frac{100\text{ km}}{\frac{50\text{ km}}{50\text{ km/h}} + \frac{50\text{ km}}{100\text{ km/h}}}$$

$$= 67\text{ km/h}$$

If the truck spent equal time at 50 km/h and 100 km/h, avg speed = 75 km/h.

Check for Understanding

Average Speed and Velocity Ask students the following: You walk down to the store 0.5 km away and then walk straight back. If the whole trip took 20 min, what was your average velocity? Your average speed? average velocity = 0; average speed = 3 km/h **OL**

Extension

Constant Velocity Divide the class into small groups, and ask each group to design a quick experiment to determine whether a person is walking with a constant velocity. Collect the experimental designs and then set up some of the most common suggestions at the beginning of the next class to have the class evaluate and test. **AL** **Interpersonal**

Reading Checks and Caption Checks

Caption Check, *student text p. 46*

3 m farther

Reading Check, *student text p. 47*

An upward slope above the *x*-axis means the velocity is positive, and the object is moving away from the origin. An upward slope below the *x*-axis means the velocity is positive, and the object is moving toward the origin. A downward slope above the *x*-axis means the velocity is negative, and the object is moving toward the origin. A downward slope below the *x*-axis means the velocity is negative, and the object is moving away from the origin.

Caption Check, *student text p. 47*

The slope would be positive.

Reading Check, *student text p. 49*

Average velocity is determined by dividing the difference in the final and initial velocities by the time interval. It does not take into account any changes in velocity during the time interval. Instantaneous velocity is the velocity of an object at a specific moment in time.

Reading Check, *student text p. 49*

Lengths of velocity vectors indicate the speed of the object compared to the speed of other objects.

Practice Problems

student text p. 48

27. a. 0.3 m/s

b. The average velocity is the slope of the line, including the sign, so it is −0.3 m/s or 0.3 m/s north.

28. The ship is moving north at a speed of 0.3 m/s.

29. −1.2 cm/s

30. a. 0.7 km/min

b. 0.7 km/min in the positive direction

31. The bicycle is moving in the positive direction at a speed of 0.7 km/min.

32.

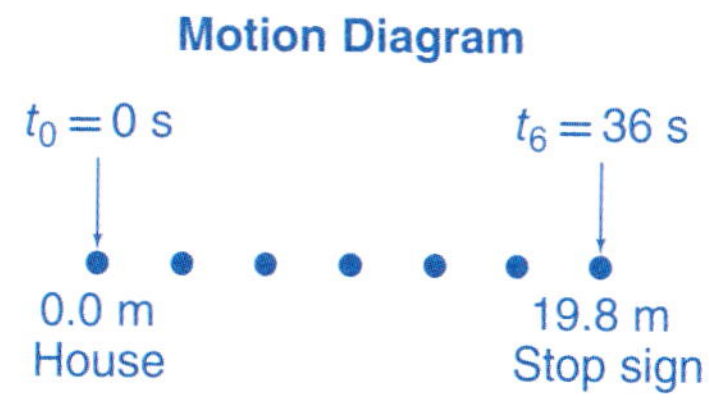

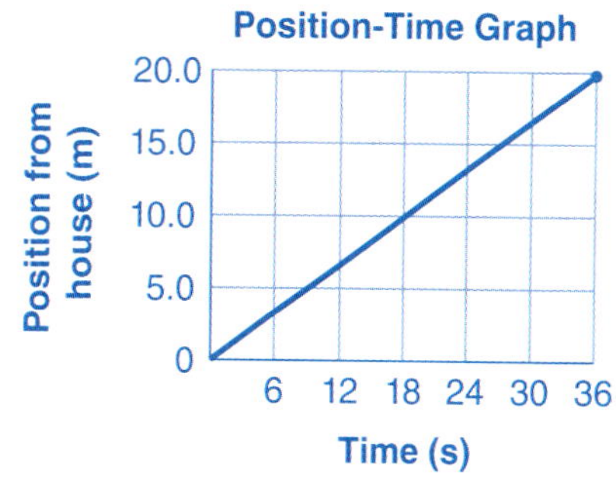

student text p. 51

33. 88 km east

34. 1.1×10^2 km east

35. 17 km west

36. 52 km west and 4.0×10^1 km west

Section 4 Review

37. An object's velocity is the rate of change in its position.

38. A, B, C = D

39. The magnitude of the average velocity of A is greater than that of B, but the average velocity of A is negative, and the average velocity of B is positive. The magnitudes of the average velocities of C and D are equal, but the average velocity of D is positive, and the average velocity of C is negative.

40. A, C, B, D. Yes; the ranking from greatest to least distance would be A, C, D, B.

41. Average speed is the absolute value of the average velocity if an object is in uniform motion.

42. The red car has traveled 8 km east of point B. The blue car has traveled 12 km west of point B. The red car's position is 14 km east of the origin. The blue car's position is 6 km west of the origin.

43. 23 km

44. Answers will vary. Drawing the models helps you organize the problem situation. It is difficult to write the proper equation if you don't have a clear picture of how things are situated and/or moving. Also, you choose the coordinate system in this step, and this is essential in making sure you use the proper signs on the quantities you will substitute into the equation later.

A CLOSER LOOK

Got the time?

Time Dilation

Purpose

Students will learn how motion can affect the rate at which time passes.

Background

The relativity of time, also called time dilation, is explained by the special theory of relativity. Other aspects of the special theory of relativity include length contraction, the limit of the speed of light, and mass-energy equivalence. The general theory of relativity expands on the special theory of relativity to explain how gravity affects space and time.

Teaching Strategies

- Students are probably used to the idea that one hour of time for one person is the same as one hour of time for another person. Reinforce the idea that this is an approximation that is only valid at slower speeds.
- Students often develop misconceptions regarding the relativity of time. For examples, a student might think that only actual clocks run faster or slower, or that a person "feels" fast or slow depending on her point of view. Reinforce with students the idea that time actually passes slower or faster depending on your point of view; it is not just an illusion.

GOINGFURTHER >>>

Expected Outcomes Earth's gravity affects the passage of time on a GPS satellite more than the satellite's motion does. A GPS satellite gains approximately 45 μs/day due to its position relative to Earth's surface in Earth's gravitational field. This is counter to the 7 μs/day loss due to the satellite's motion, resulting in a net gain of approximately 38 μs/day relative to Earth's surface.

SECTION 1

Mastering Concepts

45. A motion diagram gives you a picture of motion that helps you visualize displacement and velocity.

46. An object can be treated as a point particle if internal motions are not important and if the object is small in comparison to the distance it moves.

SECTION 2

Mastering Concepts

47. Position and displacement are different from distance because position and displacement contain information about the direction in which an object has moved, but distance does not. Distance and displacement are different from position because they describe how an object's location has changed during a time interval, but position indicates exactly where an object is located at a precise time.

48. Read the clock at the beginning and end of the interval and subtract the beginning time from the ending time.

SECTION 3

Mastering Concepts

49. Draw the two graphs on the same set of axes. One in-line skater will pass the other if the lines representing each of their motions intersect. The time coordinate of the point where the lines intersect is the time at which the passing occurs.

SECTION 4

Mastering Concepts

50. $\bar{v} \equiv \frac{\Delta x}{\Delta t} = \frac{x_f - x_i}{t_f - t_i}$

51. Both are straight lines that start at the same position, but the slope of the runner's line is steeper.

52. velocity

53. It is possible to calculate the average velocity from the information given, but it is not possible to find the instantaneous velocity.

Mastering Problems

54. 2.0×10^{1} m

55. 1.5×10^{11} m

56. Answers will vary. A possible form of the correct answer is, ".... If it takes the butterfly 7 seconds, what is its average speed?"

57. 18 min

58. 1.8 min

Applying Concepts

59. Moolinda, Dolly, Bessie, Elsie

60. If the rabbit ran twice as fast, the slope of the graph would be twice as steep. If the rabbit ran in the opposite direction, the magnitude of the slope would be the same, but it would be negative.

61. None of the combinations has the correct units, meters per second. In addition, $\Delta x + \Delta t$ increases when either term increases. The sign of $\Delta x - \Delta t$ depends upon the relative sizes of Δx and Δt. $\Delta x \times \Delta t$ increases when either increases. $\Delta t/\Delta x$ decreases with increasing displacement and increases with increasing time interval, which is backward from velocity.

62. The football can be treated as a particle if its rotations are not important and if the distance it moves is much larger than the football.

63. a. Runner A has a head start by four units.

b. Runner B is faster, as shown by the steeper slope.

c. Runner B passes runner A at point P and is ahead of runner A beyond that point.

Mixed Review

64. a.

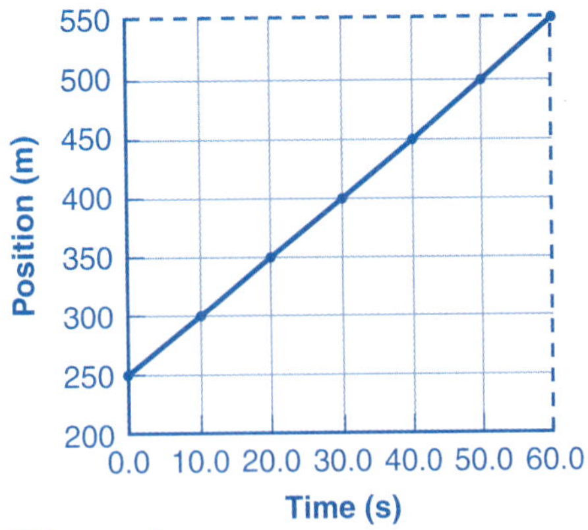

b. 550 m west

c. 3.0×10^2 m

65.

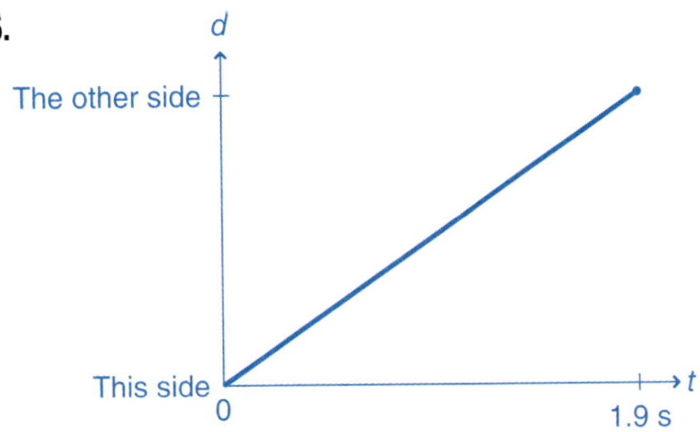

The equation is $\Delta x = \bar{v}\Delta t$.

66. a. 1.0 h

b. 45 min

c. from 6.0 to 9.0 km from the origin

67. a. See Online Solutions Manual.
Car A: 150 km
Car B: 170 km

b. Car A: 1.6 h
Car B: 1.4 h

68.

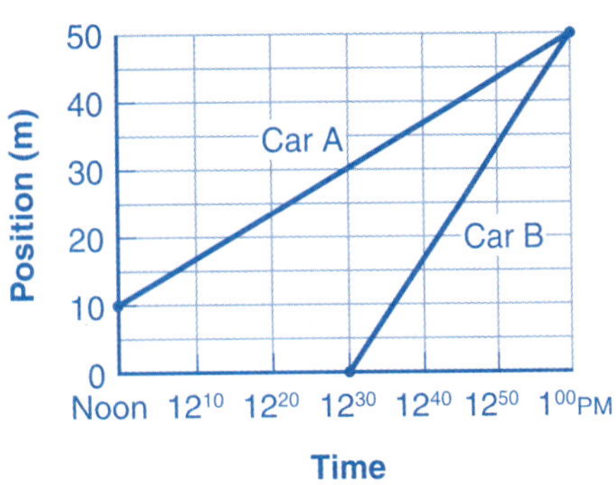

Both cars arrive at the beach at 1:00 P.M.

69. a.

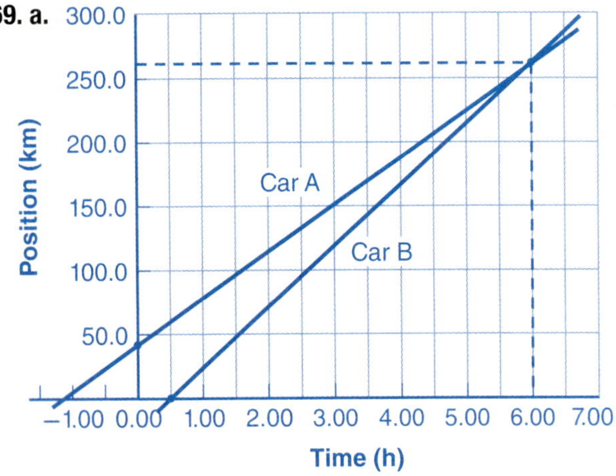

6.0 h

b. 2.6×10^2 km

c. 7.3 h

70. a. Answers will vary, but a correct form of the answer is, Jim walks 6 m in 7 s. He stops for 16 s. He walks 6 m in 9 s and stops for 5 s. He turns around and walks back toward the origin. He walks 9 m in 9 s and stops for 5 s. He then walks away from the origin again for 3 m in 1 s. He stops again for 5 s, and he turns around and walks 6 m back toward the origin in 6 s.

b. from 7.0 to 23.0 s, momentarily at 43.0 s, and from 52.0 s to 57.0 s

c. $\Delta t = 32.0\text{ s} - 0.0\text{ s} = 32.0\text{ s}$

d. -1.00 m/s

Thinking Critically

71.

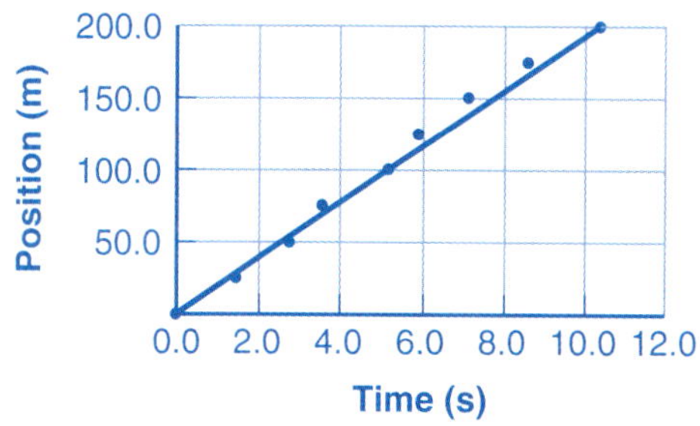

The slope of the line and the speed of the car are about 19.7 m/s.

72. 720 km/h; no

73. Possible answers: 1) Get several people together and give everyone a watch. Synchronize the watches and stand along the street separated by a consistent distance, maybe 10 m or so. When the motorcycle passes, have each person record the time (at least to an accuracy of seconds) that the motorcycle crossed in front of the person. Plot a position-time graph, and compute the slope of the best-fit line. If the slope is greater than 25 mph, the motorcycle is speeding. 2) Get someone with a driver's license to drive a car along the street at 25 mph in the same direction as you expect the motorcycle to go. If the distance between the motorcycle and the car decreases, the motorcycle is speeding. If the distance between them stays the same, the motorcycle is driving at the speed limit. If the distance increases, the motorcycle is driving less than the speed limit.

74. It is possible to have a horizontal line as a position-time graph. This would indicate that the object's position is not changing, or in other words, that it is not moving. It is not possible to have a position-time graph that is a vertical line, because this would mean the object is moving at an infinite speed.

Writing in Physics

75. Answers will vary. Galileo attempted to determine the speed of light but was unsuccessful. Danish astronomer Ole Roemer successfully measured the speed of light in 1676 by observing the moonrise of one of Jupiter's moons. His estimate was 140,000 miles/second (225,308 km/s). Many others since have tried to measure it more accurately using rotating toothed wheels, rotating mirrors, and the Kerr cell shutter.

76. Answers will vary. Examples of animals that have high endurance to outlast predators or prey include mules, bears, and coyotes. Animals that have the speed to quickly escape predators or capture prey include cheetahs, antelopes, and deer.

Cumulative Review

77. a. 5.8×10^{-8} s

b. 4.6×10^{7} s

c. 9.27 s

d. 1.23×10^{4} s

78. a. 4

b. 5

c. 3

d. 3

79. a. 7.4 mm

b. 49.6 m^2

c. 70.4 kg

Standardized Test Practice

student text p. 57

Multiple Choice

1. C
2. A
3. B
4. B
5. A

Free Response

6. $x = \bar{v}t + x_i$

$= (12.8\ \text{cm/s})(3.10\ \text{s}) + 0\ \text{s}$

$= 39.7\ \text{cm}$

The rat is 39.7 cm north of its starting point.

Rubric

The following rubric is a sample scoring device for free response questions.

Points	Description
4	The student demonstrates a thorough understanding of the physics involved. The response may contain minor flaws that do not detract from the demonstration of a thorough understanding.
3	The student demonstrates an understanding of the physics involved. The response is essentially correct and demonstrates an essential but less than thorough understanding of the physics.
2	The student demonstrates only a partial understanding of the physics involved. Although the student may have used the correct approach to a solution or may have provided a correct solution, the work lacks an essential understanding of the underlying physical concepts.
1	The student demonstrates a very limited understanding of the physics involved. The response is incomplete and exhibits many flaws.
0	The student provides a completely incorrect solution or no response at all.

CHAPTER 3

Accelerated Motion

About the Photo

Have students examine the photograph. Have students identify when the car experiences acceleration during a car race. Possible answers: when the car starts the race, when the car finishes the race and stops, whenever the car changes speed, whenever the car goes through a turn. Ask students to discuss how the car driver would know that the car is accelerating. Possible answers: The driver will feel a push or pull while inside the car, the driver might notice a change in the speedometer reading.

Chapter Overview

The chapter introduces the concept of acceleration as the rate of change of velocity. Velocity-time graphs and position-time graphs are used to develop motion equations and an understanding of the concept of motion with constant acceleration. Students use these equations to solve problems involving motion with constant acceleration. The chapter ends with a discussion of free fall as an example of this type of motion.

Before students study the material in this chapter, they should study:

- Adding vectors in one dimension
- Position-time graphs
- Uniform motion in one dimension
- Vectors vs. scalars

To solve the problems in this chapter, students will need a strong grasp of:

- Graphing data
- Significant figures
- Slope
- Solving linear equations
- Solving quadratic equations

Introduce the BIG IDEA

Demonstrate the use of a yo-yo in class. Define acceleration as a change in speed, and ask students to describe when the speed of the yo-yo is changing while in use. The speed of the yo-yo is changing nearly the entire time, both while falling and while returning to the hand.

Go online!
connected.mcgraw-hill.com

Your online experience includes:

- **Video**
- **Concepts In Motion**
- **Virtual Investigations**
- **All Lab, Applying Practices activities, and PBL worksheets**
- **Online Practice**
- **Personal Tutor**
- **Multilingual eGlossary**
- **Presentation**

Use the LaunchLAB

In *Graphing Motion,* students compare the graphs of an object moving at constant speed and an object moving with increasing speed.

Use Physics T.V.

Video

SKATEBOARD PHYSICS How does a trip to your local skate park involve physics? You might be surprised! Explore acceleration as skateboarders show off their best moves.

(l)Colin Anderson/Photographer's Choice/Getty Images, (r)Fuse/Getty Images

1 Introduce

Bellringer

A New Kind of Motion Show students an example of accelerated motion, such as a windup toy that slows down quickly, a runner just leaving the starting block, or a cart rolled onto sandpaper. Avoid using free fall as an example because the accelerated motion is difficult to observe. Have students describe differences between this type of motion and the uniform motion discussed in the previous chapter. **BL** **Visual-Spatial**

Tie to Prior Knowledge

Motion Diagrams and Graphs Sketch a motion diagram and a displacement-time graph of a person moving at a constant velocity, and have students interpret them. Ask students what quantity the slope of the line of the *x-t* graph physically represents. velocity (rate of change of position) Point out that they will use slope-analysis technique to describe motion that has a constant rate of change of velocity. **OL** **Visual-Spatial**

2 Teach

Nonuniform Motion Diagrams

student text pp. 60–61

Concept Development

Describing Velocity Changes Phrases such as *speed up* and *slow down* describe motion with specific changes in the velocity vector. If an object speeds up, consecutive velocity vectors in its motion diagram increase in length. If an object slows down, consecutive velocity vectors decrease in length.

Use Figure 2

Have students hold a plane mirror perpendicular to the page and look at **Figure 2** and its reflection. Ask students why the top motion diagram in the figure and its mirror image both show motion that is speeding up even though the motion is in opposite directions. The velocity vectors are increasing in length. Have students confirm that the bottom diagram and its mirror image both show motion that is slowing down in opposite directions because the velocity vectors are decreasing in length. **BL** **Visual-Spatial**

Use Figure 3

Have students recall how they determined $\Delta \mathbf{x}$ before. Point out that the same technique is used to determine $\Delta \mathbf{v}$ in **Figure 3.** The definition of $\Delta \mathbf{v}$, $\mathbf{v}_f - \mathbf{v}_i$, is rewritten as $\mathbf{v}_f + (-\mathbf{v}_i)$ where $(-\mathbf{v}_i)$ is a vector equal in magnitude to $\mathbf{v}_i$ but in the opposite direction. The vector sum of $\mathbf{v}_f + (-\mathbf{v}_i)$, that is, $\Delta \mathbf{v}$, is a vector whose base is located at the base of $\mathbf{v}_f$ and whose tip is located at the tip of $-\mathbf{v}_i$. **OL**

Reinforcement

Direction of $\Delta \mathbf{v}$ Emphasize that $\Delta \mathbf{v}$ is the change in velocity from $\mathbf{v}_i$ to $\mathbf{v}_f$. In **Figure 3,** have students note that as the speed of the object increases to the right, the length of the velocity vector stretches from $\mathbf{v}_i$ to $\mathbf{v}_f$ by an amount equal to the length Δv. Point out that the direction of the vector $\Delta \mathbf{v}$ is in the direction of motion. Have students draw the motion diagram of an object slowing to the right, noting that the velocity vector reduces in length from $\mathbf{v}_i$ to $\mathbf{v}_f$ by the amount Δv. Because the vector length has decreased, the direction of the change in the velocity vector, $\Delta \mathbf{v}$, is to the left, opposite the direction of motion. **OL** **Visual-Spatial**

Direction of Acceleration

student text p. 62

Use Figure 4

Point out that in the first and third motion diagrams, the lengths of the velocity vectors are increasing, indicating that the object is gaining speed. Also point out that $\mathbf{v}_i$, $\mathbf{v}_f$, $\Delta \mathbf{v}$, and $\mathbf{a}$ have the same direction. Tell students that they can predict that an object will speed up if its acceleration is in the same direction as its motion. Have students use analogous arguments to predict under what conditions an object will slow down. **OL**

Identify Misconceptions

Positive and Negative Acceleration Students often associate positive acceleration solely with speeding up and negative acceleration with slowing down. Ask students what effect a positive acceleration will have on an object moving in the positive direction. The object will speed up. Then, ask them what effect a negative acceleration will have on an object moving in the negative direction. The object will speed up. **OL**

Reinforcement

MAINIDEA Have students draw a motion diagram for the following situation, assuming the forward direction is positive. A stopped car (a) backs up, (b) stops, (c) moves forward, and then (d) stops. Ask students to identify which motions have positive accelerations and which have negative accelerations. Motions (b) and (c) have positive accelerations; (a) and (d) have negative accelerations. BL Visual-Spatial

Velocity-Time Graphs

student text p. 63

Discussion

Question A classmate is skateboarding along a level surface into a very strong wind, which slows her and eventually moves her backward. As her motion changes direction, what is her velocity? Is her acceleration as her motion changes positive, negative, or zero?

Answer The instant she changes direction, her instantaneous velocity is zero. If her initial direction of motion is positive, then her acceleration is negative, and vice versa. OL

Average and Instantaneous Acceleration

student text p. 64

Use the MiniLAB

In *Steel Ball Race,* students can compare the motion of two balls that move down a ramp from different heights.

Use the PhysicsLab

In *Acceleration,* students can use a photogate and timer to measure the motion of a rolling ball, and then calculate the acceleration of the ball.

Use the PhysicsLAB

In *Tossed-Ball Motion,* students can use a motion detector to collect data about a ball thrown straight upward, and then analyze graphs of the data.

Calculating Acceleration and Acceleration with Constant Speed

student text pp. 65–67

DIFFERENTIATED INSTRUCTION

Struggling Students For each student, prepare an index card with a different *v-t* graph sketched on it. Set up a coordinate system in the classroom and have each student demonstrate the motion represented by the graph. Have classmates volunteer to sketch on the board the graph they think represents the motion demonstrated by the student. BL Kinesthetic EL

3 Assess

Assess the MAINIDEA

Acceleration While Walking Ask students to describe the acceleration they might experience when they are walking in the hallway between classes. accelerating whenever they are speeding up, slowing down, or changing direction Ask them to describe when they are NOT accelerating. whenever they are walking in a straight line at a constant speed or standing still

Reteach

Motion and Vectors Draw two consecutive velocity vectors of increasing length and the same direction on the board, labeling them v_1 and v_2, respectively. Explain that the two velocities are positive. Have students explain how they would use the velocity vectors to find the direction of the acceleration vector. Find Δv by subtracting v_1 from v_2. The direction of a is that of Δv. OL Visual-Spatial

Extension

Experiences with Accelerated Motion Ask students to describe experiences that they have had within the past day or so involving accelerated motion. Have students estimate the accelerations or explain how they might measure them. OL

Reading Checks and Caption Checks

Reading Check, *student text p. 61*
Increasing length indicates the object is speeding up. Decreasing length indicates the object is slowing down.

Caption Check, *student text p. 61*
No; the acceleration would be zero because there is no change in velocity.

Reading Check, *student text p. 62*
The object is slowing down.

Reading Check, *student text p. 63*
The object's speed decreases to zero.

Caption Check, *student text p. 64*
2 m/s²

Reading Check, *student text p. 64*
Instantaneous acceleration is the change in velocity at an instant of time, but average acceleration is the change in velocity during a time interval divided by that time interval.

Reading Check, *student text p. 67*
The object can accelerate by changing direction.

Practice Problems

student text p. 66

1.

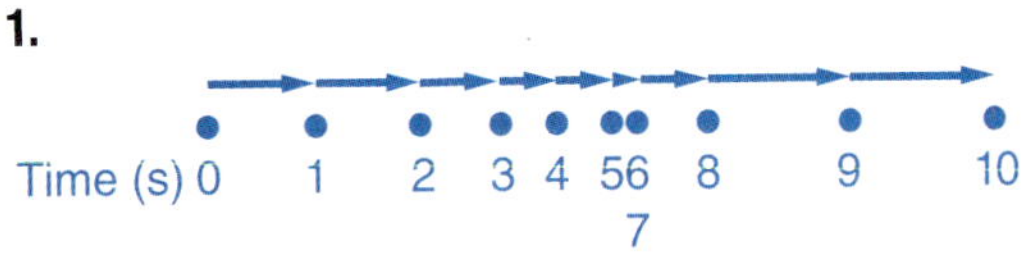

2. a. 5.0 to 15.0 s

b. 0.0 to 5.0 s

c. 15.0 to 20.0 s

3. a. 2.0 m/s²

b. −1.2 m/s²

c. 0.0 m/s²

4.

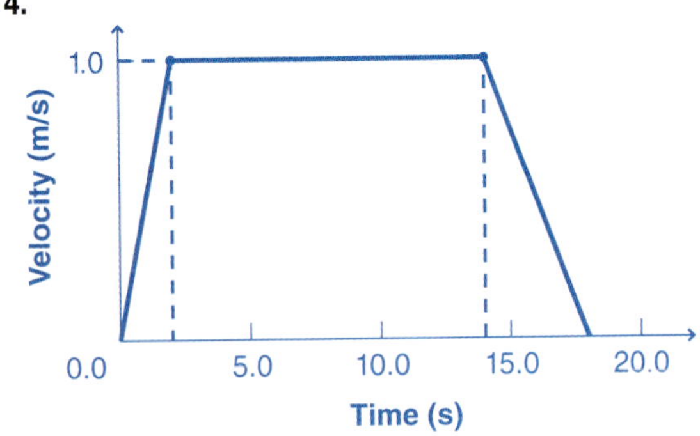

student text p. 67

5. 8.0 m/s² forward

6. 7.0 m/s² backward

7. a. 8.3 m/s² east

b. half as great (−4.2 m/s² east)

8. 3.0 m/s²

9. 0.28 m/s² west

10. 0.5 cm/y² in the direction opposite of the drift

Section 1 Review

student text p. 67

11. speed up, slow down, change direction

12. a. Both lines would have the same slope, but they would rise from the *x*-axis at different points, +15 m, and −15 m.

b. Their velocity-time graphs would be identical.

13.

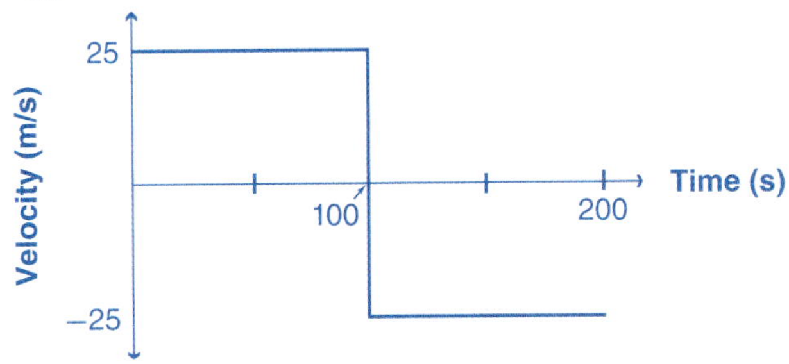

14. a. 1 m/s downstream

b. −0.75 m/s² downstream

15. See Online Solutions Manual. No; the cars had the same position, not velocity. To have the same velocity, they would have had to have the same relative position for a length of time.

1 Introduce

Bellringer

Incline Motion Construct two U-channel inclines at visibly different angles. Have students observe as you release a steel ball from rest on each incline. Ask students to identify evidence that the balls had different accelerations. It took different times for the balls to move the same distance from rest. **BL** **Visual-Spatial**

Tie to Prior Knowledge

Slope and Average Acceleration Remind students that in the first section they determined that the slope of a velocity-time graph is the average acceleration.

2 Teach

Position with Constant Acceleration

student text p. 68

Quick DEMO

Average and Final Velocity

Estimated Time 15 minutes

Materials constant-velocity vehicle, 100-cm U-channel incline, ball

Procedure Incline the U-channel so that the ball and vehicle travel 100 cm in the same time. Point out that the vehicle and the ball will both move 100 cm in the same time. Release the ball from rest at the top of the incline just as the vehicle begins its 100-cm trip. Have students observe that both reach the end of the ramp at the same time. Be sure to have a student stop the ball before it moves off the ramp. Guide students in understanding that the magnitude of the average velocity of each vehicle was the same. Point out to students that they can relate the final velocity of the ball to its average velocity. Repeat the demonstration, but this time let the ball roll off the ramp. Stop the ball and the vehicle simultaneously, and show that the ball has moved twice as far from the end of the ramp as the vehicle. Ask how fast the ball was moving compared to the vehicle. twice as fast because it went twice as far from the end of the ramp Explain that the final velocity of an object moving with constant acceleration from rest is twice its average velocity during the time interval.

Use the PhysicsLAB

In *Measuring Acceleration,* students can use a motion detector to collect data about a moving cart and then analyze graphs of the data.

PHYSICS CHALLENGE ACTIVITY

Deriving $x_f = \frac{1}{2}at_f^2$ Remind students that the Quick Demo showed that the final velocity of an object moving with constant acceleration from rest equals twice its average velocity during that time interval. Have students use this conclusion and the definition of $\bar{a}$ to show that $x_f = \frac{1}{2}at^2$ for an object initially at rest. If $v_i = 0$, $t_i = 0$, and $x_i = 0$, then $v_f = at = 2v_{ave} = \frac{2x_f}{t_f}$ or $at_f = \frac{2x_f}{t_f}$. Solving for x_f yields $x_f = \frac{a(t_f)^2}{2} = \frac{1}{2}at_f^2$.

AL **Logical-Mathematical**

ADDITIONAL IN-CLASS EXAMPLE

Use with Example Problem 3.

Problem The *v-t* graph below represents the motion of a car backing out of a driveway. What is the car's displacement at $t = 4.0$ s?

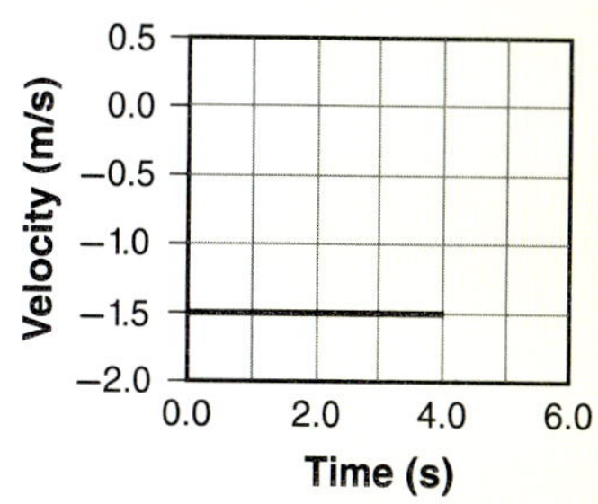

Response

$x = vt = (-1.5 \text{ m/s})(4.0 \text{ s}) = -6.0 \text{ m}$

Identify Misconceptions

Area Under a v-t Graph In Example Problem 3, students might consider the height of the area to be 75 m/s − 70 m/s. Remind students that v is the height of the plotted line above the t-axis, which is understood to intersect the v-axis at $v = 0$.

Reinforcement

Area of a v-t graph Most students have calculated the surface area of a block by multiplying two perpendicular length measurements. Point out that any rectangular area has a dimension that is the product of the dimension displayed along the x-axis and the dimension displayed along the y-axis. In a v-t graph, the x-axis displays time and the y-axis displays a rate (velocity). The dimension of this area is (m/s)•s = m, which is a physical quantity. Thus, the area represents a physical quantity. Have students consider another rate-time graph—one in which pay rate ($/h) is plotted on the y-axis and daily hours worked in a week (h) is plotted on the x-axis. Ask students what the area of this graph represents and what is its dimension. weekly pay, dollars **BL** **Visual-Spatial**

Critical Thinking

Positive and Negative Areas of a v-t Graph On the board, sketch the graph below.

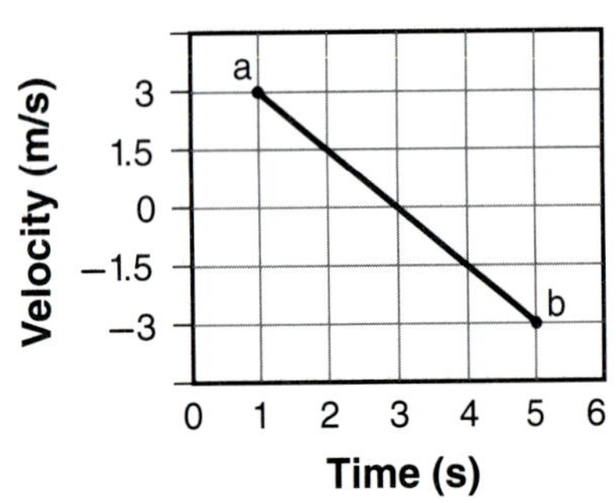

Have students explain what the displacement is for the time interval t_a–t_b. The displacement is 0 m. The areas bounded by v and the t-axis during the first half and second half of the time interval are equal, which indicates equal distances traveled. However, the first displacement is positive, while the second is negative. The total displacement for the interval is the sum of two equal-sized displacements in opposite directions, which is 0 m. **AL** **Visual-Spatial**

Applying Physics

Explain to students that when a racecar driver is trying to achieve maximum acceleration, the weight distribution and the traction of the vehicle are affected. Draw a graph that shows as acceleration increases, the weight on the front wheels decreases, while the weight on the rear wheels increases. The weight that must be placed on the front wheels in order for the tires to grip, called the traction limit, also increases during acceleration. Thus, as the force increases during acceleration, the reaction force and traction decrease in the front of the car and increase in the rear. The limit to acceleration is the point at which the traction is reached, or when the front wheels lift off the ground and there is a loss of directional control.

Careers

Motion Control Engineering Many manufacturing companies make use of robotics to automatically fabricate and package their products along advanced conveyor systems. Motion control engineers are responsible for designing robotic automation systems, integrating the operation of a variety of mechanical, optical, and electronic devices. To make sure products are manufactured properly, motion control engineers address such questions as how fast and how far an object must move on a conveyor, when the object should be moving, and where it must be at any given time. The precise determination of acceleration, velocity, and position of an object at any given time during system operation is an essential aspect to this work.

Discussion

Problem Draw students' attention to the position with average acceleration equation. Ask, *What change should you make to this equation if you want to apply it to a time interval that does* not *have an initial time of zero?*

Response Replace t_f^2 with Δt^2. **OL**

Velocity with Average Acceleration

student text pp. 69–71

Identify Misconceptions

Average Acceleration and Constant Acceleration Students might think that they can only apply the average acceleration equation for a certain time interval if the acceleration during the time interval remains unchanged. Remind students that an acceleration that does not change is constant acceleration. Average acceleration is different because it might change many times during the time interval. **OL**

An Alternative Equation

student text pp. 71–74

ADDITIONAL IN-CLASS EXAMPLE

Use with Example Problem 4.

Problem Joni jogs at a velocity of 2.50 m/s. If she then accelerates at a constant –0.10 m/s^2, how fast will she be jogging when she has moved 10.0 m?

Response

$v_f^2 = v_i^2 + 2a(x_f - x_i)$; $v_f^2 = (2.50\ \text{m/s})^2 + 2(-0.10\ \text{m/s}^2)(10.0\ \text{m} - 0\ \text{m})$ or $v_f = 2.1$ m/s

Quick DEMO

Relating Vectors *a* and *v*

Estimated Time 10 minutes

Materials U-channel, steel ball

Procedure Arrange the U-channel to form a trough with sides of equal slant. Before releasing a steel ball from rest down the left incline (facing the students), have students predict how far the ball will roll up the right incline. Release the ball and have students observe that within experimental limits, the distances are equal. Use the demonstration to review the algebraic sign of the ball's acceleration on each incline if motion to the right is positive. positive on the first incline and negative on the second Ask students why they can use the equation $v_f^2 = 2ax + v_i^2$ to show that the magnitude of the ball's acceleration on each incline is the same. The ball's final velocity on the first incline equals its initial velocity on the second incline.

PHYSICS CHALLENGE ACTIVITY

Equal but Opposite Accelerations Have students collect displacement data from the Quick Demo. From the displacement data, have them calculate the magnitudes of the two accelerations. They will discover that the magnitude of the acceleration on the first incline is equal to the magnitude of the acceleration on the second incline. As an alternative method, have them show that the two accelerations are equal in magnitude by considering the time for the ball to traverse the first incline and then the second. **AL** **Logical-Mathematical**

Concept Development

Motion with Different Accelerations Remind students that motion problems must be broken into parts whenever the acceleration changes. For example, in Example Problem 5, *a* is 0 m/s^2 for reacting and 8.5 m/s^2 for braking.

ADDITIONAL IN-CLASS EXAMPLE

Use with Example Problem 5.

Problem A cat runs at 2.0 m/s for 3.0 s and then slows to a stop with an acceleration of –0.80 m/s^2. What is the cat's displacement during this motion?

Response

$x_i = (2.0\ \text{m/s})(3.0\ \text{s}) = 6.0\ \text{m}$;

$x_f = (v_f^2 - v_i^2)/2a =$

$((0.0\ \text{m/s})^2 - (2.0\ \text{m/s})^2)/2(-0.80\ \text{m/s}^2) = 2.5\ \text{m}$;

$x_{total} = x_i + x_f = 6.0\ \text{m} + 2.5\ \text{m} = +8.5\ \text{m}$

3 Assess

Assess the MAIN IDEA

Acceleration Ask students to identify the acceleration of both graphs in **Figure 12.** The graph on the left represents an acceleration of 5.0 m/s^2, and the graph on the right represents an acceleration of zero. What would change on the graphs if both acceleration values were doubled? The slope of the graph on the left would double, and the slope of the graph on the right would remain at zero.

Check for Understanding

Motion Equations Write the equation $v_f = v_i + at_f$ on the board. Ask students to explain whether the equation could be used to calculate the answer to the following problem: Calculate the final velocity of a car that has a constant acceleration of 2.0 m/s^2 for 4.0 s. It can't be used because both the direction of the acceleration and the car's initial velocity are unknown. **BL**

Reteach

Motion Equations State the above problem again, and explain that the information can be used only to calculate the size of Δv. Point out that because neither the direction of the acceleration nor Δv is known, you cannot determine whether the car is speeding up or slowing down. **OL** **Visual-Spatial**

Reading Checks and Caption Checks

Reading Check, *student text p. 68*
parabola

Caption Check, *student text p. 69*
5.00 m/s²

Practice Problems

student text p. 69

16. a. 1.0 m/s

b. −1.0 m/s

c. The ball's velocity decreased in the first case. In the second, the ball slowed to a stop and then began rolling back down the hill. See Online Solutions Manual.

17. 67 km/h east

18. 5.1 s

19. 9.0 s

student text p. 70

20. $\Delta x_A = 9.0$ m north; $\Delta x_B = 8.0$ m north

21. $\Delta x_C = 8.0$ m south; $\Delta x_D = 4.0$ m south

22. a. See Online Solutions Manual.

b. $\Delta x = 150$ m west

c. See Online Solutions Manual.

d. The displacement was the same for both cars. For the second car, then, $v = \frac{\Delta x}{t} = \frac{150\text{ m}}{12\text{ s}} = 13$ m west (rounding to the correct number of significant figures).

student text p. 72

23. 8.8 s

24. 360 m

25. 6.3 s

26. 0.94 m/s north

27. 17 m/s west

student text p. 74

28. 32 m east

29. 4.3×10^2 m

30. 81.0 m, 1.16×10^3 m

31. 0.077 m/s² in the positive direction

32. 27 m east

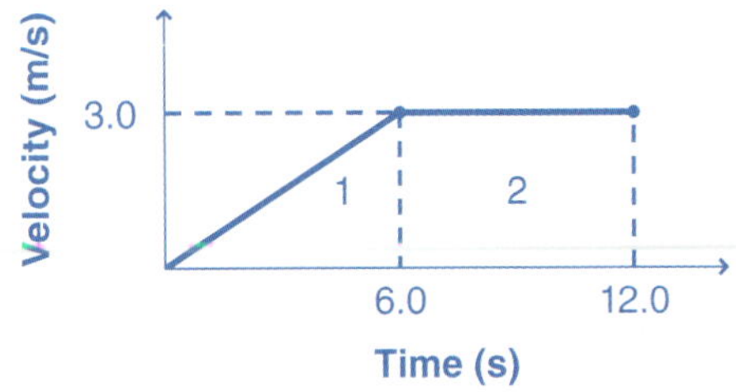

Section 2 Review

student text p. 74

33. $v_f^2 = v_i^2 + 2a\Delta x$

34. 1.3 m/s² east

35. a. 1.35×10^3 m east

b. 90.0 m/s

36. 34 m

37. 71 m/s north

38. 7.0×10^1 m/s south

39.

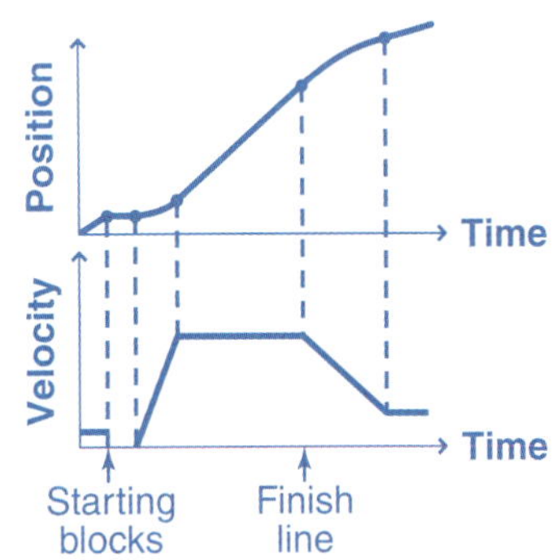

40. One person reads a stopwatch and calls out time intervals. Another person reads the speedometer at each time and records it. Plot speed versus time and find the slope.

1 Introduce

Bellringer

Vertical Incline Show students a U-channel inclined at about 30°, and then raise it to about 60°. Ask them on which incline would a rubber ball have a greater constant acceleration. the steeper one Hold the U-channel vertically and drop a ball along it. Have students draw upon the first two examples and ask them if they think that the ball most likely has a constant downward acceleration. OL Visual-Spatial

Tie to Prior Knowledge

Analyzing and Describing Free Fall Explain to students that all graphical methods of analyzing motion with constant acceleration and the relevant equations developed in Sections 1 and 2 can be applied to free fall, which they will study in this section.

2 Teach

Galileo's Discovery

student text p. 75

Use Figure 19

Have students compare the motion of the hammer and the feather in **Figure 19.** Ask them how the motion would differ if the objects were dropped near Earth instead of the Moon. The feather would fall much slower because of air resistance.

Free-Fall Acceleration

student text pp. 76–78

PHYSICS CHALLENGE ACTIVITY

Motion Photographs The photograph of the falling object in **Figure 20** was made with a camera using multiflash photography. This technique allows you to study various aspects of an object's motion, such as its speed or frequency of vibration. The camera causes the object to appear to slow down or even stop by producing pictures at an interval of about 0.06 s. Provide students with several images of objects in free fall, similar to the photograph of the falling object. Also provide them with data regarding position and time interval, and have them calculate the velocity and acceleration of the object. After the activity, point out to students that this is an example of how people in a variety of careers might use physics. AL Logical-Mathematical

Use Figure 20

To help students understand the relationship between the multiflash photo of an object falling and a ball thrown upward, have them draw the displacement over time of the ball thrown upward. Students' drawings should look similar to the photo.

Use the MiniLAB

In *Free Fall,* students can use the motion of falling objects to estimate free-fall acceleration.

Identify Misconceptions

Graphs and Paths Some students might think that the parabolic line of the position-time graph is the path of the moving ball. Point out that the ball is moving vertically in a straight line. The graph does not have this shape. Explain that the parabolic shape is that of the equation relating position and time. To emphasize this point, refer students to **Figure 20.** Ask them what shape would be produced if each successive image of the object were moved slightly to the right and a line were drawn to connect them. half a parabola Point out that the *x-t* graphs shown in **Figure 22** represent a series of snapshots separated by time of an object moving in a straight line. OL

Use Models

Vertical Motion Diagrams Have students sketch a motion diagram of an object moving horizontally with a constant acceleration in the direction of motion. Have them rotate their sketches so that the velocity vectors point downward. Point out that they now have a model for objects in free fall—the acceleration is always downward (and on Earth, approximately 9.8 m/s^2). OL Visual-Spatial

Use the PhysicsLAB

In *Free-Fall Acceleration,* students can use data they collect with a spark timer to calculate free-fall acceleration.

Reinforcement

Direction of Free-Fall Acceleration An object in free fall always accelerates downward, but the acceleration used in solving problems may be positive or negative, depending on which direction has been chosen as positive. Divide the students into two groups and have one group calculate the time for an object to fall 2.0 m from rest, assuming upward motion is positive. Have the other group do the same calculation assuming downward motion is positive. Have each group explain the calculation, and have the class compare the two answers. BL Interpersonal

Critical Thinking

MAINIDEA Toss a ball in the air and ask students what the acceleration of the ball is at the very top of its flight. The acceleration of the ball throughout its flight (even at its highest point) is 9.8 m/s^2 down. If students have difficulty with this concept, ask them to describe the change in speed and acceleration as the ball rises and then falls, including when the ball is at its highest point. At its highest point, the speed of the ball is 0 m/s. Does the acceleration ever change? No—it is always 9.8 m/s^2 down. OL

Content Background

Motion with Nonconstant Acceleration Although free-fall acceleration is assumed constant near Earth, it varies inversely with the square of the distance from the center of Earth. Thus, free-fall acceleration cannot be considered constant over large changes in altitude. On a much smaller distance scale, the non-uniform motion of a body on a spring, a falling raindrop, or a charged particle are other examples of motion with non-constant acceleration. The equations developed in this chapter cannot be used to model this motion. Such motion can be modeled, however, by calculus and by the use of specialized computer programs.

Concept Development

Special Case of Constant Acceleration To emphasize that free fall is just a special case of motion with constant acceleration, always write the general form of the constant-acceleration equation first, before substituting in any values (such as 9.8 m/s^2 for a). Encourage students to do the same.

3 Assess

Assess the MAINIDEA

Motion of a Thrown Ball Stand on a chair or desk (using appropriate safety precautions) and throw a ball towards the ground. Ask students to describe when and how the ball is accelerating. Gravity is causing the ball to accelerate towards the ground at a rate of 9.8 m/s^2 in the downward direction during the entire fall. Ask students what the acceleration would have been if you had thrown the ball up instead of down. The acceleration would have been the same, of 9.8 m/s^2 in the downward direction.

Check for Understanding

Free-Fall Motion and Initial Conditions Draw a diagram on the board similar to the motion shown in **Figure 20.** Tell students that it represents the motion of a ball in free fall. Have students identify two different sets of initial conditions that would produce such a diagram. A ball falling from rest and a ball moving (being thrown) upward. OL

Reteach

Free-Fall Motion Using the diagram above (described in Check for Understanding), draw velocity vectors for the ball falling from rest. Indicate that upward is chosen as the positive direction. Point out that free-fall acceleration is negative because the velocity vectors are increasing in length downward, and that its value is -9.8 m/s^2. Guide students as they analyze in a similar way the motion of the ball moving upward in free fall. OL

Reading Checks and Caption Checks

Caption Check, *student text p. 76*
The line would start at −20.0 m/s and slope upward.

Reading Check, *student text p. 77*
Its velocity is zero. Its acceleration is 9.8 m/s^2 downward.

Caption Check, *student text p. 79*
2.5 m/s^2

Practice Problems

student text p. 78

41. a. 39 m/s downward
b. The brick falls 78 m.

42. a. −39 m/s upward
b. The brick still falls 78 m.

43. 8.3 m/s

44. a. 26 m
b. 4.6 s

45. a $\boldsymbol{v}$ = 0 m/s; $\boldsymbol{a}$ = 9.8 m/s^2 downward
b. 2.2 m/s
c. 0.45 s

46. a. The velocity-time graph would be straight line segments that start at the origin and then rise, fall, and rise again.
b. The graph would start at the origin and have an inverted parabolic shape.

Section 3 Review

student text p. 79

47. Free fall is the motion of an object when gravity is the only significant force on it. Air significantly affects the paper but not the book.

48. 9.2 m/s downward

49. The people fall 26 m during the 2.3-s time period.

50. a. The maximum height would be three times higher on Mars.
b. Flight time is three times longer on Mars.

51. Velocity decreases at a constant rate as the ball travels upward. At the ball's highest point, velocity is zero. As the ball begins to drop, the velocity begins to increase in the negative direction until it reaches the height from which it was initially released. At that point, the ball has the same speed it had upon release. The acceleration is constant throughout the ball's flight.

52. Sample Answer: The ball is accelerating; its velocity is changing. Take a multiflash photo to measure its position. From photos, calculate the ball's velocity.

PHYSICS THAT'S ENTERTAINMENT!

Going Down

Amusement Park Thrill Rides

Background

Students may think that the fun of a thrill ride is derived simply from speed. Have them consider this question: Which gives you a bigger jolt of excitement, riding in a fast car at a steady speed for an hour or riding in a slower-moving car and stopping suddenly without notice? The unexpected backward acceleration is more likely to increase your heart rate. Explain that it's the change in speed and direction—the acceleration—that provides most of the thrill of amusement-park rides.

Teaching Strategies

- Point out that thrill-ride owners enforce size and health restrictions so that people whose bodies do not fit well into safety equipment or cannot handle extreme sensations are not harmed.
- Encourage students to go to an amusement park, if possible, and map out points of acceleration for a thrill ride. If going to a park is not feasible, have them examine a diagram or observe a video of a thrill ride.
- Prepare students for the concept of angular acceleration, studied in a future chapter, by pointing out locations on a thrill ride where the passenger is experiencing acceleration while turning a corner or moving along a curve in a track or swing.
- To connect to the future chapter on gravity, have students find out about *g* force and how positive and negative *g* affects the thrill of a roller-coaster ride.

GOING FURTHER >>>

Expected Outcomes Students should show accelerations that involve an increase in velocity, such as when the passenger is accelerating forward due to a motorized force or free falling downward. Additionally, they should show accelerations that involve a decrease in velocity, such as when the passenger is coasting up a hill, experiencing breaks, or slowing down during a bungee ride.

CHAPTER 3 ANSWERS

SECTION 1

Mastering Concepts

53. Acceleration is the change in velocity divided by the time interval in which it occurs. It is the rate of change of velocity.

54. a. if forward is the positive direction, a car moving backward at decreasing speed

b. in the same coordinate system, a car moving backward at increasing speed

c. a car that is moving on a circular track at a constant speed

55. The car starts from rest and increases its speed. As the car's speed increases, the driver shifts gears.

56. When the velocity-time graph is a line parallel to the *t*-axis, the acceleration is zero.

Mastering Problems

57. From least to greatest magnitude of acceleration: $C < E < A < D < B$.

58. a. Possible answer: "… and blocks a shot. Her hand is in contact with the ball for 0.3 s. If the basketball initially was traveling toward the basket at 1.3 m/s and then travels away from the basket at 2.0 m/s, what acceleration did she give to the ball?"

b. Possible answer: "… and is standing, dribbling the basketball at the top of the key. Once the play is set, she moves in a straight line a distance of 5.0 m in 3.0 s. What was her average speed during the move?"

59. a. 6.0 m/min^2 east

b. 0.0 m/min^2 east

c. 2.0 m/min^2 west

d. 4.0 m/min^2 west

60. See online Solutions Manual for graph.

a. speeding up from 0.0 s to 4.0 s; and 10.0 s to 12.0 s; slowing down from 5.0 s to 10.0 s

b. at 10.0 s

c. The magnitude of acceleration is the same, but the direction of acceleration is opposite.

61. 7.00×10^4 m/s forward

62. Car B has the greatest acceleration of 6.4 m/s^2. Using significant figures, car A and car C have the same acceleration, 4.5 m/s^2.

SECTION 2

Mastering Concepts

63. displacement

64. Sample answer: "A car has a forward acceleration of 5 m/s^2. After 3 s, the driver realizes that he is getting close to his desired speed and reduces his forward acceleration to 1.5 m/s^2 for 5 s. At that time, he sees a sign indicating an upcoming construction zone and slows down at a rate of 2.0 m/s^2 for 2 s."

Mastering Problems

65. a. 43 m up the hill

b. 43 m up the hill

66. 9.2×10^2 m north

67. a. 1.4×10^2 m

b. 550 m; this is about 4 times farther than stopping a car going half the speed.

68. a. 88 m

b. 75 m

c. 13 m

d. 288 m

SECTION 3

Mastering Concepts

69. All objects of the same size accelerate toward the ground at the same rate.

70. Student answers will vary. Examples of falling objects for which air resistance can be ignored are a steel ball, a rock, and a person falling small distances. Examples of falling objects for which air resistance cannot be ignored are sheets of paper, parachutes, leaves, and feathers.

Mastering Problems

71. 1.2 s

72. a. 78 m/s downward

b. 3.1×10^2 m downward

73. 2.0×10^1 m/s downward; 2.0×10^1 m downward

74. 7.3 m/s

75. 7.3 m/s

76. a. 5.9 m

b. 11 m/s upward

Applying Concepts

77. No, they have opposite signs.

78. (1) Walk in the positive direction at a constant speed. (2) Walk in the positive direction at an increasing speed for a short time; keep walking at a moderate speed for twice that amount of time; slow down over a short time and stop; remain stopped; and turn around and repeat the procedure until the original position is reached.

79. Draw a velocity-time graph and see whether the graph is a straight line, or calculate accelerations using $a = \Delta v/\Delta t$ and compare the answers to see if they are the same.

80. Velocity increases rapidly at first, then more slowly. Acceleration is greatest at the beginning but is reduced as velocity increases. Eventually, it is necessary for the driver to shift into second gear. The acceleration is smaller just before the gear change because the slope is less at that point on the graph. Once the driver shifts and the gears engage, acceleration and the slope of the curve increase.

81. Both objects traveled the same distance. The object that is shot straight upward rises to the same height from which the other object fell.

82.

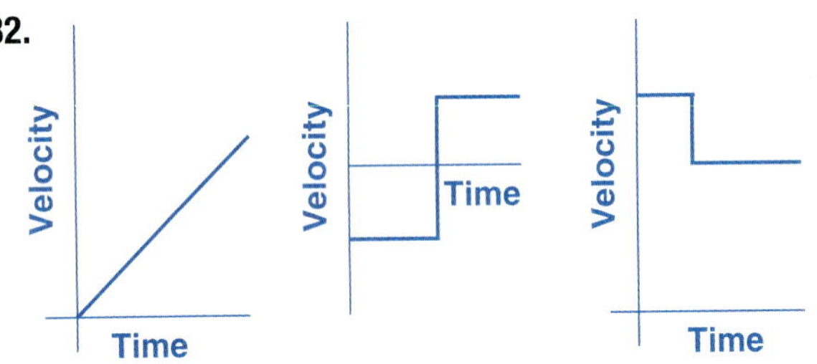

83. a. The ball would hit the Moon with a lesser speed because the acceleration due to gravity is less on the Moon.

b. The ball will take more time to fall.

84. a. Let J = Jupiter, E = Earth and a_{grav} = gravitational acceleration. At maximum height, $v_f = 0$, so

$$x_J = \frac{v_i^2}{2a_{grav,\,J}} = \frac{v_i^2}{2(3a_{grav,\,E})} = \frac{1}{3}x_E$$

b. With $v_f = 0$, the value x_f is directly proportional to the square of initial velocity, v_i. That is,

$$x_f = \frac{v_f^2}{2g} - \frac{(3v_i)^2}{2g}$$

On Earth, an initial velocity that is three times greater results in a ball rising nine times higher. On Jupiter, however, the height of nine times higher would be reduced to only three times higher because of x_f's inverse relationship to a_{grav} that is three times greater.

85. a. Rock B hits the ground with a greater velocity.

b. They have the same acceleration, the acceleration due to gravity.

c. rock A

Mixed Review

86. 1.14×10^3 m

87. You need to know the time between flashes and the distance between the first two images and the distance between the last two. From these, you get two velocities. Between these two velocities, a time interval of *t* seconds occurred. Divide the difference between the two velocities by *t*.

88. 8.0 m

89. a. 2.8×10^2 m downward

b. 7.5 s

90. a.

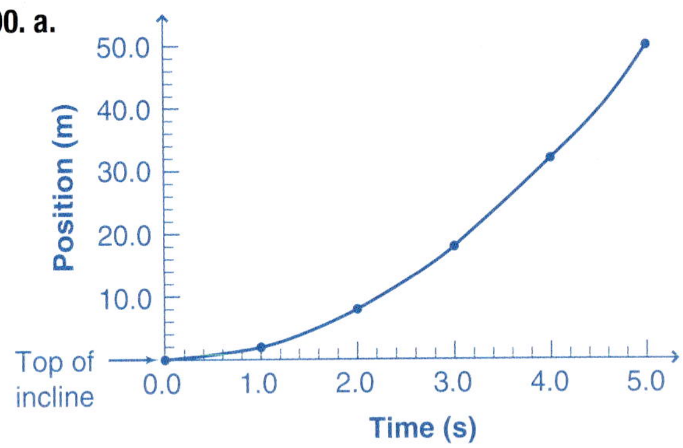

b. After 2.2 seconds, the ball has rolled approximately 10 m.

91. a. 3.1×10^8 m/s^2

b. 11 μs

92. 15 m

93. 29 times the acceleration due to gravity

94. a. 25 times free-fall acceleration

b. 21 times free-fall acceleration

95. a.

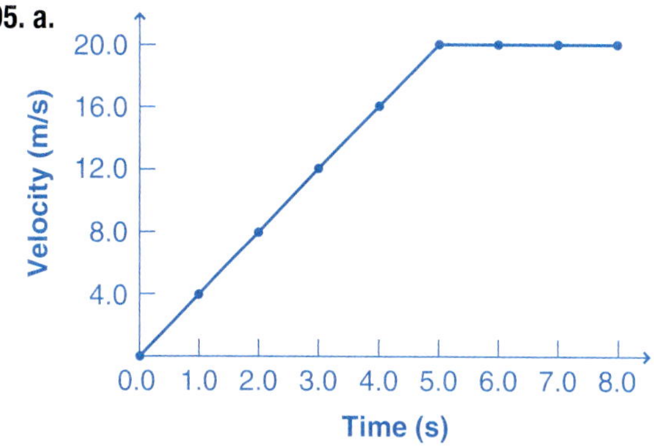

b. 8.0 m in the forward direction

c. 32 m in the forward direction

d. 110 m in the forward direction

e. 4.0 m/s^2 in the forward direction; acceleration

f. 0.0 m/s^2; constant velocity

96. 180 m from the stoplight

97. a. Upward is the positive direction. The fist moves at about 13 m/s for about 4 ms. It then suddenly comes to a halt (accelerates).

b. 3.7×10^3 m/s^2 upward

c. about 380 times free-fall acceleration

d. The area can be approximated by a rectangle: $(-13\ m/s)(0.006\ s) = -8$ cm.
This is in agreement with the position-time graph where the hand moves from +8 cm to 0 cm, for a net displacement of −8 cm.

98. a. 15 m/s downward

b. The bag has fallen 1.0×10^1 m

c. The bag is 1.0×10^1 m below the origin and 2.0×10^1 m below the helicopter.

Thinking Critically

99. Students' labs will vary. Students should find that a change in the mass over the edge of the table will not change the distance the cart moves because the acceleration is always the same: 9.8 m/s^2 downward.

100. The change in velocity is the same.

101. a. Express: 216 m; local: 232 m; on this basis, no collision will occur.

b. See Online Solutions Manual.

Writing in Physics

102. Student answers will vary. Answers should include Galileo's experiments demonstrating how objects accelerate as they fall. Answers might include his use of a telescope to discover the moons of Jupiter and the rings of Saturn, and his reliance on experimental results rather than authority.

103. Answers will vary. Because humans can experience negative effects, such as blackouts, the designers of roller coasters need to structure the downward slopes in such a way that the coasters do not reach accelerations that cause blackouts. Likewise, engineers working on bullet trains need to design the system in such a way that allows the train to rapidly accelerate to high speeds, without causing the passengers to black out.

Cumulative Review

104. a. 6.3×10^{-3} m

b. 8.4×10^8 km

c. 1.69×10^4 cm^2

d. 6.45×10^{-13} m/s

105. The graph and motion diagram indicate constant-velocity motion with a forward velocity of 35.0 m/s and an initial position of 5.0 m. The problems students create will vary. Sample problem: An object starts at a position 5.0 m west of a point and travels east at a constant velocity of 35.0 m/s. What is the object's position 10.0 s after it starts moving? Answers will vary for the create-a-problem part.

Position-Time Graph

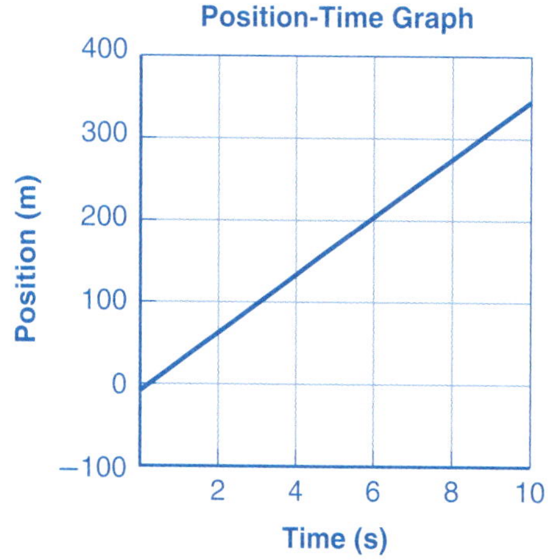

Motion Diagram

Standardized Test Practice

student text p. 87

Multiple Choice

1. C
2. B
3. A
4. A
5. C
6. C
7. D
8. B
9. D

Free Response

10. Slope = (36.9 m/s – 8.10 m/s)/6.00 s
= 4.80 m/s^2 Acceleration
= 4.80 m/s^2 Displacement
= area under graph
Area under graph = area of rectangle
+ area of triangle = (8.10 m/s×12.00 s)
+ (1/2)(12.00 s×57.6 m/s) = 443 m
See Online Solutions Manual.

Rubric

The following rubric is a sample scoring device for free response questions.

Points	Description
4	The student demonstrates a thorough understanding of the physics involved. The response may contain minor flaws that do not detract from the demonstration of a thorough understanding.
3	The student demonstrates an understanding of the physics involved. The response is essentially correct and demonstrates an essential but less than thorough understanding of the physics.
2	The student demonstrates only a partial understanding of the physics involved. Although the student may have used the correct approach to a solution or may have provided a correct solution, the work lacks an essential understanding of the underlying physical concepts.
1	The student demonstrates a very limited understanding of the physics involved. The response is incomplete and exhibits many flaws.
0	The student provides a completely incorrect solution or no response at all.

CHAPTER 4

Forces in One Dimension

About the Photo

Forces on a Ship Have the students examine the photograph. Ask students to list the forces the large ship might be experiencing. Possible answers: the ship's motor and propeller might be running, the ship might be getting pulled by the tugboat, the water current, or the wind. Ask students why a small tugboat can be useful to help a ship maneuver in a harbor environment. The tugboat can pull the ship in any direction and turn the ship very sharply.

Chapter Overview

Exerting a net force on an object causes that object's velocity to change. Forces can be exerted either by direct contact with another object or in some cases via a field, such as gravity. Newton's laws describe how force affects velocity.

Before students study the material in this chapter, they should study:

- Accelerated motion in one dimension
- Adding vectors in one dimension
- Uniform motion in one dimension
- Vectors vs. scalars

To solve the problems in this chapter, student will need a strong grasp of:

- Graphing data
- Scientific notation
- Significant figures
- Slope
- Solving linear equations

Introduce the BIG IDEA

Bounce a rubber ball. Have students describe the forces on it while it moves. Gravity pulls the ball down. Air resistance exerts small forces on it, but not as much as gravity. Have students describe the forces on the ball just as it hits the floor and explain why it bounces. When it hits, gravity pulls the ball down and the floor pushes it back up. The floor force pushing the ball up is greater than gravity pulling it down; the net force changes the motion of the ball.

Go online!
connected.mcgraw-hill.com

Your online experience includes:

- **Video**
- **Concepts In Motion**
- **Virtual Investigations**
- **All Lab, Applying Practices activities, and PBL worksheets**
- **Online Practice**
- **Personal Tutor**
- **Multilingual eGlossary**
- **Presentation**

Use the LaunchLAB

In *Forces in Opposite Directions,* students can investigate what happens when more than one force acts on an object.

Use Physics T.V.

Video

APPARENT WEIGHT Have you ever had the sensation of feeling weightless while on a free-fall ride? An elevator acts in much the same way, giving you the sensation of being lighter when you're accelerating down and heavier when you're accelerating up.

1 Introduce

Bellringer

Forces Have students put a small, relatively flat object on their desks, such as a coin or paper clip. Tell them to use different methods to move the object around on their desktop without lifting the object off the desk. After students do this for one minute, ask them to describe how they were able to move the object. All of the things they did will fit into two categories: pushes and pulls. One way of describing a force is as a push or pull.
BL Kinesthetic

Tie to Prior Knowledge

Forces and Acceleration Students have learned how to describe motion with constant acceleration using kinematics. This chapter introduces force, the cause of acceleration. This chapter answers the question of why objects accelerate.

2 Teach

Force

student text pp. 90–92

Identify Misconceptions

Force and Acceleration Many students confuse the concepts of force, velocity, and acceleration. A force is defined as a push or a pull that causes a change in motion. A net force causes an acceleration, which is a change in velocity. Whenever an object in motion changes direction, speeds up, slows down, or stops, it does so because an unbalanced force is acting on it. Clarify to students that a change in velocity can be caused by a change in the speed and/or direction of an object. However, also be clear that an object can have a velocity without a force acting on it, as in the case of a spaceship traveling in a straight line through deep space. Ask them if it has a velocity. yes Is there still a force acting on it? no Is it possible in this case that the spaceship could accelerate? not unless a force is acting on it Point out that as long as an unbalanced force is applied to an object, that object continues to accelerate. When a force no longer acts on an object, however, that object, as in the case of the spaceship, continues to move at a constant velocity. OL

Use Figure 2

Identifying forces For each of the figures, ask students to identify the agents and the directions of all the forces acting on the system.

Left figure—The agents are the table pushing up on the book, your hand pushing the book to the right, and Earth's mass (gravity) pulling down on the book.

Middle figure—The agents are the string pulling up on the book and Earth's mass pulling down on the book.

Right figure—The only agent is Earth's mass pulling down on the book (neglecting air resistance).

Point out that only in the middle figure are the forces equal in magnitude and opposite in direction. Thus, the net force is zero, and there is no change in velocity.
OL Visual-Spatial

DIFFERENTIATED INSTRUCTION

Struggling Students Help students make physical models of the forces acting on an object. Ask them to each draw a free-body diagram of a book on top of a table. Remind students that the dot in a free-body diagram represents the object. Before students draw their free-body diagrams, ask them to identify the system and where the system comes in contact with the external world. The textbook is the system, and it comes in contact with the external world on the tabletop. Ask them to tell you what other force is acting on the textbook. gravity or Earth's mass on book Remind them that each arrow represents the direction of the force.
BL Visual-Spatial

Concept Development

Force and Velocity Demo To counter the notion that a force causes a velocity, rather than a *change* in velocity, roll a ball or push a cart across a table. While the object is in motion, push it in some direction so that there is an obvious change in its velocity. Prompt students to identify this as an acceleration. Ask them if they notice any relation between the direction of the force you exerted and the direction of the resulting acceleration. The force and acceleration are in the same direction. BL

Teacher to Teacher

Falling Objects To illustrate field forces and how all objects—in the absence of air resistance—fall at the same rate, have one student hold a book parallel to the floor, while another student holds a feather at the same distance above the floor. On command, they drop their objects simultaneously. Have students note how the feather falls more slowly. This is a good place to discuss air resistance. Repeat the exercise with the feather on top of the book. The book and the feather fall at the same rate because the book removes the air resistance. **BL** **Kinesthetic**

Concept Development

Forces Have students compare forces on an object. Ask them to consider a bucket that is suspended in the air by a rope that is tied to it and run over a pulley. Ask them to compare the forces acting on the bucket. The force of gravity and the force from the rope are equal and opposite. Now tell them that the bucket begins to fall and to pull the rope through the pulley. Again, ask them to compare the forces. Because the bucket is accelerating downward, the downward force of gravity on the bucket must be greater than the upward force of the rope on the bucket.

Vectors For many students, the concept of a force vector may be baffling. Explain that vectors represent important physical quantities. Explain that vectors have both a magnitude and direction. For example, "10 km northwest" can be thought of as a vector. The magnitude is 10 km and the direction is northwest. Another example of a vector is wind velocity. Ask students, if the wind is blowing at 8 km/h toward the east, what are the magnitude and the direction? 8 km/h, east **OL** **Logical-Mathematical**

Content Background

Falling Objects Newtonian physics holds that neither size nor mass affects the velocity of a falling object. This only applies to objects falling in a vacuum where there is only one force—gravity—acting on an object. A skydiver falling through the air is also acted on by air resistance, which is also called *drag.* Students will learn more about drag force and terminal velocity at the end of this chapter.

Combining Forces

student text pp. 92–93

Critical Thinking

Forces Have students consider a cart rolling along a track at a constant velocity. Because the velocity is not changing, ask students if this means that there are no forces acting on the cart. No; Earth and the track both exert forces on the cart. Ask students why, if identifiable agents are exerting forces on the cart, there is no change in velocity. Assuming there is no air resistance, the *net* force on the cart is zero. **OL**

Content Background

MAIN IDEA The vector sum of all forces acting on an object is the net force. If the net force acting on an object is zero, then the object will not accelerate. For example, if a slowly gliding ice skater is pushed from behind, he or she will accelerate and move faster due to the unbalanced force from the push. After the push, he or she glides at a relatively constant speed because the net force is very close to zero.

Acceleration and Force

student text pp. 94–95

Content Development

Motion and Forces Divide the class into groups. Present the following statement to the class: "Motion at a constant speed requires a constant force in the direction of motion." Ask students to correct the statement and give an example of when it is and is not true. The net force acting on an object moving at a constant velocity is zero. Not true: Consider an asteroid hurtling through the solar system, for example. It may have been catapulted into space eons ago, and yet the force that initially launched the asteroid is no longer acting on it. **OL** **Interpersonal**

Newton's Second Law

student text pp. 96–97

Content Background

Correspondence Principle and Special Relativity An important concept in physics is the correspondence principle, which requires theories that may seem quite different in extreme situations to predict the same results in less extreme situations. When applied to special relativity, the correspondence principle states that, at velocities much less than c, the speed of light, Einstein's theory of special relativity agrees with Newton's laws.

REAL-WORLD PHYSICS

Applying Newton's Laws The motion of an object depends on the forces exerted on it. Engineers at NASA use Newton's laws to analyze complex, real-world situations involved in launching a spacecraft. Discuss some of the complicating factors that NASA engineers face when using Newton's laws. Consider, for example, the variables that change when calculating the acceleration of the spacecraft, such as the firing of a rocket. Force and acceleration are continually changing as fuel is burned and the mass of the spacecraft decreases. To model this situation, blow up a balloon and hold the stem closed. At arm's length, release the balloon and observe its change in motion.
CAUTION: Some students may be allergic to latex.

DIFFERENTIATED INSTRUCTION

Struggling Students Have groups come up with a paired example that will help them remember Newton's second law. In each pair, the mass, acceleration, or force should be kept constant. For example, if you had two identical bowling balls and pushed one with a large force and the other with a smaller force, the one you push with the larger force will experience the larger acceleration.
BL COOP LEARN Interpersonal

Newton's First Law

student text pp. 98–99

Quick DEMO

Mass

Estimated Time 10 minutes

Materials empty 2-liter bottle, rubber band, string, metric ruler, water

Procedure Tie the string to the rubber band. Place the rubber band around the empty bottle. The bottle should be upright with the cap on. Pull the string until the bottle starts to move. Use the ruler to measure the amount that the rubber band stretches. Record the measurement. Add some water to the bottle and repeat the process. Compare measurements. Once the bottle accelerates, the net force on the bottle is no longer zero. When more water is added to the bottle, the rubber band must exert more force to produce the same acceleration because of the greater mass of the bottle.

Reinforcement

Concept Map Ask students to draw a concept map including force, mass, acceleration, and velocity. Students should describe the relationships between items on the lines connecting them to each other.
BL Linguistic

Activity

Label Forces Have students work in small groups to come up with examples of different types of forces. Have them draw a free-body diagram for each one and then label the force. Examples could include: a ball rolling across gravel, a book on top of a table, a golf club striking a golf ball, a skydiver in free fall, a bell suspended from a rope, air escaping from a balloon, and a person jumping on a pogo stick.
OL COOP LEARN Interpersonal

3 Assess

Assess the MAIN IDEA

Forces Hang a weight from a spring or a spring scale. Ask students to identify the forces that the weight is experiencing. Gravity is pulling the weight down, and the spring is pulling the weight up. Now have a student put one hand underneath the weight and lift it until the spring is in compression. Ask students to identify all forces on the weight a second time. Gravity is pulling the weight down, the hand is pushing the weight up, and the spring is now pushing the mass down.

Check for Understanding

Free-Body Diagrams Have students draw free-body diagrams for several inanimate objects that are in contact with other stationary objects, such as a chair with a person sitting on it. Make sure students draw only the forces exerted on each object, not the forces exerted by the object. OL Visual-Spatial

Reteach

Forces Demo Have a student stand with one arm outstretched, and place a textbook on his or her hand. Ask students whether the hand exerts a force on the book. Make sure you ask the student who is exerting this force what he or she thinks as well. Then have the student close his or her eyes. Suddenly remove the book from the hand. The student's hand will accelerate upward, showing that it was indeed exerting an upward force. BL Kinesthetic EL

Reading Checks and Caption Checks

Reading Check, *student text p. 90*

Unbalanced forces cause all accelerations.

Caption Check, *student text p. 91*

$\mathbf{F}_{\text{table on book}}$ and $\mathbf{F}_{\text{hand on book}}$ are contact forces, $\mathbf{F}_{\text{Earth's mass on book}}$ is a field force.

Reading Check, *student text p. 92*

They are in the same direction.

Caption Check, *student text p. 95*

The acceleration of two carts is half the acceleration of one cart.

Reading Check, *student text p. 95*

The force must be reduced by half.

Practice Problems

student text, p. 93

1. See Online Solutions Manual for motion diagram and free-body diagram.
2. See Online Solutions Manual for motion diagram and free-body diagram.
3. See Online Solutions Manual for motion diagram and free-body diagram.
4. See Online Solutions Manual for motion diagram and free-body diagram.
5. See Online Solutions Manual for motion diagram and free-body diagram.

Practice Problems

student text, p. 96

6. 3.90×10^{2} N in the direction of the two forces
7. 6.0×10^{1} N in the direction of the larger force
8. 24 N east

Practice Problems

student text, p. 97

9. 4.2 m/s^2
10. 22 N
11. **a.** See Online Solutions Manual for free-body diagram.
    b. 144 N
    c. 111 kg

Section 1 Review

student text p. 99

12. **a.** push of a hand, friction, air resistance, spring force
    b. gravity
    c. mass, inertia, acceleration

13.

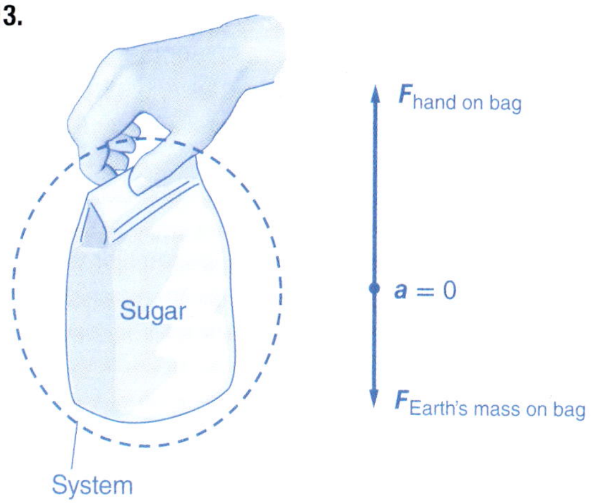

14.

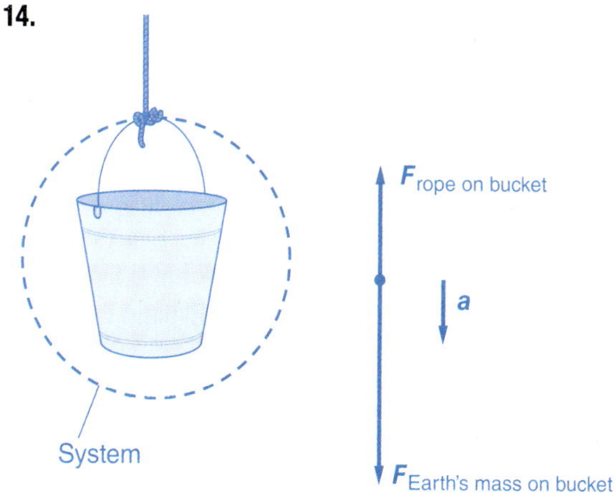

15. Because $m = F/a$ and the forces are the same, the mass of the second block is one-third the mass of the first block.

SECTION 2 Weight and Drag Force

student text pp. 100–105

1 Introduce

Bellringer

Mass and Weight Show the class several different objects (metal cylinders, etc.). Ask which object will weigh the most and why this is so. Hang the objects from spring scales to verify student predictions. Ask if the objects would weigh as much on another planet or on the Moon. No; the objects would have different weights because the local force of gravity would be different. **BL** **Visual-Spatial**

Tie to Prior Knowledge

Forces on Objects Students should be familiar with acceleration and the concept of force. In this section, they further explore Newton's first and second laws.

2 Teach

Weight

student text pp. 100–102

Concept Development

MAINIDEA Students have previously learned that free-fall acceleration near Earth's surface is 9.8 m/s^2. Point out that free-fall acceleration is the same as *g*, which is the strength of the gravitational field.

Units Make sure students understand that 9.8 m/s^2 and 9.8 N/kg are the same quantities. Explain that 1 newton (N) is equivalent to 1 kg•m/s^2.

Discussion

Problem Suppose you are standing in an elevator that is accelerating upward. Is the magnitude of the normal force exerted on you by the floor of the elevator the same as, larger than, or smaller than the magnitude of your weight?

Response The normal force must be larger in magnitude than the weight. The net force must be in the upward direction because you are accelerating in that direction. This means that the vector sum of the weight force and the normal force must be a force pointing in the upward direction, i.e., the direction of the normal force. **OL**

ADDITIONAL IN-CLASS EXAMPLE

Use with Example Problem 2.

Problem Arnold needs to lift a 35.0-kg rock. If he exerts an upward force of 502 N on the rock, what is the rock's acceleration?

Response $F_{net} = F_{Arnold\ on\ rock} - F_{Earth's\ mass\ on\ rock}$

$= 502\ N - (35.0\ kg)(9.8\ N/kg)$

$= 502\ N - 343\ N = 159\ N$

$a = \frac{F_{net}}{m}$

So, $a = \frac{159\ N}{35.0\ kg} = 4.54\ m/s^2$

$a = 4.54\ m/s^2$

PHYSICS CHALLENGE ACTIVITY

Investigation of Force and Acceleration Have students record video of themselves standing on a scale while riding in an elevator. Have them play back the video frame by frame and determine the maximum and minimum forces they applied to the scale. Ask students what the accelerations were (magnitude and direction) corresponding to these forces and at which points during the ride did the elevator have no acceleration. **AL** **Kinesthetic**

Use the PhysicsLAB

Have students perform the lab, *Forces in an Elevator,* to investigate forces that act on an object in an elevator.

Use the MiniLAB

Have students perform the lab, *Mass and Weight,* to investigate the relationship between mass and weight.

Identify Misconceptions

Cause of Apparent Weight Some students might think that apparent weight is connected with motion at a constant velocity. Rather, apparent weight is observed when an object experiences a vertical acceleration. Remind students that whether the object appears to be lighter or heavier depends upon the direction of the acceleration, not the object's velocity.

ADDITIONAL IN-CLASS EXAMPLE

Use with Example Problem 3.

Problem You are in an elevator, standing on a bathroom scale. You notice that the scale reads a weight that is less than your actual weight. (Assume that the scale is correctly calibrated.)

a. Is the elevator moving at constant velocity, or is it accelerating?

b. If it is accelerating, what is the direction of the acceleration?

Response **a.** accelerating **b.** downward

Drag Force

student text pp. 104–105

Use the PhysicsLAB

Have students perform the lab, *Terminal Velocity,* to the effects of air resistance on objects in free fall.

Use the MiniLAB

Have students perform the lab, *Upside-Down Parachute,* to investigate how terminal velocity depends on mass.

Quick DEMO

Drag Force

Estimated Time 10 minutes

Materials one marble—or metal sphere—and one cube (or metal cylinder) of equal mass, one large beaker, motor oil, lamp, thermometer, digital timer

Procedure Fill the beaker with oil at room temperature. Record the temperature of the oil in the beaker. Then, holding the marble and the cube just above the surface of the oil, release them at the same time into the beaker. (If you are using a metal cylinder, be sure to drop it flat end first.) Have students watch the two objects as they descend through the fluid in the beaker. Ask volunteers to record the rate at which each object falls. Set the beaker under a lamp and let the oil warm, to change the property of the fluid. Repeat the procedure, first recording the temperature of the fluid. Ask volunteers to record the rate at which each object falls. Draw a table on the board for comparing data. Students will see evidence of drag force.

Identify Misconceptions

Force of Air Some students might think that air only exerts a downward force. In reality, air pushes in all directions.

3 Assess

Assess the MAINIDEA

Falling Objects Throw a ball vertically into the air and allow it to fall to the ground. Ask students to explain why the ball changed direction and fell to the ground. Gravity exerted a downward force on the ball, causing acceleration and pulling toward the ground. The acceleration toward the ground eventually overcame the initial upward velocity. Ask students to describe the acceleration due to gravity while the ball was travelling upward, and then downward. The force due to gravity was the same while the ball was travelling both up and down, and so the acceleration was the same.

Check for Understanding

Terminal Velocity Show three motion diagrams indicating objects at different terminal velocities. Ask students the following questions: If the three objects were a raindrop, a piece of hail, and a snowflake, which diagram corresponds to which object? OL Visual-Spatial

Reteach

Air Resistance Give students several types of balls, including a table-tennis ball or foam ball. Ask them to rank the balls according to the time needed to fall from a given height. Have students do the experiment to see what the real ranking is. BL Kinesthetic

Reading checks and Caption Checks

Caption Check, *student text p. 100*

The forces are $\boldsymbol{F}_{\text{scale on you}}$ and $\boldsymbol{F}_{\text{Earth's mass on you}}$.

Reading Check, *student text p. 102*

The scale initially reads your weight when the elevator is at rest, then reads greater than your weight as the elevator accelerates upward, reads your weight when traveling at a constant speed, reads less than your weight as it slows down, and finally reads your weight again when the elevator is at rest.

Practice Problems

student text, p. 101

16. 39 N

17. 10.5 N/kg

18. 4.9 N; it would move at a constant speed.

19. 252 N; the bag does not hold.

20. a. 60 kg

b. 95.5 N

21. a. 735 N

b. 885 N

c. 585 N

d. 735 N

e. upward

Physics Challenge

student text p. 104

1. 7.8 s

2. −76 m/s

3. 4100 N

Section 2 Review

student text p. 105

22. Yes; for a while the diver is accelerating upward because there is an additional upward force due to air resistance on the parachute. The upward acceleration causes the diver's downward velocity to decrease. Newton's second law says that a net force in a certain direction will result in an acceleration in that direction ($F_{net} = ma$).

23. 16.2 N

24. $-0.5\ \text{m/s}^2$; the acceleration is $0.5\ \text{m/s}^2$ downward.

25. See Online Solutions Manual for free-body diagrams. Apparent weight and real weight are the same when you are traveling either up or down at a constant velocity. Apparent weight is less than real weight when the elevator is slowing while rising or speeding up while descending. Apparent weight is greater when speeding up while rising or slowing while going down.

26. $0.14\ \text{m/s}^2$ in the direction of the force applied by his friend

27. Answers may vary. One possible answer is the following: You can neglect resistance if you do all your maneuvering on the roller belt. Because you know the weight of the 1000 N crate, you can use it as your standard. Pull on the 1000 N crate with a particular force for 1 s, estimate its velocity, and calculate the acceleration that your force gave to it. Next, pull on a crate of unknown mass with as close to the same force as you can for 1 s. Estimate the crate's velocity and calculate the acceleration your force gave to it. The force you pulled with on each crate will be the net force in each case.

$$F_{\text{net 1000-N crate}} = F_{\text{net unknown crate}}$$

$$(1000\ \text{N})(a_{\text{1000-N crate}}) = (m_{\text{unknown}})(a_{\text{unknown}})$$

$$m_{\text{unknown}} = \frac{(1000\ \text{N})(a_{\text{1000-N crate}})}{a_{\text{unknown}}}$$

1 Introduce

Bellringer

Interaction Forces Have two students on inline skates and wearing proper safety gear stand at the front of the room. Ask one student to push the other. The class will see that both students move, even though only one made the effort to push the other. **BL** **Kinesthetic** **EL**

Tie to Prior Knowledge

Free-Body Diagrams In this section, students apply their skills in developing free-body diagrams. The diagrams allow students to study forces in interaction pairs. Diagramming also helps students connect Newton's second and third laws, and it increases their understanding of the normal force.

2 Teach

Interaction Pairs

student text pp. 106–108

Identify Misconceptions

Interaction Pairs There is a deeply held notion that the more powerful entity in an interaction—the faster, bigger, or stronger one—will exert a larger force. According to Newton's third law, this is not true. The objects in an interaction exert forces of identical magnitude on each other.

Use the PhysicsLAB

Have students perform the lab, *Newton's Third Law,* to investigate the force interactions between train cars.

REAL-WORLD PHYSICS

Martial Arts Karate, meaning "empty hand," was developed in Okinawa, Japan, during the early 17th century as a means of self-defense because weapons were outlawed by rulers. It can take years of practice to learn the skills of karate, but with careful training even "empty hands" can break concrete blocks. A trained expert can break a concrete block 3.8 cm thick by moving his or her hand at 11 m/s to create 3069 N of force. Of course, the block exerts the same amount of force on the hand. Amazingly, the bones in the human hand can withstand up to 40 times more force than concrete.

ADDITIONAL IN-CLASS EXAMPLE

Use with Example Problem 4.

Problem You are walking along when you slip on some ice and fall. For a moment you are in free fall. During this time, what force do you exert on Earth if your mass is 55 kg? Earth exerts a force on you: $F = mg = (55\ \text{kg})(9.8\ \text{N/kg}) = 540\ \text{N}$

Response The force you exert on Earth is equal in magnitude: $F = mg = (55\ \text{kg})(9.8\ \text{N/kg}) = 540\ \text{N}$

DIFFERENTIATED INSTRUCTION

Struggling Students Ask students how they know that for every force there is an equal and opposite force. For the student who answers this readily and correctly, ask that student to model his or her understanding for the class. For example, the student may demonstrate that when he or she tosses a heavier ball up in the air and catches it, he or she can feel a greater acceleration than for a ball that weighs less. If the force of Earth on the ball were not equal in magnitude, then there would be no difference in the weight the student sensed when catching the ball. And, if the force of Earth on the ball were not opposite in direction, the ball would not be brought back to Earth.
BL **Kinesthetic**

Tension

student text pp. 109–110

Use Figure 17

Tell students that the rope in **Figure 17** is guaranteed not to break if the tension in it is less than 500 N. Ask students to estimate the tension on the rope if they attach a 300-N bell instead of the bucket. The tension would equal the weight of all the objects hanging from it, 300 N. What would happen if they then added an additional 300-N weight to the rope? The rope would break. **OL** **Visual-Spatial**

Concept Development

Tension Demo Set up a spring scale horizontally between two pulleys. Attach two equal masses with ropes, one to each end of the scale, and run the ropes over the pulleys, so the masses are suspended in the air. Make sure the face of the scale faces away from the class. Ask the students what they think the scale reads. After a good debate, turn the apparatus so that they can see the scale reading. Some students might expect the spring scale to read the sum of the two weights. In fact, the scale reads the magnitude of tension; the weights exert equal yet opposite forces on the spring scale. Thus the tension in the scale equals the force with which each weight pulls. **OL**

PHYSICS PROJECT ACTIVITY

Tug-of-War Have students research the history of tug-of-war. There is ample evidence that the game has existed for several millennia as it spread from Africa to Asia and then to Europe. Depictions of such contests have been found painted inside Egyptian tombs dating back 4,500 years. There are also records from the courts of ancient Chinese emperors on rules of engagement and contest results. One variety of tug-of-war includes multiple ropes. Have students present the rules for tug-of-war and explain why they might be necessary. Students could also demonstrate various forms of tug-of-war as well as the rules in play. **OL** **Kinesthetic**

Discussion

MAIN IDEA Two students on inline skates each hold the end of a rope. If one student pulls on the rope, what will happen?

Response Both students will accelerate toward each other. The rope will exert an equal and opposite force on each student, so the student with less mass will accelerate more than the student with more mass. **OL**

ADDITIONAL IN-CLASS EXAMPLE

Use with Example Problem 5.

Problem You are fishing and catch a fish with a mass of 6 kg. If the fishing line can withstand a maximum tension of 30 N, what is the maximum acceleration you can give the fish as you reel it in?

Response $F = ma$

$30\text{ N} = (6\text{ kg})(a)$

$a = 5\text{ m/s}^2$

The Normal Force

student text p. 111

Reinforcement

Normal Force Have students each draw a free-body diagram to represent the forces on a cart on a frictionless ramp. Two forces act on the cart: the normal force, drawn perpendicular to the ramp's surface, and the weight, drawn down and perpendicular to Earth's surface. **OL** **Visual-Spatial**

3 Assess

Assess the MAIN IDEA

Forces on Objects Place two small model cars or wheeled carts on a desk. Place an extra weight on one of the carts so that there is a significant difference in weight between the two carts. Place a spring between them and push the carts together so that the spring is compressed. Let the carts go. Ask students to describe the motion of the carts. The carts move away from each other, but the heavier cart accelerates less than the light cart. Ask students to describe the force on each cart. The forces on each cart were equal and opposite.

Check for Understanding

Normal Force Tie a string around a box and set the box on a table. Ask students what the normal force on the box is. The table is exerting a normal force on the box that has a magnitude equal to the box's weight. Push down on the box and ask students how the normal force has changed. It has increased. Pull up on the string (box should still be in contact with the table) and ask students to describe the normal force. The normal force is less than the box's weight. Reinforce that the normal force on an object does not always equal the object's weight. **OL** **Visual-Spatial**

Extension

Tension Forces The team that wins a tug-of-war must exert a greater force on the opposing team than the opponents exert on them. Have students evaluate this statement. It is not possible for one team to exert a greater force on the opposing team than the opponents exert on them. This would be a violation of Newton's third law. Instead, one team must ensure that the *net* force on itself is away from the center, while the *net* force on the opposing team is toward the center. **OL**

Reading Checks and Caption Checks

Reading Check, *student text p. 106*

The force applied on you would be 15 N directed to the right.

Reading Check, *student text p. 107*

Earth's acceleration is usually very small because the Earth's mass is so large compared to the object's mass.

Practice Problems

student text, p. 108

28. The forces on the ball are the force of your hand and the gravitational force of Earth's mass. The ball exerts a force on your hand and a gravitational force on Earth. All these forces are exerted on your hand, on the ball, or on Earth.

29. The only force acting on the brick is the gravitational attraction of Earth's mass. The brick exerts an equal and opposite force on Earth.

30.

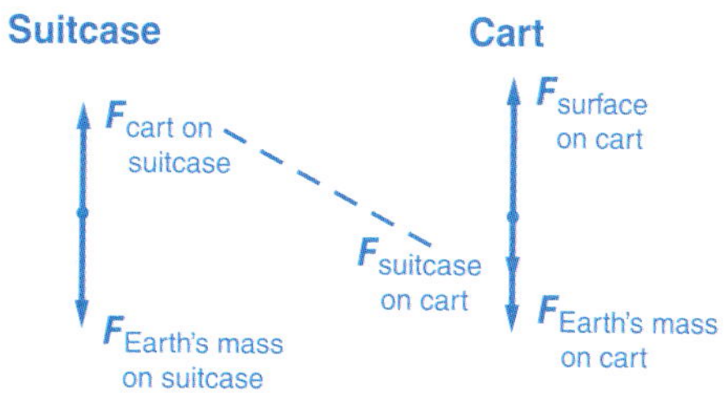

The only force interaction pair in these diagrams is $\boldsymbol{F}_{\text{cart on suitcase}}$ and $\boldsymbol{F}_{\text{suitcase on cart}}$.

31.

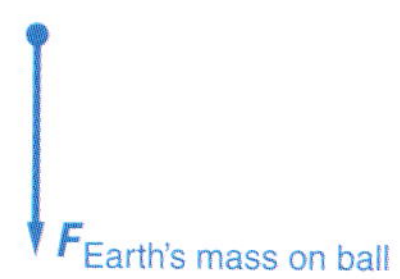

The only force acting on the ball is the force of Earth's mass on the ball, when ignoring air resistance. The ball exerts an equal and opposite force on Earth.

Practice Problems

student text, p. 110

32. 54 N

33. 0.91 m/s^2

Section 3 Review

student text p. 111

34. The forces on the book are the downward force of gravity due to the mass of Earth and the upward force of the hand. The force of the book on Earth and the force of the book on the hand are the other halves of the interaction pairs.

35. See Online Solutions Manual for free-body diagrams.

36. For the bottom rope with the positive direction upward: 49 N. For the top rope with the positive direction upward: 98 N.

37. For the bottom rope with the positive direction upward: 29 N. For the top mass with the positive direction upward: 3.5 kg.

38. The tension would be 500 N. The rope is in equilibrium, so the net force on it is zero. The team and the tree exert equal forces in opposite directions.

FRONTIERS IN **PHYSICS**

Supersonic?

Supersonic Skydiving

Background

In 1957 during Project Manhigh, Air Force Colonel Joseph Kittinger piloted a pressurized capsule and balloon to 29.5 km. Such balloon missions provided critical tests of pressure suits, life-support systems, and parachute technology used in the development of NASA's Mercury Space Program. On August 16, 1960, as part of Project Excelsior, Kittinger set the world records for highest parachute jump, longest parachute fall, and the fastest speed by a human in free fall. Kittinger is now retired but has worked as a consultant to a group of scientists, engineers, and stunt parachute jumpers seeking to break Kittinger's records.

Teaching Strategies

- Ask students which would fall faster on the Moon: a hammer or a feather? Galileo first theorized that in the absence of atmosphere, all objects would fall at the same rate regardless of mass. This was demonstrated in 1971 by astronaut David Scott during the Apollo 15 mission when he dropped a feather and a hammer on the Moon's surface. Both fell at the same rate. Videos of this event are widely available on the Internet. Refer students to Example Problem 2 to see calculations showing equal free-fall accelerations for objects with different masses.
- Ask students where they think space begins. Most will probably say where the atmosphere dissipates into pure space (this area is often described as outer space). The atmosphere does not have an abrupt boundary, so defining where space begins is somewhat open to interpretation. Kittinger describes anything above 19.2 km as space because above this altitude a human cannot survive without a pressure suit. NASA classifies anyone who has traveled into the thermosphere, which begins at an altitude of 80 km, an astronaut. Several other space agencies regard 100 km as the beginning of space.

GOING**FURTHER** >>>

Expected Outcomes During the *Excelsior I* jump, the helmet on Kittinger's pressure suit began to pull away from the suit. In addition, the first stage of his parachute system deployed too early. The chute's early deployment put Kittinger in danger of getting tangled in the parachute lines and free falling all the way to Earth. For the record-setting *Excelsior III* jump, the right glove on Kittinger's pressure suit sprung a leak during the balloon ascent. His hand swelled to double its normal size as a result of the extremely low pressure. Despite great pain, Kittinger did not immediately report the leak to mission control because he did not want to abort the mission.

SECTION 1

Mastering Concepts

39. Newton would disagree. There was an interaction between your foot and the ball. A force was exerted by your foot on the ball, which accelerated it. When moving across the field there was an interaction between the ball and the grass. A force was exerted on the ball that caused it to accelerate: it slowed down.

40. A large force is required to accelerate the mass of the bicycle and rider. Once the desired constant velocity is reached, a much smaller force is sufficient to overcome the ever-present frictional forces.

Mastering Problems

41. 9.8 N

42. 0.12 m/s^2

43. 6.9×10^3 N

44. 13 m/s^2

SECTION 2

Mastering Concepts

45. No, it only means the forces acting on it are balanced and the net force is zero. A book at rest on a table is not moving but the force of gravity pulls down on it and the normal force of the table pushes up on it. These forces are balanced so the net force is zero.

46. Yes, its velocity changed direction; thus, it was accelerated and a force is required to accelerate the basketball. The agent is the floor.

47. Joanne is correct. The force to the right means that the acceleration is to the right. If it is moving to the right it is speeding up; if moving to the left it is slowing down.

48. a. Because the force of air resistance suddenly becomes larger, the velocity of the diver drops suddenly.

b. The force of air resistance and the gravitational force are equal. Their sum is zero, so there is no longer any acceleration. The skydiver continues downward at a constant velocity.

49. a. Balloon, basketball, shot put.

b. Shot put, basketball, balloon.

c. They are inverses of each other.

Mastering Problems

50. Answers will vary depending on the mass. A person with a mass of 68 kg weighs 670 N.

51. a. 200 kg

b. 0.20 m/s^2; free-body diagram should show $\boldsymbol{F}_{\text{winch on people}}$ slightly longer than $\boldsymbol{F}_{w}$.

c. 11 s

52. a. 5.2×10^2 N

b. 410 N

c. 410 N

d. 5.2×10^2 N

e. 652 N

53. a. 22 N

b. 2.1 N

54. a. 14.0 m/s

b. -3.2×10^3 N

SECTION 3

Mastering Concepts

55. The rock does pull on Earth, but Earth's enormous mass would undergo only a minute acceleration as a result of such a small force. This acceleration would go undetected.

56. If you draw a free-body diagram for any point on the rope, there will be two tension forces acting in opposite directions. $F_{\text{net}} = F_{\text{up}} - F_{\text{down}} = ma = 0$ (because it is massless). Therefore, $F_{\text{up}} - F_{\text{down}}$. According to Newton's third law, the force that the adjoining piece of rope exerts on this point is equal and opposite to the force that this point exerts on it, so the force must be constant throughout.

57.

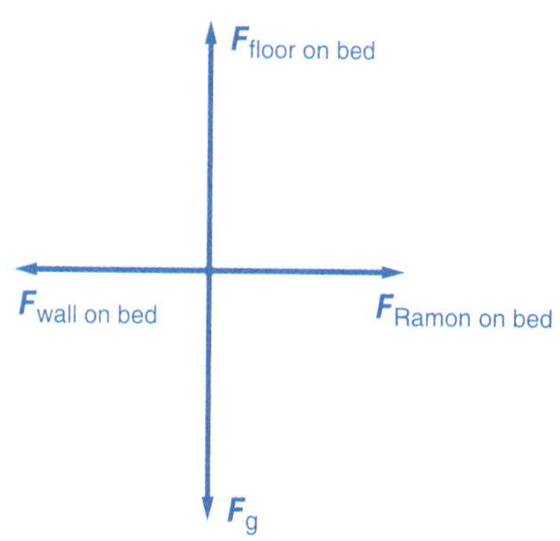

58.

$F_{\text{ball on bat}}$ and $F_{\text{bat on ball}}$ are an interaction pair.

59. From greatest to least: right box > left box > center box.

Mastering Problems

60. a. 59 N; the direction is upward.

b. 59 N downward

61. 2.10×10^{-5} N

62. 9.5×10^{2} N forward force difference, 9.3×10^{2} N normal force difference

63. 5.3×10^{4} N

64. Normal force on top block: 45 N. Normal force on middle block: 57 N. Normal force on botton block: 93 N.

Applying Concepts

65. a. The car is suddenly accelerated forward. The seat accelerates your body, but your neck has to accelerate your head. This can hurt your neck muscles.

b. The headrest pushes on your head, accelerating it in the same direction as the car.

66. The ounces tell you the weight in English units. The grams and kilograms tell you the mass in metric units. The label would need to read "Mass 0.85 kg" to be correct on the Moon. The grams and kilograms would remain unchanged.

67. The lighter, air-filled table tennis ball reaches terminal velocity first. Its mass is less for the same shape and size, so the friction force of upward air resistance becomes equal to the downward force of $m \times g$ sooner. Because the force of gravity on the massive, water-filled table-tennis ball (more mass) is larger, its terminal velocity is larger, and it strikes the ground first.

68. It means that on Earth's surface, the weight of 1 kg is equivalent to 2.2 lb. You should compare masses to masses and weights to weights. Thus 9.8 N equals 2.2 lb.

69. a. See Online Solutions Manual for free-body diagram.

b. 0 m/s

c. Because the only force acting on it is the gravitational attraction of Earth, the balls fall with free-fall acceleration of 9.8 m/s^2.

70. The hands will reduce the velocity of the ball to zero. By moving them in the direction of the moving ball, the amount of time over which the acceleration will occur will be longer, thus reducing the acceleration. Reducing the acceleration reduces the force needed to stop the ball.

Mixed Review

71. 31.89 m/s^2, 160.2 m/s

72. 1.80×10^{2} N

73. 139 m

74. a. 45 m/s^2

b. 3.9×10^{4} N

c. 3.1×10^{3} N

75. 222 m/s; 126.6 m/s is slower than the final velocity based on constant acceleration, so the acceleration cannot be constant.

76. -2.1 m/s^2

77. a. -6.0×10^{3} m/s^2

b. -8.7×10^{2} N

c. same magnitude, opposite direction (in direction of velocity of ball)

78. 8.0×10^{3} N in the direction of motion

79. a. 48 m/s^2

b. 43 m/s

80. a. 4.0 m/s^2 up

b. 40 m/s up

c. just the instrument weight, -49 N down

d. 4.1 s after release

81. 0.78 m/s^2

82. a. 2.3 m/s^2 to the right

b. 12 N to the right

c. According to Newton's third law, this should be equal and opposite to the force found in part b, so the force is 12 N to the left.

83. a. 1.17 m/s^2

b. -0.633 m/s^2

c. stopping, because the magnitude of the acceleration is less and $t = v/a$

84. a. 37 N

b. 2.4 m/s^2

Thinking Critically

85. Answers will vary, but a correct form of the answer is, "You wish to give a small child on ice skates, total mass 23 kg, a push so that she has an acceleration of 1.8 m/s^2. With what force must you push?"

86. a.

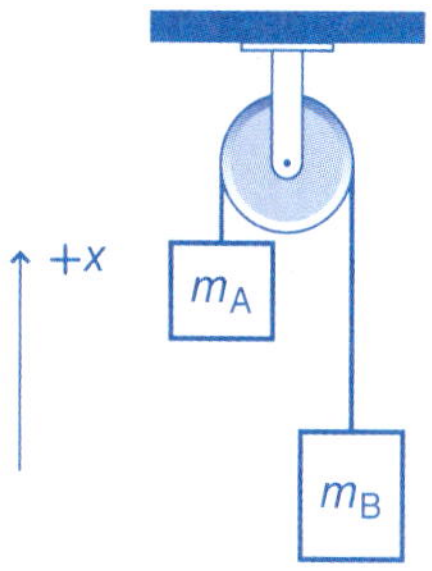

b.

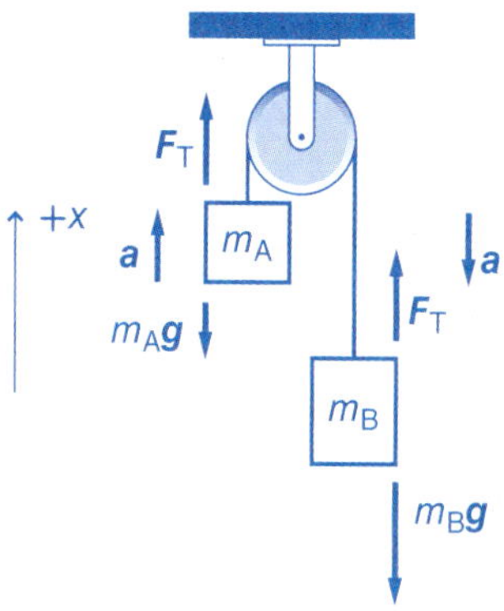

c. 2.0 m/s^2 upward

87. 5.88 m/s^2 downward

88. a. 3.0 m/s^2

b. 18 N

89. See Online Solutions Manual for diagram and possible solution.

90. See Online Solutions Manual for an example solution.

91. Student labs will vary with equipment available and designs. *x-t* graphs and *v-t* graphs should reflect uniform acceleration. The gravitational field should be close to 9.8 N/kg.

92. a. "…If he pushes the crate with a force of 40 N, what is the acceleration of the crate?"

b. "…If he pushes the crate with a force of 40 N, what force does the crate exert on him?"

Writing in Physics

93. Answers will vary. Newton's contributions should include his work on light and color, telescopes, astronomy, laws of motion, gravity, and perhaps calculus. One argument in favor of his three laws of motion being his greatest accomplishments is that mechanics is based on the foundation of these laws. His advances in the understanding of the concept of gravity may be suggested as his greatest accomplishment instead of his three laws of motion.

94. Answers will vary. Newton's first law of motion involves an object whose net forces are zero. If the object is at rest, it remains at rest; if it is in motion, it will continue to move in the same direction at a constant velocity. Only a force acting on an object at rest can change it to being in motion. Likewise, only a force acting on an object in motion can cause an object to change direction or speed. The two cases (object at rest, object in motion) could be viewed as two different frames of reference. This law can be determined, but it cannot be proven.

95. The gravitational force is a long-range force between two or more masses. The electromagnetic force is a long-range force that acts on electric charges and magnets. The weak nuclear force plays a role in beta decay. During the earliest moments of the universe, when the universe was very hot and dense, the electromagnetic force and the weak nuclear force were unified in one force called the electroweak force. The strong nuclear force has a very short range and is what holds protons and neutrons together in the nucleus of an atom.

Cumulative Review

96. 39 min; the skier should pass your house at 9:04 A.M.

97. a. 3 s, 8 s

b. car A

c. 5 s

d. none

e. ~3 s to 10 s

98. a. 0 m/s

b. ~0 m/s

c. ~1 m/s

Standardized Test Practice

student text p. 119

Multiple Choice

1. D
2. B
3. B
4. C
5. B
6. B
7. D
8. B
9. D

Free Response

10. See Online Solutions Manual. Elevator accelerates upward: apparent weight of dog increases
$F_{net} = F_g + F_{scale\ on\ dog}$.
Elevator at constant speed downward: apparent weight of dog is not changed
$F_{scale\ on\ dog} = F_g$.
Elevator falls freely downward: apparent weight of dog is zero
$F_{net} = F_g + F_{scale\ on\ dog}$ but $F_{net} = F_g$ so
$F_{scale\ on\ dog} = F_{net} - F_g = 0$.

Rubric

The following rubric is a sample scoring device for free response questions.

Points	Description
4	The student demonstrates a thorough understanding of the physics involved. The response may contain minor flaws that do not detract from the demonstration of a thorough understanding.
3	The student demonstrates an understanding of the physics involved. The response is essentially correct and demonstrates an essential but less than thorough understanding of the physics.
2	The student demonstrates only a partial understanding of the physics involved. Although the student may have used the correct approach to a solution or may have provided a correct solution, the work lacks an essential understanding of the underlying physical concepts.
1	The student demonstrates a very limited understanding of the physics involved. The response is incomplete and exhibits many flaws.
0	The student provides a completely incorrect solution or no response at all.

CHAPTER 5

Displacement and Force in Two Dimensions

About the Photo

Forces on Bridges Have students examine the photo and identify parts of the bridge that experience force. Possible answers: the roadbed, the towers, the wires Have students describe how the roadbed might be subject to force in more than one dimension. Possible answers: gravity pulls the roadbed down, wires suspend the roadbed from the top at nonvertical angles, and support pillars provide upward normal force.

Chapter Overview

This chapter extends the discussion of Newton's laws to two dimensions. The first section reviews formal vector addition in one dimension and extends it to two dimensions. The second section introduces kinetic and static friction and illustrates how to include friction in Newtonian analyses. Finally, additional two-dimensional situations, including inclined planes, are discussed, as is the concept of an equilibrant.

Before students study the material in this chapter, they should study:

- Acceleration in one dimension
- Adding vectors in one dimension
- Mass and weight
- Newton's laws of motion
- The normal force
- Uniform motion in one dimension

To solve the problems in this chapter, students will need a strong grasp of:

- Significant figures
- Solving linear equations
- Sine, cosine, and tangent
- Slope

Introduce the BIG IDEA

Ask students to identify the forces, including directions, acting on a tightrope walker standing in the middle of a sagging (slack) rope. Gravitational force acts down and tension in the rope acts at an angle. Explain that physical situations often involve forces acting in several different directions.

Go online!
connected.mcgraw-hill.com

Your online experience includes:

- **Video**
- **Concepts In Motion**
- **Virtual Investigations**
- **All Lab, Applying Practices activities, and PBL worksheets**
- **Online Practice**
- **Personal Tutor**
- **Multilingual eGlossary**
- **Presentation**

Use the LaunchLAB

In *Adding Vectors,* students will observe the vector nature of forces and use their observations to add vectors.

Use Physics T.V.

Video

FORCES AND MOTION A hurried commuter leaves a full coffee cup on top of his car before leaving for work. As the car starts moving, what will happen to the coffee cup? How long will it stay on the car before falling off?

1 Introduce

Bellringer

Displacement of a Person Ask students to consider a person who walks 100 m due north and then loses all sense of direction. Without knowing the direction, the person walks another 100 m. Ask students what the magnitude of displacement could be, relative to the original starting point. They must consider the straight-line distance from the starting point. Suggest that students draw arrows to represent the two 100-m vectors as they analyze the problem. The magnitude of the displacement of the person could be anywhere between 0 m and 200 m. **OL**

Tie to Prior Knowledge

Vectors, Forces, Acceleration Students studied how to add and subtract vectors in one dimension. Students also studied what forces were and how to apply Newton's laws in one dimension. Although the emphasis in this chapter is on forces in two dimensions, knowledge of displacement, velocity, and acceleration also is necessary to analyze some of the situations presented.

2 Teach

Vectors in Two Dimensions

student text pp. 122–124

Use Models

Rolling Ruler Demo To help students better visualize and understand the permissible ways to move vectors without changing them, use a rolling ruler to demonstrate a subset of the valid ways to move vectors. These rulers are available in stores that sell drawing and drafting equipment.

Concept Development

Designating Vectors Explain the method you will use to indicate vector quantities on the board or overhead transparency and on handouts. Usually, this is done by placing an arrow above the symbol that identifies the quantity.

Reinforcement

Magnitude Have students describe what the word *magnitude* means in physics. In physics, it means the size of a quantity, often represented by the length of an arrow. Discuss the need to always specify both the magnitude and direction of vector quantities.
OL **Linguistic**

Use Figure 2

Add Vectors Graphically **Figure 2** shows one of two correct ways to graphically add two vectors. Discuss with students that vector addition is commutative, which in this case means that the two vectors can be added in any order. You could put the tail of the eastward vector to the tip of the northward vector. The resultant will be the same as that shown in **Figure 2,** although the diagrams will appear different. **OL**

Quick DEMO

Vector Addition

Estimated Time 10 minutes

Materials two or three bungee cords with arrow heads, pegboard, metric ruler

Procedure Use the pegboard and bungee cords to demonstrate vector addition. Create a number of vector addition problems. Use the bungee cords to represent different vectors. Vectors can be started from the same origin or can be stacked to show addition visually. Use the ruler to measure the length of each "vector" as well as the resultant.
Visual-Spatial **EL**

DIFFERENTIATED INSTRUCTION

Struggling Students The text describes vector addition using the tip-to-tail method. If this does not seem to work for some students, have them try the parallelogram method of vector addition. In this method, the student draws the two vectors (**A** and **B**) being added with their tails at the same starting point. This shows the student the two directions and magnitudes of the vectors. The student then draws a parallelogram starting from the two vectors **A** and **B**. The student draws a vector (**A′**) of the same length as vector **A**, from the tip of vector **B** and parallel to vector **A**. Then the student draws a copy (**B′**) of vector **B** from the tip of vector **A** and parallel to vector **B**. This produces a parallelogram. Finally, the student connects the common point of the bases of vectors **A** and **B** with the common point of the tips of vectors **A′** and **B′** to get the resultant displacement vector.
BL **Visual-Spatial** **EL**

ADDITIONAL IN-CLASS EXAMPLE

Use with Example Problem 1

Problem Find the magnitude of the sum of two forces, one 20.0 N and the other 7.0 N, when the angle between them is 30.0°.

Response Use the law of cosines:

$$\begin{aligned} R^2 &= A^2 + B^2 - 2AB\cos\theta \\ &= (20.0\text{ N})^2 + (7.0\text{ N})^2 \\ &\quad - 2(20.0\text{ N})(7.0\text{ N})\cos 150.0° \\ &= 400\text{ N}^2 + 49\text{ N}^2 - 280\text{ N}^2(-0.866) \\ &= (400 + 49 + 242.49)\text{ N}^2 \\ &= 691.49\text{ N}^2 \\ R &= \sqrt{691.49} \\ &= 26.3\text{ N} \end{aligned}$$

This answer is consistent with this problem's vector diagram, which shows that the resultant should indeed be slightly greater in magnitude than the 20.0-N force.

Note that $\theta = 180.0° - 30.0° = 150.0°$.

Some students might too quickly conclude that θ equals 30.0°, but a vector diagram plus some basic geometry can show that θ equals the supplement of 30.0°, that is, 150°.

Identify Misconceptions

Pythagorean Theorem Students might want to apply the Pythagorean theorem whenever they add two vectors. In fact, the Pythagorean theorem only applies when they are adding two vectors at right angles to each other.

Reinforcement

Vector Addition It might be easier for students to understand vector addition and subtraction when it applies more directly to their own motion. Choose a location such as a gym, a playground, or a parking lot where the site can be marked off with a grid pattern along the edges. Have students move from one position to another using the grid to help them locate themselves on the surface. Then have them plot their motion on a piece of graph paper. Moving to a second position and plotting that movement and position on the graph paper, they can then measure their displacement from the starting point to the end point and compare that to the measurement on the graph paper. BL Kinesthetic EL

PHYSICS CHALLENGE ACTIVITY

Vector Operations Some students might ask if it is possible to multiply and divide vectors. Refer them to a university-level physics text that covers the multiplication of vectors to form a scalar (or dot) product, and a different type of multiplication to form a vector (or cross) product. Have students solve one of the physics problem sets in the textbook. Have students compare their answers and solutions with each other.
AL Logical-Mathematical

Critical Thinking

Permissible Operations Ask students which arithmetic operations are permitted between a vector and a scalar. It is not possible to add or subtract a combination of vectors and scalars. It is possible to multiply a vector by a scalar. Discuss that if you multiply a vector by a positive scalar, the product is a vector that points in the same direction as the vector that was multiplied. Explain that the opposite is true of a negative scalar; its product points in the opposite direction. If the scalar has an absolute value of 1, then the product is a vector with the same magnitude as that of the vector that was multiplied. But if the scalar has an absolute value greater than 1, then the magnitude of the product is greater than that of the vector that was multiplied. And when the scalar is less than 1, but greater than 0 the magnitude of the product is less than the vector that was multiplied. An example of multiplication by a negative scalar occurs in one of the steps to carry out vector subtraction, $\mathbf{A} - \mathbf{B}$, described in this textbook. To carry out this operation, you add vector $\mathbf{A}$ to vector, $-\mathbf{B}$ which is just vector $\mathbf{B}$ multiplied by the scalar −1. OL

Reinforcement

Vector Addition Students may perform ordinary addition when presented with word problems involving vectors. If this occurs, ask students to imagine a person at a corner of a park that is 50-m square and wants to walk to the opposite corner. Have students imagine the person walking along the edge of the 50-m-long park to its end, then turning 90° and walking another 50 m. Ask how far the person would be from the starting point. If they answer 100 m, ask what the shortest way between opposite corners would be. The person would cut across, going from the beginning point to the end point on a diagonal. Ask how far the person would be according to vector addition. 71 m BL

DIFFERENTIATED INSTRUCTION

Visually Impaired The qualitative aspects of vector addition can be communicated to students who have limited vision by using cardboard or plastic arrows of varying lengths. Fasteners can be used to place two vector arrows tip to tail, and a third arrow can be used to represent the resultant. Alternatively, pencils can be taped to a flat surface so students can feel the geometry. When moving arrows or pencils you can translate them (move up, down, right, or left), but do not rotate them. Rotation does change the value of the vector, whereas translating does not change its value. **BL** **Kinesthetic** **EL**

Critical Thinking

Adding Three Vectors Remind students that they have explored addition of two vectors. Ask them how this might solve a problem involving three vectors. Add two vectors, then add the third vector to the resultant. **OL**

DIFFERENTIATED INSTRUCTION

Struggling Students Consider a right triangle whose sides are vectors and whose hypotenuse is the vector sum of the legs. Although vector ***c*** is the vector sum of vectors ***a*** and ***b*** (that is, $\mathbf{c} = \mathbf{a} + \mathbf{b}$), the magnitude of ***c***, that is, *c*, the hypotenuse's measured length, is not equal to the algebraic sum of the magnitudes of vectors ***a*** and ***b***, that is, $c \neq a + b$.

For this activity, you need a large space, some masking tape, a meterstick, and a calculator. Clear a space large enough to mark out a relatively large right triangle on the floor with masking tape.

Have two students practice taking the same size steps (easier with a tiled floor). Then have both students stand at the vertex of one of the acute angles. Starting at the same time, have one student move to the other acute vertex along the hypotenuse, while the other student moves along the two legs. Make sure that both students take the same sized steps at the same pace. The student who moves along the hypotenuse will get there first. Remind the students that vectors have to be placed tip-to-tail in order to apply the Law of Cosines. **BL** **Kinesthetic** **EL**

Vector Components and Algebraic Addition of Vectors

student text pp. 125–129

Reinforcement

Component Sign Check It might be worth reminding or suggesting to students that, if a physics problem requires them to calculate the *x*- or *y*-components of a vector, they check that the sign of the component makes sense. If a student finds that a component's sign does not make sense, then that suggests that the student should recheck his or her work.

For instance, the solution given for the following Additional In-Class Example includes calculations of the *x*- and *y*-components of three vectors, ***A***, ***B***, and their resultant, ***R***. To check whether the signs of the five components that have signs (one component is zero, which has no sign), a student could draw the three vectors, with the correct directions and relative lengths and with their tails coinciding with the origin of an *x*-*y* coordinate grid.

From the vector diagram, the student should be able to see at a glance that all of the signs indeed make sense. They should also be able to see that all of the calculated magnitudes and directions make sense, as well.

ADDITIONAL IN-CLASS EXAMPLE

Use with Example Problem 2.

Problem Add the following two vectors via the component method: ***A*** is 4.0 m south and ***B*** is 7.3 m northwest.

Response Treat east as $+x$ and north as $+y$.

$A_x = (4.0\text{ m})\cos 270° = 0$

$B_x = (7.3\text{ m})\cos 135° = -5.16\text{ m}$

$A_y = (4.0\text{ m})\sin 270° = -4.0\text{ m}$

$B_y = (7.3\text{ m})\sin 135° = 5.16\text{ m}$

$R_x = A_x + B_x = 0 + (-5.16\text{ m}) = -5.16\text{ m}$

$R_y = A_y + B_y = (-4.0\text{ m}) + (5.16\text{ m}) = 1.16\text{ m}$

$R^2 = R_x^2 + R_y^2 = (-5.16\text{ m})^2 + (-1.16\text{ m})^2$

$R = 5.3\text{ m}$

$$\text{Direction: } \theta = \tan^{-1}\left(\frac{R_y}{R_x}\right)$$

$$= \tan^{-1}\left(\frac{1.16\text{ m}}{-5.16\text{ m}}\right)$$

$$= 167°\text{, or } 13°\text{ north of west}$$

3 Assess

Assess the MAIN IDEA

Vector Components Have students imagine themselves walking on city streets that are laid out on a grid. They walk in one direction 4 blocks, then turn the corner and walk 5 blocks down a street that is at right angles to the first. Ask students how they could show their displacement with vectors and a coordinate grid. Use a coordinate grid to plot their path on the two streets and calculate the diagonal path from where they started to where they stopped.

Check for Understanding

Vector Addition Ask students to make up a problem involving vector addition in one dimension. Have them explain how to solve it. Then have them make up another problem involving vector addition in two dimensions, where the vectors are at right angles, and ask them to outline the solution. Finally, have them create a problem involving two-dimensional vector addition without right angles and explain its solution.
OL Linguistic

Reteach

Vector Addition Review the methods of adding vectors, both graphically and algebraically. Emphasize when the Pythagorean theorem does and does not apply. Draw a number of sets of vectors on the board. Have a student use a meterstick to measure the vectors and determine the resultant vector. Then have the students determine the resultant for each set using trigonometry.
OL Visual-Spatial EL

Reading Checks and Caption Checks

Reading Check, *student text p. 123*

Vectors are added by placing the second vector's tail on the first vector's tip. The vector drawn from the tail of the first vector to the tip of the second vector is the sum of the two vectors.

Caption Check, *student text p. 125*

A vector whose *y*-component is zero is a horizontal vector.

Reading Check, *student text p. 126*

The direction is measured counterclockwise from the positive *x*-axis.

Caption Check, *student text p. 126*

The vector is in the fourth quadrant, so the *x*-component is positive and the *y*-component is negative.

Practice Problems

student text p. 124

1. $R = 141$ km

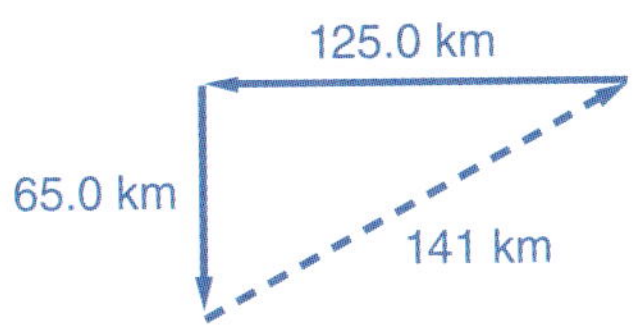

2. $R = 257$ m

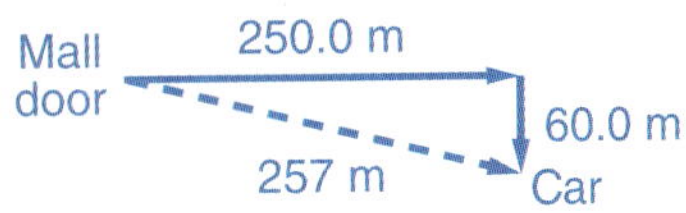

3. 1.0×10^{1} km

4. 8.3 cm

Practice Problems

student text p. 129

5. 0.87 km at 77° west of north

6. 6.0 km

7. The *x*-component is positive for angles less than 90° and for angles greater than 270°. It's negative for angles greater than 90° but less than 270°.

8. It could never be shorter than one of its components, but if it lies along either the *x*- or *y*-axis, then one of its components equals its length.

9. The force will be straight up. Because the angles are equal, the horizontal forces will be equal and opposite and cancel out. 4.4 N upward

10. The net force is 0.8 N in the upward direction.

Section 1 Review

student text p. 129

11. $\mathbf{M}_x = 4.0$ to the right

$\mathbf{M}_y = 3.0$ upward

12. Both vectors are horizontal, so they do not have any *y*-component.

$\mathbf{K}_x = -4.0$, $\mathbf{K}_y = 0$

$\mathbf{L}_x = 6.0$, $\mathbf{L}_y = 0$

13. $\mathbf{R} = 6.7$ at 27°

14. $6.0 - (-4.0) = 10.0$ to the right

15. a. Both ways of adding $\mathbf{M}$ and $\mathbf{L}$ result in the same vector $\mathbf{R}$.

b. Addition and multiplication are commutative.

Examples: $3 + 4 = 4 + 3$
$2 + 5 = 5 + 2$

Subtraction and division are not commutative.

Examples: $10 - 3 \neq 3 - 10$
$8 \div 2 \neq 2 \div 8$

16. Not necessarily; for example, you could walk around the block (one km per side). Your displacement would be zero, but the distance that you walk would be 4 km.

17. The resultant can't be zero if there are two displacements. If there are three displacements, the sum can be zero if the three vectors form a triangle when they are placed tip to tail. Also, the sum of three displacements can be zero without forming a triangle if the sum of two displacements in one direction equals the third in the opposite direction.

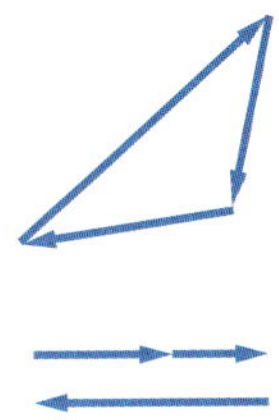

1 Introduce

Bellringer

Surface Friction Push a low-friction object, such as ice, across a table. Then push a higher-friction object, such as a book, across the same surface. Ask students to describe the difference in the behaviors of the two objects and to suggest a possible reason for this difference. Students probably will say that friction was a factor. Guide students to articulate what they mean by *friction,* so that you will be able to help guide them to the scientific definition later. BL Visual-Spatial EL

Tie to Prior Knowledge

Unbalanced Forces Shove a book across a table and ask students how they know an unbalanced force acted on the book as it came to rest. it accelerated Explain the unbalanced force is friction. Ask students to describe other instances in which they observe friction affecting an object's motion.

2 Teach

Kinetic and Static Friction

student text pp. 130–135

Use an Analogy

Investigating Friction A simplified analogy for the friction force between two objects is the interaction of two pieces of hook-and-loop tape with minute hooks. On a microscopic level, the two surfaces in contact do partially mesh into each other. Ask students to research how this tape works and draw some diagrams to help explain it. OL

Use Figure 10

Friction and Force Sketch the pictures in **Figure 10** on the board. Draw force vectors showing the applied force and the friction resistance in each case. Explain that in the left picture the static friction will match the applied force to prevent motion. Only when the applied force exceeds the resistance force will the object move, as in the right picture.

Critical Thinking

High or Low Friction Ask students whether it is better to have low friction between surfaces or high friction between surfaces. It depends on the situation. Have students give examples of each. Low-friction situations include pistons in engines and skis on snow. Examples of situations where high friction is desired include erasers on paper or a flag twirler's glove gripping the flagpole. OL Logical-Mathematical

Use the PhysicsLAB

In *Coefficient of Friction,* students will determine the coefficient of static and kinetic friction.

ADDITIONAL IN-CLASS EXAMPLE

Use with Example Problem 3.

Problem A child drags a heavy, rubber-soled shoe by its laces across a sidewalk at a constant speed of 0.35 m/s. If the shoe has a mass of 1.56 kg and the coefficient of kinetic friction is 0.65, what is the horizontal component of the force exerted by the child? (Assume the sole is in contact with the sidewalk, not bouncing around.)

Response $F_{net} = ma = 0$

$F_{net} = F_x - F_f \quad ma = 0$

Substituting yields, $F_x = F_f$

$F_x = \mu F_n = \mu mg = (0.65)(1.56\text{ kg})(9.8\text{ N/kg}) = 9.9\text{ N}$

Discussion

Problem When is it good to have a high coefficient of friction between two surfaces?

Response Possible examples are running shoes on a track, sandpaper on wood, car brakes on wheels, and car tires on roads. OL

Quick **DEMO**

Friction

Estimated Time 10 minutes

Materials nonstick cookie sheet, felt-covered wood block (6 in × 3 in × 1 in)

Procedure Have students predict which side of the block will require the greatest force to be pushed across the cookie sheet at a constant speed. Then have some students slide the largest side of the felt-covered block along the nonstick cookie sheet. Repeat with the smallest side of the block. Have students summarize their results for the class. The force of friction is the same in both cases, regardless of the area of the surfaces in contact.

Concept Development

Friction and Normal Force Demo Have a student push an empty box or crate at constant speed across a level floor. Then put a heavy load in the container. Have the student push the box across the floor again. Have the student report the qualitative change in the force required. Have students explain why the force required is greater in the second case. The required force increases because an increased normal force produces increased friction. OL Kinesthetic EL

Reinforcement

Static v. Kinetic Friction Demo Use a demonstration-sized spring scale so the class can read the scale as you pull a book across a table. Show that it takes much more force to get the book started than it does to keep it moving. Ask students to explain their observations in terms of static and kinetic frictional forces. This activity could also be done in small groups with smaller spring scales. OL Visual-Spatial EL

Applying Physics

Surfaces As surface friction varies, so does the ability of certain materials to adhere to the surface. To prevent graffiti in public places, the walls often are coated with a material whose surface makes it impossible to apply any pen, permanent marker, or pencil.

REAL-WORLD PHYSICS

Hydroplaning The coefficients of friction between rubber tires and the concrete of a road are much less when the road is wet than when it is dry. If water pools on the surface of a road, there can be enough of a layer of water between the road and the tires that there is almost no contact between the tires and the road, so that the tires are essentially riding on a layer of water.

ADDITIONAL IN-CLASS EXAMPLE

Use with Example Problem 4

Problem If the child in the previous Additional In-Class Example pulls with only half the force as before, what will happen?

Response If the new force is half of the original force (5.0 N), then it is matched by the force of static friction and the shoe will not move.

Problem If the student pulls with an extra 2.0 N in the horizontal direction, what will be the acceleration of the shoe?

Response

$$a = \frac{F_{net}}{m} = \frac{2.0\ \text{N}}{1.56\ \text{kg}} = 1.3\ \text{m/s}^2$$

3 Assess

Assess the MAIN IDEA

Friction Have students describe the force acting on a box of books at rest on the floor while someone applies a horizontal force on the box. Static friction acts on the box in the direction opposite to the applied force; gravitational force pulls down on the box; the floor pushes up on the box. Ask students how the forces on the box change if it starts to move. Gravitation force pulling down and the floor force pushing up do not change; a kinetic friction force acts in the direction opposite to the applied force.

Check for Understanding

Friction Coefficients Present a problem that requires the use of only one coefficient of friction to solve, but provide both coefficients. After students solve the problem, ask which coefficient they used and why. OL Logical-Mathematical

Reteach

Static Friction Review that a static friction force can vary and the coefficient of static friction gives the maximum possible static friction force. Pull on a heavy object with a spring scale with various forces and have students describe the variation in forces. OL

Reading Checks and Caption Checks

Caption Check, *student text p. 130*
Kinetic friction acts on the couch when it is in motion.

Reading Check, *student text p. 131*
The materials sliding past each other and the normal force between the two objects affect friction forces.

Caption Check, *student text p. 131*
Surfaces in increasing coefficient of kinetic friction: highly polished table < rough table < sandpaper.

Practice Problems

student text p. 133

18. 0.69

19. 74 N

20. 78 N

21. 0.39

Practice Problems

student text p. 135

22. 0.13

23. 0.15

24. 0.50 s

25. 66 m so he hits the branch before he can stop.

26. 6.7 m; the disk will stop in the 10-point section.

Section 2 Review

student text p. 135

27. Both result from two surfaces rubbing against each other. Both are dependent on the normal force between these two surfaces. Static friction applies when there is no relative motion between the two surfaces. Kinetic friction is the type of friction when there is relative motion. The coefficient of static friction between two surfaces is greater than the coefficient of kinetic friction between those same two surfaces.

28. 37 N

29. 1.3 m/s

30. 1.7×10^{2} N

31. All you can conclude about the coefficient of static friction is that it is between 0.16 and 0.20.

32. Friction between the dresser and the truck accelerates the dresser forward. If the force of the truck on the dresser exceeds $\mu_s mg$, the dresser will slide backward.

1 Introduce

Bellringer

Forces at an Angle Tie a string between two supports so it is nearly horizontal. Hang a small mass from the center of the string so the string dips noticeably. Ask students to identify the forces acting on the mass. gravitational force acts downward and tension in the string acts along the direction of the string, not purely vertical nor horizontal Ask students how they know the net force on the mass is zero. it is not accelerating Tension must have a vertical component that balances the downward gravitational force. Ask students to describe other situations in which several forces act on an object but the object remains in equilibrium. OL Visual-Spatial EL

Tie to Prior Knowledge

Equilibrium Students should be familiar with the concept of equilibrium from previous study. However, so far they have applied it in only one dimension.

2 Teach

Equilibrium Revisited

student text pp. 136–137

Use Figure 15

Canceling Force The figure shows that three forces are added to produce a resultant of zero. The same result may be obtained by canceling $\boldsymbol{F}_A$ and the vertical component of $\boldsymbol{F}_B$, and then canceling $\boldsymbol{F}_C$ and the horizontal component of $\boldsymbol{F}_B$. OL

Concept Development

Equilibrium Demo Attach an object to two spring scales using strings of equal length. Hold up the object using the spring scales. Demonstrate to the students that equilibrium can be achieved in many different ways by arranging the spring scales so that the angle between the strings varies and the forces exerted by each string vary.

Identify Misconceptions

Equilibrant In trying to find the equilibrant of two forces, students will often calculate just the resultant of the two vectors and call it the equilibrant. To help students avoid this mistake, ask them in what direction the equilibrant acts, reminding them that the equilibrant is a vector of the same magnitude but opposite direction of the resultant. OL

Use the MiniLAB

In *Equilibrium,* students will find the equilibrant of two forces using string-tension forces on a hanging spring scale.

Critical Thinking

Force Components Have students look at pictures or videos of weightlifters and determine how the material in this section pertains to successful weightlifting technique. Drawing force diagrams may help them articulate their explanation. Only the force exerted perpendicular to the bar by the lifter can actually elevate the bar. Weightlifters will use a grip that enables them to maximize this perpendicular force. OL Visual-Spatial EL

Content Background

Igloo Construction and Equilibrants The Inuit people use equilibrant forces to build igloos. An igloo is a dome-shaped structure built of blocks of ice. As gravity pulls the blocks closer together, each block pushes against the blocks on either side, above, and below it. The top center block is the most important one. Until the center block is in place, the top row of blocks (in particular) will move because none of them are in equilibrium. However, once the top block is inserted, all the blocks in the igloo stay in place because the sum of the forces results in a net force of zero.

REAL-WORLD PHYSICS

Equilibrium and Static Balance One of the most important applications of force vectors is made by architects and engineers in achieving equilibrium, known as static balance, in the structures they design. Whether the structure is a bridge, a building, or a highway, equilibrium is vital. Buildings with peaked roofs must be constructed so that the outward force where the roof joins the wall is counterbalanced. This can be done through an external buttress or an internal truss. In either instance, a horizontal inward force is applied to counterbalance the outward component of the vector of the roof.

Inclined Planes

student text pp. 138–141

Discussion

Problem If an object slides down an inclined plane, does its acceleration depend on its mass? The coefficient of kinetic friction between the two surfaces? The angle of inclination?

Response It depends on the angle and the coefficient of friction, but not on the object's mass.
$A = g(\sin\theta - \mu_k \cos\theta)$, for θ large enough to allow sliding. OL

Use the MiniLAB

In *Forces on a Plane,* students will measure the magnitude of the force needed to pull an object at a constant speed up an inclined plane.

Use the PhysicsLAB

In *Friction on a Plane,* students will investigate the acceleration of an object sliding down a ramp. They will compare this to their experimentally determined value of the acceleration magnitude.

Use the PhysicsLAB

In *Perpendicular Forces,* students determine by experiment how two different forces, when acting separately, make a certain object accelerate.

ADDITIONAL IN-CLASS EXAMPLE

Use with Example Problem 5.

Problem Jeff, who weighs 640.0 N, sits on the slope of a hill that descends at an angle of 35.0° from horizontal. What are the components of his weight parallel and perpendicular to the surface of the hill?

Response Choosing axes such that $+y$ is downward and normal to the slope, and $+x$ is down the slope, yields:

$F_{gx} = (640.0\ \text{N})(\sin 35.0°) = 367\ \text{N}$

$F_{gy} = (640.0\ \text{N})(\cos 35.0°) = 524\ \text{N}$

ADDITIONAL IN-CLASS EXAMPLE

Use with Example Problem 6.

Problem A child slides from rest at the top of a playground slide. If the slide is inclined at 30.0° and the coefficient of kinetic friction between the slide and the child's pants is 0.18, what is the child's acceleration?

Response $F_{gx} - F_{fx} = ma_x$

$mg(\sin\theta) - \mu mg(\cos\theta) = ma_x$

$a_x = g(\sin\theta) - \mu g(\cos\theta)$

$= (9.80\ \text{N/kg})(\sin 30.0°) - 0.18(9.80\ \text{N/kg})(\cos 30.0°)$

$= 3.4\ \text{m/s}^2$

3 Assess

Assess the MAINIDEA

Motion on a Sloping Board Ask students how they would set up axes for the motion of sliding a toolbox down a sloping board. Set the *y*-axis perpendicular to the board and the *x*-axis pointing parallel to its surface. Have students describe how the normal force and the angle of the board are related. As the angle increases, the normal force decreases ($F_N = mg \cos \theta$).

Check for Understanding

Equilibrants Ask how two forces of 6.0 N and 8.0 N could be applied to an object to give a resultant force of 10.0 N. They could be applied at an angle of 90° to each other. Sketch the two forces. Then ask how a third force could be added to produce equilibrium. The third force should have a magnitude of 10.0 N and point in the direction opposite that of the 10.0-N resultant. Remind students that this third force is the equilibrant.
OL Logical-Mathematical

Reteach

Vector Resolution One of the most common vector resolutions is the one for the weight of an object on an inclined plane. Repeat this resolution step by step, stressing why each component goes where it does, and that neither component can be greater than the total weight of the object. Drawing a large triangle helps.

Reading Checks and Caption Checks

Caption Check, *student text p. 136*

The vertical component of $\mathbf{F}_A$ is equal in magnitude but opposite in direction from the weight of the mass hanging from the ring.

Reading Check, *student text p. 137*

The equilibrant is equal in magnitude but opposite in direction from the resultant vector.

Caption Check, *student text p. 138*

As the angle increases, the component of the girl's weight parallel to the incline also increases.

Reading Check, *student text p. 138*

Acceleration will be parallel to the incline, so choosing the x-direction to be parallel to the incline means the velocity and acceleration will be in the x-direction, not a combination of the x- and y-directions. This simplifies the solution.

Practice Problems

student text p. 139

33. See Online Solutions Manual for free-body diagram.

34. $F_{g,\text{ parallel}} = 1.1$ N; $F_{g,\text{ perpendicular}} = 4.2$ N

35. 345 N

36. 26.6° relative to the horizontal

Practice Problems

student text p. 141

37. 4.90 m/s^2; 19.6 m/s

38. 0.75

39. 5.2 m/s^2

40. a. $\phi_{\text{average}} = \dfrac{21° + 17° + 21° + 18° + 19°}{5} = 19°$

b. 0.34

c. 0.20

Physics Challenge

student text p. 137

$F_{1x} = 58.3$ N
$F_{1y} = 17.8$ N
$F_{2x} = 16.7$ N
$F_{2y} = 34.2$ N
$F_{3x} = -7.52$ N
$F_{3y} = 53.5$ N
$F_{4x} = -74.3$ N
$F_{4y} = 56.0$ N
$F_{5x} = -60.7$ N
$F_{5y} = -23.3$ N
$F_{6x} = -26.4$ N
$F_{6y} = -98.5$ N
$F_{7x} = 0.0$ N
$F_{7y} = -26.0$ N
$F_{8x} = 28.8$ N
$F_{8y} = -71.4$ N
$F_{9x} = 27.8$ N
$F_{9y} = -42.8$ N
$F_{10x} = 81.7$ N
$F_{10y} = -7.15$ N
$F_x = 44.38$ N
$F_y = -107.65$ N

$$F_R = \sqrt{(F_x)^2 + (F_y)^2} = \sqrt{(44.38\text{ N})^2 + (-107.65\text{ N})^2} = 116\text{ N}$$

$$\theta_R = \tan^{-1}\left(\frac{F_y}{F_x}\right) = \tan^{-1}\left(\frac{-107.65\text{ N}}{44.38\text{ N}}\right) = -67.6°$$

The equilibrant's angle must be opposite the resultant's angle, so add 180°:

$F_{\text{equilibrant}} = 116$ N at 112°

Section 3 Review

student text p. 141

41. 3.2 m/s^2, up the incline

42. See Online Solutions Manual for diagram. The vectors shown in the free-body diagram indicate that even a small force perpendicular to the rope can increase the tension in the rope (F_T) enough to overcome the friction force. Since $F_{\text{person on rope}} = 2F_T \sin\theta$ (where θ is the angle between the rope's original position and its displaced position),

$$F_T = \frac{F_{\text{person on rope}}}{2\sin\theta}$$

For smaller values of θ, the tension, (F_T), will increase greatly.

43. 1.31×10^3 kg

44. The sum of the vectors is zero. If the vectors represent forces, the object is in equilibrium. This means the object is not accelerating.

45. F_T gets smaller as θ gets larger, and θ is larger in the bottom image.

46. No, because if the child moves up the slide, both the frictional force opposing the motion of the child and the component of Earth's gravity parallel to the slope point down the slide, not up.

ON THE **JOB**

Out on a Limb

Background

Architecture involves the careful balancing of forces to ensure a safe, stable structure. In most buildings, support beams provide the upward force required to balance the downward gravitational force. Architects must ensure that the beams can support not only the weight of the building materials but also the weight of the furniture, people, and other objects inside the building once that building is constructed.

Teaching Strategies

- Students may have heard the term *load-bearing* used to describe certain columns or walls used in buildings. Explain how this designation indicates that these structures are those primarily responsible for balancing the gravitational force in the floors and structures above them. Removing a load-bearing wall removes a structure providing an upward force—the forces are then unbalanced and the building will collapse.
- Students can think of the upward forces and the downward gravitational force as two cards leaning against each other. Removing either card causes acceleration; when both cards are in place, they balance each other and the net force is zero.

GOING**FURTHER** >>>

Expected Outcomes The most stable free-standing arch structures are catenary arches, named for having a curve identical to a hanging chain—although upside down. Individual blocks in an arch are being pushed both down and out as a result of the weight from the blocks above them.

Each block is also pushed up and in by the normal force exerted by the block underneath. For the arch to be stable, the sum of these forces must be zero.

The force exerted on a block from above increases moving down the arch, since each block has not just its own weight, but also the weight of the blocks above it. A principal source of the catenary arch's stability is the fact that the outward force on each block is constant. This is achieved by carefully shaping each block in the arch so the normal force becomes increasingly vertical toward the bottom of the arch.

The total normal force exerted by the ground is equal to the total weight of the arch.

CHAPTER 5 ANSWERS

SECTION 1

Mastering Concepts

47. Make scale drawings of arrows representing the vector quantities. Place the arrows for the quantities to be added tip to tail. Draw an arrow from the tail of the first to the tip of the last. Measure the length of that arrow and find its direction.

48. allowed: moving the vector without changing length or direction

49. The resultant is the vector sum of two or more vectors. It represents the quantity that results from adding the vectors.

50. It is not affected.

51. Reverse the direction of the second vector and then add them.

52. ***A*** is the symbol for the vector quantity. *A* is the magnitude (length) of the vector.

53. *a* and *b* represent the lengths of two vectors that are at right angles to one another. *c* represents the length of the sum of the two vectors.

54. The angle is measured counterclockwise from the *x*-axis.

Mastering Problems

55. 2.0×10^{1} km, east

56. a. $E_x = 3.5$, $E_y = 3.5$

b. $F_x = -3.5$, $F_y = -3.5$

c. $A_x = -3.0$, $A_y = 0.0$

57. a.

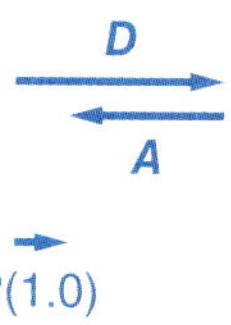

b.

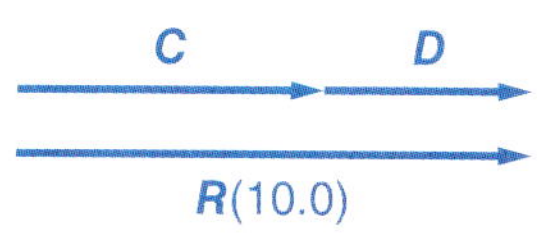

c.

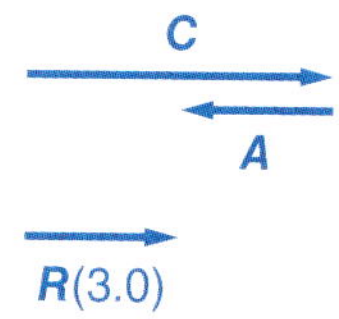

d.

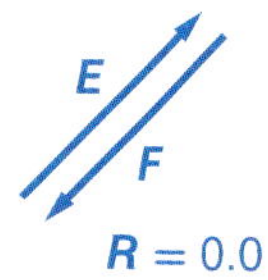

58. a.

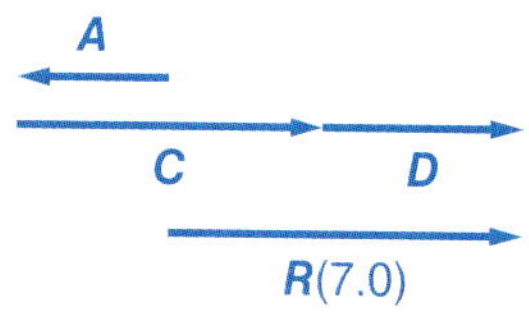

b.

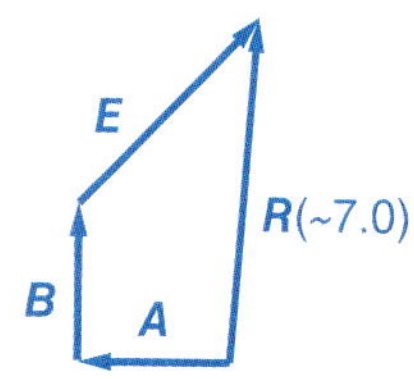

c.

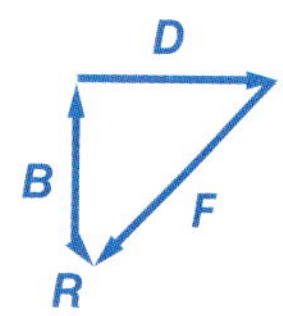

59. A < D < E < B < C

60. 40 m, 45° east of south

61. a.

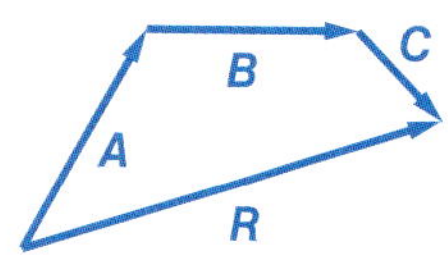

b.

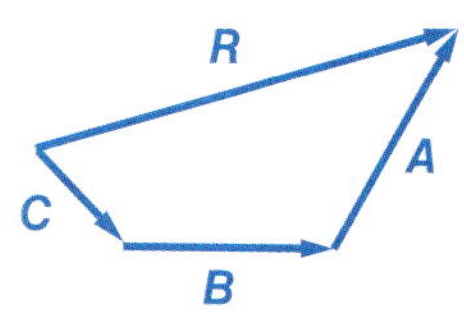

c. The resulting displacements are always the same. Adding vectors is commutative.

62. The net force is 640 N at 101°.

63. a. 6.5 m/s

b. 58° from horizontal, which is 32° from vertical

c. 42 s

64. The net force is 79 N at 54°.

65. 509.9 km, 78.69° south of west

66. 5 km, 53° south of east

SECTION 2

Mastering Concepts

67. The frictional force is greater than the normal force. You can pull the object along the surface, measuring the force needed to move it at constant speed. Also measure the weight of the object.

68. It would make no difference. Friction does not depend upon surface area.

Mastering Problems

69. 0.255

70. 1.2 m/s^2

71. a. 8.0 m/s^2

b. 1.0×10^1 N

c. 0.20

72. 180 N

73. 0.400

SECTION 3

Mastering Concepts

74. One axis is vertical, with the positive direction either up or down.

75. The two axes must be at right angles. The positive y-axis points 30 degrees away from the vertical so that it is at right angles to the x-axis.

76. For motion on a hill, the vertical (y) axis is usually set up perpendicular, or normal, to the surface of the hill.

77. The net force acting on the book is zero.

78. Yes, Newton's first law permits motion as long as the object's velocity is constant. It cannot accelerate.

79. a. Set up the y-axis perpendicular to the surface of the table and the x-axis pointing uphill and parallel to the surface.

b. One component is parallel to the inclined surface and the other is perpendicular to it.

80. a. As you increase the angle the table makes with the horizontal, the component of the book's weight along the table increases.

b. When the angle increases, the component of the weight normal to the table decreases and the friction force decreases.

Mastering Problems

81. 74.4 N, 253°

82. 34 N, 223°

83. 433 N, right

84. a. 1.5×10^2 N

b. 1.1×10^2 N

85. 123 N

86. a. The coefficient of static friction between the patient and the bed's sheets.

b. 1.00

87. a. 4.0 m/s^2

b. 93 N

88. $F = \left(\frac{g}{\mu_s}\right)(m + M)$

Applying Concepts

89. 15 km

90. 10 mm

91. The largest is 7 m, the smallest 1 m. See Online Solutions Manual.

92. The resultant increases.

93. a. A is longer.

b. B is longer

c. A and B are equal in length.

94. The northward component (y) is longer.

95. The Pythagorean theorem can be used only if the two vectors to be added are at right angles to one another.

96. See Online Solutions Manual. One component is in the negative *x*-direction, the other in the negative *y*-direction, assuming that the positive direction points upward, perpendicular to the hill.

97. The equal and opposite forces referred to in Newton's third law are acting on different objects. The horse would pull on the cart, and the cart would pull on the horse. The cart would have an unbalanced net force on it (neglecting friction) and would thus accelerate.

98. When stretching the net between the two posts, there is no perpendicular component upward to balance the weight of the net. All the force exerted on the net is horizontal. Stretching the net to remove the last bit of slack requires great force in order to reduce the flexibility of the net and to increase the internal forces that hold it together.

99. a. 45°

b. 0°

100. The component perpendicular to the ground is larger because the angle between the guy wire and horizontal is greater than 45°.

Mixed Review

101. 284 N

102. a. 166 N

b. 3.6 km

103. a. $-4.9\ \text{m/s}^2$

b. 24 m

Thinking Critically

104. 49 m

See Online Solutions Manual for complete solution.

105. a. $a = g(\sin\theta - \mu_k\cos\theta)$, so the acceleration is independent of the mass. They will tie, so Kako is correct.

b. They will reach the bottom at the same time.

106. A possible form of the correct answer would be, "…he moves the card north for 125 cm, where he places it on a shelf 115 cm above the ground. What is the total displacement of the card?"

107. A correct form of the answer is, "A 10-N carton is pushed across a floor by exerting a force of 10 N at angle of 20° below the horizontal. If the coefficient of kinetic friction is 0.13, what is the resulting acceleration of the carton?"

Writing in Physics

108. Answers will vary and may include lubricants and reduction of the normal force to reduce the force of friction.

109. Possible topics include sprinters' speed suits, swimmers' full-body, low-drag swim suits, and skiers' aerodynamic helmets, goggles, and poles.

Cumulative Review

110. a. 90.0 g

b. 1.68 km

c. 128.6 kg

d. 12 N

e. 2 m/s

111. 10 km/h

112. $13\ \text{m/s}^2$

Standardized Test Practice

student text p. 149

Multiple Choice

1. C

2. B

3. B

4. C

5. C

6. B

7. B

8. C

Free Response

8. 5.5×10^2 m

10. 1.8×10^2 N

Rubric

The following rubric is a sample scoring device for free response questions.

Points	Description
4	The student demonstrates a thorough understanding of the physics involved. The response may contain minor flaws that do not detract from the demonstration of a thorough understanding.
3	The student demonstrates an understanding of the physics involved. The response is essentially correct and demonstrates an essential but less than thorough understanding of the physics.
2	The student demonstrates only a partial understanding of the physics involved. Although the student may have used the correct approach to a solution or may have provided a correct solution, the work lacks an essential understanding of the underlying physical concepts.
1	The student demonstrates a very limited understanding of the physics involved. The response is incomplete and exhibits many flaws.
0	The student provides a completely incorrect solution or no response at all.

CHAPTER 6

Motion in Two Dimensions

About the Photo

Ask the students what forces are acting on the young man while he is airborne. gravity What determines his trajectory? his initial velocity and the net force that acts on him When he is at maximum height, is his center of mass still moving or is it momentarily stationary? It is not moving vertically but is moving at a constant speed in the horizontal direction because no force is acting in this direction.

Chapter Overview

In this chapter, the concepts of kinematics and dynamics that students have previously studied are extended to motion in two dimensions. The chapter analyzes projectile motion by applying kinematics of motion with constant horizontal velocity and constant vertical acceleration. Students learn to analyze circular motion by applying Newton's laws. The chapter closes with a discussion of relative velocity.

Before students study the material in this chapter, they should study:

- Adding vectors
- Mass v. weight
- Newton's laws of motion
- Uniform motion in one dimension
- Vectors v. scalars

To solve the problems in this chapter, student will need a strong grasp of:

- Graphing data
- Significant figures
- Sine, cosine, and tangent
- Solving linear equations
- Solving quadratic equations

Introduce the BIG IDEA

Armed with vectors and Newton's laws, a large variety of motion can be analyzed. This chapter looks at projectile motion, circular motion, and relative velocity. By applying vectors and Newton's laws, the future velocity, position, and acceleration of objects can be predicted with great accuracy.

Go online!
connected.mcgraw-hill.com

Your online experience includes:

- **Video**
- **Concepts In Motion**
- **Virtual Investigations**
- **All Lab, Applying Practices activities, and PBL worksheets**
- **Online Practice**
- **Personal Tutor**
- **Multilingual eGlossary**
- **Presentation**

Use the LaunchLAB

In *Projectile Motion,* students can use words and a graph to describe the horizontal and vertical motion of a projectile.

Use Physics T.V.

Video

PROJECTILE PHYSICS Have you ever seen a catapult or trebuchet in action? Discover the physics of launching projectiles!

1 Introduce

Bellringer

Projectile Motion Have two students toss a softball (or similar ball) underhand back and forth in front of the class. Ask students to focus on the horizontal and vertical motions of the ball by having them first describe the motion as seen by an observer above the motion and then describe the motion as seen by one of the students doing the tossing. BL Kinesthetic

Tie to Prior Knowledge

Kinematics The models students developed in previous chapters to analyze constant-velocity motion and constant-acceleration motion will apply to a projectile's horizontal and vertical motions.

2 Teach

Path of a Projectile

student text p. 152

Use Figure 1

Draw students' attention to the difference between the shapes of the paths in the two photos. Emphasize that the initial direction is the same as the direction of the force that launches a projectile, but the gravitational force changes the shape of the path. Ask students to name other examples of both types of motion.

Independence of Motion in Two Dimensions

student text p. 153

Use the MiniLAB

In *Over the Edge*, students can investigate how mass affects the motion of a projectile.

Use the MiniLAB

In *Projectile Path*, students will analyze the vertical and horizontal motion of a projectile.

Quick DEMO

MAINIDEA

Estimated Time 5 minutes

Materials two identical marbles, table

Procedure To show the independent nature of simultaneous horizontal and vertical motions, hold one marble next to the edge of a tabletop and place an identical marble on the edge of the table. Drop the first marble and at the same time give the second marble a horizontal push. Repeat the experiment with different horizontal velocities for the first marble. For all experiments, students should hear and see the two marbles both hit the floor at the same time.

Concept Development

Independence of Velocities Emphasize again that a projectile's horizontal motion is constant in the absence of air resistance. The projectile's vertical velocity changes as the force of gravity accelerates the projectile.

DIFFERENTIATED INSTRUCTION

Struggling Students Provide students with a copy of a picture similar to **Figure 2.** Using a ruler, have the students draw a horizontal line between each side-by-side pair of objects. Using a square, have students verify that the line is straight with respect to the left or right edge of the image. Reinforce for the students that the vertical position of the two objects is the same at each interval and, therefore, the objects are falling at the same rate regardless of the horizontal component of motion.
BL Visual-Spatial

Horizontally Launched Projectiles

student text pp. 154–156

Use the PhysicsLAB

In *Launch an Investigation,* students use physics to determine the launch parameters of a projectile.

Use the PhysicsLAB

In *On Target*, students can design an investigation to determine what factors affect the path of a projectile.

Activity

Constant Horizontal Velocity Place a large bottle with about a 5-cm-wide mouth on the floor where students can walk past it. Give each student a ball that will easily fit through the mouth of the bottle. Have students walk past the bottle at a constant speed, holding the ball at their side, and drop the ball into the bottle. After the activity, ask students at what point they should drop the ball so that it falls into the container. Students should release the ball before it is above the opening.
OL **Kinesthetic**

Angled Launches

student pp. 156, 157

Identify Misconceptions

Apex Acceleration Activity Have students draw a free-body diagram of a projectile at the apex of its trajectory. Some students might believe that a projectile's acceleration at the top of its trajectory is zero and will indicate no force acting on the projectile at that time. Point out that a single force acts on the projectile; namely, the force of gravity. Because gravity acts downward, the projectile must always have a downward acceleration until another force acts to counter gravity.
OL **Visual-Spatial**

DIFFERENTIATED INSTRUCTION

Visually Impaired To help students get an idea of the shape of a trajectory, attach pieces of string at equal intervals along a meterstick. The intervals represent equal time intervals. The string lengths should have a ratio of 1:4:9:16:25, and so on. The lengths of the pieces of string represent vertical distances traveled. Attach a small washer to the end of each string. Students can then "feel" what the trajectory "looks" like. By holding the stick at different angles, students can simulate the trajectories of projectiles with different launch angles. **OL** **Kinesthetic**

ADDITIONAL IN-CLASS EXAMPLE

Use with Example Problem 2.

Problem Courtney kicks a soccer ball that is at rest on level ground and gives it an initial velocity of 7.8 m/s at an angle of 32° above the ground. Assume that forces due to air drag on the ball are insignificant.

a. How long will the ball be in the air?

b. How high will the ball go?

c. What will be its range?

Response

$v_{yi} = v_i \sin\theta = (7.8\ \text{m/s}) \sin 32°$

$v_{yi} = 4.1\ \text{m/s}$

$v_{xi} = v_i \cos\theta = (7.8\ \text{m/s}) \cos 32°$

$v_{xi} = 6.6\ \text{m/s}$

a. At landing, $y = 0$

$0 = 0 + v_{yi}t - \frac{1}{2}gt^2$

$t = 2v_{yi}/g = 2(4.1\ \text{m/s})/(9.8\ \text{m/s}^2) = 0.84\ \text{s}$

b. $y_{\text{max}} = v_{yi}\left(\frac{1}{2}t\right) - \frac{1}{2}g\left(\frac{1}{2}t\right)^2$

$= (4.1\ \text{m/s})(0.42\ \text{s}) - \frac{1}{2}(9.8\ \text{m/s}^2)(0.42\ \text{s})^2$

$= 0.86\ \text{m}$

c. $R = v_{xi}t = (6.6\ \text{m/s})(0.84\ \text{s}) = 5.5\ \text{m}$

Reinforcement

Projectiles, Kinematics, and Vectors Have students make lists of the concepts from previous chapters needed to analyze projectile motion. Such lists should include velocity, acceleration, freefall, vector resolution, and the independence of perpendicular vectors. **OL** **Linguistic**

Critical Thinking

Projectile Motion Explain to students that three objects of equal mass are shot straight upward with the same initial velocity. One of the objects is on the Moon, one is on Earth's surface, and one is on Earth but submerged in a pool of water so deep that the projectile does not leave the water. Have students compare and contrast the shapes of the objects' trajectories. Each trajectory will have a range of zero. The trajectories, ranked from highest to lowest, are Moon, Earth, and underwater. **OL**

PHYSICS CHALLENGE ACTIVITY

Trajectories Have students watch a video of a football kickoff. Have them use video analysis software to measure the time of flight and the distance that the ball is kicked. From these measurements, have them calculate the initial horizontal and vertical velocities, the initial velocity (magnitude and angle), and the maximum height. Have students select another sport in which projectile motion is evident, such as volleyball or soccer. They should analyze the motion of the projectile and compare the motion of the two objects. **AL** **Visual-Spatial**

Center-of-Mass Trajectory

Have the students estimate where the center of mass is for the young man in the Chapter Opener Photo. somewhere near his bellybutton Using tracing paper, connect this point between the different photos. Make a time axis (going to the right) under the trace and a vertical distance axis to the left of the trace. What does the trajectory of the center of mass look like? a parabola.

Forces from Air

student text p. 158

Use Figure 6

Have students compare the paths of the water in the two lower photos with the path of the water in the upper photo. Remind them that it is the net force on an object that changes the object's velocity.

3 Assess

Assess the MAIN IDEA

Imagine that gravity can be turned on and off with a flick of a switch. If you switch gravity off just at the moment a punter kicks a football, how will the football's trajectory change? Gravity acts in the vertical direction (downward), so only the vertical component of the football's trajectory will change. Because no forces (neglecting air resistance) act on the football in the vertical direction, it will not accelerate vertically and so continue forever upward with its initial vertical velocity. Its horizontal motion is not affected by gravity, so the football will move horizontally as it does when gravity is on.

Check for Understanding

Velocity and Acceleration Demo Throw a ball vertically upward and ask students the following questions. How does the velocity change as the ball rises? The upward velocity decreases. What is the velocity of the ball at its highest point? Zero How does the velocity change as the ball falls? The downward velocity increases. What is the acceleration of the ball as it rises? The free-fall acceleration is 9.8 m/s^2 downward. What is the ball's acceleration at its highest point? 9.8 m/s^2 downward What is the ball's acceleration as the ball falls? 9.8 m/s^2 downward **OL**

Reteach

Independence of Velocities Give students the horizontal and vertical components of the initial velocity of a projectile. Ask them to compute the velocity components at later times in the projectile's flight. Stress that the vertical velocity is continuously changing whereas the horizontal velocity remains constant. **OL** **Logical-Mathematical**

Reading Checks and Caption Checks

Caption Check, *student text p. 153*
−9.8 m/s

Reading Check, *student text p. 153*
The horizontal and vertical velocities of the dropped object are independent, so the vertical velocity does not depend on the initial horizontal velocity.

Caption Check, *student text p. 154*
$-9.8\ m/s^2$

Reading Check, *student text p. 154*
If we ignore air resistance, there are no forces acting in the horizontal direction. Therefore, there is no acceleration in the horizontal direction so the horizontal velocity is constant.

Practice Problems

student text p. 156

1. a. 4.0 s
   b. 2.0×10^1 m
   c. $v_x = 5.0$ m/s, $v_y = 39$ m/s
2. 1.1 m/s
3. 0.60 cm

Reading Checks and Caption Checks

Reading Check, *student text p. 156*
at the top

Practice Problems

student text p. 157

4. a. 2.8 s
   b. 9.3 m
   c. 65 m
5. hang time = 4.8 s; distance = 65 m; maximum height = 28 m
6. 32 m/s at 82° above horizontal

Section 1 Review

student text p. 158

7. The faster ball is in the air a shorter time and thus gains a smaller vertical velocity.
8.

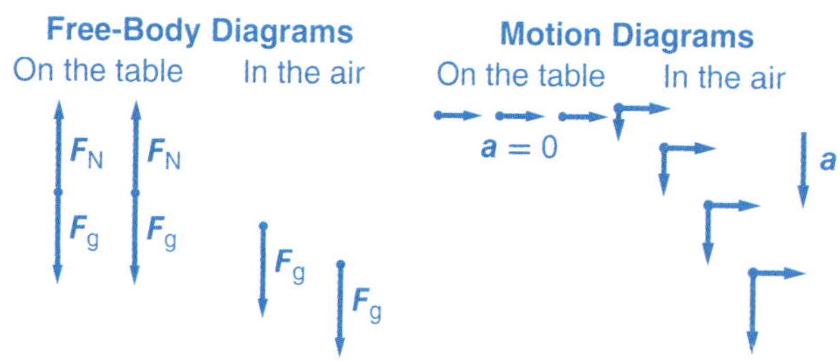

9. 27 m
10. 3.6 m
11. no change in horizontal velocity, time of flight is longer on the Moon, maximum height is larger on the Moon, horizontal distance is larger on the Moon

1 Introduce

Bellringer Activity

Centripetal Force Draw a large circle (with a diameter of at least 50 cm) on a large piece of paper. Have students observe as you have one or more volunteers try to make a ball roll along the circumference of the circle by only tapping the ball. Have students identify each tap as a force. Ask them what they noticed about the direction of each force. Each force should be directed toward the center of the circle. OL

Visual-Spatial

Tie to Prior Knowledge

Vector Quantities Remind students that velocity and acceleration are vector quantities because they have both magnitude and direction.

2 Teach

Describing Circular Motion

student text p. 159

Identify Misconceptions

Acceleration in Circular Motion Students might think that an object in circular motion can only be accelerating if its speed is changing. Explain that the word *acceleration* is often used in this way in everyday language, but in science an object also accelerates if the direction of its motion changes, even if its speed remains constant.

Centripetal Acceleration

student text pp. 160–162

Use Figure 9

Explain to students that the diagram at the bottom of **Figure 9** makes use of the definition of $\Delta v = v_f - v_i$ by drawing it in the form $v_i + \Delta v = v_f$. OL

Concept Development

Centripetal Force Point out that, if there is a centripetal acceleration, then the net force must be radially inward.

Use the PhysicsLAB

In *Centripetal Force*, students can investigate centripetal acceleration and circular motion.

Quick DEMO

Centripetal Acceleration

Estimated Time 5 minutes

Materials rotating stool, accelerometer

Procedure

1. Have a student sit on the stool and slowly spin while holding an accelerometer at arm's length.
2. Have the student hold the accelerometer tangentially to the circular path and observe the accelerometer.
3. Repeat step 2 but with the accelerometer held radially.
4. Have the students compare the two situations.

The accelerometer will indicate no acceleration in step 2 and an acceleration directed toward the center in step 3.

Identify Misconceptions

MAINIDEA Students might believe that, if the net force supplying a centripetal acceleration is removed from an object, the object will continue to move in a circular path. Whirl a foam ball on the end of a string above your head then release the string. The ball moves in a straight line at a tangent to its original circular path. Explain that the ball moves in a circle because the string provides a centripetal force to cause the centripetal acceleration. Once the string is released, the ball continues in a straight line because there are no horizontal forces acting on it.

REAL-WORLD PHYSICS

***g* Forces** Pilots who fly combat and aerobatic aircraft undergo the strain of what is called a "*g* force," which is a measure of an apparent increase in gravitational force due to the force exerted by the pilot's seat. In a tight 4- or 5-*g* turn, the pilot's apparent body weight is 4 or 5 times its normal weight, so a 900-N pilot would feel like she weighed 4500 N. The harder the turn, the greater the force applied by the seat, so the greater the *g* force. At about 9 *g*, eyesight begins to fail and the pilot begins to experience tunnel vision. The pilot momentarily experiences a "grayout," in which everything appears to be black and white.

Discussion

Question In circular motion, what aspects of the velocity and acceleration vectors are similar? What aspects are different?

Answer Both have constant magnitude and changing direction. The direction of the velocity is tangential to the orbit, whereas the direction of the acceleration is radially inward. OL

ADDITIONAL IN-CLASS EXAMPLE

Use with Example Problem 3.

Problem Jay is on a fishing trip and is having such a good time that he decides to swing a fishing-line weight around on the end of a piece of fishing line. The weight has a mass of 0.028 kg, the length of the fishing line between his hand and the weight is 0.75 m, and the weight makes one revolution in 1.2 s. What is the magnitude of the force exerted by the string on the weight?

Response 0.59 N

$a_c = 4\pi^2 r/T^2$

$F_T = ma_c = 4\pi^2 mr/T^2$

$= 4\pi^2 (0.028\text{ kg})(0.75\text{ m})/(1.2\text{ s})^2 = 0.58\text{ N}$

Reinforcement

Net Force Activity Glue some tubing in the shape of a semicircle onto a piece of wood or heavy cardboard. Roll a marble around the inside of the channel and have students observe that, when it emerges from the channel, it follows a straight line that is the tangent of the channel at the point of exit. OL Visual-Spatial EL

Centrifugal "Force"

student text p. 163

PHYSICS CHALLENGE ACTIVITY

Banked Curves Have students either find the specifications for a banked curve on a highway or estimate the value of the specifications. From the angle and the suggested speed limit for the curve, have them determine the minimum value of the coefficient of static friction needed between the road and a car's tires to keep the car from slipping. Curves may be banked at angles ranging from several degrees for public roads to several tens of degrees for racetracks. Assume that, if friction were not present, the car would slide up the roadway (away from the center of the turn). In this case, friction must act down the banked road to prevent this motion. Applying Newton's second law gives

$$F_{net} = ma_{centripetal}$$

$$F_{N_c} + F_{f_c} = m\frac{v^2}{r}$$

$$F_N \sin\theta + F_f \cos\theta = m\frac{v^2}{r}$$

$$mg\cos\theta\sin\theta + \mu mg\cos^2\theta = m\frac{v^2}{r}$$

$$\mu g\cos^2\theta = \frac{v^2}{r} - g\cos\theta\sin\theta$$

$$\mu = \frac{\frac{v^2}{r} - g\cos\theta\sin\theta}{g\cos^2\theta}$$

Inserting $\theta = 5.0°$, $v = 29$ m/s (65 mph), and $r = 100$ m gives $\mu = 0.78$

AL Logical-Mathematical

Critical Thinking

Banked Curves Ask students why curves on highways are banked. The inclined roadway produces a horizontal component of the normal force. This component is in the same direction as the desired centripetal acceleration and so contributes to accelerating the car to make it round the curve. OL

Discussion

Question When riding in a vehicle that is turning, why do you slide toward the outside of the curve?

Answer If the force of static friction between you and the seat is not sufficient then, as the car turns beneath you, the inertia of your body will make your body continue to move in a straight line until it comes in contact with the car door. OL

3 Assess

Assess the MAIN IDEA

Why do airplanes bank to turn? By banking, a component of the lift force of the wings becomes horizontal and perpendicular to the plane's velocity. This force causes the plane to accelerate centripetally (i.e., turn).

Check for Understanding

Uniform Circular Motion Ask students to describe an object's speed, velocity, and acceleration as it moves in uniform circular motion. Its speed is constant, but the direction of the velocity changes constantly to follow a circular path. The magnitude of the acceleration is constant, but the direction of the acceleration changes so that it is always oriented toward the center of the circle. OL

Extension

Space Station Activity Have students view one of the scenes from the movie *2001: A Space Odyssey* that shows the space station rotating. According to the film's scientific consultant, the diameter of the station was supposed to be 305 m (1000 ft). Using the video, have students measure the station's period of rotation and calculate the centripetal acceleration of a person in the station. Have students compare this centripetal acceleration to the free-fall acceleration on Earth. To achieve the free-fall acceleration of Earth, the station's period of rotation would have to be $T = 2\pi\sqrt{r/g} = 2\pi\sqrt{(152.5\ \text{m})/(9.8\ \text{m/s}^2)} = 25\ \text{s}$

Reading Checks and Caption Checks

Caption Check, *p. 159*
The velocity vectors have the same length.

Caption Check, *p. 161*
The horizontal acceleration would be zero, and the horizontal velocity would remain constant.

Practice Problems

student text p. 162

12. 3.1 m/s^2, friction

13. 8.1 km

14. 1.2×10^2 N

15. 0.24 N

16. 8.6 m/s^2, 0.88

Section 2 Review

student text p. 163

17. The ball accelerates toward the center of the circle because of the centripetal force.

18. The force is toward the center of the tub. The walls exert the force on the clothes.

19. There is an acceleration because the direction of the velocity is changing. There must be a net force toward the center of the circle. The road supplies that force, and the friction between the road and the tires allows the force to be exerted on the tires. The seat exerts the force on the driver towards the center of the circle. The note should also make it clear that centrifugal force is not a real force.

20.

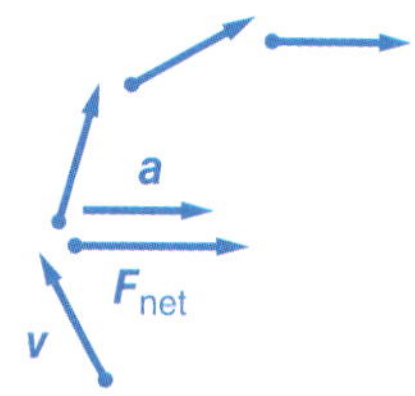

a. right

b. right

c. the car's seat

21. 4.7 m/s^2

22. 0.32 N

23. 55 m/s^2, 15 m/s

24. 61 N

25. Earth's gravity supplies the force that accelerates you. The scale will record a lower weight if you are in uniform circular motion.

SECTION 3 Relative Velocity

student text pp. 164–167

1 Introduce

Bellringer Activity

Moving Sidewalks Place a constant-speed car on a long board on a tabletop and start it. Ask students to describe two situations in which the moving car has zero velocity. (1) Pull the board while walking with the same speed as the car but in the opposite direction (so that the velocity of the car relative to the tabletop is zero). (2) Walk with the same speed and in the same direction as the car (so that the velocity of the car relative to the walker is zero). OL

Tie to Prior Knowledge

Velocity and Vector Addition Students extend their understanding of velocity to include relative velocity. Students apply vector addition to velocity vectors.

2 Teach

Relative Motion in One Dimension

student text pp. 164, 165

Concept Development

Reference Frames Explain that $v_{a/b}$ is the velocity of object a as measured in observer b's reference frame, and $v_{b/c}$ is the motion of observer b's reference frame as measured in observer c's reference frame. Thus, $v_{a/c}$ is the velocity of the object as measured in observer c's reference frame. Note that, by canceling the repeated subscripts, the correct order of addition is assured (e.g., $v_{a/b} + v_{b/c} = v_{a/c}$).

Reinforcement

Adding Vectors The sum of two vectors acting in one dimension is their algebraic sum and not their arithmetic sum.

Use the PhysicsLAB

In *Moving Reference Frame,* students can explore motion from different points of view.

Using an Analogy

Net Motion and Net Wage Draw students' attention to the vector showing their net motion relative to the street in the diagram on the right in **Figure 14.** To help them understand this vector, use the following analogy: If your hourly wage is $12/h and you must pay $2/h for parking while you work, your net hourly wage is $10/h.

Relative Motion in Two Dimensions

student text pp. 165–167

Identify Misconceptions

Relative Velocity Path Activity Ask students to sketch the displacement with respect to the water of the marble discussed in Example Problem 4. The path is a straight line in the direction of $v_{m/w}$. Some students may draw the path as a curve. Help students model relative motion using the strategy found on the next page in Using Models.
OL Visual-Spatial

ADDITIONAL IN-CLASS EXAMPLE

Use with Example Problem 4.

Problem Lalei places her lunch tray on a cafeteria conveyor belt that moves westward at 0.150 m/s. With respect to the tray, a ladybug on the tray crawls northward at 0.050 m/s. What is the ladybug's velocity with respect to the ground?

Response 0.16 m/s, 18° north of west

$$\boldsymbol{v}_{l/g} = \boldsymbol{v}_{l/t} + \boldsymbol{v}_{t/g}$$

$$v_{l/g}{}^2 = v_{l/t}{}^2 + v_{t/g}{}^2$$

$$v_{l/g} = \sqrt{v_{l/t}{}^2 + v_{t/g}{}^2}$$

$$= \sqrt{(0.150\ \text{m/s})^2 + (0.050\ \text{m/s})^2}$$

$$= 0.16\ \text{m/s}$$

$$\theta = \tan^{-1}(v_{l/t}/v_{t/g})$$

$$= \tan^{-1}((0.050\ \text{m/s})/(0.150\ \text{m/s}))$$

$$= 18° \text{ north of west}$$

Content Background

Jet Streams and Air Travel Jet streams are fast-flowing, confined air currents found in the atmosphere at an altitude of around 12 km. In the northern hemisphere, the streams are most commonly found between latitudes 30°–70° and between latitudes 20°–50°. The wind speeds vary according to the temperature gradient; they average 55 km/h in summer and 120 km/h in winter, although speeds of over 400 km/h are known to occur. The location of the jet stream is an extremely important datum for airlines. In the United States, for example, the time needed to fly from the east coast to the west coast can decrease or increase by about 30 minutes depending on whether the airplane flies with or against the jet stream, respectively.

Using Models

MAINIDEA To simulate a constant-velocity boat moving in a river, use a constant-velocity toy car and a long sheet of paper. The motion of the car represents $v_{b/w}$ and the sheet can be pulled at constant speed to simulate the motion of the current, $v_{w/g}$. Have students observe that the velocity of the boat relative to the ground ($v_{b/g}$) is a straight line.

3 Assess

Assess the MAINIDEA

Consider **Figure 2.** If the camera were moving with the same velocity as the initial velocity of the purple ball with respect to the ground, $\mathbf{v}_{p/g,\,i}$, what would the picture look like? Consider the reference frames of the ground (g) and the camera (c). In g, the camera moves horizontally at the same velocity as the initial velocity of the purple ball (i.e., $\mathbf{v}_{p/g,\,i} = \mathbf{v}_{c/g}$). Therefore, the purple ball has no horizontal velocity in c ($\mathbf{v}_{p/c,\,i} + \mathbf{v}_{c/g} = \mathbf{v}_{p/g,\,i} \Rightarrow \mathbf{v}_{p/c,\,i} = 0$), so it will fall straight down in the photo. In c, the initial velocity of the red ball will be opposite to the initial velocity of the purple ball in g: $\mathbf{v}_{r/c,\,i} + \mathbf{v}_{c/g} = \mathbf{v}_{r/g,\,i} \Rightarrow \mathbf{v}_{r/c,\,i} = -\mathbf{v}_{c/g} = -\mathbf{v}_{p/g,\,i}$ (note that $\mathbf{v}_{r/g,\,i} = 0$). Therefore, in c, the initial velocity of the red ball is opposite that of the purple ball in g, so the red ball will execute the same parabolic trajectory as the purple ball does in g, except it will be to the left instead of to the right.

Check for Understanding

Relative Velocity Equation Review the general form of the relative velocity equation. Then ask students to explain why it is important that the equation involves velocities rather than speeds. Speed may be used only if the reference frames and the objects inside of them are all moving in the same direction (e.g., horizontally or vertically). If any of these are moving in different directions, you have to consider the speed in the different dimensions (e.g., horizontal and vertical) to describe the motion. A quantity that requires two scalars to describe it is called a vector. **OL**

Reteach

Relative Velocity Ask each student to write a relative-velocity problem on a sheet of paper and to write its solution on the back of the sheet. (You may want to specify the number of dimensions.) Have students exchange sheets and solve the problems. The students may then present their solutions to the class, and the class can arbitrate any differences in solutions.
OL **Interpersonal**

Reading Checks and Caption Checks

Caption Check, *student text p. 164*

speed

Physics Challenge, *student text p. 165*

$$x = \sqrt{\frac{2Trh}{ma_{grav}}}$$

Yes, the expression changes if Phillipe is walking 0.50 m/s relative to the ground. If the stone travels in the same direction as Phillipe is walking, the velocity of the stone relative to the ground will be greater, resulting in a larger value of *x*.

Caption Check, *student text p. 165*

The air vector would be in the opposite direction, and the placement of the resultant vector would change in both length and direction.

Reading Check, *student text p. 165*

yes

Reading Check, *student text p. 166*

Resolve the velocity vectors into their *x*- and *y*-components. Each component gives the speed in the corresponding direction relative to a given reference frame.

Practice Problems

student text p. 167

26. 6.0 m/s

27. 0.73 m/s

28. 2.0 m/s, against the boat

29. a. 4.0 m/s

b. 19° south of east

30. 1.7×10^{2} km/h

31. a. 250.0 km/h

b. 150.0 km/h

Section 3 Review

student text p. 167

32. 260 km/h west

33. 5 m/s with the current, 0 m/s against the current

34. 14 m/s, 69° west of north

35. a. 8.0 m/s east

b. 1.0×10^{1} m/s

36. 3.8 m/s, 9.3° north of east

37. 1.9×10^{2} km/h, 64° south of east

38. 2.9×10^{2} km/h, 81° north of east

39. Choose the component of your velocity along the direction of the river to be equal and opposite to the velocity of the river.

ON THE JOB

Need for Speed

Race-Car Driver

Background

The weight of a race car is not distributed evenly among all four tires during a turn. A driver can manipulate the distribution of downward force on the tires by braking or accelerating. Braking increases the downward force on a car's front wheels, which increases steering control. Of course, braking also decreases the car's speed. Drivers must maintain a strategic balance between speed and control during the turns to achieve a competitive position in the race.

Teaching Strategies

- Tell students that the extreme forces at work on the race car even affect the body of the driver inside the cockpit. Race-car drivers must have strong necks, legs, and arms to maintain control of the cars during large accelerations. Drivers must also be prepared to make a quick exit should the vehicle catch fire during a crash.
- Have students find out how the improved safety features of a car style introduced to NASCAR in 2007 have changed the sport and the physics of stock-car racing.
- Encourage students to find out how race tracks with banking differ from flat tracks and how they affect the forces at work when the race cars turn.

GOING FURTHER >>>

Expected Outcomes Students should choose two auto-racing events and describe how the car body styles differ, how the track shapes differ, and any difference in regulations. Then students should explain how these differences affect the speeds the cars achieve and the forces that affect their motion. For example, drag races take place on straight tracks that have no turns, while stock-car races take place on oval tracks where much of the skill lies in managing the forces at work during turns.

SECTION 1

Mastering Concepts

40. It is not correct. A throw, kick, or other force is a contact force, and once there is no contact, there is no force.

41. a. E

b. The horizontal velocity is the same at all points.

c. B and C

d. The acceleration is the same everywhere.

42. After they are launched, the only force acting on both objects is gravity. Both objects immediately begin to accelerate downward, but the object launched upward at an angle has an initial upward velocity, causing it to go upward and then curve downward. The object launched horizontally immediately curves downward.

43. Answers will vary, but a correct form of the answer is, "A baseball is thrown horizontally at an initial speed of 1.5 m/s. How far does it travel horizontally before it hits the ground 8 m below?"

44. The plane will be directly over the crate when the crate hits the ground. Both have the same horizontal velocity. To an observer on the ground, the crate will look like it is moving horizontally while falling vertically.

Mastering Problems

45. 29 m

46. 3.2 m

47. a. 0.50 s

b. 0.80 m/s

48. 33 m, 7.3 m

49. a. 31 m

b. 2.1×10^2 m

50. 31 m/s at 45°

51. a. 14 s

b. 5.0×10^2 m

52. 6.5 m/s

53. 12 m/s

54.

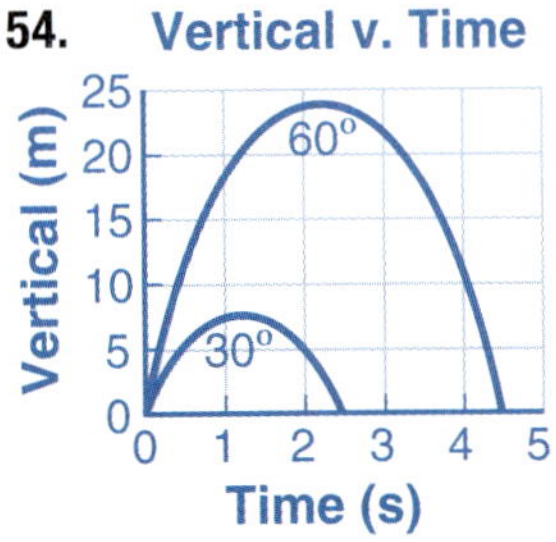

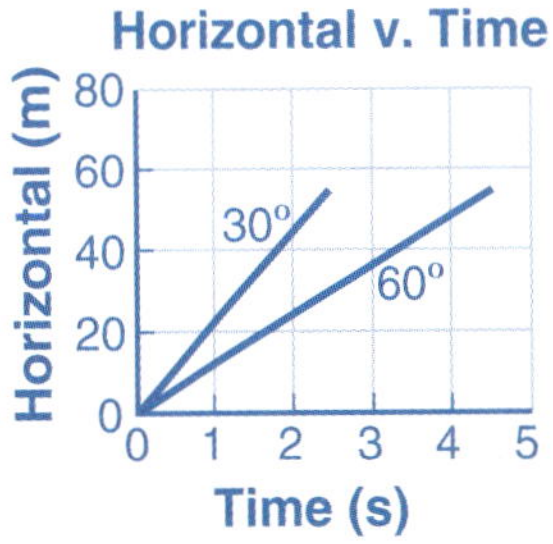

SECTION 2

Mastering Concepts

55. a. No; going around a curve changes the direction of velocity. Thus, the acceleration cannot be zero.

b. No; the magnitude of acceleration may be constant, but the direction of the acceleration changes.

56. The net force is directly proportional to the square of the speed of the moving object.

57. a. The force is along the string toward the center of the circle that the yo-yo follows.

b. The string exerts the force.

c. If the string were released, the velocity of the yo-yo would not change. According to Newton's first law of motion, it would move tangent to the circle in the direction it had been moving. There would be a gravitational force on it, so according to Newton's second law of motion, it would also have a downward acceleration. It would act like a horizontally launched projectile.

Mastering Problems

58. a. 9.65 m/s^2

b. 5.94×10^3 N

59. $A < C < B = D < E$

60. 71 m/s^2; 5.0×10^2 N

61. a. 2.64×10^7 m/s^2

b. 2.6×10^4 N

62. 13 m/s

63. 1.5×10^3 m

SECTION 3

Mastering Concepts

64. The magnitude of the relative velocity of that car to your car can be found by adding the magnitudes of the two cars' velocities together. Since each car is probably moving at close to the speed limit, the resulting relative velocity will be larger than the posted speed limit.

Mastering Problems

65. Odina wins.

66. a. 5.0 m/s, 53° from the shore

b. 3.0 m/s, 4.0 m/s

67. 1.6×10^2 km/h, 18° west of south

68. Answers will vary. A possible form of the correct answer is, "She wishes to reach a camp on the east bank that is 75 m downstream. If she rows at a speed of 5 m/s, at what angle should she direct the boat to go straight to the camp?"

Applying Concepts

69. Horizontal motion is uniform because there are no forces acting in that direction (ignoring friction). Vertically, there is acceleration due to the force of gravity. The projectile-motion equations in this book do not hold when friction is taken into account. Projectile motion in both directions will be impacted when air drag is taken into consideration because air drag is a friction force.

70. 20 m/s downward

71. Because of the acceleration due to gravity, the baseball falls a greater distance during the second $\frac{1}{4}$ s than during the first $\frac{1}{4}$ s.

72. **a.** The time does not change.

 b. A higher horizontal speed produces a longer horizontal distance.

73. yes

74. 6.0 s

75. Both speed and angle of launch matter, so height does make a difference. Maximum range is achieved when the resultant velocity has equal vertical and horizontal components—in other words, a launch angle of 45°. For this reason, height and speed affect the range.

76. The relative speed of two cars going in the same direction is less than the relative speed of two cars going in the opposite direction. Passing with the lesser relative speed will take longer.

77. **a.** in you hand

 b. The ball will land beside you, toward the outside of the curve.

78. The tension in the string is doubled, since $F_T = ma_c$.

79. See Online Solutions Manual for free-body diagram. The acceleration is toward the center of the track.

 a. The component of the normal force acting toward the center of the curve and, depending on the car's speed, the component of the friction force acting toward the center both contribute to the net force in the direction of acceleration.

 b. yes

Mixed Review

80. **a.** 464 m/s

 b. 3.3 N

 c. 9.5×10^2 N

 d. 9.5×10^2 N

81. 1157 m/s

82. −1.50 km/s

83. 24 N

84. 3.0×10^2m

85. 35°, 49°

86. **a.** 15 N

 b. 0.69 m/s

87. 53 m or 4.0×10^1 m

88. 8.5 m/s

Thinking Critically

89. The vertical gravitational force changes the speed of the cars, so the motion is not uniform circular motion.

90. **a.** Yes; the ball clears the wall by 2.1 m.

 b. 41 m/s

 c. 25° to 73°

91. 3.0×10^8 m/s, $\frac{4}{5}$c

92. It is not uniform circular motion. Gravity increases the speed of the ball when it moves downward and reduces the speed when it is moving upward. Therefore, the centripetal acceleration needed to keep it moving in a circle will be larger at the bottom and smaller at the top of the circle. At the top, tension and gravity are in the same direction, so the tension needed will be even smaller. At the bottom, gravity is outward while the tension is inward. Thus, the tension exerted by the string must be even larger.

Writing in Physics

93. Answers will vary. Students should explain that the clothoid shape reduces the centripetal acceleration experienced by the riders, making the ride safer.

94. Answers will vary. Students might explain that a pendulum ride swings riders in an arc, with the centripetal acceleration acting opposite gravitational acceleration at the top of the arc. Riders on a carousel are in circular motion at a constant speed, but because their direction changes, they experience centripetal acceleration.

Cumulative Review

95. a. 2×10^{16} m^2

b. 1.4×10^{-7} km^2

c. 2.8 kg/m^3

d. 1.7×10^{-3} m/s

96. 8 m/s

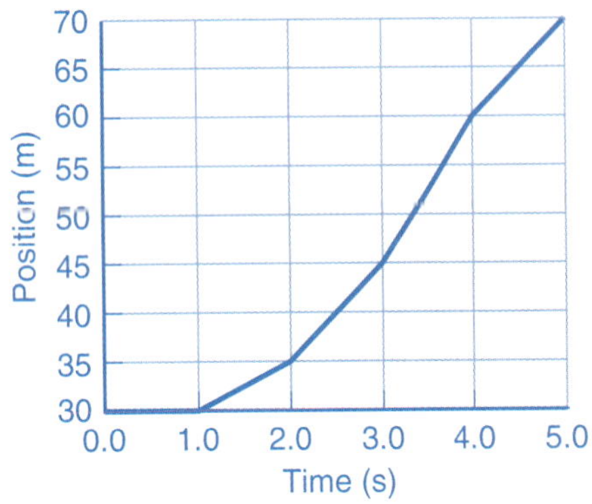

or

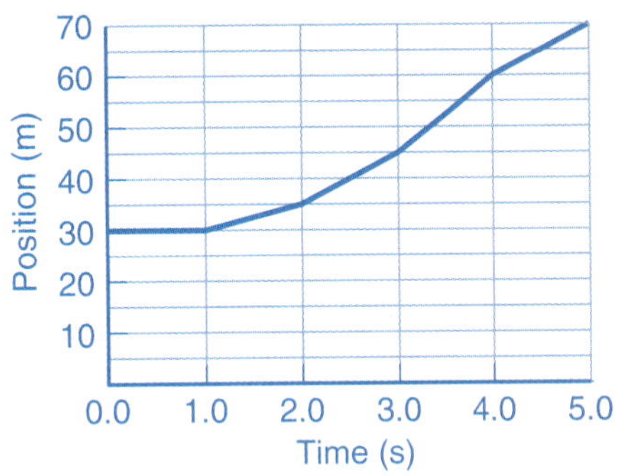

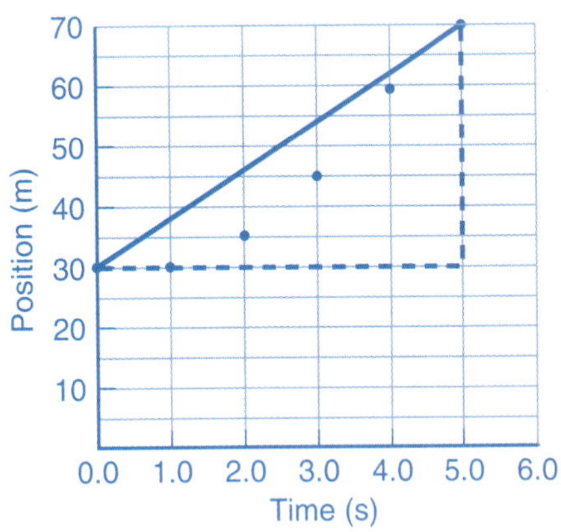

97. a. 5.9 N

b. 3.4 N

Standardized Test Practice

student text p. 175

Multiple Choice

1. C
2. B
3. B
4. C
5. B
6. B
7. D

Free Response

8. 82 m, so the ball lands just outside the ring. They should adjust the cannon to shoot a little lower.
9. 59 N

Rubric

The following rubric is a sample scoring device for free response questions.

Points	Description
4	The student demonstrates a thorough understanding of the physics involved. The response may contain minor flaws that do not detract from the demonstration of a thorough understanding.
3	The student demonstrates an understanding of the physics involved. The response is essentially correct and demonstrates an essential but less than thorough understanding of the physics.
2	The student demonstrates only a partial understanding of the physics involved. Although the student may have used the correct approach to a solution or may have provided a correct solution, the work lacks an essential understanding of the underlying physical concepts.
1	The student demonstrates a very limited understanding of the physics involved. The response is incomplete and exhibits many flaws.
0	The student provides a completely incorrect solution or no response at all.

CHAPTER 7
Gravitation

About the Photo

Where was the photo taken? From the *International Space Station (ISS).* Describe what living in the *ISS* might be like. Students might say that the astronauts and other objects are floating and that there is "zero gravity" in space. What keeps the *ISS* from crashing to the ground? The *ISS* is traveling about 17,500 mph sideways as it is being pulled downward by Earth's gravitational force. The net effect is that the path of the *ISS* follows the curve of the surface of Earth. Estimate the orbital period of the *ISS*. about 90 minutes

Chapter Overview

This chapter introduces the laws governing planetary motion from a historical perspective. Kepler's laws are explained and discussed in terms of Newton's law of universal gravitation. Weight and apparent weightlessness in orbit also are discussed. This chapter introduces the concept of gravitational fields and Einstein's theory of general relativity.

Before students study the material in this chapter, they should study:

- Accelerated motion in one dimension
- Adding vectors in two dimensions
- Circular motion
- Mass and weight
- Projectile motion
- Vectors vs. scalars

To solve the problems in this chapter, students will need a strong grasp of:

- Graphing data
- Scientific notation
- Significant figures
- Slope
- Solving linear equations
- Solving quadratic equations

Introduce the BIG IDEA

What do an electron, Jupiter, a building, Earth, a human, and a car have in common? mass, gravitational fields Is gravity repulsive or attractive? attractive Objects with mass exert attractive forces on other objects with mass.

Go online!
connected.mcgraw-hill.com

Your online experience includes:

- **Video**
- **Concepts In Motion**
- **Virtual Investigations**
- **All Lab, Applying Practices activities, and PBL worksheets**
- **Online Practice**
- **Personal Tutor**
- **Multilingual eGlossary**
- **Presentation**

Use the LaunchLAB

In *Model Mercury's Motion,* students can plot Mercury's orbit based on data and determine whether the plotted curve is a circle.

Use Physics T.V.

Video

PLANETARY MOTION What can looking through a telescope tell you about gravitation? Explore the physics of amateur astronomy and find out!

1 Introduce

Bellringer

Scale Hold up a bowling ball and tell students that it represents the Sun. Ask them to find an object that would represent the size of Earth on the same scale. Table 1 on page 173 will help. The scale is about 109:1, so Earth should be about 2 mm in diameter. Have objects, such as a tennis ball, an acorn, a hazelnut, a peanut, peppercorns, and pinheads available. Explain to students that in later activities they will use the ball and peppercorns to construct a model of the solar system.
OL Visual-Spatial EL

Tie to Prior Knowledge

Gravitation Students will apply Newton's second law of motion to a new force, gravitation, which gives rise to the centripetal acceleration of a satellite in its orbit.

2 Teach

Early Observations and Kepler's Laws

student text pp. 178–181

Concept Development

MAIN IDEA Separate students in groups of four, and give each group several pairs of objects and a meter stick. Object combinations could include soccer ball-acorn, tennis ball—tennis ball, potato-cork, paper clip-bowling ball, book-soda can, quarter-penny, etc. Ask students to separate each pair by 1-m of distance. Next ask them to rank the pairs by greatest to least gravitational force between the pairs. Have a triple beam balance on hand in case students want to measure any masses. Check rankings at groups. Ask students if their ranking would change if all of the distances were halved. No. What happens to the force between the objects when they are closer? Increases.

Identify Misconceptions

The Sun-Centered Solar System Students might think that the Sun-centered solar system was proven by the time of Copernicus, Galileo, or Newton. Actually, the observational proof was not accepted until the 1830s. Copernicus's system was eventually accepted because it was a simpler system than the Earth-centered model and because Newton's laws could describe the motion of planets in Copernicus's system. Newton's laws could not be used to account for the elaborate orbits of planets around Earth in an Earth-centered system.

Use the PhysicsLAB

In *Modeling Orbits,* students determine the orbital shape of planets and satellites in the solar system.

Concept Development

Constructing a Model of Planetary Motion Find out what students know about Earth, the Sun, and the solar system. Start with what they have observed, such as the apparent motion of the Sun around Earth, details of how the Sun moves across the sky, and the motion of stars through the night and the year. Ask students to construct a model that explains these observations. If they choose a system with the Sun at the center, ask them which observations support this placement.
OL Visual-Spatial EL

PHYSICS PROJECT ACTIVITY

Eccentric Orbits Have students research various historic models of the solar system to explain how they might have influenced Kepler. Ask them why it was so important to understand that planetary orbits are elliptical. In fact, even Copernicus had worked with the assumption that planets moved in circles. Have students work together to depict and explain the more eccentric orbits of some of the planets and Pluto, which is now classified as a dwarf planet. Ask them if Pluto is always farther than Neptune from the Sun. No, in fact, sometimes, due to their eccentric orbits and where they are—perihelion or aphelion—Pluto, can be closer to the Sun than Neptune. OL Visual-Spatial EL

Identify Misconceptions

Elliptical Orbits Draw a circle with a 1-m diameter on the board and place the Sun 1 cm from the center of the circle. The difference between the circular and elliptical orbits is about 0.07 mm. Ask where Earth is in relation to the Sun during the North American winter. Earth is located closest to the Sun in the month of January. Point out that Earth's seasons are caused not by Earth's position relative to the Sun, but by the angle of Earth's axis. When the Northern Hemisphere is tilted away from the Sun, the angle of insolation is not as direct––especially at higher latitudes––hence the cold. OL

Extension

Stellar Parallax The positions of nearby stars change as Earth's location around the Sun changes. Have students complete this activity to model the change. You will need two 2-lb coffee cans with translucent plastic lids, two lightbulbs (one in the center of the classroom and one at the other end), and masking tape. Punch a small hole in the center of the bottom of each coffee can. Tape the two cans together side by side. The cans represent telescopes: one at the location of Earth in January, the other in July. One lightbulb represents a star nearby, while the other represents a distant star. With the two cans at one end of the classroom and the can bottoms toward the stars (lights), darken the room and note the relative positions of the images of the two stars (lightbulbs) on the plastic lids. The positions on the two lids of the lightbulb nearby will be shifted slightly with respect to those of the lightbulb farther away. This greatly exaggerates the effect that astronomers see. OL Kinesthetic EL

REAL-WORLD PHYSICS

Astronomy in Ancient Culture Astronomy was a major part of many cultures and religions. Knowledge of the length of the year was crucial for any culture that depended on farming. Ancient peoples of Central America made accurate observations of Venus. An early astrolabe, a medieval instrument for measuring the positions of stars and planets, was made in Baghdad in the eighth century. The Chinese people observed the explosion of the Crab Nebula in 1054, an event not recorded in the West.

REAL-WORLD PHYSICS

Telescopes Brahe and Kepler did all of their work without telescopes. Galileo used a crude telescope when he discovered Jupiter's moons, Saturn's rings, and the phases of Venus. These discoveries certainly helped support the Sun-centered model of the solar system. Ask students how telescopes are an improvement over the naked eye. Work with a local astronomy club to have a star party. Have students summarize what they saw and how the telescope made this possible. The telescope collects more light than the eye can and magnifies the image, making much fainter objects visible. BL

ADDITIONAL IN-CLASS EXAMPLE

Problem Europa, a satellite of Jupiter, has a period of 3.55 days. How many units is its radial distance?

Response

$$r_E{}^3 = (4.2\text{ units})^3\left(\frac{3.55\text{ days}}{1.8\text{ day}}\right)^2 = 288\text{ units}^3$$

$$r_E = 6.6\text{ units}$$

Reinforcement

Kepler's Laws Compare the orbits of Io and Callisto as modeled in the Example Problem. Ask students how Galileo could have applied Kepler's third law. Galileo could treat Jupiter as if it were the Sun, and the moons as if they were orbiting planets. OL

DIFFERENTIATED INSTRUCTION

Struggling Students Review the material on circular motion, using motion diagrams to show how acceleration corresponds to a change in velocity (its magnitude, its direction, or both). Draw a circular orbit and show that the gravitational force (which we will reasonably assume is the net force on the planet) and the acceleration are toward the center of the circle. Next draw an elliptical orbit with a relatively high eccentricity (deviation from the shape of a circle, which is an ellipse with zero eccentricity), with the Sun at one focus. Draw lines from the Sun to selected points on the ellipse, noting that the gravitational force on and acceleration of the planet, at any given point on the orbit, are directed along that line and pointing in the direction of the Sun. At the points of closest and farthest distances, the planet's acceleration is perpendicular to the direction of its orbital velocity (and so the planet's direction changes but its speed does not). At other points in the orbit, the planet's acceleration is not perpendicular to its orbital velocity, so there is a component of acceleration parallel to the planet's orbital velocity (and so the planet's speed increases or decreases). BL Visual-Spatial EL

Newton's Law of Universal Gravitation

student text pp. 182–185

Critical Thinking

Inverse-Square Proportionality Newton's law of gravitation involves examples of direct and inverse-square proportionality. Using numbers, explain how the force calculated by the law of universal gravitation changes when one mass and then the other is cut in half, doubled or tripled. Repeat for number changes in distance. If the mass of one object is cut in half, the force is cut in half; if the mass is doubled, then the force is doubled; if the mass is tripled, the force is tripled, and so on. If the distance between objects is cut in half, the force increases fourfold; doubling the distance, the force is reduced by one-fourth; tripling the distance reduces the force by one-ninth. OL

Quick DEMO

Newton's Law of Universal Gravitation

Estimated Time 5 minutes

Materials two golf balls

Procedure Hold a golf ball in each hand. Hold one about 1 m above the floor and the other about 2 m above the floor. Ask students to look at the equation for the law of universal gravitation: $F = \frac{Gm_1m_2}{r^2}$ and compare the forces on the two balls. The force is about the same because *r* is measured from the center of Earth.

Concept Development

Newton's Law of Universal Gravitation It is not obvious that the attraction of a very large object, such as Earth, on an apple can be explained by Newton's universal law of gravitation. Newton spent 20 years developing integral calculus to prove his idea. Sketch Earth and an apple. Ask students what would happen if Earth were made of separate rocks. Draw force arrows from representative rocks. Show how symmetry averages force to the left and to the right. Within a small cone of angles, there is much more material farther away than close to the apple, which compensates for the much weaker force of the parts of Earth that are far from the apple. OL Visual-Spatial EL

ADDITIONAL IN-CLASS EXAMPLE

Problem Make a transparency similar to the illustration from the children's book *Le Petit Prince* (the prince standing on his personal asteroid). Ask students to estimate the value of *g* at the surface. The radius of the asteroid is 1.5 m, and its density is that of the asteroid Vesta, 3.3×10^3 kg/m^3. What would happen if the prince jumped upward?

Response

$m = (3.3\times10^3\ \text{kg/m}^3)(14\ \text{m}^3) = 4.6\times10^4\ \text{kg}$

$g = \frac{F}{m} = \frac{Gm}{r^2} = \frac{(6.67\times10^{-11}\ \text{N·m}^2/\text{kg}^2)(4.6\times10^4\ \text{kg})}{1.5\ \text{m}^2}$

$= 1.4\times10^{-6}\ \text{m/s}^2$

He would jump very high. OL

Use Figure 8

Cavendish Balance Explore photographs of other gravitation torsion balances and computerized Cavendish balances. The gravitational force exerted by the large spheres on the smaller ones causes the wire to twist. The amount of twist can be measured by observing the deflection of the light beam. The wire exerts a torque proportional to the angle through which it is twisted. The proportionality constant is calculated from a measurement of the period of oscillation. The equilibrium angle is measured, and the gravitational force is deduced. OL

Concept Development

The Value of G Derived from Newton's law of universal gravitation, the gravitational constant, (G), is the number used to determine the force of gravity. In the 1980s, the accepted value of G was 6.67260×10^{-11} N·m^2/kg^2, with a relatively high uncertainty of 0.01%. More recently, the accepted value is $G = 6.67390\times10^{-11}$ N·m^2/kg^2, with an uncertainty of 0.0014%. Scientists are still learning about the nature of gravity. In deep space, for example, gravity does not exist as a direct force. Instead of determining G, scientists consider the pressure exerted by space on the matter/energy that is displacing it. To calculate gravity more accurately now, scientists multiply the Relative Space Warp (RSW) by the Space Constant (SC).

Content Background

Gravitational Force Ask students to consider the following questions: How universal is Newton's law of universal gravitation? Does the gravitational force depend only on mass and not on the substance? Could it, for example, depend on the relative numbers of protons and neutrons in the substance? The earliest tests were made by Lóránd Eötvös, a Hungarian, born in 1848, who invented a very sensitive torsion balance. He compared the gravitational force on objects with the same inertial mass. Using a variety of woods and metals, Eötvös found that the forces were the same to five parts in a billion.

3 Assess

Assess the MAIN IDEA

Ranking Gravitational Forces On the board, draw five diagrams (labeled A–E) of various pairs of masses separated by different distances. Have students rank the gravitational forces on the left-side objects from highest to lowest. Ask students to rank the gravitational forces on the right-side objects from highest to lowest. Is there a difference? No, the gravitational forces in each pair are equal but in opposite directions, the equation used is the same with the same variables.

Check for Understanding

Sketching a Free-Body Diagram Draw a circular orbit of a satellite around Earth. Show the satellite in two locations. Ask students to copy the sketch and draw a free-body diagram for the satellite at both locations. Ask them to identify the force(s) and show the direction of the acceleration. There should be only one force, F_g. It should be directed toward Earth. The acceleration vector should be in the same direction. Discuss why there are no additional forces. Nothing is touching the satellite, so there are no contact forces. The only long-range force is that of gravity. So, there is only one force. Note that students will commonly say that centripetal forces or centrifugal "forces" are additional forces. If students do so, you might clarify for them the meaning of whichever of these terms they use. In the present example, the gravitational force acting on the satellite is a *centripetal force,* which is any force directed toward the center of a circle along which an object is moving.

Regarding centrifugal force, suppose that the satellite is observed from a reference frame that rotates about Earth's center at the same rate as does the satellite. With respect to this rotating reference frame, the satellite would be motionless. According to Newton's first law of motion, if an object is at rest, there must be no net force acting on it. We know that a gravitational force is pulling the object toward Earth's center. Newton's first law will not be valid in this rotating reference frame unless there is, acting on the satellite, a force equal in magnitude to that of the gravitational force and opposite in direction, that is, pointing away from Earth's center. A centrifugal force is that force which would make Newton's first law valid in a rotating reference frame (in the sense described above). It differs from the gravitational force in that, whereas there is a law associating the gravitational force on an object with a certain mass with the relative positions of other objects with mass, centrifugal force has no comparable physical source. **OL** **Visual-Spatial** **EL**

Reteach

Mass Demo Drop an object—a ball, for example. Ask students if the object's acceleration tells them anything about either the object's mass or Earth's mass. Ask whether their response regarding the object that was dropped applies as well to a satellite in orbit around Earth. An object's freefall acceleration is independent of its mass (to an approximation, for example neglecting air resistance), so measuring the acceleration tells you nothing about the object's mass. But $a_{\text{freefall}} = \frac{F_{\text{gravitational}}}{m_{\text{object}}} = \frac{Gm_{\text{object}}m_{\text{Earth}}}{r^2 m_{\text{object}}} = \frac{Gm_{\text{Earth}}}{r^2}$. So measuring the acceleration magnitude allows you to calculate Earth's mass (assuming you know the values of G and r). The same results apply to satellites, that is, you cannot calculate the satellite's mass, but you can calculate Earth's mass, from a knowledge of the satellite's freefall acceleration. **OL**

Reading Checks and Caption Checks

Reading Check, *student text p. 179*

The distance between points 1 and 2 is longer than the distance between points 6 and 7. Earth is closer to the Sun and travels faster between points 1 and 2 than between points 6 and 7.

Caption Check, *student text p. 179*

The equal time areas are shaped differently because the Sun is located at a focus of the ellipse, which is offset from the ellipse's center.

Reading Check, *student text p. 184*

The amount of horizontal rotation of the rod is used to determine the force of attraction between the spheres.

Practice Problems

student text p. 181

1. 11 units
2. 2.8 y
3. $0.724r_E$
4. $19r_E$
5. 684 days
6. **a.** 89 min

   b. 3.2×10^2 km
7. 4.3×10^4 km

Physics Challenge

student text p. 183

1. For planet B, $\frac{r_B{}^3}{T_B{}^2} = 9.6\times10^{-6}\ \text{AU}^3/\text{days}^2$

   For planet C, $\frac{r_C{}^3}{T_C{}^2} = 9.77\times10^{-6}\ \text{AU}^3/\text{days}^2$

   For planet D, $\frac{r_D{}^3}{T_D{}^2} = 9.82\times10^{-6}\ \text{AU}^3/\text{days}^2$

   Kepler's third law is obeyed.

2. For the Earth-Sun system,

   $\frac{r^3}{T^2} = \frac{(1.000\ \text{AU})^3}{(1.000\ \text{y})^2} = 1.000\ \frac{\text{AU}^3}{\text{y}^2}$

   For the planet C-Upsilon system,

   $\frac{r^3}{T^2} = 9.77\times10^{-6}\ \text{AU}^3/\text{days}^2$

   $= (9.77\times10^{-6}\ \text{AU}^3/\text{days}^2)(365\ \text{days/y})^2$

   $= 1.30\ \text{AU}^3/\text{y}^2$

The star's mass is 1.30 times that of the Sun.

Section 1 Review

student text p. 185

8. 1.2×10^{-7} N; 8.4×10^3 parts per billion of the weight.
9. 6.02×10^4 days
10. The value of *g* would increase.
11. It would be the same, because the same value of G is used to describe the attraction of bodies having diverse chemical compositions: the Sun (a star), the planets, and satellites.
12. No. A scientific law is a statement of what has been observed to happen many times. A theory explains scientific results. None of these statements offers explanations for why the motion of planets is as it is or for why gravitational attraction acts as it does.
13. **a.** Horizontal throwing requires the same effort because the inertial character, $F = ma$, of the rock is involved. The mass of the rock depends only on the amount of matter in the rock, not on its location in the universe. The path would still be a parabola, but it could be much wider because the rock would go farther before it hits the ground, given the smaller acceleration rate and longer time of flight.

    b. Assume the rocks would be dropped from the same height on Earth and on the Moon. It will hurt less because the smaller value of *g* on the Moon means that the rock strikes the toe with a smaller velocity than on Earth.

1 Introduce

Bellringer

Projectile Motion ***WARNING:* Wear goggles.** Hold a golf ball at arm's length. Drop it and catch it at the top of its bounce. Ask students how far the ball dropped in the first second of its fall $\frac{(9.8\text{ m/s} - 0\text{ m/s})}{2} = 4.9\text{ m/s}/1\text{ sec} = 4.9\text{ m}$. Ask them to consider what would happen if the ball were thrown sideways at 10 m/s, 50 m/s, and 500 m/s. How far would the ball drop in the first second? 4.9 m How far would the ball move horizontally (sideways) in the first second? 10 m, 50 m, 500 m What would its path look like? curved Ask students to relate this activity to **Figure 11.**
BL Visual-Spatial EL

Tie to Prior Knowledge

Motion of Satellites Have students apply Newton's law of gravitation to the motion of satellites. They will need to review the meaning of *weight* and *mass.* OL

2 Teach

Concept Development

MAINIDEA Bring in copies or go online to show students images of orbital debris around Earth that can be found on NASA's Orbital Debris Program Office website. Allow students to explore the images of orbital debris at Low Earth Orbit (LEO) and Geostationary Orbit (GEO). Ask students why the debris does not fly off into space. Earth's gravitational force keeps the debris in orbit. Ask students how far the Earth's gravitational field extends. To infinity.

Orbits of Planets and Satellites and Free-Fall Acceleration

student text pp. 186–189

Concept Development

Orbits Start students with the fact that in 1 s an object at or near Earth falls 4.9 m. Develop a table for the horizontal distance traveled in that second for various horizontal speeds.

Diagramming Help students develop an appreciation for the curvature of Earth by drawing a diagram and explaining the following. Show students that in traveling a horizontal distance x, an object has subtended a central angle given by $\tan\theta = \frac{x}{r_E}$, where r_E is Earth's radius, which is 6.37×10^3 km. Over that angle, the distance that Earth has "fallen away" from the horizontal is approximately given by $y = r_E(1 - \cos\theta)$. For example, when $x = 8$ km, $\tan\theta = 1.3\times10^{-3}$ and $y = 5$ m. OL

Applying Physics

Remote-sensing Satellites Two remote-sensing satellites called *GOES* (Geostationary Operational Environmental Satellite) cover the western hemisphere. *GOES-East* is stationary above 75° W longitude and *GOES-West* is above 135° W longitude. Each has a mass of 2100 kg and was launched by an Atlas Centaur rocket. *GOES-12* has replaced *GOES-8* as *GOES-East* after more than 6 years of service. A third satellite is in orbit and can be moved into position if either of the two working satellites fails. Updated information can be found at http://goespoes.gsfc.nasa.gov/goes.

Reinforcement

Orbital Speed Ask students to explain how the speed of an object in a circular orbit depends on the radius of the orbit. If you double the radius, what happens to the speed? The speed will be $\frac{1}{\sqrt{2}}$, or 70.7% of its original speed. Repeat the question for the period of the orbit. The speed is $\sqrt{2^3}$, or 2.8 times its original speed. OL

Using Models

Where is Mars at midnight? On a large sheet of paper, have students draw circles to scale that represent the orbits of Earth and Mars. Students might draw a 15-cm radius for Earth and a 23-cm radius for Mars. (If the circles have a common center, select a point on both circles to indicate conjunction, otherwise there will be an actual measurement. This is your starting point.) Label the monthly locations of Earth on its orbit and research the dates when Mars is at conjunction (closest to Earth) or opposition (farthest from Earth). Use the period of Mars (684 Earth days) to mark and label the location of Mars once every Earth month. Knowing that at midnight, the night sky is the direction away from the Sun, ask students to find the months when Mars is visible and when it is in the east, south (for inhabitants in the northern hemisphere), and west. OL Visual-Spatial EL

Careers

Astronomer A professional astronomer studies the origin and evolution of structure in the universe. Others study phenomena that, decades ago, would have been considered topics for a science fiction book: light echoes around exploding stars, and gravitational lenses. Not all astronomers are at universities. Many work in business and industry. The best way to prepare for a career in astronomy is to acquire an excellent education in mathematics. Math is the best preparation for the study of physics, which is an integral part of any astronomer's education. It is not necessary to pursue a Ph D in astronomy, but one should be motivated to pursue studies through Master's level.

ADDITIONAL IN-CLASS EXAMPLE

Problem Engineers are planning to place the International Space Station (ISS) into orbit at an altitude of 450 km above Earth's surface. What would be the orbital speed and period of the ISS?

Response

Known: $h = 4.50\times10^{5}$ m

$r_E = 6.38\times10^{6}$ m

$m_E = 5.97\times10^{24}$ kg

$G = 6.67\times10^{-11}$ N•m^2/kg^2

Determine orbital radius. Add height of satellite's orbit to Earth's radius.

$$r = h + r_E = 4.50\times10^{5}\text{ m} + 6.38\times10^{6}\text{ m} = 6.83\times10^{6}\text{ m}$$

Solve for speed.

$$v = \sqrt{\frac{Gm_E}{r}} = \sqrt{\frac{(6.67\times10^{-11}\text{ N·m}^2/\text{kg}^2)(5.97\times10^{24}\text{ kg})}{6.83\times10^{6}\text{ m}}} = 7640\text{ m/s or } 27{,}500\text{ km/h}$$

Solve for the period.

$$T = 2\pi\sqrt{\frac{r^3}{Gm_E}} = 2\pi\sqrt{\frac{(6.83\times10^{6}\text{ m})^3}{(6.67\times10^{-11}\text{ N·m}^2/\text{kg}^2)(5.97\times10^{24}\text{ kg})}} = 5620\text{ s}$$

This is approximately 94 min.

DIFFERENTIATED INSTRUCTION

Visually Impaired NASA has produced *Touch the Universe: A NASA Braille Book of Astronomy* by Noreen Grice. It uses Braille and large-print captions for 14 pages of *Hubble Space Telescope* photos. Each image is embossed with lines, bumps, and other textures. The raised patterns translate colors, shapes, and other intricate details of the cosmic objects, allowing visually impaired people to feel what they cannot see. Braille and large-print descriptions accompany each of the book's 14 photographs, making the design of this book accessible to readers of all visual abilities. The book starts readers at Earth, goes through the solar system, and ends with the most distant images from *Hubble*.

Identify Misconceptions

Weightlessness Many people believe that Earth's gravity stops at the edge of the atmosphere. This misconception is reinforced by misuse of the terms *weightlessness* and *microgravity*. When discussing acceleration due to gravity far from Earth, point out that Newton's development of the law of universal gravitation rested on his realization that Earth's gravity extends to the Moon and far beyond.

Discussion

Problem How is it possible to use pools to model apparent weightlessness, similar to what astronauts experience on the Moon or on a space station?

Response Due to the upward, buoyant force of the water, one's apparent weight feels less. By working in the pool, one can experience a sense of being somewhat weightless. **OL**

The Gravitational Field and Two Kinds of Mass

student text, pp. 190–191

Content Background

The Gravitational Field Action at a distance is a worrisome idea. Suppose the Sun ceased to exist. If gravity acts at a distance, it must do so instantaneously. Thus, at the moment the Sun disappeared, Earth would begin to move in a straight-line path. The concept of a field makes all action local. Earth's mass does not respond to the Sun's mass, but to the Sun's gravitational field at Earth's location. An experiment has purported to compare the speed of the gravitational force with the speed of light. The scientists found that the speed of gravity is equal to 1.06 times the speed of light (with an uncertainty of 20%). Other scientists dispute this interpretation, saying that the experiment does nothing more than measure the speed of light. For references regarding this issue, see the webpage http://physics.wustl.edu/cmw/SpeedofGravity.html.

Use the MiniLAB

In *Weightless Water,* students put water in a cup with a hole in it. In a kitchen or a restaurant, students would rightly consider this a bad idea. But for students who have predicted that the cup would, or would not, leak while in freefall, and now want a definitive verdict, putting water in a cup with a hole in it is actually a very good idea.

Use the MiniLAB

In *Weight in Free Fall,* students can observe what happens to an object's weight with respect to a reference frame in free fall.

Quick DEMO

Measuring Inertial Mass

Estimated Time 15 minutes

Materials a hacksaw blade

Procedure Clamp one end of the hacksaw blade to a desk so that the blade can vibrate horizontally. Clamp objects with various masses to the other end. Measure the period of oscillation. Note that the period depends on the square root of the mass. Only the inertial mass is involved.

PHYSICS CHALLENGE ACTIVITY

Scales and Balances Collect as many devices as possible on which an object can be weighed. Determine how each works. For example, a spring scale measures the deflection of the spring by the force (weight) exerted on it. An electronic scale also uses deflection but uses electrical resistance to measure it. A beam balance compares the force of gravity on the object with the force on a calibrated glider. An inertial-mass balance determines the period of oscillation that depends on the inertial mass of the object and the force exerted by the balance spring. AL Kinesthetic EL

Use the PhysicsLAB

In *How Can You Measure Mass?, students use an inertial balance to measure mass.*

Use the PhysicsLAB

In *Inertial Mass and Gravitational Mass,* students determine the relationship between inertial mass and gravitational mass.

Content Background

Particles and Objects Newton's law of gravitation gives us the gravitational force between two particles, both of which have mass but, as with geometric points, no volume. What, then, is the gravitational force between two *objects* (1 and 2), which do have volume? Each object is modeled as a large number of particles (or many objects as continuous distributions of matter, the particle model). Each particle in Object 1 exerts a gravitational force on each particle in Object 2. The total gravitational force on Object 1 due to Object 2 is the vector sum of all the gravitational forces on Object 1 due to Object 2, each of which is determined by Newton's law of gravitation. (If the objects are continuous matter, then the total force is a vector integral, rather than a vector sum.) If the sizes of Objects 1 and 2 are very small relative to the distance between them, then the objects simply are point particles to which Newton's law is applied.

But if the sizes of 1 and 2 are not small relative to the distance between them, such as the case in which each of the two objects have spherical symmetry (as does, for example, a uniformly dense, solid sphere), then the above-mentioned vector sum of forces on Object 1 is the same force that would result if Object 1 is replaced with a single particle with Object 1's mass and located at the center of mass and Object 2 is replaced with a single particle with Object 2's mass at the center of Object 2's mass. That is, replace the two-object force problem with a much simpler two-particle force problem. For the latter, the magnitude of the force on Particle 1 (Object 1) is determined by the scalar equation $F_g = \frac{Gm_1m_2}{r^2}$.

The direction of the force is along the line connecting Particles 1 and 2. Thus, the net gravitational force on Object 1 due to Object 2 is a single force directed from the center of Object 1 to the center of Object 2. AL

Critical Thinking

Earth and Satellite One way to encourage students to think about the ideas in the Content Background is to apply those ideas to a satellite orbiting Earth. We can model the Earth as a sphere consisting of a great number of particles, and the satellite, which is small relative to the Earth, as a single particle. Sketch the satellite in orbit around Earth, and draw a line connecting the satellite and Earth's center ("the Earth-satellite line").

Ask students to explain why the direction of the net gravitational force on the satellite is toward Earth's center, even though most of Earth's material lies on one side or the other of the Earth-satellite line. *Hint: Assume that Earth is made up of separate equal-size, equal-mass, and uniformly distributed rocks, each attracting the satellite.* Draw two dots representing points lying in a plane inside Earth, one on each side of and equidistant from the Earth-satellite line and from the satellite. Next draw, at the point representing the satellite, the equal-magnitude force vectors pointing toward the two chosen points in Earth. Resolve each force vector into a component along the Earth-satellite line and a component perpendicular to that line. Ask students what they notice in comparing like components of the two force vectors. The components perpendicular to the Earth-satellite line are equal in magnitude and opposite in direction, whereas the components along the Earth-satellite line are in the same direction. Ask them what the vector sum of the perpendicular components is. zero This can be expressed by saying that those components "cancel each other out." Prior to selecting a second set of points, ask students if they have any intuition or reasoning about whether the perpendicular components will cancel for a new pair of points. The perpendicular vector components should cancel no matter which pair of points is selected, as the cancellation didn't depend on "where" the two points were, but only on their symmetry about the Earth-satellite line. What then accounts for the net gravitational force on the satellite pointing toward Earth's center? Because of the symmetrical distribution of Earth's material about the Earth-satellite line, the only gravitational force vector components that do not cancel are the ones that point toward Earth's center. Since this shows the force of gravity in a particular situation, students need to look for symmetry in a problem to simplify the solution to a particular problem. AL Visual-Spatial

Einstein's Theory of Gravity

student text, pp. 192–193

Concept Development

Spacetime and Its Curvature Students might want to pursue some of the ideas in the student text.

Spacetime: An *event* is any occurrence that happens at a certain place at a certain time—the clap of your hands and the bounce of a ball being two examples. Space is said to be 3-dimensional because one needs three coordinates to specify the location of an event. Time is said to be 1-dimensional because only one coordinate can specify when an event takes place. In both Newtonian and relativistic mechanics one can think of space and time as being part of a single 4-dimensional ($3 + 1 = 4$) structure called *spacetime,* the "arena" in which events take place (this viewpoint is especially useful and important in relativistic mechanics). Think of marking an event by drawing a dot in a 3-dimensional *x-y-t* coordinate grid, called a *spacetime diagram,* with the *t*-axis usually drawn as the vertical axis. The fourth coordinate, *z* in this case, cannot be accommodated on such a diagram, so it is not drawn.

The Curvature of Spacetime: General relativity is (at a minimum) a theory of gravitation, whereas special relativity is a special case of general relativity in which gravitation is relatively weak. In general relativity, matter affects both space and time measurements (as described below). These effects are often referred to as the "curvature," "warping," or "distortion" of spacetime, where the term *matter* is shorthand for everything in spacetime—objects, radiation, etc., by virtue of those systems' energy and momentum. To say that space is "curved" ("warped" or "distorted") means that the geometry of space does not correspond to Euclidean geometry. That is, for example, the sum of the measures of the three angles of a triangle does not equal 180° (where the triangle is formed by connecting three points in space with the shortest possible lines).

There are different ways to think about or visualize such a space. One way is to imagine that 3-dimensional space is literally curved as it "sits" in a higher-dimensional space (6 dimensions would actually be required), much as the 2-dimensional surface of Earth is curved as it "sits" in a 3-dimensional space. This interpretation takes literally the term "curved space." Another way is to imagine that 3-dimensional space is itself "really" Euclidean, that is, not curved, but that metersticks are distorted by the gravitational field in just such a way that the results of measurements are described by non-Euclidean geometry. In this interpretation, metersticks expand or contract depending on their location, orientation, and the time at which the measurement is made. For example, metersticks that pointed along a radial direction with respect to Earth's center contract the closer they are brought toward Earth.

Neither the theory of general relativity nor experiments nor observations can determine which of these two interpretations is "correct." And so they are just that—interpretations, which can be chosen based on usefulness, convenience, or preference. To say that time is "curved" means that the rate at which clocks "tick" depends on where and "when" they are. For example, the higher a clock is from Earth's surface, the faster it runs, relative to clocks on the ground, an effect that has been measured. This effect must be taken into account in the design and operation of the GPS system, in which signals are transmitted between satellites and ground receivers.

Further Reading: Spacetime Physics by Edwin Taylor and John A. Wheeler; *Black Holes and Time Warps: Einstein's Outrageous Legacy* by Kip Thorne; and "Curve Space," an edited transcript of a lecture by Richard Feynman, available in Feynman, Leighton, Sands, *The Feynman Lectures on Physics* (Vol. II, Ch. 42) or Feynman, *Six Not-So-Easy Pieces: Einstein's Relativity, Symmetry and Space-Time* (Ch. 6).

Use an Analogy

Curved Rubber Sheet-Curved Spacetime

Analogy Point out to students the ways in which the curved rubber sheet-curved spacetime analogy, shown in **Figure 16,** is useful and the ways in which it is misleading. For example: *Useful:* (1) Just as the yellow ball curves the rubber sheet, so too does the Sun curve spacetime; (2) Just as the red ball's path is affected by the shape of the rubber sheet, so too is the path of a planet in space affected by the curvature of spacetime. *Misleading:* (1) The yellow ball curves the rubber sheet, which is a two-dimensional *spatial* surface, or two-dimensional *space.* The Sun curves not only space, but space *and time,* that is, spacetime. In fact, it is the curvature of time rather than that of space that is most important in accounting for the motion of the planets around the Sun. (2) The analogy may suggest that the Sun curves space in the same way or to as great an extent as the yellow ball curves the rubber sheet. In fact, the curvature of space around the Sun is so small that this curvature has not yet been detected. (3) The red ball accelerates toward the yellow ball because the red ball is on a hill, that is, the rubber sheet that is bent by the yellow ball. (Remind students that even if the red ball moves in a circular path around the yellow ball without moving closer, it is nonetheless accelerating toward the yellow ball, though it is simultaneously moving with a certain velocity in a direction tangent to the path.) The red ball accelerates toward the yellow ball because the sloped rubber surface exerts a contact or normal force on the ball, one component of which points toward the yellow ball. Students might be misled to believe that a similar mechanism is at work according to the curved spacetime view of the planet's motion. Actually, there is no "hill exerting a force" on the planets. Indeed, no gravitational force is invoked at all. Rather the planets are moving on a "natural path" in curved spacetime.

Content Background

Curvature in Newtonian Theory Although it is correct that, in thinking about gravitation, Newton used a Euclidean space, whereas Einstein used a non-Euclidean spacetime (often called "curved spacetime"), that is actually not what distinguishes the two theories. The Newtonian theory of gravitation can, in fact, be expressed in the language of curved space, as well. There are more than one substantive difference between the Newtonian theory and general relativity, but one important difference is that in Newtonian theory, time is absolute and universal, whereas in general relativity it is not.

Use Figure 16

Ask students what would happen to the large ball if the small ball were more massive. Increasing the mass of the small ball would reduce the effect of the large ball on the small ball's path. As the mass of the small ball is increased, it will begin to have a greater effect on the large ball such that eventually, when the masses are equal, the larger ball would begin to move relative to the small ball. This would also be relative to the distance between the two balls. The farther apart they are, the less the interaction. OL

Content Background

General Theory of Relativity Many of the predictions of Einstein's general theory of relativity have been confirmed. The ability of a massive object such as a galaxy to act as a lens has been spectacularly confirmed by images gathered by the *Hubble Space Telescope*. In addition, rapidly spinning neutron stars, or pulsars, have been found to slow their spin in a manner that is consistent with general relativity. Gravitational radiation causes the pulsars to slow their spin. The Laser Interferometer Gravitational-Wave Observatory (LIGO) experiment has been searching for gravitational radiation from supernovae and pulsars since 2002.

3 Assess

Assess the MAIN IDEA

Comet Impact Have students go online to look for videos and photographs of comet Shoemaker-Levy 9 that impacted Jupiter in July 1994. Why did the comet hit Jupiter and why was it broken into pieces? The comet was captured by Jupiter's strong gravitational field about 20–30 years before it impacted Jupiter. This was the first comet ever seen orbiting a planet before it hit. The comet was broken into pieces by a close approach to the planet where tidal forces tore it apart. When the pieces of the comet hit Jupiter, explosions occurred, and scars more prominent than the Great Red Spot were visible at the impact sites for months afterward.

Check for Understanding

Gravitational Field Review with students how they can calculate the weight of an object by using the relationship $F_g = mg$.

F_g = the weight of an object in newtons (N)

m = the mass of the object in kilograms (kg)

g = the gravitational field strength in newtons/kilogram (N/kg)

Ask students to calculate the strength of the gravitational field around Earth. They should find g [in newtons per kilogram (N/kg)] for distances nr_E, where $n = 1, 2, 3, 4,$ and 5. They should then find their own weight (in newtons) at these locations using their measured mass (in kilograms). OL

Reteach

Weightlessness Review methods of measuring inertial and gravitational mass and weight. Discuss three situations in which your apparent weight approaches zero: being so far from any planet, satellite, or star that there is no gravitational force on you; being supported by a force such as buoyancy; or accelerating at the rate of g along with the scale and all other possible supports. Ask students what it would feel like for them to experience such a situation. They would experience a sense of weightlessness. BL

Reading Checks and Caption Checks

Caption Check, *student text p. 186*
The example does not consider the effects of air resistance.

Reading Check, *student text p. 187*
The mass of a satellite affects neither the satellite's orbital speed nor its period.

Caption Check, *student text p. 190*
The gravitational field (g) is found by $g = \frac{F_g}{m}$. The gravitational force (F_g) will never reach zero because it is inversely proportional to the distance between two objects squared. When *r approaches 0, the force is at maximum. When r approaches infinity,* F_g *approaches 0 but never reaches that value due to the* $\frac{1}{r^2}$ relationship.

Caption Check, *student text p. 193*
On Earth, we experience light traveling in straight lines.

Practice Problems

student text p. 188

14. a. 7.75×10^3 m/s
b. slower
c. The speed is slower because r is larger. The satellite is farther from Earth's center.

15. 6.70 km/s, 1.4 Earth days

16. a. 7.8×10^3 m/s
b. 5.3×10^3 s or 88 min

17. a. 2.86×10^3 m/s
b. 1.65 h

Section 2 Review

student text p. 193

18. a. $\frac{g_S}{g_E} = 2.2$
b. 8.5×10^{19} N/kg

19. Yes. The chairs are weightless but not massless. They still have inertia and can exert contact forces on your toe.

20. 1.5 N/kg

21. a. When the orbital radius is large, the period also will be large. Thus, the one at 160 km will have the larger period.
b. The one at 150 km, because the smaller the orbital radius, the greater the speed.

22. Newton's law describes how to calculate the force between two massive objects. Einstein's theory explains how an object, such as Earth, attracts the Moon.

23. 7.35 N/kg

24. No. Because the speed and period of the orbit don't depend at all on the mass of the satellite, the scientific advisors would not have been able to calculate the mass of the satellite.

25. Earth rotates toward the east, and its velocity adds to the velocity given to the satellite by the rocket, thereby reducing the velocity that the rocket must supply.

FRONTIERS IN **PHYSICS**

No Escape

Is a BLACK HOLE really a hole?

Purpose

This article describes how the high escape velocity of a black hole prevents everything, even light, from escaping.

Background

The event horizon—boundary of influence—of a black hole is a sphere that surrounds the black hole in space. This event horizon marks the distance at which nothing, not even light, can escape. The only variable determining the radius of the sphere is the mass of the black hole. As objects pass through the event horizon, the black hole becomes more massive, and the event horizon expands outward.

Teaching Strategies

An analogy for a black hole is a "drain hole," a vortex in water. The idea is that nothing can move through the water faster than the speed of sound. At some point close to the drain hole, the water is moving toward the drain hole so fast that no sound can escape. Discuss with your students some of the unique properties of such a drain hole. What sorts of signals could or could not escape such a sound horizon?

GOING**FURTHER** >>>

Expected Outcomes While the speed of the light beam does not change, Albert Einstein's general theory of relativity shows that gravity affects light in an unusual way. Gravity robs light of some of its energy. For a light beam, energy equals color. As the light beam loses energy, its color moves down the spectrum, from blue to red and beyond. For a light beam projected from a black hole, the energy of the beam has dropped to zero.

SECTION 1

Mastering Concepts

26. Answers will vary. A possible form of the correct answer would be, ". . . if the planet's average orbital radius is 9.50×10^8 km, what would you expect its period to be?"

27. The path of Io is an ellipse, with Jupiter at one focus.

28. Because Earth moves more slowly in its orbit during summer, by Kepler's second law, it must be farther from the Sun. Therefore, Earth is closer to the Sun in the winter months.

29. No. The equality of the area swept out per unit of time applies to each planet individually.

30. Newton knew that the Moon followed a curved path; therefore, it was accelerated. He also knew that a force is required for acceleration.

31. He carefully measured the masses, the distance between the masses, and the force of attraction. He then calculated *G* using Newton's law of universal gravitation.

32. According to Newton, $F_g \propto \frac{1}{r^2}$. If the distance is doubled, the force is cut to one-fourth.

33. Because $\frac{T^2}{r^3} = \frac{4\pi^2}{Gm_S}$, if the mass of the Sun, m_S, is doubled, the ratio will be halved.

Mastering Problems

34. 12 y

35. 246 y

36. 4.16×10^{23} N

37. 9.11×10^{-31} kg

38. 6.5×10^{-8} N

39. 6.1×10^{-9} N

40. **a.** 489 N

b. 4.90×10^{2} N

41. **a.** 6.0×10^{24} kg

b. 5.5×10^{3} kg/m^3

42. 5.84×10^{-10} N

43. 8.0×10^{-10} N

44. The mass of the star is 1.91 times the mass of the Sun.

45. 23 years

46. $\frac{F_E}{F_S} = \frac{1.0}{2.3}$

47. D < A < E < C < B

48. 0.37 kg, 0.75 kg

49. 101 N

50. 5.65×10^{26} kg

51. 18 AU

52. **a.** 2.2×10^{15} m^2/s

b. 2.0×10^{11} m^2/s

53. 79 days

SECTION 2

Mastering Concepts

54. Its speed; it is falling all the time.

55. Speed depends only on b, the distance from Earth and c, the mass of Earth.

56. gravitational attraction to the central body

57. A force of 5*g* means that an astronaut's weight is five times heavier than it is on Earth. The force exerted on the astronaut is five times the force of Earth's gravitational force.

58. Einstein's view is that gravity is an effect of the curvature of space as a result of the presence of mass, whereas Newton's view of gravity is that it is a force acting directly between objects. Thus, according to Einstein, the attraction between Earth and the Moon is the effect of curvature of space caused by their combined masses.

59. $\frac{N}{kg} = \frac{kg \cdot m/s^2}{kg} = \frac{m}{s^2}$

60. The value of *g* would double.

Mastering Problems

61. a. 3.07×10^3 m/s or 3.07 km/s

b. 8.66×10^4 s or 24.1 h

62. a. 0.2 N/m

b. 20 N

63. a. 2.03×10^{20} N

b. 2.80×10^{-3} N/kg

64. Answers will vary, but a correct form of the answer is, "A satellite is in a circular orbit around Earth. If it moves with a speed of 8.3×10^3 m/s, what is its orbital radius?"

65. a. 1.80×10^3 N

b. 8.00×10^2 N

c. 2.92×10^2 N

66. 2.64×10^3 km

67. a. 1.6×10^3 kg

b. 1.3×10^{-10} m/s^2

68. 8.3×10^{-9} N

69. 7.3×10^{22} kg

70. 1.60 N/kg

71. 3.0×10^{-47} N

72. a. 1.7×10^{-10} N

b. 1.7×10^{-12} N

73. 241 N

74. a. 29 N/kg

b. 1.1 N/kg

c. 4.9 N/kg

Applying Concepts

75. The acceleration is independent of the object's mass. This is because more massive objects require more force to accelerate at the same rate.

76. You must know the period and orbital radius of at least one of its satellites.

77. Orbital motion of an object does not depend on its mass and cannot be used to find the mass. Newton's version of Kepler's third law is used to find an object's mass when a satellite is known to orbit the object.

78. $\frac{1}{4}g$

79. Nothing. G is a universal constant, and it is independent of Earth's mass. However, the force of attraction would double.

80. It will also double.

81. Only the lower right orbit is possible. The top two orbits do not have the Sun at a focus, and in the lower left orbit, the planet is not in orbit around the Sun.

82. No. The forces constitute an action-reaction pair, so under Newton's third law, they are equal and opposite.

83. A satellite with a small orbital radius has the greater velocity.

84. If the orbital radius increases, so will the period.

85. g on Jupiter is three times that on Earth.

86. As the mass of the planet increases, the period of the satellite decreases. Because Earth has a larger mass than Mars, Earth's satellite will have a smaller period.

87. a. Your mass increased.

b. The ratio remained constant because it is equal to the gravitational field at the location.

88. To "drop" an object down to Earth, you would have to launch it backward at the same speed at which you are traveling in orbit. With respect to Earth, the object's speed perpendicular to Earth's gravity would be zero, and it could then "drop" down to Earth. However, the object is likely to burn up as a result of friction with Earth's atmosphere on the way down.

89. The satellite is positioned as close to the equator as possible so it doesn't appear to have much north-south movement. Because it is placed at that distance, a satellite has a period of 24.0 h. If it were positioned any closer, its period would be less than 24.0 h and it would appear to move toward the east. If it were positioned any farther, its period would be longer than 24.0 h.

Mixed Review

90. 2.01×10^{30} kg

91. a. 1.7×10^{3} m/s

b. 6.5×10^{3} s

92. $r\geq7.8\times10^{1}$ m

93. a. 1.2×10^{2} min

b. 1.6×10^{3} m/s

94. a. 0.707 months

b. 1.26 times the present orbital radius of the Moon

c. The length of a year on Earth would not be affected. It does not depend on Earth's mass.

95. $0.35T_M$

96. 84.5 min

Thinking Critically

97. at sea level: $c = 4.0\times10^{8}$ units, $y = 9.77$ m/s^2
on top of Mt. Everest: 9.74 m/s^2
in a typical satellite orbit: 9.47 m/s^2
in a much higher orbit: 9.18 m/s^2

98. About 8 min

99. a. $F_{Sm} = (5.90\times10^{-3}\text{ N})m$; $F_{Mm} = (3.40\times10^{-5}\text{ N})m$

b. The Sun pulls approximately 100 times stronger on the waters of Earth.

c. $(2.28\times10^{-6}\text{ N})m$

d. $(1.00\times10^{-6}\text{ N})m$

e. the Moon

f. The tides are primarily due to the difference between the pull of the Moon on Earth's near side and Earth's far side.

Writing in Physics

100. One of the earliest, crude measurements was made by James Bradley in 1732. The answers should also discuss measurements of the transits of Venus done in the 1690s.

101. Astronomers measure the star's tiny velocity due to the gravitational force exerted on it by a massive planet. The velocity is calculated by measuring the Doppler shift of the star's light that results from that motion. The star moves back and forth as the planet orbits the star, allowing calculation of the planet's period. From the size of the velocity, they can estimate the planet's distance and mass. By comparing the distances and periods of planets in solar systems with multiple planets and using Kepler's third law, astronomers can better separate the distances and masses of stars and planets.

Cumulative Review

102. 4.0×10^{2} km

103. 610 N

Standardized Test Practice

student text p. 201

Multiple Choice

1. C
2. D
3. A
4. C
5. D

Free Response

6. 8×10^5 km

Rubric

The following rubric is a sample scoring device for free response questions.

Points	Description
4	The student demonstrates a thorough understanding of the physics involved. The response may contain minor flaws that do not detract from the demonstration of a thorough understanding.
3	The student demonstrates an understanding of the physics involved. The response is essentially correct and demonstrates an essential but less than thorough understanding of the physics.
2	The student demonstrates only a partial understanding of the physics involved. Although the student may have used the correct approach to a solution or may have provided a correct solution, the work lacks an essential understanding of the underlying physical concepts.
1	The student demonstrates a very limited understanding of the physics involved. The response is incomplete and exhibits many flaws.
0	The student provides a completely incorrect solution or no response at all.

CHAPTER 8
Rotational Motion

About the Photo

Circular Rides Most students have probably ridden a swing ride at an amusement park. These rides consist of chairs attached by chains to a vertical pole that rotates. Ask students to describe the ride in terms of the angular velocity from the beginning to the end of the ride. What causes the swing's angular velocity to change during the ride? Aid the students in understanding that all objects that change their angular velocity are subject to a net torque.

Chapter Overview

Like linear motion, rotational motion obeys Newton's laws. But there is an added twist: different parts of an object being rotated experience different velocities and accelerations. New concepts, like torque and moment of inertia, and several new symbols are required to describe this common type of motion. Torque also is needed to determine whether an object is in rotational equilibrium.

Before students study the material in this chapter, they should study:

- Accelerated motion in one dimension
- Adding vectors in two dimensions
- Newton's laws of motion
- Newton's law of universal gravitation
- Uniform motion in one dimension

To solve the problems in this chapter, students will need a strong grasp of:

- Graphing data
- Scientific notation
- Slope
- Solving linear equations

Introduce the BIG IDEA

Linear and Circular Motion Predict what will happen to a dropped yo-yo while holding the end of the string. Yo-yo will spin while falling. Will it accelerate at 9.8 m/s^2? No. Drop yo-yo to confirm. The upward tension in the string provided a torque causing the angular velocity to change. Upward tension is the reason the yo-yo didn't accelerate at 9.8 m/s^2 but at a lesser value.

Go online!
connected.mcgraw-hill.com

Your online experience includes:

- **Video**
- **Concepts In Motion**
- **Virtual Investigations**
- **All Lab, Applying Practices activities, and PBL worksheets**
- **Online Practice**
- **Personal Tutor**
- **Multilingual eGlossary**
- **Presentation**

Use the LaunchLAB

In *Rolling Objects,* students can explore objects that have different moments of inertia.

Use Physics T.V.

Video

DANGEROUS CURVE Have you ever seen a road sign depicting a truck tipping over? You will usually find them near curves. They warn truck and bus drivers to slow down while rounding the curve. What can happen if these drivers do not slow down on the curve?

Describing Rotational Motion

student text pp. 204–207

1 Introduce

Bellringer

Rotating Football Using a football, first ask students when it can be treated as a point particle. A football can be treated as a point particle when its rotational motion is small compared to its linear motion. As you hold it in your hands and casually rotate it in all possible orientations, ask them to brainstorm all the quantities that would have to be determined to describe where the football is, how to orient it, and how it is moving. Quantities should include not only *x*, *y*, and z (and the three velocities) for the center of the ball, but also the ball's orientation about three rotational axes (one vertical and two horizontal) and how fast it is rotating about those three axes. Briefly discuss other objects whose rotations are important to them.
OL **Visual-Spatial** **EL**

Tie to Prior Knowledge

Linear Motion Students will use the quantities that describe linear motion (position, velocity, and acceleration) and the geometry of the circle to develop equations for rotational motion.

2 Teach

Angular Displacement

student text pp. 204–205

Concept Development

Greek Symbols The symbols used for the quantities involved in rotational motion—θ (theta), ω (omega), α (alpha), and τ (tau)—will be unfamiliar to most students. Point out that these symbols are used to distinguish between linear and rotational motion.

Radians Angle measurements in radians (rad) are based on the ratio between the arc length and the radius of a circle. Point out that in calculations, radian units are dimensionless.

Reinforcement

Degrees and Radians To help students become accustomed to measurements in radians draw a circle diagram showing frequently used angles (30°, 45°, 60°, 90°, 120°, 180°, etc.), and their associated radian measures ($\frac{\pi}{6}, \frac{\pi}{4}, \frac{\pi}{3}, \frac{\pi}{2}, \frac{2\pi}{3}, \pi$, etc.) Show students where 1 radian fits. **BL** **Visual-Spatial**

Concept Development

MAIN IDEA On a turntable affix a small figurine to the outer edge. Place the figurine at the 12 o'clock position on the turntable. Give one student a stopwatch to time the motion, and another student a soft cloth measuring tape. Turn the turntable on and allow the student to time the motion of the figurine for several revolutions. Have the student measure the total distance the figurine traveled by using the cloth measuring tape. Ask students if there might be an easier way of measuring the distance traveled? Students might say it would be easier to count the number of revolutions and measure the radius of the circular path to determine this value from the circumference. Inform students that the angular displacement θ is related to the linear distance as measured by the cloth measuring tape, and determine the angular displacement in radians. Next, ask students to determine the average angular velocity ω in rad/s.

Critical Thinking

"Natural" Angles To begin, ask students what they think is a natural way to measure angles. Next, on the board draw two radii in a large circle that are about 60° ($\frac{\pi}{3}$ radians) apart in angle. Carefully obtain measurements for the radius and intercepted arc length. Ask students to compute the ratio of arc length to radius and to determine in which form the measure of the angle numerically is closest to this ratio. The ratio should be about 1, and $\frac{\pi}{3}$ radians is then closest. On this basis, radians are the natural form of angle measure. **OL** **Visual-Spatial**

Quick DEMO

Angular Displacement

Estimated Time 10 minutes

Materials bicycle wheel, meterstick, and cloth measuring tape

Procedure Measure the radius of the wheel. Mark one point on the sidewall of the wheel at the outer edge. Place the wheel on the ground so that the marked point is at the bottom. Roll the wheel along the ground by one rotation and measure the distance along the ground that the wheel traveled. Use this information to relate θ to distance moved. The distance traveled will be $2\pi r$. **AL** **Kinesthetic**

Angular Velocity and Angular Acceleration

student text pp. 205–207

Identify Misconceptions

Average and Instantaneous As in the case of linear motion, average and instantaneous values of an object's angular velocity and angular acceleration often are confused. If the angular velocity is changing, one can always define an average angular velocity, but for each instant, one also can define an instantaneous angular velocity. If the angular position is plotted as a function of time, then the instantaneous angular velocity would be the tangent to the graph. If the angular velocity is changing at a constant rate, then the average and instantaneous angular acceleration will be the same. **OL** **Logical-Mathematical**

Use Table 1

Angular and Linear Displacement Create a drawing on the board that illustrates the connection between angular and linear displacement and showing how x increases with r for fixed θ. Illustrate, using motion diagrams, the similar relationship between v and r for fixed ω. **OL** **Visual-Spatial**

3 Assess

Assess the MAIN IDEA

Motion of a Second Hand Bring into the class a large clock with a second hand. Have students determine the angular displacement in radians (starting at 12) of the second hand when it reaches 9. $\theta = \frac{3\pi}{2}$ radians. Have students determine the angular velocity of the second hand. $\omega = 0.10$ rad/s. Have students determine the angular acceleration of the second hand. $\alpha = 0$ rad/s^2.

Check for Understanding

Demonstrating Angular Velocity Use a large wheel, preferably one with a smaller pulley attached. Mark a spot on the outer edge. Wrap a string around the wheel. Pull the end of the string at a constant velocity. Ask students to observe the angular velocity of the wheel. Repeat pulling the string at about the same constant velocity as before. This time have it wrapped around the smaller pulley. But first, ask them to predict whether the angular velocity will change. Angular velocity: $\omega = \frac{v}{r}$. Because the radius is smaller, angular velocity must be larger.

Extension

Determining Angular Acceleration Ask students to consider a wheel that is suspended several feet above the ground with its axle parallel to the ground. A massive object is hung from a string wrapped around the axle many times. How can probeware be used to measure the angular acceleration of the wheel? A rotational motion sensor can be connected to the apparatus. Ask whether the linear acceleration of a point on the circumference will be different if a pulley is used. $a = \frac{\alpha}{r}$ The relationship depends on r. When the pulley is used, a will be much smaller.

Reading Checks and Caption Checks

Caption Check, *student text p. 205*

The variable r represents the distance a point is from the center of the rotating object. The variable x represents the distance that the point moves. The variable θ represents the angle of rotation.

Reading Check, *student text p. 205*

4π

Reading Check, *student text p. 205*

Angular velocity is an object's angular displacement divided by the time it takes the object to make the angular displacement.

Reading Check, *student text p. 206*

Angular velocity and angular acceleration both describe the motion of an object, but angular acceleration is equal to the change in angular velocity divided by the time required to make the change.

Practice Problems

student text p. 207

1. **a.** -120π rad or -377 rad

   b. -2π rad or -6.28 rad

   c. $-\frac{\pi}{6}$ rad or -0.524 rad

2. **a.** 6π rad

   b. 2π rad/min

   c. negative acceleration because the toy slows down until it stops moving

3. 0.707 m

4. **a.** The changes in velocity are the same, so the linear accelerations are the same.

   b. Because the radius of the wheel is reduced from 35.4 cm to 24 cm, the angular acceleration will be increased. $\alpha_1 = 5.23$ rad/s^2; $\alpha_2 = 7.7$ rad/s^2

5. Because $\omega = \frac{v}{r}$, if r is increased, ω will decrease. The number of revolutions will also decrease.

Section 1 Review

student text p. 207

6. **a.** 2.36×10^6 s

   b. 2.66×10^{-6} rad/s

   c. 4.63 m/s

   d. The speed on Earth's equator is 464 m/s, or about 100 times faster.

7. **a.** $\Delta\theta = -\frac{\pi}{3}$ rad

   b. $\Delta\theta = -4\pi$ rad

   c. $\Delta\theta = -240\pi$ rad

8. -8.3 rad/s^2

9. angular displacement—yes; linear distance—no, because it is a function of the radius

10. **a.** 52 rad/s or 5.0×10^2 rev/min

    b. 25 rad/s or 2.4×10^2 rev/min

    c. -5.9×10^{-3} rad/s^2

1 Introduce

Bellringer

Lever Arm Use a classroom door or other massive object that pivots about one edge. Have students explore how the location and direction of the force they exert changes the rotation of the door. Show how the greatest rotation for the least force occurs when the force is exerted perpendicular to the door as far from the hinge as possible. OL Kinesthetic EL

Tie to Prior Knowledge

Force Students will use the concept of angular motion, Newton's second law, and the geometry of the circle to develop Newton's second law for rotational motion.

2 Teach

Force and Angular Velocity

student text pp. 208–210

Concept Development

Maximum Torque When a string is wrapped around an object, it is always tangent to the circle and thus perpendicular to the radius. The torque exerted when you exert a force F on the string is $\tau = rF$. OL

PHYSICS CHALLENGE ACTIVITY

MAIN IDEA Bring a football into class. Give the football to a student and have him/her stand in front of the room. Draw a "goal line" on the floor with some masking tape. Have the student with the football stand a few feet from the "goal line." How should the player be tackled in order to prevent the player from being carried over the "goal line"? Students might respond that the player should be tackled below the waistline. Be sure to explain that a tackle below the waistline creates a torque on the player that causes his body to rotate with some angular velocity. This angular motion will cause the player to hit the ground sooner than if the player were hit at the waistline. Hitting at the waistline might carry the player several feet before falling, and therefore, over the "goal line." For safety, remind students that this is only a demonstration and do not allow them to tackle the demonstrator.

Use Figure 6

Perspectives on Torque Equation Explore the equation for torque: $\tau = Fr \sin \theta$. Note that it can be written $\tau = r(F \sin \theta)$ or $\tau = F(r \sin \theta)$. In the first case, it can be interpreted as a force, reduced because of the angle at which it is exerted, times a distance. In the second, it is a force times a lever arm, which depends on the distance and the angle at which the force is exerted. OL Logical-Mathematical

Discussion

Problem At what point between the hinged side and outside edge of a door would a given force perpendicular to the door produce the same torque as a force of equal strength but at a 30° angle along the outer edge of the door?

Response The torques would be equal if the force were applied halfway across the door. OL

ADDITIONAL IN-CLASS EXAMPLE

Use with Example Problem 1.

Problem What force is applied when the same wrench (as in the example problem) is used with a 35-N•m torque applied at an angle 75° from the perpendicular?

Response Compared to the 60° from the perpendicular in the example problem, 75° from the perpendicular is closer to a right angle with the radius arm. Therefore, the force needed to achieve a torque of 35 N•m should be less than the force found in the example problem.

$$F = \frac{\tau}{r \sin \theta} = \frac{(35\ \text{N}\cdot\text{m})}{(0.25\ \text{m})\sin(75)} = 1.4\times10^{2}\ \text{N}$$

DIFFERENTIATED INSTRUCTION

Visually Impaired Feel the torque! Pivot a broomstick around one end. Screw hooks into the bottom at a variety of distances from the pivot. Let a student with visual impairment test-lift a heavy object. Hang the object from a selected hook while the student lifts the free end of the broomstick. Ask the student to describe the exerted force on the broomstick. Describe to the student the location of the hook. Repeat, using a different hook, and ask the student to relate the torque and the lever arm. Alternatively, have a student pull down on a hook, using a scale to exert a fixed force but adjusting the angle to vary the torque. OL Kinesthetic EL

Critical Thinking

Lever Arm Graph Ask students how you can graphically represent the lever arm. To visualize how exerting a force at an angle reduces the lever arm, students should use the example in **Figure 6** to draw the lever arm when a force is exerted on a wrench of length 25 cm, at angles of 15°, 30°, 45°, 60°, and 75°. How long are these lever arms? 6.5 cm, 12 cm, 18 cm, 22 cm, and 24 cm OL

Logical-Mathematical

Finding Net Torque

student text pp. 211–213

Use an Analogy

Averages The balance point, as determined by where all contributions to net torque sum to zero, can be compared to other kinds of mathematical averages. For example, consider student scores on a science exam. Adding a score far above or far below the class average will move the average more than adding a score that is closer to the initial average.

REAL-WORLD PHYSICS

The Torque Wrench Applying the correct torque when tightening a bolt is very important—too little torque and the bolt will not exert enough force to keep the two objects together. On the other hand, if you apply too much torque you can break the bolt. A torque wrench is designed to keep a person from applying too much torque. The simplest models have a flexible handle and a thin indicator that converts the amount that the handle is flexed into the torque. The more expensive models have a device that produces an audible click when the correct torque has been applied. Many electric and pneumatic wrenches have built-in torque indicators to prevent workers from applying excessive torque.

ADDITIONAL IN-CLASS EXAMPLE

Use with Example Problem 2.

Problem Suppose that Aysha is replaced by the 52-kg Mario. Where should Mario sit?

Response $F_{gM} = m_M g = (52\text{ kg})(9.8\text{ N/kg}) = 5.1\times10^2\text{ N}$. Replace 4.2 by 5.1 in the final equation to obtain

$$r_M = \frac{(5.5\times10^2\text{ N})(1.75\text{ m})}{(5.5\times10^2\text{ N} + 5.1\times10^2\text{ N})} = -0.91\text{ m}$$

Discussion

Problem Tell students to imagine a very heavy, uniform plank of length l and mass m_1 placed so it extends a distance r over the edge of a small platform. The plank is not fastened to the platform in any way. Ask students what factors determine the maximum value of r so a person of mass m_2 can walk out to the end of the plank. Have students develop an equation for this maximum distance.

Response The length of the plank, the positioning of the center of the plank with respect to the edge of the platform, and the mass of the person determine the maximum distance.

Develop the equation by first recognizing that the plank is uniform, so the mass can be considered to act at its center. Then assume that the person is at the end of the plank while it is positioned for maximum distance r just before the plank tips over. Net torque is zero. Setting clockwise torque equal to counterclockwise torque and then solving for r yields

$$m_1 g\left(\frac{l}{2} - r\right) = m_2 g r,$$

$$r = \left(\frac{m_1}{m_1 + m_2}\right)\frac{l}{2}$$ OL

Content Background

Seesaw All changes in rotational motion are the result of net torque. When considering the case of the seesaw, you might need to remind some students that torques are vector quantities in rotational dynamics that add. They can be balanced or unbalanced. If torques do not balance on a seesaw, then the board rotates. Two people on opposite ends of a seesaw exert torques in opposite directions. If the torques they exert are equal in magnitude and opposite in direction, the board does not rotate. If the two people have equal weights, then the board balances when they sit equal distances from the axis—or fulcrum, in this case. If they are not of equal weights, then the heavier person must sit closer to the axis to balance the board.

Use the PhysicsLAB

In *Leverage,* students can demonstrate how a lever can reduce the effort needed to create large forces.

Use the PhysicsLAB

In *Torques,* students can measure forces that produce torque.

The Moment of Inertia

student text pp. 213–216

Concept Development

Distance and Torque Point out to students that torques add if they are in the same direction and subtract if they are in opposite directions. Have students consider pushing open a door. Where they push and in what direction they push are important factors in determining the ease with which they will be able to push open a door. Some might have watched young children figure out that the farther away from the hinges (point of rotation) they push on the door, the easier it is to open it. In other words, increasing radial distance increases torque. Students might understand the moment of inertia after they understand the importance of distance in the definition of torque. BL Visual-Spatial

Use Table 2

Distribution of Mass Have students compare the moments of inertia of a sphere, a cylinder, and a ring that have the same mass and radius. Remind students that the farther mass is from the axis, the greater the moment of inertia is. Ask students which object has a larger fraction of its mass away from the axis. The ring has all its mass at the edge; the cylinder is intermediate; the sphere has the most mass near the axis. OL

Concept Development

Integration of Mass Calculation of the moment of inertia requires the use of integral calculus. Conceptually, the object is broken into tiny blocks of mass. The tiny mass is multiplied by the square of its distance from the axis. Those products are summed over all the tiny mass blocks.

Reinforcement

Demonstrate If two objects have the same mass but different shapes, the one with its mass distributed farther from its axis of rotation will have a larger moment of inertia. Show students pairs of objects of the same mass but different shapes, such as a ring and a disk. Have them pick the one that should have the larger moment of inertia. Students should be able to say that the ring shape has a larger moment of inertia.
BL Kinesthetic EL

DIFFERENTIATED INSTRUCTION

Struggling Students Have students investigate how the moment of inertia affects the velocity of an object. Have students use a hoop and a disk that have the same diameter, width, and mass. Ask students which object has its mass located farthest from its center. the hoop What you are really asking students is to identify how each object's mass is distributed around its axis of rotation. Ask students to predict which object will reach the bottom of an incline first. Have them test their predictions. Instruct students to place the disk and the hoop on an incline and to release them simultaneously. The solid disk will reach the bottom first.
BL Kinesthetic

ADDITIONAL IN-CLASS EXAMPLE

Use with Example Problem 3.

Problem If the baton in Example Problem 3 is rotated about a point one-fourth of the distance from one round object, how does the moment of inertia compare to the two examples?

Response One round object is one-fourth the length of the rod from the axis; the other is three-fourths the length. Therefore, $I = m\left(\frac{1}{4}l\right)^2 + m\left(\frac{3}{4}l\right)^2 = m\left(\frac{5}{8}\right)l^2$.
So, $I = (0.30\text{ kg})\left(\frac{5}{8}\right)(0.65\text{ m})^2 = 0.08\text{ kg}\cdot\text{m}^2$; which is intermediate between the two other cases.

Quick **DEMO**

Distributed Model Moment of Inertia and Resistance to Motion

Estimated Time 10 minutes

Materials duct tape, quarters, meterstick, stopwatch

Procedure

1. Tape quarters at the 40-cm and 60-cm marks on a meterstick. ***WARNING:*** **Everyone present should wear goggles.**
2. Have one student hold the meterstick horizontally at the 50-cm mark with an extended arm. Have another student use a stopwatch to record the amount of time required for the first student to oscillate the meterstick back and forth, from horizontal to vertical, five times, as quickly as possible.
3. Repeat with quarters set at the 25-cm and 75-cm marks and again at the 10-cm and 90-cm marks. Ask students what happens to the period of oscillation (time required per oscillation) as the mass is distributed farther from the center of rotation. The period should increase quickly, with the second increment larger than the first, provided the student uses a similar twisting effort in all three trials. This illustrates that as point masses are distributed farther from the axis of rotation, their resistance to rotation increases faster than a linear relationship would suggest. OL Kinesthetic EL

PHYSICS CHALLENGE ACTIVITY

Test the Validity of Models Have students calculate the added moment of inertia of a quarter due to its rotation. If the quarter is a distance *r* from the axis of rotation, then its moment of inertia as a point mass is $I = mr^2$. Its moment of inertia as a thin disk is $I_{disk} = m\left(\frac{1}{12}h^2 + \frac{1}{4}r^2\right)$, where *h* is its thickness and *r* is its radius. Compute the added moment of inertia due to its rotation. What fraction is this of its point-particle moment of inertia when it is 10 cm from the axis? 2 cm from the axis?
$m = 5.5$ g, $r = 12$ mm, $h = 1.3$ mm.
$I_{disk} = 2.0\times10^{-7}$ kg•m^2. At 2 cm,
$I = 2.2\times10^{-6}$ kg•m^2. At 10 cm,
$I = 5.5\times10^{-5}$ kg•m^2. So, at 2 cm, the added moment of inertia is about 9 percent, at 10 cm, it is about 0.4 percent. AL Logical-Mathematical

Use the MiniLAB

In *Balancing Torques,* students can find the equilibrium condition and calculate torques.

Newton's Second Law for Rotational Motion

student text pp. 216–218

Quick **DEMO**

Newton's Second Law of Rotational Motion

Estimated Time 5 minutes

Materials two 1-m lengths of 3/4″ PVC water pipe, four 0.3-m lengths of steel reinforcing rods (rebar)

Procedure Before class, drive two rebar rods into one length of pipe so they meet at the center. Drive the other two rebar rods into the other pipe length so they are flush with the pipe ends. Have students lift the pipes to feel that their masses are the same. Have students grasp each pipe in the center and rotate them back and forth. The torque needed to accelerate the pipes is easily felt in the wrists.

Content Background

Describing Rotational Motion Rotational motion concerns the movement of a rigid body around a point, namely, the axis of rotation. To describe it, we use a coordinate system that allows us to measure changing angles. Angular acceleration ($\alpha = \frac{\Delta\omega}{\Delta t}$, rad/s^2) is the rate of change of angular velocity ($\omega = \frac{\Delta\theta}{\Delta t}$, rad/s). Angular velocity and angular acceleration are vector quantities. The angular acceleration vector is determined in relation to the angular velocity vector, which points along the axis of rotation. In rotational motion, because the moment of inertia (I) tells how the mass of an object is distributed about the axis of rotation, torque (τ, tau) also can be described as a product of the moment of inertia and angular acceleration ($\tau = I\alpha$).

ADDITIONAL IN-CLASS EXAMPLE

Use with Example Problem 4.

Problem The wheel in the example problem is accelerated from rest to 8.00 rev/s in only 5.00 s. How would the force required be affected?

Response Only one known quantity (t) is changed. In part a, the angular acceleration is now given by $\alpha = \frac{2\pi(8.00 \text{ rev/s})}{5.00 \text{ s}} = 10.1 \text{ rad/s}^2$. It is a factor of three greater. Torque and force are also proportional to angular acceleration, so they also increase by a factor of three. Thus, the force is 16.5 N.

Identify Misconceptions

Inertia and Torque Moment of inertia for a point mass has two factors, the value of the mass and the square of its distance from the axis of rotation. Because torque involves force times just the first power of distance, confusion may result. The following hint might help students sort this out. First, remind students that angular acceleration is related to linear acceleration, $a = r\alpha$. Next, students should apply Newton's second law for linear motion, $\frac{F}{m} = r\alpha$. Now have them multiply both sides of this equation by r, $\frac{Fr}{m} = r^2\alpha$. So, $\alpha = \frac{Fr}{mr^2} = \frac{\tau}{I}$. Angular acceleration, torque, and moment of inertia, as defined, make Newton's second law work for rotational motion. OL

PHYSICS CHALLENGE ACTIVITY

Can Competition Have students race two soup cans, a liquid broth soup versus a thick paste soup, down an inclined plank. Ask students which one wins. Why? Have students run the race with cans of different radii, analyze the results, and present them to the class. The key to understanding is to pick the axis of rotation where the can touches the incline. Sum the moment of inertia of a point object at the edge of the can to the moment of inertia of the can, so, $I = mr^2 + I_{can}$. Then, Newton's second law leads to $a = \frac{g \sin \theta}{1 + \frac{I_{can}}{mr^2}}$. The liquid broth soup can acts like a hoop because the liquid does not rotate in the short race time. The paste soup can acts as a solid cylinder, so it wins. The radius of the can does not affect acceleration. AL Kinesthetic

3 Assess

Assess the MAIN IDEA

Friction as a Torque Spin a bicycle wheel or other disk-like object until it is moving fairly fast. Ask students to list different ways of stopping the wheel or disk with a chalkboard or white-board eraser. Test a few of the methods. Which method works best? Students may answer that applying a force to the edge of the wheel with as much force as possible is the best. Explain how the angular velocity is reduced the fastest when the greatest torque is applied. Maximizing the torque requires applying a force as far from the center of the wheel as possible.

Check for Understanding

Demonstrate Have a bicycle available, mount a wheel, or draw a bicycle gear with a pedal on the board. Ask students which position of the pedal they would choose to start the bike from rest. Ask them to use torque and Newton's second law for rotational motion in their answer. The torque for a given force is greatest when the pedal is horizontal. If one starts with the pedal 30° to 45° above the horizontal, the torque is reduced to only 87 percent, or 71 percent of its maximum, but the pedal can go farther before torque goes to zero at the bottom. OL

Reteach

Moment of Inertia Review moment of inertia and the fact that it can be expressed as the ratio of the torque to the angular acceleration. Have a variety of objects with equal mass but different moments of inertia—from pipes stuffed with rebar to coins on pencils or rings and solid cylinders. Have students devise ways of comparing moments of inertia of two items. OL

Reading Checks and Caption Checks

Caption Check, *student text p. 209*
The sin must be used when the angle (θ) is not equal to 90°.

Reading Check, *student text p. 209*
The variable L is the length of the lever arm. The variable r is the distance from the axis of rotation to the point where the force is exerted. The variable θ is the angle between the force and the axis of rotation to the point where the force is applied.

Reading Check, *student text p. 209*
The variable τ represents torque. The variable F represents force. The variable r is the distance from the axis of rotation to the point where the force is exerted. The variable θ is the angle between the force and the axis of rotation to the point where the force is applied.

Caption Check, *student text p. 214*
Less torque is needed when your hands are placed in the middle of the book because the average distance of the book's mass from the rotational axis is much less in this case.

Reading Check, *student text p. 214*
$I = mr^2$

Practice Problems

student text p. 210

11. 1.4×10^2 N

12. 0.407 m

13. a. 36.6°

b. 30.2°

14. 94 N·m

15. 1.1×10^2 N·m; 0.0 N·m

student text p. 213

16. 1.5 m

17. A torque of +2.70 N·m must be exerted.

18. 0.056 kg

19. 0.042 kg

20. 789 N

student text p. 216

21. When r is doubled, I is multiplied by a factor of 4.

22. When more mass is located far from the center, moment of inertia will be greater. Thus, the hollow ball has a greater value of I.

23. a. 4.0 kg·m^2

b. 2 kg·m^2

c. 1.6 kg·m^2

24. a. The moments of inertia are different. If the spacing between spheres is r and each sphere has mass m, then rotation about sphere A is $I = mr^2 + m(2r)^2 = 5mr^2$. Rotation about sphere C is $I = mr^2 + mr^2 = 2mr^2$. The moment of inertia is greater when rotating around sphere A.

b. about sphere A: 0.020 kg·m^2; about sphere C: 0.008 kg·m^2

student text p. 218

25. 16 rev/s

26. 9.0 N

27. 5.5 N

28. 4.3 N

29. 8.99 kg·m^2

30. 7.7 N

Physics Challenge

student text p. 216

B > C > D > A

Section 2 Review

student text p. 218

31. To produce a torque with the least force, you should push as close to the edge as possible and at right angles to the door.

32. $\frac{F_{friend}}{F_{you}} = 1.8$. That is, your friend pushes almost twice as hard as you do.

33. a. 0.44 N·m

b. 1.6 N

34. 29 N·m

35. Sphere < solid disk < wheel. The less the moment of inertia, the less torque needed to give an object the same angular acceleration.

36. 5.99 kg·m^2

37. Torque: $\tau = Fr\sin\theta$. The force is due to friction, and the torque causes the ball to rotate clockwise. If the surface is friction-free, then there is no force parallel to the surface, no torque, and thus no rotation. Remember, forces acting on the pivot point (the center of the ball) are ignored. See Online Solutions Manual for free-body diagram.

1 Introduce

Bellringer

Balance Point Make and demonstrate a simple mobile with one rigid rod or stick, some string, and three weights. Two of the weights should be about equal, and the third should be about half as heavy. Tie different combinations of two of the three weights to the stick. In each case, try to hold up the mobile by another piece of string attached to a balance point proposed by students. Keep trying their suggestions until the mobile balances. Let students get a sense of where the balance point should be, given the different combinations. OL Visual-Spatial

Tie to Prior Knowledge

Equilibrium Students will combine the concepts of net force and net torque to analyze equilibrium situations. Then, they will extend their knowledge of angular velocity and linear velocity to situations where motion occurs within a rotating reference frame.

2 Teach

The Center of Mass

student text pp. 219–220

Concept Development

Importance of Center of Mass Any object or group of objects has a center of mass. This is the point at which the total mass can be considered to be acting. For example, ask students to locate the center of mass of a 10-kg object and a 20-kg object that are 3 m apart. In this case, the center of mass is on a line between the objects at a point 1 m from the 20-kg object and 2 m from the 10-kg object. OL Visual-Spatial

Reinforcement

MAIN IDEA Describe and draw this movie scene on the board. In a popular movie, two scientists are trying to hold a door closed from a raptor dinosaur on the other side. One scientist is pushing on the door near the handle while the other is pushing near the hinge. The scientist pushing near the hinge claims that she cannot leave the door to grab a weapon because the other scientist couldn't hold the door closed alone. Why should the scientist by the hinge stop pushing and grab the weapon? Students might respond that the scientist by the hinge is supplying little to no torque in this situation and should either push closer to the door handle or get the weapon.

Explain that the torque is maximized when the force is farthest from the hinge. For quite a period of time, the scientist pushing by the handle is supplying enough torque to counteract the raptor's torque. The door will remain in this state of equilibrium where all net forces and net torques are zero while the scientist pushing near the hinge leaves to retrieve the weapon.

Quick DEMO

Center of Mass

Estimated Time 10 minutes

Materials large piece of foam (10 cm thick, about 30–50 cm), two chemiluminescent glow sticks (approximately 15 cm long)

Procedure

1. Activate the glow sticks. Punch a hole through the foam near an edge, and push a glow stick through. Find the center of mass of the foam plus the glow stick. Mark the center of mass, punch a hole through, and press the second glow stick through the hole. If the glow sticks are insufficiently heavy, add weights such as lead sinkers.
2. Toss the foam across the room, making it spin. Ask what path it took. Did all parts follow the same path? an arc; no
3. Then turn off the room lights. Toss the foam again. Ask students to describe the path. The glow stick at the center of mass followed a parabolic path; the other executed circles around it.

REAL-WORLD PHYSICS

High Jumps If students were able to see computer tracking of a high jumper's body clearing the bar, they would see a path plotted according to the jumper's center of mass, or center of gravity—a perfect parabola. Search the web for a video of high jumps and video analysis software. Help students analyze the high jumps using the software.

Center of Mass and Stability

student text, pp. 220–221

Concept Development

Tipping Obtain boxes of different shapes. Fill some with moderately heavy and stable materials that can't move inside the box. Have students try tipping the boxes over (carefully) and analyze when they are stable and when they are not. Challenge them to find the center of mass of the boxes. OL Kinesthetic EL

Content Background

Center of Mass and Center of Gravity For a solid object, one can replace the entire mass of the object with a point mass equal to the object's mass located at the center of the solid object. The object will rotate about its center of mass. The center of mass also is the balance point of the object; that is, the object could be suspended from its center of mass and it would not rotate. It would not be out of balance. Likewise, all lines of suspension of a mass must pass through the center of mass because there can be no net torque on the object when it is hanging in equilibrium. This is the basis of the method used for finding center of mass through intersecting lines of suspension. The terms *center of mass* and *center of gravity* are used interchangeably.

Use the MiniLAB

In *Spinning Tops,* students can develop the concept of center of mass and Newton's first law as it relates to rotating objects (rotational inertia).

Use the PhysicsLAB

In *Equilibrium,* students can collect and organize data about forces acting on scaffolding.

Reinforcement

Stability Review stability and equilibrium by asking students how they stand to keep from falling over. Have a student kneel on the floor. Place the elbows against the knees and forearms along the floor. Have someone place a small box at the person's fingertips. Then, moving the hands behind the back, try to touch the box with the nose. Generally, a male is unstable because he must move his higher center of mass in front of his knees. Females can usually do this trick.

Conditions for Equilibrium

student text, pp. 221–223

Quick DEMO

Balance and the Center of Mass

Estimated Time 5 minutes

Materials none

Procedure Have a student stand with his or her toes against a wall and attempt to stand on tiptoe. He or she will find it extremely difficult, if not impossible. Discuss why this is.

PHYSICS PROJECT ACTIVITY

Stability of Automobiles The stability of automobiles, especially those with high centers of mass, such as sport utility vehicles, is of great public interest and debate. Have students research the viewpoints of the National Highway Traffic Safety Administration, the automakers, and independent consultants. Note that the discussion of stability in this chapter concentrated on static stability. If, on the other hand, an object is in the process of rolling, additional torque is needed to reduce the angular velocity. Boxes can be used to simulate these situations. Students should present their results to the class. **OL** **Linguistic**

ADDITIONAL IN-CLASS EXAMPLE

Use with Example Problem 5.

Problem Suppose a person lifts the end of the ladder that is nearest to sawhorse B. What force would the person have to exert? What force would the remaining sawhorse exert?

Response In this case, $l_B = 0$. The unknowns are F_A and F_B. The only change in the solution is r_B, which is now 1.20 m. Therefore,

$$F_B = \frac{(0.30\ \text{m})(56.84\ \text{N})}{1.20\ \text{m}} = 14\ \text{N}.$$

$$F_A = \left(\frac{1 - 0.30\ \text{m}}{1.20\ \text{m}}\right)(56.84\ \text{N}) = 43\ \text{N}.$$

Critical Thinking

Static Equilibrium Tell students that *static* means "unchanging state." Then ask if this definition also means that there must be no forces acting on an object in static equilibrium. No. Emphasize that there are forces acting on objects in static equilibrium and that these forces are balanced so their net force is zero. The word *equilibrium* comes from two Latin words meaning "even balance."

Concept Development

Center of Mass Ask students where they would find the center of mass for a ladder. The center of the center rung, assuming a constant density. Ask students to draw a free-body diagram of the forces on a ladder resting on two sawhorses. Designate forces: the left sawhorse is F_A, and the right sawhorse is F_B. These forces are parallel to each other and act in the upward direction. Ask students to write the net force acting on the ladder for the ladder to be in equilibrium. $F_A + F_B - F_g = 0$, or $F_A + F_B = F_g$ OL Visual-Spatial

Careers

Architects and Engineers Statics can be used to analyze forces acting on objects in equilibrium. Architects and civil engineers use statics to determine the structural integrity of their designs. In Munich, Germany the BMW Welt building is a conference facility built so that the roof looks like a floating cloud. Statics allowed architects to use just 11 columns to support a steel roof that stretched as far as $2\frac{1}{2}$ football fields placed end to end.

Concept Development

Torque Recall for students that torque is defined as the product of the force (F) and the lever arm $\tau = rF$. In the example problem, F_A, F_B, and F_g are each perpendicular to the ladder. Ask students to define the lever arm of each force. It is the distance along the ladder from the axis of rotation to the point where each force is acting. OL

Critical Thinking

Rotational Equilibrium Refer students again to the In-Class Example for Problem 5. Point out that the equation $F_A + F_B - F_g = 0$ (or, $F_A + F_B = F_g$) shows the net force acting on the ladder. Ask them how to find F_A and F_B. Use the second condition of static equilibrium: The ladder must be in rotational equilibrium. Ask students what the requirement is for the ladder to be in rotational equilibrium. The net torque on it must be zero. OL

DIFFERENTIATED INSTRUCTION

Struggling Students Equilibrium problems usually involve an equation for the net force—a sum that must be zero for equilibrium—and another for the net torque, which likewise must be zero. Most of the problems students confront involve an object with uniform mass distribution, a separate weight that acts at some particular point along the object, and two points of support. They should understand that choosing the axis of rotation at a point where one of the forces acts or at the center of mass of the distributed object will simplify the problem. Make sure students can use these principles to set up the correct equations. Then, address difficulties they might have completing the algebraic solution. BL

Rotating Frames of Reference, Centrifugal "Force," and The Coriolis "Force"

student text, pp. 224–225

Identify Misconceptions

Apparent Forces Students might believe they are feeling a force pushing them backward in a forward-accelerating car or outward when turning a corner. Reemphasize that Newton's first law applies only in reference frames that are not accelerating. Thus, when a car is accelerating forward, the student seems to feel a "force" pushing backward. Actually, a force is pushing the student forward with the car, which causes the student's body to sink backward in the seat. This feels like a force pushing the student backward, but it is a result of the inertia of the student's body resisting the acceleration. The same is true of centrifugal "forces."

Reinforcement

Apparent Forces Centrifugal and Coriolis "forces" seem to arise in rotating reference frames, but in each case, the effect is actually subjective from the standpoint of the person or object within the reference frame. The centrifugal "force," for example, is a major public misconception, precisely because it feels so real. An understanding of apparent forces is nonetheless important because Newton's laws hold only in fixed reference frames. OL Logical-Mathematical

3 Assess

Assess the MAIN IDEA

Balancing Soda Cans Have students each bring in an empty soda can. Ask students to try to balance the empty soda cans on the bottom rim of the can. Why won't an empty can balance on the rim? The center of mass of the can produces a torque around the pivot point, which causes the can to rotate or tip over. Next, have the students slowly add water in small amounts to the cans to determine whether they can balance the can on its bottom rim. Why are the cans now able to balance? Because the center of mass of the can is directly over the pivot point and produces no net torque, which means the can will not tip.

Check for Understanding

Center of Mass Set up a ring stand. Clamp a bar to the top of the ring stand so that it forms a support parallel to the lab table. Have string, scissors, masking tape, marker, and a small, irregularly shaped object available. Have students guide your actions to find the center of mass of the small irregularly shaped object. Suspend the object from the support. When it stops moving mark where the string falls with masking tape or a marker. Suspend the object again so that the string falls in a different location on the object. Mark the line made by the string. Where the two lines intersect on the object is the object's center of mass.

Reteach

Equilibrium Demo Ask students what is needed to keep at rest an extended object that is at rest. When force components in all three directions sum to zero, the object will not accelerate. Demonstrate the disruption of equilibrium by balancing a meterstick on your finger, then rolling the finger slightly so the stick rotates. Discuss torques and what forces are at work when the meterstick is in equilibrium.

Reading Checks and Caption Checks

Reading Check, *student text p. 219*
The center of mass is the point on the object that moves in the same way that a point particle would move.

Reading Check, *student text p. 221*
When the center of mass is above the base of the object, it is the most stable.

Reading Check, *student text p. 224*
Centrifugal force is an apparent force that does not really exist that seems to push an object outward in a rotating frame of reference.

Reading Check, *student text p. 225*
You would calculate the angle in the westward direction that is needed to compensate for the rotation of the Earth.

Practice Problems

student text p. 223

38. $F_B = 32$ N; $F_A = 8.0 \times 10^1$ N

39. a. counterclockwise: (0.51 m)F_B;
clockwise: $-$(0.66 m)F_A

b. (0.51 m)$F_B = -$(0.66 m)F_A

c. 31 N

d. F_A would become greater, and F_B would be less.

40. $F_{center} = 2.4 \times 10^2$ N; $F_{end} = 0$ N

41. $F_{end} = -8.3 \times 10^2$ N; $F_{center} = 1.8 \times 10^3$ N

Section 3 Review

student text p. 225

42. a. a book falling without rotating

b. an unbalanced seesaw that rotates until one person's feet hit the ground

43. Yes, an object moves as if all its mass is concentrated at the center of mass. There is nothing in the definition that requires any or all of the object's mass to be at that location.

44. The center of mass of the vehicle will be raised, but the size of its base will not be increased. Therefore, it needs to be tipped at a smaller angle to get the center of mass outside the base of the vehicle.

45. It is in the middle of the roll, in the open space.

46. Possible answer: Obtain a piece of string and attach a small weight to it. Suspend the string and the weight from one corner of the book. Draw a line along the string. Suspend the two from another corner of the book. Again, draw a line along the string. The point where the lines cross is the center of mass.

47. Earth's mass exerts a downward force. The turntable's surface exerts both an upward force to balance gravity and an inward force due to friction that gives the penny its centripetal acceleration. There is no outward force. If it were not for friction, the penny would move in a straight line.

48. Clockwise in the northern hemisphere and counterclockwise in the southern hemisphere; although wind provides the force that generates surface ocean currents, Earth's rotation significantly influences the movement of these currents. Because of Earth's eastward rotation on its axis, ocean currents are deflected to the right (east) in the northern hemisphere and to the left (west) in the southern hemisphere.

HOW IT WORKS

Spin Cycle

Centrifuges

Purpose

to see the application of rotational motion principles in a common piece of laboratory equipment—the centrifuge

Background

From a physics perspective, the laboratory centrifuge is not named well. A centrifuge uses the absence of centripetal force within a liquid mixture to separate that mixture. Circular motion requires force. The force on the liquid as a whole is provided by the walls of the centrifuge tube, but within the tube, the various components of a mixture separate by density because they are unable to produce sufficient centripetal force upon one another to hold their original configuration.

Teaching Strategies

A common carnival ride uses similar principles. This ride mirrors the physics of a centrifuge by spinning riders rapidly until they stick to the wall, even when the floor has fallen from beneath their feet. Riders feel as if the ride is tilting, even though it remains horizontal. This illusion occurs because centripetal acceleration (the wall pushing in) mimics the normal force (the ground pushing up). Model the carnival ride by having the class make reasonable estimates of the revolution rate and friction present within the ride.

GOINGFURTHER >>>

Expected Outcomes Answers will vary. Centrifuges are used to separate body fluids such as blood, to isolate nucleic acids and proteins from treated tissue samples, to separate chloroplasts or other organelles from cellular tissue, and in many other applications.

SECTION 1

Mastering Concepts

49. It is constant.

50. It is zero.

51. All parts of a rigid body rotate at the same rate angular velocity but not the same linear velocity.

52. It is inward (centripetal).

Mastering Problems

53. 0.600 rad

54. 51 rad/s

55. 0.49 m

56. **a.** 197 rad/s

b. 492 rad

57. $-7.54\ \text{rad/s}^2$

58. 17.5 cm/s

59. **a.** 2.73

b. 1.65

c. $71g$

60. 1.1×10^5 rev/min

SECTION 2

Mastering Concepts

61. They are all different. The one with the most mass, farthest from the axis, has the greatest moment of inertia.

62. The angular acceleration acting on the bolt produces a force. Different torques can be exerted with wrenches of different lengths.

63. Recall that $\tau = Fr\sin\theta$. Thus, $0 = E < D < C < B < A$.

Mastering Problems

64. 23 N

65. 3.8 N•m

66. $0.050\ \text{kg}\cdot\text{m}^2$

67. **a.** 7.5×10^2 rad/s

b. 0.72 N

68. $0.048\ \text{kg}\cdot\text{m}^2$

SECTION 3

Mastering Concepts

69. When the wheel is balanced so it does not tilt (rotate) in any direction, there is no net torque on it. This means that the center of mass is at the pivot point.

70. It is directly above the line between the points where the two wheels are touching the ground. There is no net torque on the truck, so it is momentarily stable.

71. Your center of mass must be above the point of support. But your center of mass is roughly at the center of your body. Thus, while you are on your toes, about half of your body must be in front of your toes, and half must be behind them. If your toes are against the wall, no part of your body can be in front of your toes.

72. She moves her center of mass closer to her head.

73. The center of mass of the vehicle with the larger wheels is located at a higher point. Thus, it does not have to be tilted very far before it rolls over.

Mastering Problems

74. **a.** At the opposite end, she will only lift half the mass. $F_{\text{least}} = 61$ N

b. At the board's center of mass (middle), she will lift the entire mass. $F_{\text{greatest}} = 120$ N

75. 1.16 m from the front of the car

76. $F_{\text{right}} = 62$ N upward, $F_{\text{left}} = 38$ N upward

Applying Concepts

77. The teeth have identical linear velocities. Because the radii are different and $\omega = \frac{v}{r}$, the angular velocity of the larger gear is smaller.

78. You can apply a known torque and measure the resulting angular acceleration.

79. The more mass there is far from the axis, the greater the moment of inertia. If torque is fixed, then the greater the moment of inertia is, the less the angular acceleration will be. Thus, the wheel with mass mostly at the hub has the least moment of inertia and the greatest angular acceleration. The wheel with mass mostly near the rim has the greatest moment of inertia and the least angular acceleration.

80. Its rotation rate can be increased only if a torque is applied to it. The frictional force of the alley on the ball provides this force. When the ball is rolling without sliding, there is no more kinetic frictional force and thus no more torque.

81. Put an extension pipe on the end of the wrench to increase the lever arm, exert your force at right angles to the wrench, or exert a greater force, perhaps by getting two people to push on the end of the wrench.

82. The torque caused by these forces is zero because the lever arm is zero.

83. The pole increases moment of inertia because of its mass and length. The drooping ends of the pole bring the center of mass closer to the wire, thus reducing the torque on the walker. The increased moment of inertia and decreased torque reduce the angular acceleration if the walker becomes unbalanced. The walker also can use the pole to easily shift the center of mass over the wire to compensate for instability.

84. You have forward tangential velocity, so the key will leave your hand with that velocity. Therefore, you should toss it early.

85. That makes the torque caused by that force equal to zero, reducing the number of torques that must be calculated.

86. A disclike object will smaller moment of inertia than a hooplike object, so D > A > C > B > E.

Mixed Review

87. a. $\alpha = \frac{3g}{2l}$

b. No; the angle between the door and the weight is changing. Thus, the acceleration changes.

88. 5.0×10^2 rad/s^2

89. $F_{left} = 1.5\times10^2$ N; $F_{right} = 6.0\times10^2$ N

90. Suki can move 0.848 m from the support or $1.75 - 0.848 = 0.90$ m from the end.

91. -1.3×10^{-3} m/s

92. a. The center of mass is always over the point of contact with the surface for a uniform cylinder. Therefore, the center of mass moved 2.50 m.

b. 2.00 m/s

c. 8 rad/s

93. a. 21 rad/s

b. 16 rev

c. 1.0×10^2 rad

94. a. -2.2 rad/s^2

b. -1.3×10^{-2} N•m

95. Because increasing the diameter decreases the angular velocity, it also will decrease the reading of the speedometer.

96. $F_{rope} = \frac{(0.86\ m^2/s^2)M}{h - 0.25\ m}$ Note that when you pull the box at the height of its center of mass, the denominator becomes zero. That is, you can pull with any amount of force and not tip the box.

97. a. The masses are the same, so the weights are the same. Thus, the same upward force is required to lift each load.

b. The longer piece of lumber would be easier to keep from rotating because it has a greater moment of inertia.

98. a. 86 N

b. Harris has to lift 2.0 m from Paul's end of the board.

Thinking Critically

99. 4.20×10^2 N

100. a. 21 N•m

b. The tension in the rope is 64 N.

101. Answers will vary, but a correct form of the answer is, "A flywheel is spinning at an angular velocity of 20 rad/s when a constant angular acceleration of 3.5 rad/s^2 is applied to slow it down. Through what angle has it turned after 4 s have elapsed?"

102. Answers will vary. A possible form of the correct answer would be, "He places one hand 20 cm in front of the ladder's center and the other hand 40 cm behind the ladder's center. What are the forces exerted by each hand?"

103. a. when $\omega = 0.0$

b. when $\alpha = 0.0$

c. When $\omega = 0.0$ instantaneously, but α is not zero, ω will keep changing.

d. Yes, as long as ω is constant but not zero.

104. The road exerts a force on the tires that brings the car to rest. The center of mass is above the road. Therefore, there is a net torque on the car, causing it to rotate in the direction that forces the front down.

Writing in Physics

105. For a planet and a moon with identical densities, the Roche limit is 2.446 times the radius of the planet. Earth's Roche limit is 18,470 km.

106. The force exerted by the ground on the tire accelerates the car. The engine produces this force. It creates the force by rotating the axle. The torque is equal to the force on the edge of the tire multiplied by the radius of the tire. Gears in the transmission may cause the force to change, but they do not change the torque. Therefore, the amount of torque created by the engine is delivered to the wheels.

Cumulative Review

107. a. 24 N

b. 2.0 m/s^2

108. 18.4°

109. 122 km/h at 14.2° west of north

110. 972 N

Standardized Test Practice

student text, p. 233

Multiple Choice

1. C
2. B
3. C
4. D
5. B
6. C

Free Response

7. 44 N•m

Rubric

The following rubric is a sample scoring device for free response questions.

Points	Description
4	The student demonstrates a thorough understanding of the physics involved. The response may contain minor flaws that do not detract from the demonstration of a thorough understanding.
3	The student demonstrates an understanding of the physics involved. The response is essentially correct and demonstrates an essential but less-than-thorough understanding of the physics.
2	The student demonstrates only a partial understanding of the physics involved. Although the student may have used the correct approach to a solution or may have provided a correct solution, the work lacks an essential understanding of the underlying physical concepts.
1	The student demonstrates a very limited understanding of the physics involved. The response is incomplete and exhibits many flaws.
0	The student provides a completely incorrect solution or no response at all.

CHAPTER 9

Momentum and Its Conservation

About the Photo

Momentum in Sports Ask students to recall memorable athletic experiences they have had, and ask them to describe the experiences in terms of motion in one or two dimensions, including velocity, force, and acceleration. Creating a sketch may help students formulate questions that can be answered using the concepts of momentum and impulse.

Chapter Overview

This chapter discusses the changes in motion of an object by considering an object's momentum before and after impulses act on it, and what is required to conserve the momentum of a system.

Before students study the material in this chapter, they should study:

- Accelerated motion in one dimension
- Adding vectors in one dimension
- Adding vectors in two dimensions
- Circular motion
- Mass and weight
- Newton's laws of motion
- Uniform motion in one dimension
- Vectors and scalars

To solve the problems in this chapter, students will need a strong grasp of:

- Graphing data
- Scientific notation
- Significant figures
- Sine, cosine, and tangent
- Solving linear equations
- Solving quadratic equations

Introduce the BIG IDEA

Quantity of Motion Isaac Newton treated momentum as such a fundamental concept that he referred to it simply as the "quantity of motion." How might this phrase be interpreted today? Two identical particles moving at the same velocity have more "quantity of motion" than either one does on its own.

Go online!
connected.mcgraw-hill.com

Your online experience includes:

- Video
- Concepts In Motion
- Virtual Investigations
- All Lab, Applying Practices activities, and PBL worksheets
- Online Practice
- Personal Tutor
- Multilingual eGlossary
- Presentation

Use the LaunchLAB

In *Colliding Objects,* students can investigate how balls with similar size but different masses react after they collide.

Use Physics T.V.

Video

CRASH! From fender benders to bumper cars, there's a lot of physics involved in collisions. How do forces shape what happens when two objects collide?

1 Introduce

Bellringer

Collision Force Drop a heavy object, such as this book, onto your desk. Drop the book again, but this time drop it onto a pad on your desk. Have students use their prior knowledge to list what they can and cannot determine about the nature of the collision. They can find the mass of the object and should be able to calculate its velocity just as it hits the desk. They can also observe that the object's velocity is zero after hitting the desk. Without knowing the time it took for the object to stop, however, they cannot determine the acceleration of the object as it came to a stop. They also cannot determine the force of the desk on the object, or vice versa. OL Visual-Spatial

Tie to Prior Knowledge

Laws of Motion Students will apply Newton's second law of motion to explore the relationship between momentum and impulse. Students may need to review angular velocity before they learn about angular momentum.

2 Teach

Impulse-Momentum Theorem

student text pp. 236–239

Use Figure 1

Students can learn about the connection between force and impulse by examining the bell curve in **Figure 1.** Ask them to find the maximum force. 1.47×10^4 N Ask if it is possible to determine how long the collision lasted? yes How? Observe the interval of time over which the force on the baseball is non-zero ($t = -1.5$ ms through $t = 1.5$ ms, or 3.0 ms). Explain that one standard measure is the time interval over which the force was larger than half its maximum value. Have students use the graph to calculate this interval. $\Delta t \approx 0.4$ ms $- (-0.4$ ms$) = 0.8$ ms Ask students how they can approximate the magnitude of the impulse. The area under the curve is the impulse. One method is to copy the area under the curve onto graph paper and count squares, find the area of each square and multiply. They could also find the area of a triangle that has an apex at the maximum force and that touches the two half-maximum force points. Have students compare the area of that triangle with the true area, 13.1 N•s. A triangle with its apex at (0.0 ms, 1.47×10^4 N) would have its base points at (−0.8 ms, 0.0 N) and (0.8 ms, 0.0 N). Its area would be (0.5)(base)(height), which is (0.5)(0.0016 s)(14,700 N) = 12 N•s. OL Visual-Spatial

Concept Development

Momentum and Velocity Help students understand the difference between momentum and velocity by writing $\boldsymbol{p} = m\boldsymbol{v}$ on the board and having students analyze it.

Content Development

MAIN IDEA To differentiate between momentum and change in momentum (impulse), write the following on the board:
$F\Delta t = m\Delta\boldsymbol{v} = m(\boldsymbol{v}_f - \boldsymbol{v}_i) = m\boldsymbol{v}_f - m\boldsymbol{v}_i = \boldsymbol{p}_f - \boldsymbol{p}_i$.
Emphasize that both momentum and impulse are vector quantities—they have both magnitude and direction. Explain that impulse points in the same direction as the change in momentum.

Use the PhysicsLAB

In *Sticky Collisions,* students can investigate what happens to the momentum of an object during a collision.

Applying Physics

Forces on Runners Have students examine their running shoes to determine how far the soles compress when a given force is exerted on them. Suggest designing a shoe tester that could exert a measured force on an object the size of the ball of the foot and measure the distance that force compresses the shoe. Forces as large as four times the student's weight should be tested. Have students research what make and model of shoe most reduces the force on the running foot. Students can gather a variety of shoes to test which kind most reduces force on the foot.

Quick DEMO

Impulse

Estimated Time 5 minutes

Materials bedsheet, raw egg

Procedure Take your class out of the building or to a safe and easy-to-clean area. Have two assistants hold a bedsheet vertically between them. Ask students to predict whether you can break an egg by tossing it with all your might against the bedsheet. Then throw a raw egg with considerable velocity toward the middle of the sheet. The bedsheet usually stops the egg without breaking it. Explain that the bedsheet stops the egg over a much longer time interval (Δt) than a brick wall would. Emphasize that a longer Δt means that the force exerted on the egg is reduced.

Content Background

Force Function How can one develop a mathematical function for a force that is varying (such as that in **Figure 1**)? One approach is to approximate it as a constant force. Mathematically, this is equivalent to creating a rectangle with the same area as is under the curve of the *F*-*t* graph. Of course, the shape of the rectangle does not change the impulse. One choice is to make the average force equal to the maximum force and adjust the time interval to get the correct area. A second way is to choose a time interval that best represents the time over which the force acts and adjust the force to get the correct area. Without detailed measurements of the force as a function of time, there is no single correct method.

Identify Misconceptions

Momentum and Velocity Momentum is not the same as velocity. In all examples so far, the momentum and velocity are related by a fixed ratio–the mass. For this reason, some students may not see a reason to have a second quantity and will treat momentum as if it were velocity. The difference will become more clear when students study collisions.

Reinforcement

Vectors Create some subtraction exercises using initial and final momentum vectors. Emphasize that the difference will be the impulse. Include some with either initial or final momentum of zero and with the two momenta in both the same and opposite directions. BL Visual-Spatial EL

ADDITIONAL IN-CLASS EXAMPLE

Use with Example Problem 1.

Problem Assume that there is a passenger of 85-kg mass in the vehicle described in Example Problem 1. For both gentle and emergency braking situations, calculate the impulse and average force needed to bring the person to a stop along with the vehicle.

Response $p_i = (85\ \text{kg})(26\ \text{m/s})$

$p_i = 2.2\times10^3\ \text{kg·m/s};\ p_f = 0.$

So $F\Delta t = -2.2\times10^3\ \text{kg·m/s}$

When $\Delta t = 21\ \text{s},\ F = -1.0\times10^2\ \text{N}$

When $\Delta t = 3.8\ \text{s},\ F = -5.8\times10^2\ \text{N}$

PHYSICS PROJECT ACTIVITY

Seat Belts and Airbags As an extension to what students have learned about impulse and momentum, students can explore how airbags work to reduce forces exerted in automobile collisions. Have them research and download film clips that show how airbag inflation can "soften the blow" to the crash dummies. Students can each develop a demo using a laboratory cart, a clay "passenger," and various strategies to protect the passenger when the cart crashes into a barrier. They can cover the sharp edge of the cart with rubber to model a padded dashboard. Ribbons can model seat belts, and a soft balloon can be the airbag. Some students may even opt to equip their carts with foam bumpers. OL Kinesthetic

PHYSICS PROJECT ACTIVITY

Teacher To Teacher

Egg-Drop Physics Have students construct containers for eggs that will be dropped from a significant height, such as the top of a stairwell in your school. Regulate the mass of each container so that the change in momentum or impulse of impact is the same for each container. The goal is to reduce the force of impact on the egg so that the egg will not break. Students can do this by increasing the time of impact of the egg inside the container to decrease the force. OL Kinesthetic

Angular Momentum

student text pp. 240–243

Activity

Angular Impulse You can illustrate angular momentum either by using a stool that rotates freely or by spinning a bicycle wheel with a weighted tire and an extended axle for easy handling. If using the stool, it may be best if you sit on it while twirling. In either case, ask students how to get you or the wheel rotating. Apply a torque, such as shoving off the side of a desk with your hand if you are on a stool, or in the case of the wheel, by pushing off the axle. Ask students what they would do to increase the rotation speed. Apply a larger torque over a longer period—in other words, increase the angular impulse. OL Logical-Mathematical

Concept Development

Angular Momentum versus Angular Velocity Ask students to define both angular momentum and angular velocity. Angular momentum, the momentum of an object rotating about an axis, is the product of an object's moment of inertia and its angular velocity. Angular velocity is an object's rate of rotation around an axis.

Discussion

Problem How does angular momentum differ from angular velocity?

Response The angular momentum of an object is the angular velocity of that object multiplied by that object's moment of inertia. OL

Critical Thinking

Rotating Systems Lead students in brainstorming examples of rotating systems that have angular momentum. Have them draw examples from everyday life and from nature—consider, for example, the fields of astronomy, meteorology, and sports. Examples of rotating systems include galaxies, solar systems, Earth and other planets, storms like hurricanes and tornadoes, high- and low-pressure areas, whirlpools and eddies, spinning balls, swinging bats, gymnasts, dancers, divers, spinning wheels, rotating doors, drills, and rotary saw blades. AL

Activity

Applications of Rotation Students can relate what they have learned about angular momentum to everyday life by researching various models of sport-utility vehicles, jeeps, and other passenger vehicles to assess them for safety. Suggest that they consider how well the vehicles handle sharp turns when traveling at high or even moderate speeds. Ask students to choose the vehicle they believe is best designed to tilt at large angles. They also can choose a poor design. Have them determine which, if any, of these models is prone to rollover. Ask students to prepare a safety report with a model showing the vehicle's center of mass, base, point of support, and axis of rotation. If time permits, have students present their research to the class and explain how their vehicle is designed for safety and how it maintains stability when tipped or tilted.
OL Visual-Spatial

3 Assess

Assess the MAIN IDEA

A stuntman with a mass of 100.0 kg falls from a roof before hitting an air-filled cushion, which stops his fall in 0.75 s. If his velocity at the moment he reaches the cushion is −14 m/s, what is the impulse on the stuntman? 1400 N•s What is the average force exerted on the stuntman by the cushion? 1900 N

Check for Understanding

Bounce Impulse Bounce a ball on the floor. Have students diagram the initial and final momenta and the impulse. Ask students what provides the impulse. the floor Have students compare the impulses on two balls of different mass. Both balls hit the floor with the same velocity, but the ball with the larger mass has the larger momentum, so it will have the greater impulse. OL

Extension

Specific Impulse Have students interested in model rocketry explain to the class how impulse and momentum are involved in the operation of rockets. The specific impulse of a rocket propellant is a rough measure of how fast the propellant is ejected out of the back of the rocket. Because space travel is concerned with acceleration and acceleration is determined by thrust, the larger the exhaust velocity, the larger the specific impulse. In designing chemical-propellant rockets, the goal is not to minimize the amount of fuel but to maximize the thrust (force) per unit of fuel burned.
AL Logical-Mathematical

Reading Checks and Caption Checks

Caption Check, *student text p. 241*

Extending her arms moves her mass away from the axis of rotation. This increases her moment of inertia and decreases her angular velocity so that she goes straight into the water.

Reading Check, *student text p. 241*

The tuck position increases the angular velocity by bringing the diver's mass closer to her axis of rotation so that her moment of inertia decreases.

Practice Problems

student text p. 239

1.

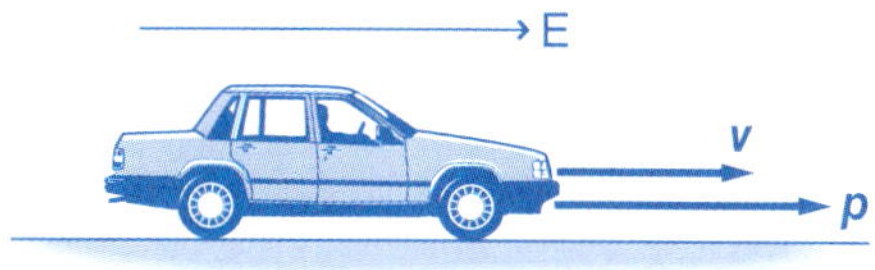

a. 2.32×10^4 kg·m/s eastward

b. 38.4 km/h eastward

2. a. 1.0×10^4 N·s directed westward

b.

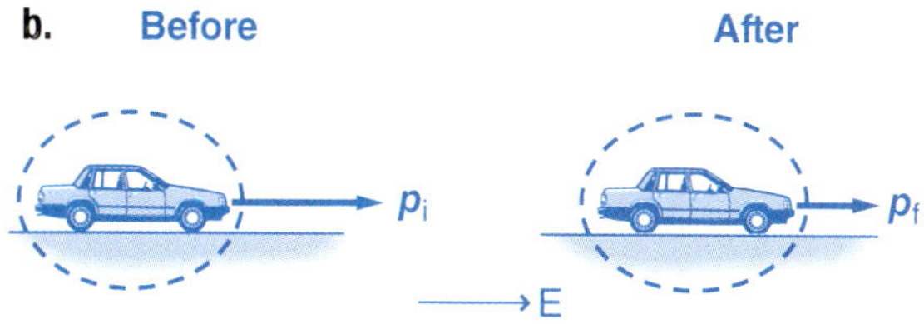

1.3×10^4 kg·m/s eastward, 65 km/h eastward

3. a. 2.7 m/s in the same direction as the original velocity

b. 1.3 m/s in the same direction as the original velocity

4. a.

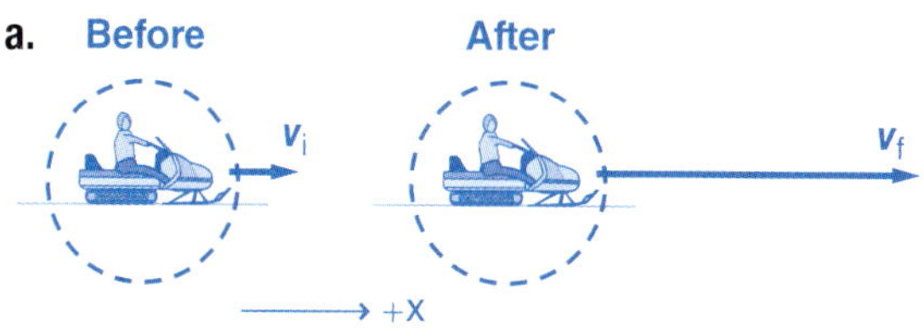

b. 5.28×10^3 kg·m/s

c. 88.0 N

5.

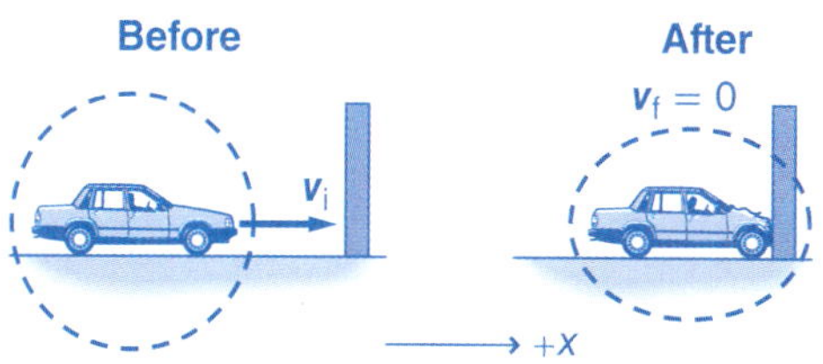

a. 7.8×10^3 N opposite to the direction of motion

b. 8.0×10^2 kg; such a mass is too heavy to lift. You cannot safely stop yourself with your arms.

Practice Problems

student text p. 243

6. 18 s

7. 63 rad/s; 0.033 N·m

8. 2.45 N·m·s; 1.02 N·m

9. 31 rad/s; 1.9 N·m·s

Section 1 Review

student text p. 243

10. A falling raindrop has more momentum. The truck's momentum is zero because its velocity is zero.

11. Yes, momentum is a vector quantity, and the momenta of the two cars are in opposite directions.

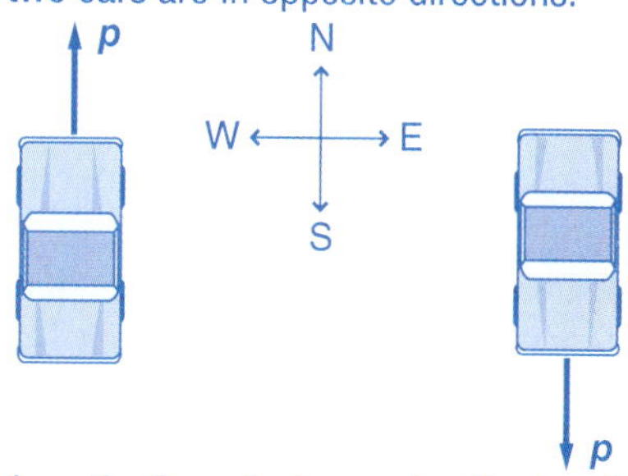

12. You reduce the force by increasing the length of time it takes to stop the motion of your body.

13. a.

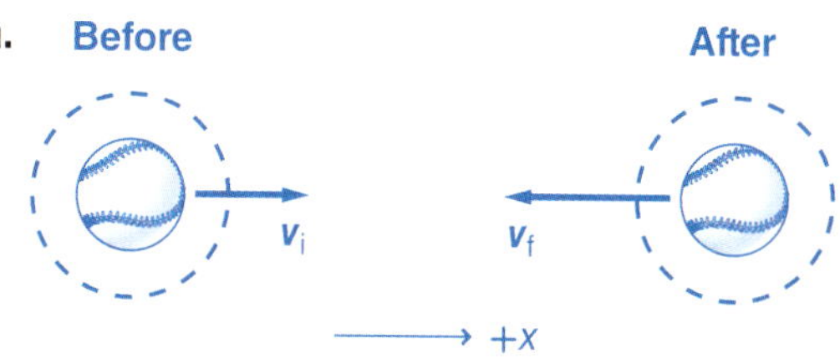

b. -11.1 kg·m/s

c. -11.1 N·s

d. -1.4×10^4 N

14. No; the change in momentum is upward. Before the ball hits the floor, its momentum vector is downward. After the ball hits the floor, its momentum vector is upward.

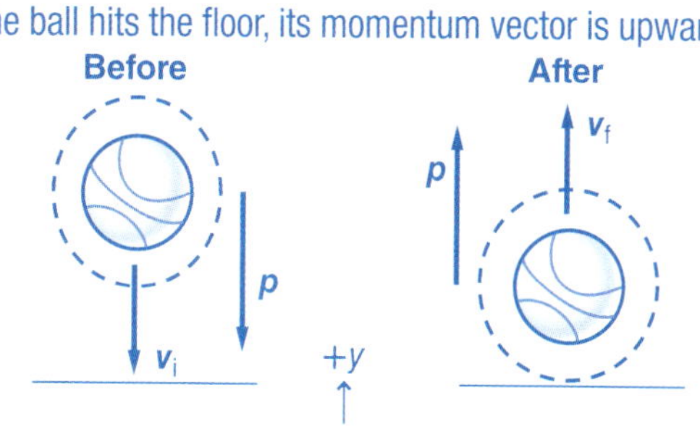

15. No torque acted on him. Drawing his arms in decreased his moment of inertia. The angular momentum did not change; his angular velocity increased.

16. The ones that bounce off produce larger impulses because they end up with some momentum in the reverse direction.

1 Introduce

Bellringer

Newton's Cradle Obtain a Newton's cradle apparatus that has six or seven steel balls suspended from two parallel bars. Pull all but two steel balls out of the way. Pull back one of the two steel balls, and let it collide into the other one. Have students describe the collision. Repeat, but this time let one steel ball collide into three steel balls. Before you release the steel ball, ask students to predict the outcome of the collision on the system. BL Visual-Spatial EL

Tie to Prior Knowledge

Newton's Laws of Motion This section relates Newton's first and third laws of motion to conservation of momentum. Two-dimensional collisions will require vector addition. Students will also use what they know about accelerated motion, angular velocity and angular momentum, and rotational dynamics to understand the law of conservation of angular momentum.

2 Teach

Two-Particle Collisions

student text pp. 244–245

Identify Misconceptions

System of Objects Ask students if it is true that a system is restricted to a single object or body. No; a system can include more than one object, even if the objects are not attached to each other. Describe the Earth-Moon system and how it moves as a single system in orbit around the Sun, even though Earth and the Moon are not physically attached and move relative to each other. The Earth-Moon system is not an isolated system because of the role of the Sun.

Momentum in a Closed, Isolated System

student text pp. 245–246

Discussion

Problem How do Newton's laws of motion relate to a closed, isolated system of two objects that collide?

Response The two objects will follow Newton's third law of motion when they collide. Each object exerts an equal and opposite force on the other during the collision. The two objects will continue to move together. If you could calculate the center of mass of the system, you would see that it moves with a constant velocity before, during, and after the collision, according to Newton's first law of motion. OL

ADDITIONAL IN-CLASS EXAMPLE

Use with Example Problem 3.

Problem A 1875-kg car going 23 m/s strikes a 1025-kg car going 17 m/s in the opposite direction. The two cars, traveling on ice, stick together. How fast do they move after the collision?

Response Let $v_{Di} = -17$ m/s. Then

$$v_f = \frac{(1875\ \text{kg})(23\ \text{m/s}) + (1025\ \text{kg})(-17\ \text{m/s})}{1875\ \text{kg} + 1025\ \text{kg}}$$

$= 8.9$ m/s. Note that this is slower than the rear-end type of collision in the example.

PHYSICS PROJECT ACTIVITY

Teacher To Teacher

Force Shield Place a massive object on your hand, then hit it with a small hammer. Show students that this does not hurt your hand. Due to conservation of momentum, the large mass of the object in your hand compared to the mass of the hammer causes the high velocity of the hammer to be converted into a low velocity of the object—so low that it does not crush your hand. OL Logical-Mathematical

Recoil
student text p. 247

Use the PhysicsLAB
In *Colliding Carts,* students can use a motion detector to investigate what happens to the momenta of two carts when they collide.

Use the MiniLAB
In *Rebound Height,* students can observe how mass and velocity affect the momentum of a bouncing ball.

REAL-WORLD PHYSICS
Sweet Spot Tennis rackets are designed to maximize the velocity that is imparted to the ball and to help the player control the direction of the ball. The design also helps reduce the forces exerted by the racket on the player's hand. Players use the term *sweet spot* to denote the location on the racket where it feels good to hit the ball. Hitting the ball at its sweet spot minimizes the high-frequency vibrations of the racket. The sweet spot also is the region where the coefficient of restitution (COR) is high. The COR is measured by dropping a ball on a rigidly held racket. It is the ratio of the magnitude of the velocity of the ball leaving the racket to the magnitude of the velocity just before it hits the racket.

DIFFERENTIATED INSTRUCTION
Visually Impaired Students can feel the recoil of a balloon. Blow up a balloon and hand it to a student. Make sure the student keeps the neck closed tightly to prevent the air from escaping. Have the student place the opposite end of the balloon on the palm of his or her hand. Then, ask the student to open the neck of the balloon to release the air. The student should feel the force of the air from the balloon on his or her hand. The force of the air on the front of the balloon pushes the balloon forward. **OL** **Kinesthetic** **EL**

Propulsion in Space
student text pp. 248–249

Identify Misconceptions
Rocket Propulsion When Robert Goddard began his rocket experiments, a prominent newspaper said he was doomed to failure because, as any high school student knew, the rocket could move only if the gases that were expelled pushed on air, and there is no air in space. How, then, do rockets move? The newspaper had it all wrong: Gases expelled from a rocket do not push on the air. The gases push on the rocket itself. It is the forward push of the gases that accelerates the rocket.

ADDITIONAL IN-CLASS EXAMPLE
Use with Example Problem 4.

Problem What if the same astronaut in Example Problem 4 had a mass of only 62 kg? What would be the final speed of the astronaut?

Response The same analysis and final equation can be used, but with a different mass:

$$\mathbf{v}_{Cf} = \frac{-m_D\mathbf{v}_{Df}}{m_C}$$

$$= \frac{-(0.035\text{ kg})(-875\text{ m/s})}{62\text{ kg}}$$

$$\mathbf{v}_{Cf} = +0.49\text{ m/s}$$

Two-Dimensional Collisions
student text pp. 249–250

DIFFERENTIATED INSTRUCTION
Struggling Students Two-dimensional collisions give students who have had trouble with forces in two dimensions a second chance to learn how to handle vectors. Most students will be tempted to add momenta as if they were scalar quantities. Some will find the graphical approach easier to understand while others, who are not visual learners, will find the algebraic approach easier. Pick the approach that fits each student's learning style. Pair two students who have different strengths and have them work on the same problem, compare answers, and teach each other how to solve it their way. **COOP LEARN**

Use Figure 12

Have students draw free-body diagrams of the collision. They can transfer the figure onto a worksheet and then measure the lengths and angles of the momentum vectors. Explain to students that they can move vectors around the page as long as they do not change the lengths and directions. Have students check to see if the vector sum of the two final momenta equals the initial momentum of ball C (because ball D starts at rest). **OL** **Visual-Spatial**

Careers

Accident Reconstruction Expert Investigating automobile accidents requires an understanding of collisions, friction, and Newton's laws of motion. Accident reconstruction experts work in a variety of ways to determine the causes of automobile accidents using evidence such as tire tracks. Some of their techniques are illustrated in the practice problems in this section. Accident reconstruction experts frequently serve as expert witnesses in court cases. In recent years, specialized data-collecting equipment and computer software have made their job easier. These experts have formed numerous professional organizations that help them improve their skills and exchange information. Both community colleges and universities have short training courses for investigators.

ADDITIONAL IN-CLASS EXAMPLE

Use with Example Problem 5.

Problem A 975-kg car, C, moving south at 22.5 m/s, collides with a 2165-kg truck, D, moving west at 17.5 m/s. They stick together. In what direction relative to south and with what speed do they move after the collision?

Response $m_C = 975$ kg, $m_D = 2165$ kg, $v_{Ci,y} = -22.5$ m/s, $v_{Di,x} = -17.5$ m/s.

$p_{f,y} = p_{i,y} = m_C v_{Ci,y} = (975\text{ kg})(-22.5\text{ m/s})$
$= -2.19\times10^4\text{ kg·m/s}$

$p_{f,x} = p_{i,x} = m_D v_{Di,x} = (2165\text{ kg})(-17.5\text{ m/s})$
$= -3.79\times10^4\text{ kg·m/s}$

$p_f = 4.38\times10^4\text{ kg·m/s}$, $v_f = p_f/(m_C + m_D)$
$= 13.9$ m/s

$\theta = \tan^{-1}(p_{f,x}/p_{f,y})$

$$= \tan^{-1}\left(\frac{-3.79\times10^4\text{ kg·m/s}}{-2.19\times10^4\text{ kg·m/s}}\right)$$

$\theta = 60.0°$ west of south

Use the MiniLAB

In *Momentum,* students can use the law of conservation of momentum to determine an object's velocity after a collision.

Use Models

Clay and Inelastic Collisions Students can use clay balls to model moving objects sticking and moving together. Have them construct balls with the same ratio of masses as the objects in Example Problem 5. Suggest that they use 10 g to represent 100 kg. Provide students with several examples of collisions, having them define the system, sketch the "before" and "after" states, and determine an unknown. **OL** **Visual-Spatial**

Conservation of Angular Momentum

student text p. 251

Concept Development

MAINIDEA The "nebular hypothesis" is a widely accepted model that attempts to explain the formation of the solar system. In this model, a huge gas cloud collapses on itself due to gravitational attraction. Because angular momentum is conserved, the overall angular velocity of the gas cloud increases as it shrinks. In fact, the conservation of angular momentum implies that almost all cosmological objects, including stars, galaxies, and black holes, are rotating.

Quick DEMO

Changing Moment of Inertia

Estimated Time 10 minutes

Materials rotating stool, two heavy blocks

Procedure Sit on the stool holding the blocks close to your body. Have a student gently spin you. Extend your arms and bring them back in. Discuss what happens in terms of conservation of angular momentum and changes in moment of inertia. How does that affect ω, angular velocity? When the moment of inertia (I) decreases—arms in closer—angular velocity (ω) increases and you spin faster. Moment of inertia increases when arms are extended; angular velocity decreases, and you spin more slowly.

Concept Development

Conservation of Angular Momentum Angular motion is different from linear motion in that the angular speed of a rotating object can be changed without changing its mass or angular momentum. Moment of inertia can be changed. Use the Quick Demo above to illustrate the concept, but be sure to define the system carefully. Make sure that torques cannot be exerted between the system and the external world. For example, a spinning skater will eventually slow and stop because of the torque exerted on her by the ice via friction. OL

Content Background

Orbiting Versus Spinning Objects Angular momentum is one characteristic used to describe an object in motion about an axis. Because angular momentum is a vector quantity, a complete description includes both a magnitude and a direction. The angular momentum of an orbiting object is mvr, where m is the object's mass, v is the magnitude of its linear speed, and r is the distance from the center of the orbit. The angular momentum can be expressed as $I\omega$ where $I = mr^2$ and $\omega = v/r$. The angular momentum of a spinning object is also represented by $I\omega$.

Reinforcement

Conservation of Angular Momentum Have students write a description of angular momentum in their own words. Their descriptions must relate angular momentum to the moment of inertia and angular speed. Their descriptions should also explain how angular momentum is conserved. Example answer: Angular momentum depends on the angular speed and the moment of inertia. Angular momentum is conserved as long as there is no net external torque acting on any object in the system. BL Linguistic

Critical Thinking

Adding Mass to a Rotating System Ask students to imagine they are conducting a conservation of momentum laboratory. They are using a disk rotating at 25 rad/s that has a moment of inertia of 2.5 kg•m^2. They drop a metal ring on the rotating disk so that its center coincides with the center of the rotating disk and observe that the new rotational speed is 18 rad/s. Assuming that angular momentum is conserved and that the metal ring initially was not rotating, ask students what its moment of inertia is. The initial and final angular momentum is (2.5 kg•m^2)(25 rad/s) = 63 kg•m^2/s. Because you know the final angular velocity, you find $I = L/\omega$ = (63 kg•m^2/s)/(18 rad/s) = 3.5 kg•m^2. Thus the added moment of inertia is 1.0 kg•m^2. AL Visual-Spatial

3 Assess

Assess the MAIN IDEA

Conservation of Momentum Imagine that two ice-skaters collide on an ice rink. What conditions must exist before they collide, in order for both of the skaters to be motionless after the collision? The skaters must have equal and opposite momentums. If both skaters have the same mass, they must be moving at the same speed in opposite directions. Could either skater be motionless before the collision? No; in this case, one or both skaters would be moving after the collision.

Check for Understanding

Momentum in Proportion Have students use the principle of conservation of momentum to estimate the amount of linear momentum that Earth acquires when a person jumps into the air. Sample answer: If a person jumps up 0.80 m, the velocity when leaving the ground is 4.0 m/s. If the person's mass is 60.0 kg, the momentum is 240 kg•m/s. Ask students the following questions. What change would that make in Earth's velocity? Earth's mass is 6.0×10^{24} kg, so its change in velocity would be 4.0×10^{-23} m/s. What change would it make in Earth's velocity if one million people in New York City jumped simultaneously? The change in velocity would be 4.0×10^{-17} m/s. OL Logical-Mathematical

Extension

The Slowing Earth Have students research how and why the angular momentum of Earth due to its rotation is changing with time. Be aware that it is a very complex problem. Changes are due to interactions with the atmosphere, oceans, melting snow, and solar flares. Students should also consider other factors affecting Earth's slowing; for example, the bulge at Earth's equator is decreasing and Earth's molten core rotates slightly faster than Earth. AL Linguistic

Reading Checks and Caption Checks

Caption Check, *student text p. 245*
The diagram shows that the forces exerted by the balls are equal and opposite. Because impulse is force multiplied by the time interval, and because the time intervals are the same, the impulses must also have the same magnitude but opposite directions.

Reading Check, *student text p. 245*
A closed system does not gain or lose mass. A system is isolated if the net external force on it is zero.

Caption Check, *student text p. 247*
Even though the forces on the boy and girl had the same magnitude, the boy had a smaller acceleration, which according to Newton's second law of motion implies that his mass is greater.

Caption Check, *student text p. 249*
The more massive ball would move less, and the less massive ball would move more.

Caption Check, *student text p. 252*
There is no unbalanced torque on it unless the top is tilted.

Reading Check, *student text p. 252*
As Earth rotates, the Sun exerts a torque on it, causing it to precess.

Practice Problems

student text p. 246

17. 1.1 m/s

18. 0.034 m/s

19. 1.2×10^3 m/s

20. 2.8 m/s

21. 6.7 m/s

22. 2.0 m/s in the opposite direction

Practice Problems

student text p. 249

23. 7.91 m/s

24. 9.0 cm/s to the right

25. 2.8 m/s in the opposite direction

Practice Problems

student text p. 251

26. 11.2 m/s, 36.6° north of west

27. 18.1 m/s, 15.9° south of east

28. 22.1 m/s; Yes, it was exceeding the speed limit.

29. 3.5 m/s, 30.0° to the right, and 2.0 m/s, 60.0° to the left

Physics Challenge

student text p. 252

1. 2.0×10^1 m/s

2. 3.0×10^1 m/s; the friend was not exceeding the 22 m/s speed limit. The other car was exceeding the limit.

Section 2 Review

student text p. 253

30. Most of the mass of the disk is located at the rim, thereby increasing the moment of inertia. Therefore, when the disk is spinning, its angular momentum is larger than it would be if more mass were near the center. With the larger angular momentum, the disk flies through the air with more stability.

31. a. 3.13 m/s

b. 1.25 m/s

32. No; because the mass of the racket is much larger than that of the ball, only a small change in its velocity is required. In addition, it is being held by a massive, moving arm that is attached to a body in contact with Earth. Thus, the racket and ball do not comprise an isolated system.

33. The vertical momentum comes from the impulsive force of Earth against the pole. Earth acquires an equal and opposite vertical momentum.

34. Because their final momentums are zero, their initial momentums were equal and opposite.

35. a. In the first case, the system is the skateboard, the ball, and you. In the second case, Earth is included in the system.

b. In the first case, the momentum is shared. In the second case, Earth's large mass is included in the system, and change in its velocity is negligible.

A CLOSER LOOK

Fire in the Sky

Background

With near-Earth objects, size matters. Anything smaller than 40 m across would likely burn up in the atmosphere and not pose a danger. However, those as big as 1 km across would have enough momentum upon impact to cause significant local destruction. Anything 2 km or greater in diameter might cause serious problems worldwide when debris kicked up into the atmosphere changed global climate. Such an "impact winter" could lead to famine when crops all over the world died from a lack of adequate sunlight.

Teaching Strategies

- To give students an idea of scale, have them look up the diameter of Earth (about 12,700 km). Then, have students calculate how large a scale model of a spherical asteroid with a diameter of 10 km would be if the model of Earth were a basketball (about 24 cm). The model asteroid would be about 0.02 cm in diameter.
- Explain that whether or not the dinosaur extinction did occur during the Chicxulub impact, one thing is certain: a collision that great caused ample destruction. Scientists estimate that the asteroid hit Earth with the equivalent effect of detonating 300 million nuclear weapons.

GOING**FURTHER** >>>

Expected Outcomes The "giant impact" hypothesis of the formation of the Moon is the most commonly accepted hypothesis, but there are others. The fission hypothesis that the material forming the Moon broke off from the Earth's crust is disconfirmed by the fact that the Moon does not orbit Earth exactly in its orbital plane. Studies of the Moon's composition indicate that it is about 100 million younger than Earth, so the impact hypothesis is more likely. The high angular momentum of the Earth-Moon system, and the fact that the Moon is moving slowly away from Earth, both favor the giant impact hypothesis.

ANSWERS

SECTION 1

Mastering Concepts

36. Yes, but for a bullet to have the same momentum as a truck, it must have a higher velocity because the two masses are not the same. $m_{bullet}\mathbf{v}_{bullet} = m_{truck}\mathbf{v}_{truck}$

37. a. The pitcher and the catcher exert the same amount of impulse on the ball, but the two impulses are in opposite directions.

b. The catcher exerts the larger force on the ball because the time interval over which the force is exerted is smaller.

38. No net force on the system means no net impulse on the system and no net change in momentum. However, individual parts of the system may have a change in momentum as long as the net change in momentum is zero.

39. Cars are made with bumpers that compress during a crash to increase the time of a collision, thereby reducing the force.

40. a. by applying an external torque

b. by changing the moment of inertia

Mastering Problems

41. 0.013 s

42. a. −14 kg•m/s

b. -3.2×10^{4} N

43. 25 m/s

44. 74 kg•m/s; 1.0×10^{1} m/s

45. a. −7.1 kg•m/s

b. -1.4×10^{4} N

46. a. 2.0×10^{4} kg•m/s

b. 1.3×10^{3} N

47. -1.2×10^{3} N

48. -6.0×10^{1} N

49. a. -1.0×10^{2} kg•m/s

b. -5.0×10^{2} kg•m/s

50. a. −5.1 kg•m/s

b. -1.0×10^{2} N

c. -1.0×10^{1} N

51. $D < A < B < C = E$

52. 4.8 N•s

53. 3.5×10^{2} kg•m/s

54. 42 m/s

55. a. $+5.2\times10^{-23}$ N•s

b. +7.8 N

56. 1.3×10^{3} s, or 22 min

57. 8.9×10^{2} kg•m/s at 44°

58. a. -2.00×10^{2} kg•m/s

b. -4.0×10^{3} N

c. 4.1×10^{2} kg

d. no

e. You would not be able to protect a child on your lap in the event of a collision.

59. Answers will vary, but a correct form of the answer is, "In a game of croquet, a 1.3-kg ball is struck by a mallet that makes contact with the ball for 0.55 s. The ball, initially at rest, is given a speed of 20.0 cm/s. What average force did the club exert on the ball?"

SECTION 2

Mastering Concepts

60. An isolated system has no external forces acting on it.

61. Momentum is conserved. The change in momentum of gases in one direction must be balanced by an equal change in momentum of the spacecraft in the opposite direction.

62. If you define the two balls as a system, the eight ball must be moving with the same velocity that the cue ball had just before the collision.

63. a. The momentum of a falling ball is not conserved because a net external force, gravity, is acting on it.

b. One such system in which total momentum is conserved includes the ball plus Earth.

64. a. The floor is outside the system, so it exerts an external force, and therefore an impulse on the ball.

b. Momentum is conserved in the system of ball plus Earth.

65. The external force of a car's brakes can bring the car to a stop by stopping the wheels and allowing the external frictional force of the road against the tires to stop the car. If there is no friction—for example, if the road is icy—then there is no external force and the car does not stop.

66. The child would have to exert a torque on it. He or she could stand next to it and exert a force tangential to the circle on the handles as they go past. He or she could also run at the side and jump onboard.

67. Answers will vary. A possible form of the correct answer is, "... skating at a speed of 4.3 m/s when he collides head-on with Gabe, mass 50.0 kg, who is skating in the opposite direction at 2.7 m/s. The two stick together. If Armando and Gabe are a closed, isolated system, what is their final velocity?"

Mastering Problems

68. 0.37 m/s

69. a. Before: $m_{FB} = 95$ kg
$v_{FB} = 8.2$ m/s
$m_{DT} = 128$ kg
$v_{DT} = ?$
After: $m = 223$ kg
$v_f = 0$ m/s

Before

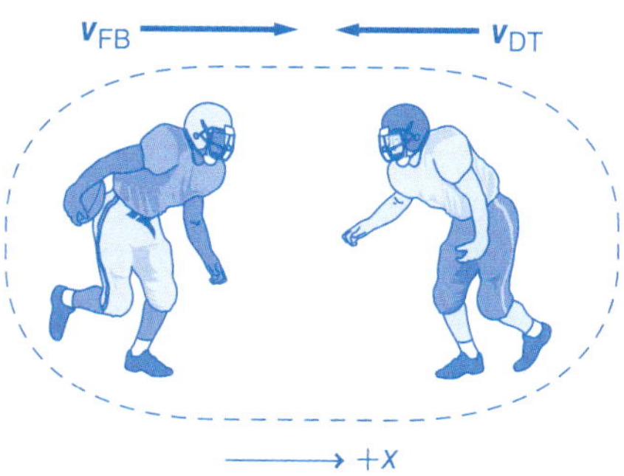

After

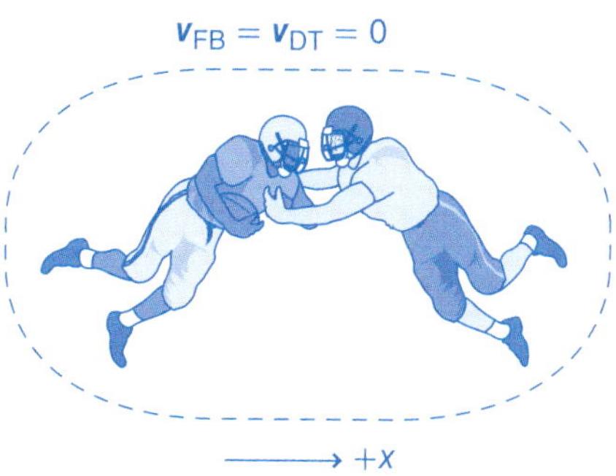

b. 7.8×10^2 kg•m/s

c. -7.8×10^2 kg•m/s

d. $+7.8\times10^2$ kg•m/s

e. -7.8×10^2 kg•m/s

f. -6.1 m/s

70. a. Before: $m_C = 5.0$ g
$m_D = 10.0$ g
$v_{Ci} = 20.0$ cm/s
$v_{Di} = 10.0$ cm/s
After: $m_C = 5.0$ g
$m_D = 10.0$ g
$v_{Cf} = 8.0$ cm/s
$v_{Df} = ?$

Before

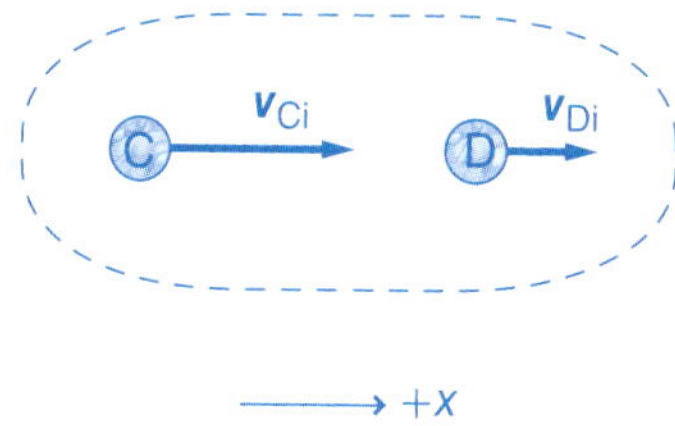

After

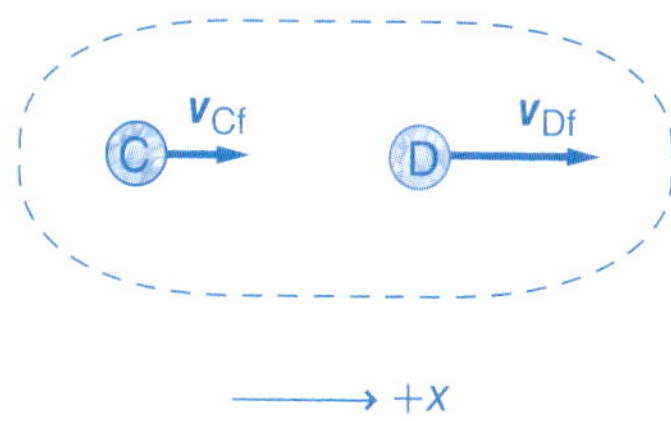

b. $p_{Ci} = 1.0\times10^{-3}$ kg•m/s, $p_{Di} = 1.0\times10^{-3}$ kg•m/s

c. 4.0×10^{-4} kg•m/s

d. 1.6×10^{-3} kg•m/s

e. 16 cm/s

71. -0.30 m/s

72. -4.94 m/s, or 4.94 m/s backward

73. 1.26 m/s in the same direction as he was riding

74. a. Before: $m_K = 60.0$ kg
$m_D = 90.0$ kg
$v_i = 0.0$ m/s
After: $m_K = 60.0$ kg
$m_D = 90.0$ kg
$v_{Kf} = ?$
$v_{Df} = ?$

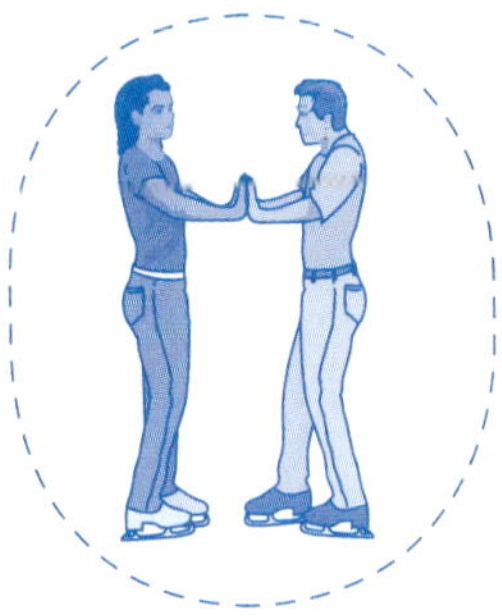

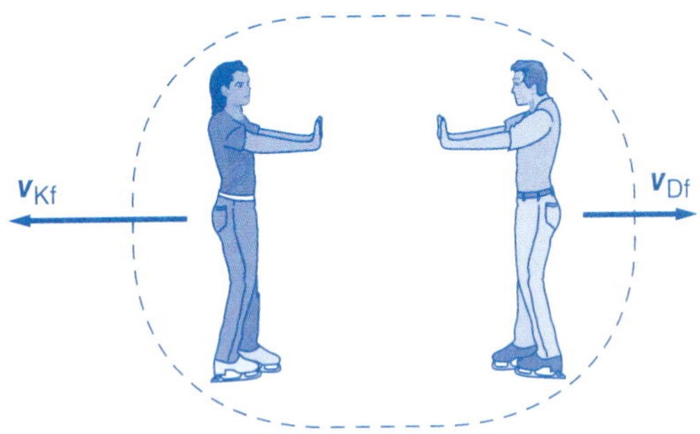

b. −1.50

c. Keshia, who has the smaller mass, has the greater speed.

d. The forces were equal and opposite.

75. cue ball: 2.8 m/s; eight ball: 2.8 m/s

76. 11 m/s

77. 0.22 m/s in the original direction

Applying Concepts

78. A force, $\mathbf{F}$, exerted on an object over a time, Δt, causes the momentum of the object to change by the quantity $\mathbf{F}\Delta t$.

79. From 4 to 6 s, the object moves with a constant, positive velocity. From 8 to 10 s, the object is at rest. After 11 s, the object moves with a constant, negative velocity.

80. Yes, if the smaller force acts for a long enough time, it can provide a larger impulse.

81. You should move your hands in the same direction the ball is traveling to increase the time of the collision, thereby reducing the force.

82. The bullet is in the rifle a longer time, so the momentum it gains is larger.

83. When the gas pistol is fired in the opposite direction, it provides the impulse needed to move the astronaut toward the spaceship.

84. Consider the system to be the ball, the wall, and Earth. The wall and Earth gain some momentum in the collision.

85. Consider the two trucks to be an isolated system. If the two trucks had equal masses, they would have moved off at half the speed of the moving truck. Thus, the moving truck must have had a more massive load.

86. In each case, consider the bullet and the block of wood to be an isolated system. Momentum is conserved, so the momentum of the block and bullet after the collision equals the momentum of the bullet before the collision. The rubber bullet has a negative momentum after impact, with respect to the block, so the block's momentum must be greater in this case.

Mixed Review

87. 60.0 kg·m/s; 20.0 m/s

88. a. 2.12×10^4 kg·m/s

b. 313 N

89. a. She spins around the center of mass of her body, first in the tuck position and then also as she straightens out.

b. giant swing (A), straight (C), tuck (B)

c. tuck (greatest), straight, giant swing (least)

90. a. 2.35×10^4 kg·m/s

b. 2.6×10^4 N

c. The force is exerted by the track through friction.

91. 8.39 m/s

92. a. Before: $m_w = 50.0$ kg
$m_c = 10.0$ kg
$v_i = 5.0$ m/s
After: $m_w = 50.0$ kg
$m_c = 10.0$ kg
$v_{wf} = 7.0$ m/s
$v_{cf} = ?$
See Online Solutions Manual for sketch.

b. −5.0 m/s, or 5.0 m/s west

93. a. 1.5×10^2 kg•m/s down

b. -1.5×10^2 kg•m/s up

c. 3.0×10^3 N

d. 5.88×10^2 N; the force is about five times the weight.

Thinking Critically

94. Consider the two balls to be an isolated system. Dotted lines show that the changes of momentum for each ball are equal and opposite: $\Delta(m_A\mathbf{v}_A) = \Delta(-m_B\mathbf{v}_B)$. Because the masses have a 3:2 ratio, a 2:3 ratio of velocity changes will compensate.

95. The student and the stool would spin slowly in the direction opposite to that of the wheel. Without friction there are no external torques. Thus, the angular momentum of the system is not changed. The angular momentum of the student and stool must be equal and opposite to the angular momentum of the spinning wheel.

96. a. Before: $m_A = 92$ kg
$m_B = 75$ kg
$m_C = 75$ kg
$v_{Ai} = 5.0$ m/s
$v_{Bi} = -2.0$ m/s
$v_{Ci} = -4.0$ m/s
After: $m_A = 92$ kg
$m_B = 75$ kg
$m_C = 75$ kg
$v_f = ?$
See Online Solutions Manual for sketch.

b. 0.041 m/s

c. Yes. The velocity is positive, so the football crosses the goal line for a touchdown.

Writing in Physics

97. The change in a car's momentum does not depend on how it is brought to a stop. Thus, the impulse also does not change. To reduce the force, the time over which a car is stopped must be increased. Using barriers that can extend the time it takes to stop a car will reduce the force. Flexible, plastic containers filled with sand often are used.

98. There are two ways an airbag reduces injury. First, an airbag extends the time over which the impulse acts, thereby reducing the force. Second, an airbag spreads the force over a larger area, thereby reducing the pressure. Thus, the injuries due to forces from small objects are reduced. The dangers of airbags mostly center on the fact that airbags must be inflated very rapidly. The surface of an airbag can approach the passenger at speeds of up to 322 km/h (200 mph). Injuries can occur when the moving bag collides with the person. Systems are being developed that will adjust the rate at which gases fill the airbags to match the size of the person.

Cumulative Review

99. −6.0 N

100. 4.3×10^7 m

101. $\alpha = 8.33$ rad/s^2; $\omega = 16.7$ rad/s; $I = 1.44$ kg•m^2

Standardized Test Practice

student text p. 261

Multiple Choice

1. B
2. C
3. D
4. D
5. C
6. C
7. A
8. A

Free Response

9. $F\Delta t = m\Delta v$

$= (12.0\text{ kg})(20.0\text{ m/s} - 0.00\text{ m/s})$

$= 2.40\times10^2\text{ kg}\cdot\text{m/s}$

$= 2.40\times10^2\text{ N}\cdot\text{s}.$

The rock's impulse on the ground is 2.40×10^2 N•s. Thus, the ground's impact on the rock is -2.40×10^2 N•s.

Rubric

The following rubric is a sample scoring device for free response questions.

Points	Description
4	The student demonstrates a thorough understanding of the physics involved. The response may contain minor flaws that do not detract from the demonstration of a thorough understanding.
3	The student demonstrates an understanding of the physics involved. The response is essentially correct and demonstrates an essential but less-than-thorough understanding of the physics.
2	The student demonstrates only a partial understanding of the physics involved. Although the student may have used the correct approach to a solution or may have provided a correct solution, the work lacks an essential understanding of the underlying physical concepts.
1	The student demonstrates a very limited understanding of the physics involved. The response is incomplete and exhibits many flaws.
0	The student provides a completely incorrect solution or no response at all.

CHAPTER 10

Work, Energy, and Machines

About the Photo

Have students examine the photo and identify all the machines. Ask students what the advantages are of machines building cars over humans doing so by hand. Students may say that machines can tighten bolts and screws better than human hands because they can apply more torque or force. Ask students to make lists of machines that they use in their everyday lives. Be sure to point out that objects such as bottle openers, pulleys, screw drivers, etc. are also considered simple machines.

Chapter Overview

The relationships among force, displacement, work, and energy are developed through hands-on activities, sketches, and equations. Work is the transfer of energy by mechanical means and is defined as the product of force magnitude and displacement in the direction of the force. Machines transfer energy with greatest advantage through the judicious trade-off of force and displacement. Ideal and real machines are differentiated and efficiencies are analyzed and calculated for simple and compound machines.

Before students study the material in this chapter, they should study:

- Accelerated motion in one dimension
- Circular motion
- Newton's laws of motion
- Normal force
- Uniform motion in one dimension
- Vectors vs. scalars

To solve the problems in this chapter, students will need a strong grasp of:

- Scientific notation
- Significant figures
- Sine, cosine, and tangent
- Solving linear equations

Introduce the BIG IDEA

Show students a video of a karate expert breaking a piece of wood with her hand. What happened to the kinetic energy of the karate expert's hand? It went into breaking the fibers of the wood.

Go online!
connected.mcgraw-hill.com

Your online experience includes:

- **Video**
- **Concepts In Motion**
- **Virtual Investigations**
- **All Lab, Applying Practices activities, and PBL worksheets**
- **Online Practice**
- **Personal Tutor**
- **Multilingual eGlossary**
- **Presentation**

Use the LaunchLAB

In *Energy and Falling,* students observe how objects of different sizes and masses that are dropped from different heights affect the size of the resulting impact crater in the sand into which the objects are dropped.

Use Physics T.V.

Video

MACHINES What happens when simple machines are combined to do work? Explore the physics of everyday gadgets and learn more about energy, work, and simple machines.

1 Introduce

Bellringer

***WARNING:* Wear goggles.**

Transfer of Energy You can mount a wheel and axle on the demo table. The next step is to hold the wheel while you attach a 500-g block to a string around the part of the axle outside the wheel. Ask students to predict what will happen when a 100-g or 200-g block is attached. It's a good idea to have plenty of string on the wheel so the smaller mass can fall a large distance. The wheel speeds up fastest with the 500-g block, much more slowly with the 200-g block, and even more slowly with the 100-g block. Explain that more work is done (and the wheel's final kinetic energy is greater) when blocks of larger mass fall through the same distance.
OL EL Visual-Spatial

Tie to Prior Knowledge

Force and Motion This chapter brings together concepts of force and motion. It starts by suggesting that a quantity other than momentum is important when objects interact. The concept of kinetic energy is developed by applying equations of motion to the definition of a new quantity called work.

2 Teach

Work and Energy

student text pp. 264–270

Concept Development

MAIN IDEA Have students collect soda cans and bring empty cans to class. Build a simple pile driver to crush the cans using a large mass that may slide on a set of vertical rails. Have students explore what happens to their cans when the mass is dropped from various heights. Determine the minimum height needed to completely crush the can.

Ask students what caused the can to become crushed. force of the falling mass Ask students for what distance this "crushing" force acted on the can. about the height of the can Inform students that the falling mass did work on the can because an applied force caused a displacement. Ask students how the height of the mass changed the crushing of the can. The higher the mass, the more it crushed. Connect with the students that the higher drop level gave the falling mass more kinetic energy, and therefore, it could be transformed into more work to crush the can.

Concept Development

Keeping Track of Work When work is done on a system, the work-energy theorem implies that there is a change in the energy of the system. For this relationship to hold, the total work done in an interaction must be considered. In addition, the system must not also do work on its environment and no other types of energy transfer should take place, such as heating or cooling. You can use the following examples to illustrate how the work-energy theorem holds in various situations.

- **Pushing Against a Wall** Force is applied when pushing, but the wall applies an equal and opposite force. There is no displacement, so work done by the pushing force, work done by the reaction force, and total work are zero. Kinetic energy does not change because the wall remains stationary.
- **Lifting at Constant Speed** The net work and change in kinetic energy are zero. In a system including the object and Earth, however, doing work to lift an object transfers energy to that system in the form of gravitational potential energy.
- **Free Fall** When the only force acting on an object dropped from rest is the gravitational force, then the work done on the object by this force equals the increase in the object's kinetic energy. This increase is given by $W = \Delta E$. The work is $W = Fd = (mg)(d) = mgd$, where d is the distance the object falls. The change in energy is $\Delta E = \Delta KE = KE_f - KE_i = \frac{1}{2}mv_f^2 - 0 = \frac{1}{2}mv_f^2$. So, $mgd = \frac{1}{2}mv^2$.

Reinforcement

Force in the Direction of Motion Activity You can draw a right triangle on the board with the legs horizontal and vertical, and the hypotenuse diagonal. It might be useful to review the definitions of the sine, cosine, and tangent of the angle θ and to establish that the horizontal and vertical legs are the perpendicular components of a force represented by the hypotenuse. If the motion is along the horizontal leg, then the force in that direction is the horizontal component of the triangle. The angle θ is adjacent to the component in the direction of motion. You can ask students which trigonometric function is equal to the adjacent side over the hypotenuse of the right triangle. the cosine function You can then ask what the magnitude of the force component is along the direction of motion. $F_x = F\cos\theta$
OL EL Visual-Spatial

Critical Thinking

Friction Force and Normal Force You can ask students to find a mathematical relationship for the magnitude of the component of force in the direction of motion when a uniform sled is dragged horizontally at constant speed by a cord attached to its center. The cord makes an angle θ with the horizontal, m is the mass of the sled, μ is the coefficient of sliding friction, F is the magnitude of the force along the cord, F_N is the magnitude of the normal force, and g is the strength of the gravitational field. The normal force between the sled and the surface is the weight of the sled minus the vertical component of force along the cord. The magnitude of the horizontal component of force along the cord must be equal to the magnitude of the force of friction when the object is sliding at constant speed. So, $F \cos \theta = \mu F_N = \mu(mg - F \sin \theta)$.

AL Logical-Mathematical

Content Background

Determining Work for a Variable Force The argument presented in the text for the equivalence of work done and change in kinetic energy assumes that force is constant. This does not need to be the case. Methods of calculating work due to a variable force include finding the area under the force magnitude-v.-displacement graph or performing an integral on the force magnitude-v.-displacement function over the range of displacement.

Discussion

Problem How much work is done by a person on a box when that person pushes the box with a force of 184 N, 10.0 m across a floor?

Response $W = Fd = (184\ \text{N})(10.0\ \text{m}) = 1840\ \text{J}$. BL

REAL-WORLD PHYSICS

Automobiles and the Work-Energy Theorem Imagine that a 1200-kg vehicle speeds up from rest to 20.0 m/s in 80.0 m on level pavement. The change in kinetic energy is

$$\frac{1}{2}mv^2 = \frac{1}{2}(1200\ \text{kg})(20.0\ \text{m/s})^2$$
$$= 2.4\times10^5\ \text{J}$$

This energy is transformed into thermal energy through friction. You can show students how to use the work-energy theorem to find the braking distance from 20.0 m/s with locked wheels if the coefficient of friction between the tires and road on dry pavement is $\mu = 0.70$. The magnitude of the force exerted by the road to stop the car is $F = (0.70)(1200\ \text{kg})(9.8\ \text{N/kg}) = 8.2\times10^3\ \text{N}$.

The stopping distance is $\frac{(2.4\times10^5\ \text{J})}{(8.2\times10^3\ \text{N})} = 29\ \text{m}$. OL

ADDITIONAL IN-CLASS EXAMPLE

Use with Example Problem 1.

Problem When a club head strikes a 46-g golf ball, the ball picks up 43 J of kinetic energy. The force of the club head on the ball is not constant. Rather, in the short collision time of the club head and ball, the force builds to a peak and then diminishes. But for the purposes of the calculation in this problem, assume that the average force magnitude, which you can treat as constant, equals 2300 N. Over what distance is the club head in contact with the ball?

Response

$$d = \frac{W}{F} = \frac{43\ \text{J}}{2300\ \text{N}} = 1.9\times10^{-2}\ \text{m}$$

Use the MiniLAB

In *Force Applied at an Angle,* students pull a block along a table for various angles between the pulling force and the table.

ADDITIONAL IN-CLASS EXAMPLE

Use with Example Problem 2.

Problem How much work is done in pushing a tall box 15 m with a force, the magnitude of which is 4.0×10^2 N, that is applied slightly upward at an angle of 10.0° from horizontal?

Response

$W = Fd \cos \theta = (4.0\times10^2\ \text{N})(15\ \text{m}) \cos 10.0° = 5900\ \text{J}$

PHYSICS CHALLENGE ACTIVITY

Graph of Force Magnitude Verse String Angle Refer to the MiniLAB *Force Applied at an Angle*. You can ask students how the friction force magnitude is related to the coefficient of friction and the string angle. The friction force magnitude equals the coefficient of friction, μ, times the normal force magnitude F_N. The normal force magnitude equals the difference of the magnitude (mg) of the block's weight and the magnitude of the vertical component of the pulling force. So, $F \cos \theta = \mu F_N = \mu(mg - F \sin \theta)$ and $F = \frac{\mu mg}{(\cos \theta + \mu \sin \theta)}$. You can then ask students to calculate the value of μ from the equation $\mu = \frac{F}{mg}$, which is obtained by solving for μ in the equation $F = \frac{\mu mg}{(\cos \theta + \mu \sin \theta)}$, into which is substituted $\theta = 0$ and the value of F corresponding to that angle, as measured in the MiniLAB. Students can then graph F as a function of θ. That is, $F(\theta) = \frac{\mu mg}{(\cos \theta + \mu \sin \theta)}$. If $\mu = 0.6$, then the graph starts at $F = 6$ N when $\theta = 0$, has a minimum close to 5 N at about 30°, and increases to $F = 10$ N when $\theta = 90°$. **AL** **Logical-Mathematical**

Use Figure 4

Constant Force and Changing Force You can ask students to give two examples of forces that do not vary with position, such as the force illustrated on the left in **Figure 4.** A car under steady acceleration and an object in free fall without air drag near Earth's surface are examples of objects under constant net forces.

Another example of a force that varies linearly with displacement is force due to a spring. The slope of the force magnitude-v.-displacement graph is the spring constant, k. You can ask students: How can the work done to displace the spring be found for any distance x? The force magnitude is related to position by $F = kx$. The work is the area under the triangle. So the general case is $W = \left(\frac{1}{2}\right)(kx)(x)$, or $W = \frac{1}{2}kx^2$. See Chapter 14, Section 1 for more details. **OL**

Power

student text pp. 271–273

Identify Misconceptions

Energy Versus Power Students often confuse energy and power. You can, first, ask students if they can tell how big and powerful a machine or appliance is by the number of joules of energy it uses. No; a machine can use any number of units of energy, given enough time. What rating of a machine or appliance is associated with its size? Its power rating, in watts or horsepower; this rating indicates how fast the device can use or transfer energy. **OL** **Logical-Mathematical**

DIFFERENTIATED INSTRUCTION

Physically Challenged Finding the mechanical power output for travel up a wheelchair-accessible ramp requires three measurements: change in elevation of the ramp, mass of the chair plus user, and time to travel up the ramp. Measure and record these values for a building entrance or other site. Time the student's travel up the ramp from the lowest to the highest point. Insist that students travel in an ordinary manner. Then have the students calculate the power output using the relationship $P = \frac{mgh}{t}$ *where* P is power in watts, m is total mass in kilograms, g is the gravitational field strength, and h is the elevation change of the ramp. **OL** **Kinesthetic** **EL**

ADDITIONAL IN-CLASS EXAMPLE

Use with Example Problem 3.

Problem A net force, the magnitude of which is 2800 N, accelerates a 1250-kg vehicle for 8.0 s. The vehicle travels 80.0 m during this time. What power output does this represent?

Response $P = \frac{W}{t} = \frac{Fd}{t} = \frac{(2800\text{ N})(80.0\text{ m})}{8.0\text{ s}} = 28\text{ kW}$

Use the PhysicsLAB

Students will calculate the work done and the power generated in *Stair Climbing and Power*. The key concept reinforced by the lab is that the same amount of work done in a shorter period of time results in greater power.

PHYSICS PROJECT ACTIVITY

Automobile Mass and Fuel Economy Several students should work together to obtain mass and fuel-economy data for at least five automobiles of varying size. Fuel economy should be converted to kilometers per liter of fuel. Each participant should carefully record distance traveled and fuel volume added for three or four fueling station visits. After a few weeks, participants should compile their average data and plot fuel economy versus vehicle mass. They should report whether mass influenced fuel economy and explain why or why not.

Results will vary. All other relevant factors being equal, the greater the mass of the vehicle, the more energy (supplied by the gasoline in its gas tank) is required for an increase in the vehicle's energy. This increase in energy can be in the form of kinetic energy (when the vehicle's speed increases) or in the vehicle-Earth system's gravitational potential energy (when, say, the vehicle climbs a hill). This study does not consider or control for other factors besides vehicle mass that could affect fuel economy. Thus, further study would be required before any confident conclusion is made regarding the relationship between vehicle mass and fuel economy. OL Kinesthetic EL

Applying Physics

Cycling You can ask students the following questions: What is the average distance traveled by and the average energy output of a Tour de France rider in 6 hours?

$d = (8.94\ \text{m/s})(2.16\times10^4\ \text{s}) = 1.93\times10^5\ \text{m} = 193\ \text{km}$;

$E = (1.00\ \text{kW})(2.16\times10^4\ \text{s}) = 2.16\times10^7\ \text{J}$. OL

3 Assess

Assess the MAIN IDEA

Forces that Work On the front board, draw a diagram of a mass being pushed up a rough surface incline with the pushing force parallel to the incline surface. Ask students to draw and label a free-body diagram of all forces on the mass. F_{push}, F_f, $mg \sin \theta$, F_N, and $mg \cos \theta$. Ask students to determine which forces do work and which do not. F_{push}, F_f, and $mg \sin \theta$ do work; the others do not. Ask students why F_N and $mg \cos \theta$ do not do work. They are perpendicular to the direction of motion.

Check for Understanding

Work and Power in Lifting José lifts a 20.0-kg block to a height of 2.0 m in 5.0 s. Sue lifts 30.0 kg to a height of 1.5 m in 8.0 s. You can ask students to compare the work and power of each student and to explain their answers. José's work is $(20.0\ \text{kg})(9.8\ \text{N/kg})(2.0\ \text{m}) = 390\ \text{J}$. Sue's work is $(30.0\ \text{kg})(9.8\ \text{N/kg})(1.5\ \text{m}) = 440\ \text{J}$. José's power is $\frac{390\ \text{J}}{5.0\ \text{s}} = 78\ \text{W}$, and Sue's power is $\frac{450\ \text{J}}{8.0\ \text{s}} = 55\ \text{W}$. OL Logical-Mathematical

Extension

Force Direction Advantage You can ask students to make a quick sketch to show why it is much easier to pull a wheelbarrow up steps than to push it up. When you push a wheelbarrow, a component of the force will be directed downward unless you can get in the awkward position necessary to push it upward from underneath. When pulling the wheelbarrow, however, a component of the force is naturally upward, which makes it easier to pull the wheelbarrow up the steps. AL Visual-Spatial EL

Reading Checks and Caption Checks

Caption Check, *student text p. 264*
Sample Answer: when you push a saltshaker across a tabletop

Reading Check, *student text p. 265*
2 J

Reading Check, *student text p. 266*
You do positive work because your force is exerted in the direction of displacement and your friend does negative work because the force applied is in the direction opposite of displacement.

Reading Check, *student text p. 266*
Sample Answer: Suppose you pull straight up on a fishing line that has bait and a sinker attached. If the bait, fishing line, and sinker have a mass of 0.15 N and you pull it straight up 8.0 m, you do 1.2 J of work.

Caption Check, *student text p. 266*
$W_{friend} - W_{you} = W_{total}$; the answer would be a positive number.

Practice Problems
student text p. 268

1. a. Because $W = Fd$, doubling the force would double the work to 1.35 J.
b. Because $W = Fd$, halving the distance would cut the work in half to 0.68 J.

2. a. 2.9×10^4 J
b. 5.8×10^4 J

3. a. 6.0×10^2 J
b. 5.9×10^3 J

4. 1.1×10^2 J

Practice Problems
student text p. 269

5. 4.92×10^3 J

6. 6.5×10^3 J

7. a. 903 J
b. −903 J

8. 6.54×10^3 J

9. a. 6.9×10^3 J
b. -1.5×10^4 J

Practice Problems
student text p. 272

10. 1.15×10^3 W; 1.15 kW

11. a. 348 W
b. 696 W

12. 0.63 kW

13. 1.3×10^5 N

14. 5.7 min

Section 1 Review
student text p. 273

15. Kinetic energy is proportional to the square of the velocity, so doubling the energy doubles the square of the velocity. The velocity increases by a factor of the square root of 2, or 1.4.

16. 8×10^2 J

17. 1.9×10^3 J; See Online Solutions Manual for force-displacement graph.

18. 3.46×10^3 J

19. 4.4 J

20. No, work is not a function of time. However, power is a function of time, so the power required to lift the book does depend on how fast you raise it.

21. 3.4×10^4 W

22. 6.0×10^2 kg

23. Both do the same amount of work. Only the height lifted and the vertical force exerted count.

24. Since work is the change in kinetic energy, calculate the work done by each force. The work can be positive, negative, or zero, depending on the relative angles of the force and displacement of the object. The sum of the three works is the change in energy of the system.

1 Introduce

Bellringer

Wind-Up Toys Try to obtain several small wind-up or battery-powered toys. Display them on the demo table. You can run each one and ask students to describe how each toy takes advantage of its power source. If possible, show how the wind-up mechanism or battery-powered motor drives the moving parts of the toy. Answers will depend on which kinds of toys are displayed. In general, students should be able to note that there is a mechanical connection between the power source and the arms, legs, or other moving parts of the toys. These connections allow the parts to move. BL EL Visual-Spatial

Tie to Prior Knowledge

Force, Distance, and Work This chapter builds on the understanding of the relationship between these three quantities established in Section 1. In the ideal case, input work is equal to output work. Students should now know that work is equal to force magnitude times distance. Simple machines trade off these factors of force magnitude and distance to their advantage when doing a mechanical task. Concepts covered during student's study of rotational motion also are helpful in describing the physics of many simple machines.

2 Teach

Benefits of Machines

student text pp. 274–276

Concept Development

MAIN IDEA Separate students into groups of four, and have students do internet research and make sketches of the following machines: levers, gears, bevel gears, worm gears, monkey wrench, toothed peg wheels, chains and belts, rack and pinion, pulley, ratchet, cam, wedge, wheel and axel. For each machine, the groups should identify whether the machine changes the magnitude of the force and/or the direction of the force exerted. Have students brainstorm for experiences in their life during which they may have used such machines.

Concept Development

Levers and Torque The fulcrum of a lever is the axis of rotation. The torque applied to the effort side is the same as the torque developed on the resistance side, $F_e d_e = F_r d_r$.

Use Figure 11

Pulley Advantage The basic way to determine the mechanical advantage of a pulley or system of pulleys is to count the number of lines supporting the load.

Identify Misconceptions

Reasons to Use Machines Students might not understand that there are more reasons for using machines than just reducing the force magnitude necessary to perform some task. Sometimes a machine simply changes the direction of an effort force without multiplying it. In fact, other machines have exactly the opposite effect—they trade a larger effort force so the displacement of the effort force is increased. You can ask students to come up with an example of each of these kinds of machines. A single pulley with one line supporting the load reverses the effort force. A broom is a type of lever where the distance the region near the handle moves is much smaller than the distance the broom head moves. The mechanical advantage of a broom might be 0.33 or less. OL Logical-Mathematical

Content Background

Ideal Mechanical Advantage Computations The ideal mechanical advantage (*IMA*) of a simple machine depends on the physical geometry of the device. This is the factor by which effort force would be multiplied to obtain load force if there were no energy losses. For example:

Lever $IMA = \frac{L_e}{L_r}$, where L_e and L_r are the lengths of the effort arm and resistance arm, respectively.

Inclined Plane $IMA = \frac{d_e}{d_r}$, where d_e and d_r are the length of the ramp and height of the ramp, respectively.

Wheel and Axle $IMA = \frac{r_w}{r_a}$, where r_w and r_a are the radii of the wheel and the axle, respectively.

Reinforcement

Mechanical Advantage You can hold up a screwdriver and ask students which factor is most important to loosen a tight screw: a longer blade, a longer handle, or a larger-diameter handle. The correct answer is a larger-diameter handle. You can increase the *IMA* and the applied force to the screw by increasing the distance from your hand to the axis of the screwdriver. OL

Use the PhysicsLAB

In *Lifting with Pulleys,* students set up and investigate pulley systems and determine their efficiencies and *IMA*s.

REAL-WORLD PHYSICS

Thomas Jefferson's Dumbwaiter Thomas Jefferson is noted for devices he created during the construction of Monticello. The illustration below is a diagram of a Monticello dumbwaiter. In this design, the food storage box would carry the contents to the dining room 4.0 m overhead by turning the handle 24 complete turns. The handle moved 0.30 m per turn. You can ask students to calculate the ideal mechanical advantage of the dumbwaiter.

$$IMA = \frac{d_e}{d_r} = \frac{(24)(0.30\ \text{m})}{4.0\ \text{m}} = 1.8$$

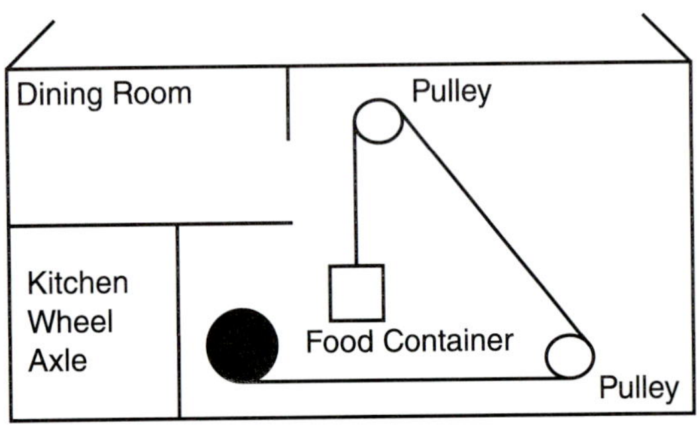

Compound Machines

student text pp. 277–280

Critical Thinking

Compound Efficiency You can ask students the following question: If two simple machines are connected in a series, how should the efficiencies of the two individual machines be combined to obtain the net efficiency of the whole machine? If the first could transmit only $e_1 = 80\%$ of the work, followed by the second transmitting $e_2 - 70\%$, then 70% of 80% of the work appears at the output. This is a net of $e = e_1e_2 = 56\%$. You can then ask, what is the net efficiency of *n* such machines in series? $e = e_1e_2e_3 \ldots e_n$

AL Logical-Mathematical

Quick DEMO

Pinching Tools

Estimated Time 5 minutes

Materials wire cutters, bolt cutters, or other long-handled tool, scrap wire or other material to test tool

Procedure You can start by holding up one of these tools and show how it easily cuts the scrap material. You can ask the class how these devices can apply so much force. Distance moved at the end of the handle is several times the distance moved by the pinching end, so resistance-force is much larger than effort-force. If the tool being demonstrated is a cutting tool, you can ask students if they can identify a second simple machine on the tool, other than the levers formed by the handles. The cutting heads on the tool are wedges. That is, they are small inclined planes that further increase force when splitting materials.

Discussion

Problem A locking pliers is a device where the long handle and a second release lever operate another lever system connected to the moveable jaw of the pliers. (If possible, obtain and demonstrate such a tool.) What are the features and advantages of this tool?

Response The set of levers driving the jaws on a pair of locking pliers can be pushed with the main handle into a position where they "snap in" under tension. They will not "snap out," relieving the user of the need to maintain force on the handle. The user pushes the release lever to push the jaw lever out of the locking position. OL

DIFFERENTIATED INSTRUCTION

Visually Impaired Most of the time, measurements of effort and resistance distances can be made by touch. A student who is visually impaired should have a Braille ruler or a plastic ruler with raised markings. The lines on most metersticks have indentations sufficient for tactile use. Spring balances could be similarly adapted for force measurements. As an activity, have students design and construct pulley systems by touch. Some guidance might be needed to get the students started. Data and analysis could involve distance and force measurements under several different load masses. Students should be able to feel and count the number of lines holding the load to determine ideal mechanical advantage. **BL** **Visual-Spatial** **EL**

Use the MiniLAB

In *Wheel and Axle,* students lift a 500-g object attached to a string wrapped around the wheel (of a wheel and axle) by pulling on another string wrapped around the axle. Students contrast the magnitude of the pulling force with the magnitude of the object's weight. Students then contrast the distance the pulling hand moves with the distance the object is lifted.

Use Models

Paper Mechanism Activity Have students construct simple models of mechanisms of their choice using stiff paper strips, paper circles, and metal fasteners. They should be limited only by their imaginations, but they should be required to make a clear argument that their models are representative of real, useful devices. They also must analyze all simple machines at work in the models by taking measurements, computing ideal mechanical advantage, and explaining the qualitative advantages of the mechanisms.

Examples of what might be constructed include a system of two or three gears or pulleys using paper circles or a scale model of a human limb using paper strips. **OL** **Kinesthetic** **EL**

Concept Development

Mounted Bicycle It is much easier for students to visualize the action of a bicycle if one is mounted for demonstration. If possible, use a stationary bike. If you use a two-wheel bicycle, then you might want to use a stand to hold the bike stationary. With a meterstick or other measuring device, you can find the gear, wheel, and pedal radii. Next you can perform a calculation of *IMA* using the appropriate equation from the student book. **OL** **Visual-Spatial** **EL**

ADDITIONAL IN-CLASS EXAMPLE

Use with Example Problem 4.

Problem A bicycle has a pedal radius of 15.0 cm, a front-gear radius of 5.57 cm, a rear-gear radius of 4.00 cm, and a rear-wheel radius of 35.6 cm.

a. What is the *IMA* of the bicycle?

b. How many times will the rear wheel turn for one complete turn of the pedal?

Response

a. $IMA = \left(\frac{4.00\text{ cm}}{5.57\text{ cm}}\right) \times \left(\frac{15.0\text{ cm}}{35.6\text{ cm}}\right) = 0.303$

b. $\frac{5.57\text{ cm}}{4.00\text{ cm}} = 1.39\text{ turns}$

PHYSICS CHALLENGE ACTIVITY

Law of the Machine You can have the students set up a simple pulley system with an *IMA* of 2. Using a range of masses from 50.0 g to 1.0 kg for loads and a spring balance for measuring effort force magnitude, you can have them carefully record the reading on the spring balance for each resistance force magnitude. Next, you can plot the effort force magnitude versus the load force magnitude on a graph. Students can then fit the data obtained to the equation $F_e = aF_r + b$, where F_e is the effort force magnitude, F_r is the resistance force magnitude, and a and b are constants. You can ask them what the meanings of the constants a and b are. The constant a is the inverse of the mechanical advantage (*MA*) and the slope of the graph, and b is the "starting" friction of the machine. This relationship is sometimes called the law of the machine. AL Logical-Mathematical

Use an Analogy

Automobile Transmission Ask students to imagine an object rolling down from the top of a series of three hills. The hills have progressively less slope, with the lowest being nearly flat. Ask students how these hills are like a car's transmission. The highest hill allows maximum downward force with the least forward motion, and the lowest hill applies the least force in the forward direction. The middle hill is an intermediate step. The highest hill is analogous to first gear, and the lowest hill corresponds to a higher gear of an automobile. OL

REAL-WORLD PHYSICS

Vehicle Jacks It seems amazing that a small person can lift part of a motor vehicle with one hand. A device called a jack makes this possible. You can show one or two kinds of vehicle jacks and ask students to explain how they could determine the mechanical advantage of the jack. An experimental method may work best, if it is practical. Here are some possible results. A person might apply a force of 250 N over a distance of 0.5 m with the car moving upward by only 0.1 m. Such a jack would have an ideal mechanical advantage of $\frac{d_e}{d_r} = \frac{0.5\text{ m}}{0.1\text{ m}} = 5$. OL

3 Assess

Assess the MAIN IDEA

Exploring Everyday Machines Assign students into groups of four, and give each group a machine to "dissect." Bring in items like an egg beater, garlic press, tongs, pulley, wheel barrow, screw driver, corkscrew, a compact can crusher, a hammer and wedge, nail clippers, scissors, a windup toy, an old windup clock, a bicycle. Have each group use/observe the machine from many angles and identify the elements of the machines as researched earlier. Students should discuss in groups how the parts relate and affect each other, and then sketch a detailed diagram of the machine. Have students add arrows and notes to indicate the directions of motion. Have the groups identify whether the machine changes the force and/or the direction of the exerted force.

Check for Understanding

Input and Output Work Ask students what is wrong with this statement: "Using a simple machine increases the amount of work that can be applied to a task." Output work from a simple machine never can exceed input work. A simple machine simply rearranges force, distance, or both so the task can be done more easily. OL

Reteach

Pulley System *IMA* You can use a pulley system with an *IMA* of 3 to lift a 1.0-kg object. You can ask students to note that the string must be pulled down by 60 cm to lift the object by only 20 cm. You can then ask students whether the *IMA* of the system can be determined without measuring the distances. Yes; the *IMA* is the same as the number of support strings. OL

Reading Checks and Caption Checks

Reading Check, *student text p. 275*
6 N

Reading Check, *student text p. 276*
$IMA = 6$

Caption Check, *student text p. 277*
Sample Answers: lever, see-saw; pulley, flag pole pulley; wheel and axle, door knob; inclined plane, wheel-chair ramp; wedge, human teeth; screw, screw-type soft drink lid

Reading Check, *student text p. 277*
Both are designed to make a task easier. Compound machines are composed of simple machines.

Reading Check, *student text p. 278*
The effort force is the force you exert on the pedal. The resistance force is the force the gear exerts on the chain.

Reading Check, *student text p. 278*
Both *MA* and *IMA* are dimensionless quantities and have no units.

Reading Check, *student text p. 280*
Gears are needed to apply different amounts of forces for different requirements.

Practice Problems

student text p. 280

25. $IMA = 0.225$

$MA = 0.214$

$F_r = 33.2$ N

$d_e = 3.15$ cm

26. a. 4.0

b. 1.5

c. 38%

27. a. 1.82

b. 91.0%

28. a. 6.0

b. 1.7×10^2 N

29. 0.81 m

Physics Challenge

student text p. 276

1. The work done in lifting is $F_g d = mgd$. Therefore, the power is

$$P_{lift} = \frac{W}{t} = \frac{F_g d}{t} = \frac{mgd}{t}$$

$$= \frac{(0.25\ m^3)(1.00\times10^3\ kg/m^3)(9.8\ N/kg)(25\ m)}{1.0\ s}$$

$$= 6.1\times10^4\ W$$

$$= 61\ kW$$

2. The work done in increasing the pump's kinetic energy is $\frac{1}{2}mv^2$. Therefore,

$$P_{KE} = \frac{W}{t} = \frac{\Delta KE}{t} = \frac{\left(\frac{1}{2}mv^2\right)}{t} = \frac{mv^2}{2t} =$$

$$\frac{(0.25\ m^3)(1.00\times10^3\ \frac{kg}{m^3})(8.5\ \frac{m}{s})^2}{(2)(1.0\ s)}$$

$$= 9.0\times10^3\ W = 9.0\ kW$$

3. $$e = \frac{W_o}{W_i} \times 100$$

$$= \frac{\frac{W_o}{t}}{\frac{W_i}{t}} \times 100$$

$$= \frac{P_o}{P_i} \times 100 \text{ so, } P_i = \frac{P_o}{e} \times 100$$

$$= \frac{(9.0\ kW + 61\ kW)}{80} \times 100$$

$$= 8.8\times10^4\ W = 88\ kW$$

Section 2 Review

student text p. 281

30. a. wheel and axle; increases the size of the force

b. lever; increases the size of the force and changes the direction of the force

c. wedge; increases the size of the force and changes the direction of the force

d. lever; increases the size of the force and changes the direction of the force

31. 5.2×10^3 N

32. 18

33. Either *MA* increases while *IMA* remains the same, or *IMA* decreases while *MA* remains the same, or *MA* increases while *IMA* decreases.

34. a. large

b. Small, because less chain travel, hence few pedal revolutions, will be required per wheel revolution.

c. smaller

A CLOSER LOOK

disAdvantage?

Prosthetic Running Legs

Background

The specific running prosthetic described here is called the Cheetah. In 2008, controversy arose when a double amputee was denied eligibility to compete in the 2008 Olympics in Beijing. The ruling was based on a study which claimed that, upon impact with a running track, the Cheetah prosthetic returned more energy than a natural human ankle joint—and thus gave the runner an advantage over non-amputees. The ban was later lifted when the International Association of Athletics Federations determined that there was insufficient evidence showing that the Cheetah gave runners an overall advantage. A second study has revealed that runners with prosthetics are at a disadvantage in terms of force production.

Teaching Strategies

- Point out the challenge of making rules that attempt to create an equal playing field for amputees with many different types of residual limbs. Have students research the different types of amputees—lower limb, upper limb, bilateral, and unilateral—and discuss how governing bodies might ensure that a race that includes all different kinds of athletes is fair.
- Encourage students to find out more about the different types of prosthetics that athletes use for different sports. The racing prosthetics described here are ideal for track and field events but are not suited for walking, running marathons, or participating in sports such as baseball or basketball.
- Have students find out about biomechatronics, a field in which scientists incorporate sensors and robotic mechanisms in artificial limbs and other prosthetics.

GOINGFURTHER >>>

Expected Outcomes Students should discuss each type of athlete's advantages and disadvantages. Students might wish to compare this debate to other sports controversies, such as the fairness of swimmers being allowed to compete in high-tech racing suits, or the use or ban of performance-enhancing drugs and supplements in all types of athletic competitions.

SECTION 1

Mastering Concepts

35. joules

36. No, the force of gravity is directed toward Earth and is perpendicular to the direction of displacement of the satellite.

37. Only gravity and an upward, normal force act on the object. No work is done because the displacement is perpendicular to these forces. There is no force in the direction of displacement because the object is sliding at a constant speed.

38. Work is the product of the force and the distance over which an object is moved in the direction of the force. Power is the time rate at which work is done.

39. $kg{\cdot}m^2/s^3$

Mastering Problems

40. 1×10^4 J

41. 59.9 kg

42. 2.75×10^4 N

43. 8.87×10^7 J

44. 126 W

45. 7.5 J

46. **a.** 9.00 kJ

b. 3.00 kW

47. 4.43×10^3 J

48. Yes, because the force is applied in the direction of mower's motion.

7.9×10^4 J

49. 36.2°

50. 2.0 kW

51. 518 J

52. 7.4×10^3 J

53. 1.20×10^4 J

54 **a.** 6.0×10^4 J

b. 7.4×10^4 J

55. 54.7 m

56. 9.0×10^1 kW

57. **a.** 8.0×10^2 J

b. 5.9×10^2 J

58. **a.** 3.4×10^2 J

b. -2.8×10^2 J

c. -1.3×10^2 J (work done against friction)

59. **a.** 25 N/m

b. 0.50 J

c. $W = \frac{1}{2}kd^2 = \frac{1}{2}(25\ \text{N/m})(0.20\ \text{m})^2 = 0.50\ \text{J}$

60. 0.80 J

61. **a.** total $= 2.6\times10^2$ J

b. 1.3×10^2 W

62. **a.** 1.10×10^5 J

b. 3.14 kW

63. 3.7×10^2 W

SECTION 2

Mastering Concepts

64. no, $e \leq 100\%$

65. Pedals transfer force from the rider to the bike through a wheel and axle.

Mastering Problems

66. **a.** 3.0×10^2 N

b. 4.0×10^1 N

c. 6.0×10^3 J

d. 6.8×10^3 J

e. 3.5

67. 98 J

68. 0.24 m

69. Answers will vary, but a correct form of the answer is, "A constant force of 12.5 N acts on a 6.0-kg object, increasing its speed from 0.05 m/s to 1.10 m/s. Over what distance did this force act?"

70. **a.** 4.00

b. 3.59

c. 89.8%

71. **a.** 3.5

b. 4.00

c. 88%

72. **a.** 61 N

b. 4.0

c. 3.3; 82%

73. 31.4 cm

74. 0.50 m/s

75. 1.64×10^4 J

76. Answers will vary. A possible form of the correct answer would be, "... if he pushes it with a force of 20 N for a distance of 7.0 m in 14.0 s, how much power did he deliver?"

77. a. $W_{i1} = W_{o1} = W_{i2} = W_{o2}$

$W_{i1} = W_{o2}$

$F_{e1}d_{e1} = F_{r2}d_{r2}$

For the compound machine

$IMA_c = \frac{d_{e1}}{d_{r2}}$

$\frac{d_{e1}}{d_{r1}} = IMA_1$ and $\frac{d_{e2}}{d_{r2}} = IMA_2$

$d_{r1} = d_{e2}$

$\frac{d_{e1}}{IMA_1} = d_{r1} = d_{e2} = (IMA_2)(d_{r2})$

$d_{e1} = (IMA_1)(IMA_2)(d_{r2})$

$\frac{d_{e1}}{d_{r2}} = IMA_c = (IMA_1)(IMA_2)$

$= (3.0)(2.0) = 6.0$

b. 150 N

c. 2.0 cm

Applying Concepts

78. Each requires the same amount of work because force magnitude times distance is the same.

79. You do positive work on the box because the force and motion are in the same direction. Gravity does negative work on the box because the force of gravity is opposite to the direction of motion. The work done by you and by gravity are separate and do not cancel each other.

80. The net work is zero. Carrying the carton upstairs requires positive work; carrying it back down is negative work. The work done in both cases is equal and opposite because the distances are equal and opposite. The student might arrange the payments on the basis of the time it takes to carry paper, whether up or down, not on the basis of work done.

81. No, the force on the box is up and the displacement is down the hall. They are perpendicular and no work is done.

82. a. Both people are doing the same amount of work because they are both climbing the same flight of stairs and they have the same mass.

b. The person who climbs in 25 s expends more power, as less time is needed to cover the distance.

83. $P = \frac{W}{t}$, but $W = Fd\cos\theta$ so, $P = \frac{Fd\cos\theta}{t}$ because $v = \frac{d}{t}$, $P = Fv\cos\theta$

84. Increase the ratio of $\frac{d_e}{d_r}$ to increase the *IMA* of a machine.

85. Assuming a circular orbit, the force due to gravity is perpendicular to the direction of motion. This means the work done is zero. Hence, there is no change in kinetic energy of the planet, so it does not speed up or slow down. This is true for a circular orbit.

86. Your hand should be as far from the head as possible to make d_e as large as possible. The nail should be as close to the head as possible to make d_r as small as possible.

87. Reduce friction as much as possible to increase the resistance force.

Mixed Review

88. Either ramp: only the vertical distance is important. If Isra used a longer ramp, she would require less force. The work done would be the same.

89. a. 5.5×10^3 J

b. $d = 0$, so no work

c. -5.5×10^3 J

d. No. He exerts no force, so he does no work.

90. a. 9.5×10^2 N

b. 1.8×10^4 J

c. 2.2 kW

91. a. 1.04×10^3 J

b. 958 J

c. 92.1%

92. a. 681 N

b. 456 N, opposite to the direction of motion

c. -1.14×10^4 J

93. 58.7°

94. a. 57 W

b. 67 W

Thinking Critically

95. a. 6.1×10^{2} W

b. 1.2×10^{3} W

96. a. 1.5×10^{3} W

b. 3.0×10^{3} W

97. $W = 1.76\times10^{4}$ J. From the graph, the maximum power is 25 W at 15 kg. The time is about 12 minutes.

98. $W_c = W_e < W_b < W_a = W_d$

Writing in Physics

99. The overall efficiency is 15–30 percent. The transmission's efficiency is about 90 percent. Rolling friction in the tires is about 1 percent (ratio of pushing force to weight moved). The largest gain is possible in the engine.

100. Answers will vary. Some examples include, the company Consumers' Power changed its name to Consumers' Energy without changing its product, natural gas. "It's not just energy, it's power!" has appeared in the popular press.

Cumulative Review

101. 82 N

102. 1.02 m

SECTION 1 The Many Forms of Energy

student text pp. 292–300

1 Introduce

Bellringer

Toys, *KE,* and Energy Sources Gather a good collection of toys—some using energy from batteries, some spring energy, and some gravitational potential energy—that all show energy transformations. Have students manipulate and observe the toys. Ask what obvious types of energy all the toys have in common. potential and kinetic energy Ask students to brainstorm some sources that supply energy that make it possible for toys to move. Possible answers: batteries, wound springs, gravity. **OL** **Visual-Spatial** **EL**

Tie to Prior Knowledge

Work-Energy Theorem The students are introduced to a monetary model to reinforce the work-energy theorem presented in the previous chapter. The monetary model is then expanded to include types of energy other than kinetic energy.

2 Teach

A Model of the Work-Energy Theorem

student text pp. 292–293

Use an Analogy

Rotational Energy Explain that the equation representing rotational kinetic energy, $KE_{rot} = \frac{1}{2}I\omega^2$, is analogous to the equation for translational kinetic energy, $KE = \frac{1}{2}mv^2$, because each part of the former corresponds to a part of the latter. The moment of inertia (I), which depends on the mass and shape of an object, corresponds to the mass (m) of a point object, and the angular velocity (ω), to the translational velocity (v). Ask students to state an analogous equation for angular momentum. $I\omega$ **OL**

Discussion

Problem According to **Figure 2,** the work you do on a ball in throwing it and the work you do on the ball in catching it are equal in magnitude. Does the figure indicate that it takes the same size force to throw a ball as to catch it?

Response Not necessarily; $W = Fd$; in catching the ball, the average force acts over a smaller distance, so the average force needed to catch the ball is usually larger than the average force needed to throw the ball. This conclusion is also supported by analysis using the impulse-momentum theorem. **OL** **Interpersonal**

Quick **DEMO**

Work and Potential Energy

Estimated Time 5 minutes

Materials strong spring, goggles

Procedure Put on goggles and an apron. Pull the ends of the spring to noticeably stretch it. Ask the class whether work was done in stretching the spring. Yes, a force produced a displacement in the direction of the force. Write $W = Fd$ on the board. Have students identify F and d. F is the force needed to stretch the spring, and d is the distance the spring stretched from its rest position. Ask students whether it took energy to stretch the spring and whether the energy is still available in the stretched spring. Yes, the teacher supplied energy to stretch the spring, and the energy is available because the stretched spring can be used to move another object.

Content Background

Hooke's Law Hooke's law states that the force necessary to elongate a spring a small distance (x) is proportional to the elongation. That is, $F \propto x$. This proportionality can be stated as the equation $F = -kx$, where k is the spring constant. (The spring constant is an indication of the spring's stiffness.) The work done in elongating the spring equals $\frac{1}{2}kx^2$. The potential energy stored in an elongated spring is given by the equation $PE_{spring} = \frac{1}{2}kx^2$.

Kinetic Energy

student text pp. 293–294

Identify Misconceptions

Energy Is Not a Vector Draw two equal masses moving in opposite directions with velocities $+v$ and $-v$. Ask students which mass has the greater momentum. the mass with the positive velocity Ask which has the greater kinetic energy. The kinetic energies of the masses are equal. Point out that squaring the velocity makes energy a positive quantity without direction. Write on the board, "Energy is not a vector." **OL** **Logical-Mathematical**

Careers

Energy Engineers In the wind-power industry, engineers are keenly aware that the kinetic energy of the wind can be transformed into electrical energy using wind turbines. The power, the rate at which mechanical energy is delivered to a turbine by wind, depends on the cube of the wind's velocity. In a wind turbine, the maximum power delivered equals $(\pi/8)\rho D^2 v^3$, where ρ is the density of the air, D is the diameter of the wind turbine, and v is the wind speed. The efficiency at which mechanical power is converted to electrical power depends on such factors as the efficiencies of the wind turbine and generator.

Potential Energy and Gravitational Potential Energy

student text pp. 294–298

Critical Thinking

Platform Diving Ask students the following questions. Is the potential energy the same for all divers on the platform? No; they have different masses. Will they all have the same kinetic energy when entering the water if they dive similarly? No; the potential energies are different. Will they have the same velocity when they enter the water if they dive similarly? Yes; objects in free-fall accelerate at the same rate. Will they each take the same amount of time to fall from the platform to the water? Yes; objects in free-fall accelerate at the same rate. **OL** **Logical-Mathematical**

Use Figure 4

Juggling Oranges Have students draw free-body diagrams of the left and center oranges in the photo. Point out that the net force acting on each orange is its weight. Have students assume that the center orange is at the peak of its trajectory, and then ask them if the values of *KE* and *GPE*, respectively, are at a maximum or minimum. The value of *KE* is minimum, and that of *GPE* is maximum. Ask students whether gravity is doing positive or negative work on the left orange. negative work Explain that it is negative work because the force on the orange is in the opposite direction of the displacement.

$W = Fd\cos\theta$

$\theta = 180° \Rightarrow \cos 180° = -1$

OL **Visual-Spatial**

DIFFERENTIATED INSTRUCTION

Struggling Students Point out that the complicated motion of an orange shown in **Figure 4** can be analyzed in four parts. (1) The upward force of the juggler's hand does positive work on the orange. Before the juggler releases the orange, the acceleration of this upward force increases the *KE* of the system. (2) As the orange moves upward, its *GPE* increases as *KE* decreases because an unbalanced force—the force of gravity—is acting on it. At the top of the orange's flight, *GPE* is at a maximum. (3) As the orange falls, *GPE* decreases as *KE* increases. (4) In catching the orange, the juggler does negative work to slow the orange to a stop before repeating the first step. **BL** **Visual-Spatial**

ADDITIONAL IN-CLASS EXAMPLE

Use with Example Problem 1.

Problem How much work does a bricklayer do to carry 30.2 kg of bricks from the ground up to the third floor (height = 11.1 m) of a building under construction? What is the gravitational potential energy of the brick-Earth system when the bricklayer reaches the third floor?

Response

$W = Fd = mgh = (30.2\text{ kg})(9.8\text{ N/kg})(11.1\text{ m}) = 3.29\text{ kJ};$

$GPE = mgh = (30.2\text{ kg})(9.8\text{ N/Kg})(11.1\text{ m}) = 3.29\text{ kJ}$

Reinforcement

Concept Map Have students work in pairs to draw a concept map that relates the following key concepts or quantities: work-energy theorem, potential energy, kinetic energy, velocity, height, gravity, and mass. **BL** **Interpersonal** **EL**

DIFFERENTIATED INSTRUCTION

Visually Impaired All students profit from feeling what is happening in an experiment. For instance, if work is being discussed, have students slide a slotted mass along a vertical meterstick. Students can then calculate how much work they did in lifting the mass. Arrange a collision between sliding blocks so that a student can catch each block in his or her hand after the collision. Such a situation gives students tactile information about the momentum of the object after collisions. **BL** **Kinesthetic**

Elastic Potential Energy

student text pp. 298–299

Concept Development

Work Done on Bow Explain to students that the work done on the bowstring is positive because the force and the displacement of the bowstring are in the same direction. The work done on the bow is also positive because the bow deforms in the direction of the pulled bowstring.

Reinforcement

MAIN IDEA Sketch or show a picture of a roller coaster. Ask students what type of energy riders possess at various locations along the track. Locations should include positions with only gravitational potential energy, only kinetic energy, and combinations of both.

Content Background

Elasticity The elasticity of all materials is a result of the electromagnetic interactions among the atoms in the material. Almost all solids can be stretched or compressed slightly, but springs are bent and designed to do so in a controlled and predictable way. Materials can be permanently stretched. This condition, called plastic deformation, occurs because the atoms actually change their relative places due to the stretching. Even a small force will cause the plastic deformation of a clay ball.

Mass and Other Forms of Energy

student text pp. 299–300

REAL-WORLD PHYSICS

To inspect food materials, the food industry relies on a variety of instruments for spectrographic analysis. Spectroscopy is based on an understanding of how energy is distributed within atoms and molecules. The total energy of a molecule comprises numerous sources of energy—electronic, nuclear, rotational, translational, and vibrational. Spectroscopy looks at the interactions between electromagnetic radiation and matter to analyze properties such as molecular composition, structure, and dynamics. Students can work in small teams to research various instruments used in food analysis and explore how physics is applied and then report their findings to the class.

3 Assess

Assess the MAIN IDEA

Exploring Forms of Energy On the board make a list of various systems that posses kinetic energy, potential energy, and combinations of both. Have students write down all types of energy each system possesses. Examples include: a compressed spring in a pinball machine (*EPE*), a juggling ball thrown vertically at its highest point (*GPE*), a football flying from the quarterback to the wide receiver (*KE* and *GPE*), a set mousetrap (*EPE*), a communication satellite in orbit around Earth (*KE* and *GPE*), a snowboarder carving midway down a mountain face (*KE* and *GPE*), an apple high up on a tree branch (*GPE*), a stretched rubber band (*EPE*).

Check for Understanding

Potential Energy Ask students to describe potential energy changes as they climb a flight of stairs and return on an escalator. In climbing the stairs, the change in potential energy equals $mg\Delta h$. In returning on the escalator, the change in potential energy equals $-mg\Delta h$ because Δh is negative. **BL** **EL**

Extension

Center of Mass In analyzing motion, all the mass of an object can be considered concentrated at one point, the center of mass. For a person, this point is usually located behind the belly button. In the high jump, athletes jump so that they go over the crossbar in a layout, or horizontal, position. Five decades ago, most jumpers went over the crossbar with the upper body upright, much like a hurdler. Have students investigate the changes in potential energy that occur in high jumping and how today's athletes apply them to jump higher. A jumper using the layout style does not have to lift his or her center of mass as high as a jumper using the upright style. Therefore, with the same initial kinetic energy, the jumper can clear a bar of higher height. **AL**

Reading Checks and Caption Checks

Caption Check, *student text p. 293*

Answers will vary; sample answers: pulling a rake while raking leaves; pushing a lawnmower while mowing the lawn; pushing and pulling a vacuum cleaner while cleaning the carpet; pushing and pulling a broom while sweeping the floor

Reading Check, *student text p. 295*

As the orange rises, gravity does negative work on the orange, which slows the speed of the orange down.

Caption Check, *student text p. 296*

Answers will vary; halfway between the highest point and the lowest point of the orange.

Reading Check, *student text p. 298*

Elastic potential energy is stored energy due to an object's change in shape.

Practice Problems

student text p. 294

1. KE changed by friction $= 160$ J; $W = 160$ J
2. $KE_i = 2.12\times10^5$ J; $KE_f = 8.47\times10^5$ J; $W = 6.35\times10^5$ J
3. $KE = 2.45\times10^{20}$ J; 5.8×10^4 bombs
4. $KE_{trans} = 200$ J, $KE_{rot} = 200$ J

Practice Problems

student text p. 298

5. -44 J
6. -2.4×10^2 J
7. 28 J
8. -1.2×10^2 J
9. To lift the box to the table: $W = 1.1\times10^2$ J.
   To slide the box across the table: $W = 0.0$ J because the height did not change, and we ignored friction.
   To lower the box to the floor: $W = -1.1\times10^2$ J.
   The sum of the three energy changes $= 0.0$ J.

Section 1 Review

student text p. 300

10. The bowling ball has zero kinetic energy when it is resting on the rack or when it is held near your shoulder. Therefore, the total work done on the ball by you and by gravity must equal zero.
11. **a.** See Online Solutions Manual for energy bar diagram. There should be three bars: one for the spring's potential energy, one for gravitational potential energy, and one for kinetic energy. The spring's potential energy is at the maximum level, and the other two are zero.

    b. See Online Solutions Manual for energy bar diagram. The kinetic energy is at the maximum level, and the other two are zero.

    c. See Online Solutions Manual for energy bar diagram. The gravitational potential energy is at the maximum level, and the other two are zero.
12. **a.** 1.0×10^5 J

    b. -4.9×10^4 J
13. At the top, $GPE = 4.0\times10^4$ J; at the bottom, $GPE = -3.5\times10^4$ J; see Online Solutions Manual for energy bar diagram.
14. The angular momentum is doubled because it is proportional to the angular velocity. The rotational kinetic energy is quadrupled because it is proportional to the square of the angular velocity. The children did work in rotating the merry-go-round.
15. **a.** Karl exerted a constant force F over a distance d and did an amount of work $W = Fd$ on the puck. This work changed the kinetic energy of the puck. $W = KE_f - KE_i = \frac{1}{2}mv_f^2$; See Online Solutions Manual for energy bar diagram.

    b. If the puck has half the mass, it still receives the same amount of work and has the same change in kinetic energy. However, the smaller mass will move faster by a factor of 1.414.

    c. The second puck has less momentum than the first puck. Because the change in momentum is equal to the impulse provided by the air hose, the second puck receives a smaller impulse.

1 Introduce

Bellringer

Conservation of Energy Have students observe the motion of a hard rubber ball as you drop it onto the floor from a height of about 1 m. Repeat several times so that students can observe that the ball does not rebound higher than the height from which it was released. Repeat with the ball falling onto a tennis racket instead of onto the floor. Then, drop the ball into a cake pan filled with sand placed on the floor. Ask what differences the students observed. Why didn't the ball return to the height from which it was released? What happened to the potential energy? The potential energy was transformed into other forms of energy, such as thermal energy and sound energy. **OL** **Visual-Spatial** **EL**

Tie to Prior Knowledge

Conservation Laws As they learn about the law of conservation of energy, students should recall the law of conservation of momentum, presented in an earlier chapter, and the law of conservation of mass during chemical reactions if they have studied chemistry.

2 Teach

The Law of Conservation of Energy

student text pp. 301–305

Concept Development

Energy and Momentum Energy often is used to describe the motion of an object, while energy and momentum together often are used to describe collisions.

Concept Development

Conservation of Total Mechanical Energy It is often necessary to identify all forms of energy that an object possesses and then to determine whether the object's situation allows its total mechanical energy to be conserved.

Reinforcement

MAIN IDEA Take a blob of clay and a rubber ball, and throw each onto the floor. What type of energy did each have just before hitting the floor? kinetic energy What happened to the kinetic energy? Transformed into sound energy, thermal energy, and some went back into kinetic energy in the bouncing ball. Was total energy conserved? yes Kinetic energy? no Ask students to develop a total energy conservation expression for both the blob of clay and the rubber ball.

PHYSICS CHALLENGE ACTIVITY

Energy Sharing Investigate how energy is transferred between two pendulums that are weakly linked to each other. From the ceiling or a high support, hang two 1-kg masses from strings of the same length. At a point about 0.5 m from the top of the strings, tie a rubber band that loosely connects one pendulum to the other. The rubber band should cause one pendulum to occasionally tug the other pendulum. Pull back one pendulum and release it. Observe the motions of the two pendulums over several oscillations and describe them in terms of the transfer of energy from one to the other. **AL** **Visual-Spatial** **EL**

Quick DEMO

Conservation of Mechanical Energy

Estimated Time 5 minutes

Materials string, clay, lab support, empty soda can

Procedure Using a fist-sized blob of clay and about 1.5 m of string, make a pendulum and suspend it from a lab support. Pull the bob to one side, and place an empty soda can in line with the bob so that the bob barely touches the side of the can. Before releasing the bob, ask students to hypothesize what will happen after you release the bob. Because of conservation of energy, the bob should swing back and barely touch the soda can.

REAL-WORLD PHYSICS

Colonial Surveyor and Inventor Benjamin Banneker was an African American inventor who lived from 1731 to 1806. George Washington, who appointed him to the commission that planned the land usage for Washington, DC, recognized his abilities in surveying and mathematics. Not only was Banneker a city planner and surveyor, but he was also a skilled carpenter. One of his projects was a clock made entirely of carved wood. The clock ran by falling weights. As gravity pulled down the weights, an intricate set of gears caused the hands to move, keeping time.

ADDITIONAL IN-CLASS EXAMPLE

Use with Example Problem 2.

Problem A 68.2-kg diver steps off a 5.0-m diving platform. Ignoring air resistance, what is the kinetic energy and velocity of the diver as she enters the water?

Response

$KE_f = GPE_i = mgh = (68.2\text{ kg})(9.8\text{ N/kg})(5.0\text{ m})$
$= 3.3\times10^3\text{ J}; KE_f = \frac{1}{2}mv^2 = 3.3\times10^3\text{ J}$
$= \frac{1}{2}(68.2\text{ kg})(v^2)$
$v = 9.8\text{ m/s}$

Concept Development

Laboratory Carts Have students compress the springs of two spring-loaded carts, place the springs together, release the springs, and then measure the distance each cart travels. Ask students to repeat this activity, using various masses on one of the carts, and have them answer the following questions: How is the distance each cart travels related to its initial velocity? The distance each cart rolled is proportional to its initial velocity. Was momentum conserved in each case? yes Does your data show that the spring released the same amount of energy each time it was released? yes EL

DIFFERENTIATED INSTRUCTION

Struggling Students Point out how gravitational potential and kinetic energies are related to concepts presented earlier. For example, if a ball is thrown upward and leaves the ground at velocity v_i, its maximum height is determined by the formula $v_f^2 = v_i^2 + 2ah$. The height is $h = \frac{-v_i^2}{2a}$, where $a = -g$. Now analyze the same situation in terms of energy. The total energy of the ball at ground level is $GPE_i + KE_i = 0 + \frac{1}{2}mv_i^2$. This equals the energy at the maximum height, $GPE_f + KE_f = mgh + 0$. Setting the two equations equal yields $\frac{1}{2}mv_i^2 = mgh$. Solving for h yields $h = \frac{v_i^2}{2g}$, the same answer as before. BL

Use the MiniLAB

In *Energy Exchange,* students can observe the kinetic energy of a ball shot upward transform into gravitational potential energy, which is maximum at the peak of the ball's flight.

Use the MiniLAB

In *Interrupted Pendulum,* students can observe the conservation of mechanical energy in a swinging pendulum bob.

Analyzing Collisions

student text pp. 306–309

Use the PhysicsLAB

In *Conservation of Energy,* students can observe the relationship between gravitational potential energy and kinetic energy of a falling object.

Use the PhysicsLAB

In *Is Energy Conserved?,* students can observe the experimental results of the law of conservation of energy and the conservation of mechanical energy.

Quick DEMO

Energy Transfers

Estimated Time 5 minutes

Materials large rubber ball, smaller rubber ball

Procedure Drop each ball separately from about chest height. Ask students to observe the heights to which the balls rebounded. Hold the small ball on top of the large ball and ask students to hypothesize what would happen if you dropped the balls as before. Then drop the balls. Have students explain their observations. During the collision, kinetic energy was transferred from the larger to the smaller ball. The larger ball rebounded less and the smaller ball rebounded to a point higher than its release point.

PHYSICS PROJECT ACTIVITY

Design a Roller Coaster Have each student design a roller coaster using bent, hard-plastic insulation tubing for ball bearings. Have students consider the characteristics of a good roller-coaster ride and how energy is used to provide these characteristics. As they design their roller coasters, ask students what they must know about rolling friction and the effect of rotational kinetic energy. Have students do preliminary tests for their designs. Each project report should show how the roller coaster was designed, what preliminary tests were done, and how the actual coaster compares with the design expectations. Have students present their results to the class. OL Kinesthetic EL

Identify Misconceptions

Momentum and Energy To differentiate *p* and *KE*, ask students to discuss the following among themselves: How can two objects have the same mass and energy but different momenta? $v_1 = -v_2$ How can two objects have the same mass and momentum but different energies? They cannot. OL Interpersonal

Use Models

Energy Transfer The model of exchanging money has been used to describe the transfer of energy from one form to another. In collisions, energy and momentum are exchanged from one object to another. Have students create a model that keeps the momentum currency separate from the energy currency. OL Visual-Spatial

Discussion

Problem Consider the collision of two skaters on ice. They are of equal mass. Is it possible that a body could change in momentum without changing in kinetic energy?

Response Yes; consider two skaters of the same mass and speed that collide elastically. For each skater, the momentum and velocity vectors would reverse directions and the kinetic energy would be unchanged. OL

Content Background

Elliptical Orbits The conservation of energy explains why a planet in an elliptical orbit changes speed. Assume that the only force acting on the planet is the gravitational force between it and the star. As the planet moves in its orbit, the system's *PE* changes because the planet's distance from the star changes. Because the sum of *PE* and *KE* is constant, the system must have its minimum *KE* when *PE* is at a maximum; this is when the planet is farthest from the star. Because *KE* depends on speed, the planet has its slowest speed at the farthest distance. Conversely, *PE* is at a minimum and *KE* is at a maximum when the planet and star are closest together. Therefore, the planet would have the greatest speed at the closest point in its orbit.

ADDITIONAL IN-CLASS EXAMPLE

Use with Example Problem 3.

Problem A 54.5-kg ice skater moving at 3.2 m/s collides with a 44.7-kg skater who is motionless. They then slide together along the frictionless ice. What is their velocity after the collision? How much kinetic energy was lost in the collision? What fraction of the original kinetic energy was lost?

Response

$m_A v_A + m_B v_B = (m_A + m_B)v_f$;

$(54.5\text{ kg})(3.2\text{ m/s}) + (44.7\text{ kg})(0\text{ m/s})$

$= (54.5\text{ kg} + 44.7\text{ kg})v_f$;

$v_f = 1.8\text{ m/s}$;

$KE_f = \frac{1}{2}(m_A + m_B)v_f^2$

$= \frac{1}{2}(54.5\text{ kg} + 44.7\text{ kg})(1.8\text{ m/s})^2 = 160\text{ J}$,

$KE_i = \frac{1}{2}m_A v_A^2 + \frac{1}{2}m_B v_B^2$

$= \frac{1}{2}(54.5\text{ kg})(3.2\text{ m/s})^2 + \frac{1}{2}(44.7\text{ kg})(0\text{ m/s})^2$

$= 280\text{ J}$, $KE_i - KE_f = 280\text{ J} - 160\text{ J} = 120\text{ J}$;

$\frac{\Delta KE}{KE_i} = \frac{120\text{ J}}{280\text{ J}} = 0.43$, or 43%

Reinforcement

Potential and Kinetic Energy

Sketch the following bar graph on the board.

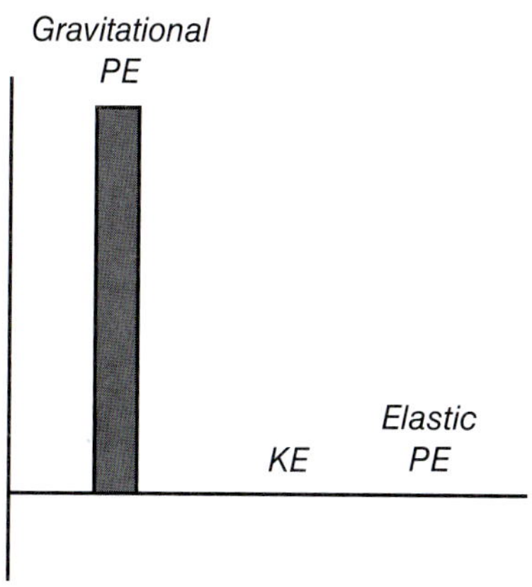

Explain that the graph shows the energy distribution of a system consisting of a gymnast, a trampoline, and Earth. The reference level is at the trampoline. Here, the gymnast is at the top of a rebound from the trampoline. Ask students to describe the graph when the gymnast just reaches the trampoline. At this height ($h = 0$) the trampoline is at its equilibrium point, so both *GPE* and *EPE* bars are at zero: $PE_{g2} = PE_{e2} = 0$, Thus, by conservation of energy $KE_2 = PE_{g1}$. **OL** **Visual-Spatial**

DIFFERENTIATED INSTRUCTION

Hearing Impaired Have a student hold up a golf ball in each hand and move them in the same direction but move one about twice the speed of the other. Display a transparency with the following questions: 1. Compare the kinetic energies of the two balls. The ball with the greater speed has the greater energy. 2. If the one ball is traveling about twice as fast as the other, is its *KE* twice that of the slower ball? No; *KE* depends on v^2, so the faster ball has about four times the *KE* of the slower ball. **BL** **Visual-Spatial**

PHYSICS CHALLENGE ACTIVITY

Collisions In their quest to drive golf balls farther and farther off the tee, golfers are using new types of club heads and golf balls. The hitting of the ball can be approximated as a free collision of the club head with the ball. Have students think about what qualities of the ball and club, such as the mass of the head, would affect the length of a shot. Have them make a list of these qualities and compare them to the descriptions in advertisements in sports magazines. Ask students which of the advertisements indicate an understanding of physics and whether physics principles support the claims made in the advertisements. **AL** **Linguistic**

3 Assess

Assess the MAINIDEA

Car Bumpers Explain that cars are designed to undergo inelastic collisions. Ask students why this is. By reducing the *KE* of the car during a collision, less work has to be done on the passenger to decrease his or her *KE* to zero. Therefore, the force to do the work on the passenger is also reduced. **OL**

Check for Understanding

Car Collisions Sketch the diagrams from **Figure 13.** Have students describe momentum and energy changes. In each case, momentum was conserved; in case 1, *KE* remained constant; in case 2, *KE* decreases because some kinetic energy is transformed to thermal energy and sound energy; in case 3, *KE* decreases because some of the kinetic energy is transformed to thermal energy and sound energy; and in case 4, *KE* increases because additional energy comes from a released spring or an explosion. **OL** **Visual-Spatial**

Reading Checks and Caption Checks

Caption Check, *student text p. 301*

Reference level is the person's hand. As the ball leaves the hand, its velocity is greatest and its kinetic energy is greatest. The *KE* bar is highest at this point, and the *GPE* bar is zero. Halfway, the *KE* and *GPE* bars are equal. Back at the hand, *GPE* is zero again and *KE* is greatest. The *KE* bar is the same height as first thrown if there is no air resistance.

Reading Check, *student text p. 302*

The final *KE* would be less because some of the ball's *PE* would be transformed to thermal energy due to friction and possibly some of the energy would be transformed to sound energy.

Caption Check, *student text p. 306*

The energy was transformed into other kinds of energy, including sound energy, thermal energy.

Reading Check, *student text p. 306*

Case 1: $p_i = p_f = 1.00$ kg•m/s; momentum is conserved.
Case 2: $p_i = p_f = 1.00$ kg•m/s; momentum is conserved.
Case 3: $p_i = p_f = 1.00$ kg•m/s; momentum is conserved.
Case 4: $p_i = p_f = 1.00$ kg•m/s; momentum is conserved.

Practice Problems

student text p. 305

16. 3.1×10^3 J; 3.7 m

17. The system of Earth, bike, and rider remains the same, but now the energy involved is not mechanical energy alone. The rider must be considered as having chemical potential energy, some of which is converted to mechanical energy.

18. a. bottom of valley: 3.0×10^1 m/s

b. top of next hill: 9.9 m/s

c. No; the angles do not have any impact.

19. 1.1 m above the platform

20. 0.8 m/s

Practice Problems

student text p. 309

21. 1.13×10^2 m/s

22. a. 4.4×10^3 J; 1.2×10^3 kg•m/s

b. 6.8 m/s

c. −200 J

23. a. See Online Solutions Manual for sketch. The system includes the suspended target and the dart.

b. Only momentum is conserved in the inelastic dart-target collision. As the dart-target combination swings upward, energy is conserved.

c. 46 m/s

Physics Challenge

student text p. 309

1. Conservation of momentum:

$$mv_1 = mv_2 + m_B v_B$$
$$m_B v_B = m(v_1 - v_2)$$
$$v_B = \frac{m(v_1 - v_2)}{m_B}$$

2. For the bullet alone:

$$KE_1 = \frac{1}{2}mv_1^2; \quad KE_2 = \frac{1}{2}mv_2^2$$
$$\Delta KE = \frac{1}{2}m(v_1^2 - v_2^2)$$

3. Energy lost to friction

$$E_{lost} = KE_1 - KE_2 - KE_{block}$$
$$E_{lost} = \frac{1}{2}mv_1^2 - \frac{1}{2}mv_2^2 - \frac{1}{2}m_B v_B^2$$

Section 2 Review

student text p. 309

24. a. Diagrams should show that all of the system's energy is gravitational potential energy.

b. Diagrams should show that all of the system's energy is elastic potential energy.

25. To simplify problems that take place over a short time, Earth is considered a closed system. It is not actually isolated, however, because it is acted upon by the gravitational forces from the planets, the Sun, and other stars. In addition, Earth is the recipient of continuous electromagnetic energy, primarily from the Sun.

26. Even though the rubber ball rebounds with little waste of energy kinetic energy would not be conserved in this case because the glob of chewing gum probably was deformed in the collision.

27. The balls, paddle, and racket are designed to match so that the maximum amount of kinetic energy is passed to the ball. A softer ball receives energy with less loss from a softer paddle or racket. Other combinations are a golf ball and club (both hard) and a baseball and bat (also both hard).

28. after three bounces

29. 720 J

30. The ball falls more slowly during the beginning part of its drop. Therefore, in the first half of the time that it falls, it will not have traveled half of the distance that it will fall. Therefore, the ball will have more potential energy than kinetic energy.

FRONTIERS IN **PHYSICS**

What's the Alternative?

Alternative Energy Debate

Purpose

To introduce students to a variety of alternative energy sources.

Background

The term *energy crisis* entered into common speech during fossil-fuel shortages in the 1970s. While fossil fuels remain the main source of energy, their fluctuating availability and cost, as well as environmental concerns, have led to research into alternatives. Some of these, such as wind, water, and biomass, are modern twists on energy sources that have been around for centuries. Others, such as photovoltaic cells and nuclear reactors, are the result of more modern physics understanding.

Teaching Strategies

- Before reading the feature, have students review the various types of energy discussed in this chapter, including kinetic, potential, radiant, chemical, nuclear, and electrical.
- Have students identify alternative energy sources that they use. This should include determining what energy source is used by local electric companies. Some possible answers: solar powered calculators, hybrid cars, hydroelectric plants, wind farms.
- Have students identify the energy transformations that occur for each alternative energy source. To guide them, show them the energy transformations that take place in a typical fossil fuel plant:

  Chemical energy → Thermal energy → Kinetic energy → Electrical energy
- The U.S. Department of Energy and The U.S. Energy Information Administration Web sites provide a wealth of information about energy sources and usage.

GOING**FURTHER** >>>

Expected Outcomes Students should consider a wide variety of factors, such as construction and operating costs, availability, efficiency, safety, environmental impact, and storage/distribution concerns.

SECTION 1

Mastering Concepts

31. The work done on an object causes a change in the object's energy. This is the work-energy theorem.

32. A force exerted over a distance does work, which produces a change in energy.

33. A system that contains a wound-up spring has elastic potential energy. As the toy goes, the elastic potential energy is converted to kinetic energy. The toy runs down when all of the energy has been used to do work or converted to thermal energy by friction.

34. a. The potential energies are different due to the different reference levels.

b. The changes in the potential energies as a result of the fall are equal because the change in h is the same for both reference levels.

c. The kinetic energies of the ball at any point are equal because the velocities are the same.

35. The kinetic energy of a baseball can never be negative because the kinetic energy depends on the square of the velocity, which is always positive.

36. The gravitational potential energy of a baseball can be negative if the height of the ball is lower than the reference level.

37. The sprinter's kinetic energy increases by a factor of 9 because the velocity is squared.

38. The pole-vaulter runs (kinetic energy) and bends the pole, thereby adding elastic potential energy to the pole. As she lifts her body, that kinetic and elastic potential energy is transformed into kinetic and gravitational potential energy. When she releases the pole, all of the energy is kinetic and gravitational potential energy.

39. A flexible, fiberglass pole can store elastic potential energy because it can be bent easily. This energy can be released to push the pole-vaulter higher vertically. By contrast, the wooden pole does not store elastic potential energy, and the pole-vaulter's maximum height is limited by the direct conversion of kinetic energy to gravitational potential energy.

Mastering Problems

40. 1.3×10^5 J

41. 203 J

42. a. 2.3×10^3 J

b. 5.6×10^2 J

c. 4/1; twice the velocity gives four times the kinetic energy. The kinetic energy is proportional to the square of the velocity.

43. 6.86×10^5 J

44. a. 4.5×10^3 J

b. 2/1

c. 2/1; the ratio of their combined kinetic energy to Katia's kinetic energy is the same as the ratio of their combined mass to Katia's mass. Kinetic energy is proportional to mass.

45. a. 2.54×10^8 J

b. 2.54×10^8 J

c. 2.54×10^8 J

d. 143 m/s

46. 66 m

47. a. −345 J

b. −345 J

c. 34.5 m

48. 2.1×10^3 J

49. 1.3×10^2 J

50. -2.78×10^3 J

51. 3.4×10^3 J

52. 0.10 kg

53. 17 J

54. 1.7×10^2 J

55. a. 8.6×10^4 J

b. 3.9×10^3 N

56. a. -1.44×10^5 J

b. -1.44×10^5 J

c. -2.88×10^5 N

SECTION 2

Mastering Concepts

57. a. The total momentum of the ball and the puck together is conserved in the collision because there are no unbalanced forces on this system.

b. The total kinetic energy is not conserved. Part of it is lost in the smashing of the clay ball and the adhesion of the ball to the puck.

58. a. See Online Solutions Manual for energy bar diagram. An ice cube, initially at rest, slides down a frictionless slope.

b. See Online Solutions Manual for energy bar diagram. An ice cube, initially moving, slides up a frictionless slope and comes momentarily to rest.

59. On a roller-coaster ride, the car has mostly potential energy at the tops of the hills and mostly kinetic energy at the bottoms of the hills.

60. On each bounce, some, but not all, of the ball's kinetic energy is stored as elastic potential energy; the ball's deformation dissipates the rest of the energy as thermal energy and sound. After the bounce, the stored elastic potential energy is released as kinetic energy. Due to the energy losses in the deformation, each subsequent bounce begins with a smaller amount of kinetic energy and results in the ball reaching a lower height. Eventually, all of the ball's energy is dissipated, and the ball comes to rest.

Mastering Problems

61. 2.0×10^{1} m

62. 26 m

63. a. 4.90×10^{3} J

b. 4.90×10^{3} J

c. 4.90×10^{3} J

64. 407 J

65. a. 2×10^{4} J

b. 2×10^{4} J

c. 40 m/s

66. a. 30 m/s

b. 20 m/s

67. a. 4.0×10^{6} kg•m/s

b. 4.0×10^{6} kg•m/s

c. before: 1.6×10^{7} J; after: 8.0×10^{6} J

d. While momentum was conserved during the collision, kinetic energy was not. The amount not conserved was turned into thermal energy and sound energy.

68. -1.2×10^{3} J

69. 39 m

70. a. Answers will vary. Possible forms of the correct answers would be: "If the cliff is 43 m high and he releases the anvil from rest, what will be its speed when it impacts the ground at the foot of the cliff?"

b. Possible answer: "What force would he need to exert on the anvil to give it an upward acceleration of 0.06 m/s^2?"

71. a. 3.4 m/s

b. -1.7×10^{2} J

72. 4.1 m/s

73. a. See Online Solutions Manual for energy bar diagram. Yes; the mechanical energy has changed; the potential energy increased by 3200 J.

b. from the internal energy of the person

Applying Concepts

74. a. If the car wheels do not skid, the brake surfaces rub against each other and do work that stops the car. The work that the brakes do is equal to the change in kinetic energy of the car. The brake surfaces heat up because the kinetic energy is transformed to thermal energy.

b. If the brakes lock and the car wheels skid, the wheels rubbing on the road are doing the work that stops the car. The tire surfaces heat up, not the brakes. This is not an efficient way to stop a car, and it ruins the tires.

75. The trailer truck has more kinetic energy, $KE=\frac{1}{2}mv^2$, because it has greater mass than the compact car. Thus, according to the work-energy theorem, the truck's engine must have done more work.

76. Elastic potential energy is stored in the wound rope, which does work on the rock. The rock has kinetic and potential energy as it flies through the air. When it hits the wall, the inelastic collision causes most of the mechanical energy to be converted to thermal and sound energy and to do work breaking apart the wall structure. Some of the mechanical energy appears in the fragments thrown from the collision.

77. The energy went into bending sheet metal on the cars. Energy was also transferred due to frictional forces between the cars and the tires in the form of thermal energy and sound.

78. The larger skater will have more kinetic energy. The kinetic energy of each skater will be dissipated by the negative work, $W = Fd$, done by the friction of the ice. Since the frictional forces are equal, the larger skater will go farther before stopping.

79. a. No work is done by the tension force on the mass because the tension is pulling perpendicular to the motion of the mass.

b. This does not violate the work-energy theorem because the kinetic energy of the mass is constant; it is moving at a constant speed.

80. The hill must be made higher by a factor of 4.

81. Even though the balls are moving in opposite directions, they have the same kinetic energy and potential energy when thrown. They will have the same mechanical energy and speed when they hit the ground.

82. a. Pushing a hockey puck horizontally across ice; system consists of hockey puck only.

b. Dropping a ball; system consists of ball and Earth.

c. Compressing the spring in a toy pistol; system consists of spring only.

d. A car speeding on a level track reduces its speed.

Mixed Review

83. 8.6 m/s

84. 1.3 m

85. Assume that the foam peanuts exert a constant force to slow her down, $W = Fd = E = mgh$. If the height is increased five times, then the depth of the foam peanuts also should be increased five times to 5 m.

86. The two players had equal and opposite momenta before the collision. The energy loss for each player was $\frac{1}{2}mv^2 = \frac{1}{2}\left(\frac{m^2v^2}{m}\right) = \frac{p^2}{2m}$. The linebacker lost more energy. Thus, $KE_{linebacker} > KE_{end}$.

87. 13 J added by the spring

88. 6.28 m

89. The work done by friction (5.0 J) is greater than the energy of the cart (3.9 J). The cart would not reach the bottom of the hill.

90. 0.67 kg

Thinking Critically

91. 73 m/s

92. a. If $m_1 = m_2$, we know that m_1 will be at rest after the collision and m_2 will move with velocity v_1. All of the energy will be transferred to m_2.

b. If $m_1 \gg m_2$, we know that the motion of m_1 will be unaffected by the collision and that the energy transfer to m_2 will be minimal.

c. The best way to stop a neutron is to have it hit a hydrogen atom, which has about the same mass as the neutron.

93. Answers will vary, but a correct form of the answer is, "A ball at ground level is thrown into the air with an initial kinetic energy of 30 J. How high above the ground will it be when its kinetic energy is reduced to 15 J?"

94. Answers will vary, but a correct form of the answer is, "A person is preparing to bowl when she accidentally drops the ball, mass 6 kg, on her foot. If the ball is released 0.92 m above her foot, how fast is it moving when it hits the foot?"

95. $$v_{A2} = \left(\frac{m_A - m_B}{m_A + m_B}\right)v_{A1} + \left(\frac{2m_B}{m_A + m_B}\right)v_{B1}$$

$$v_{B2} = \left(\frac{m_B - m_A}{m_A + m_B}\right)v_{B1} + \left(\frac{2m_A}{m_A + m_B}\right)v_{A1}$$

96. $v_A = v_E < v_B = v_C = v_D$

97. 6.7 m/s

Writing in Physics

98. The Sun transforms nuclear energy into thermal energy through nuclear fusion and releases that energy in the form of electromagnetic radiation, which is transferred through the vacuum of space to Earth. Earth absorbs that electromagnetic radiation in its atmosphere, land, and oceans in the form of thermal energy or heat. Part of the visible radiation also is converted by plants into chemical energy through photosynthesis. There are several other chemical reactions mediated by sunlight, such as ozone production. The energy then is transferred into various forms, some of which are the chemical processes that allow us to digest food and turn it into chemical energy to build tissues, to move, and to think. In the end, after we have used the energy, the remainder is dispersed as electromagnetic radiation back into the universe.

99. Potential energy is stored in the binding of the protons and neutrons in the nucleus. The energy is released when a heavy nucleus is broken into smaller pieces (fission) or when very small nuclei are combined to make bigger nuclei (fusion). In the same way, chemical potential energy is stored when atoms are combined to make molecules and released when the molecules are broken up or rearranged. Separation of electric charges produces electric potential energy, as in a battery. Electric potential energy is converted to kinetic energy in the motion of electric charges in an electric current when a conductive path, or circuit, is provided. Biological processes are all chemical, and thus, biological energy is just a form of chemical energy. Solar energy is fusion energy converted to electromagnetic radiation. (See the answer to the previous question.) Light is a wave form of electromagnetic energy whose frequency is in a range detectible by the human eye.

Cumulative Review

100. 6.0×10^{24} kg

101. a. -0.500 kg·m/s

b. -0.995 kg·m/s

c. When the bullet ricochets, its change in momentum is larger in magnitude, and so is the block's change in momentum, so the block ends up with a greater speed.

Standardized Test Practice

student text p. 317

Multiple Choice

1. D

2. D

3. D

4. C

5. C

6. C

7. C

Free Response

8. 0.50 m

9. 1520 J

Rubric

The following rubric is a sample scoring device for free response questions.

Points	Description
4	The student demonstrates a thorough understanding of the physics involved. The response may contain minor flaws that do not detract from the demonstration of a thorough understanding.
3	The student demonstrates an understanding of the physics involved. The response is essentially correct and demonstrates an essential but less-than- thorough understanding of the physics.
2	The student demonstrates only a partial understanding of the physics involved. Although the student may have used the correct approach to a solution or may have provided a correct solution, the work lacks an essential understanding of the underlying physical concepts.
1	The student demonstrates a very limited understanding of the physics involved. The response is incomplete and exhibits many flaws.
0	The student provides a completely incorrect solution or no response at all.

CHAPTER 12
Thermal Energy

About the Photo

Cooking Ask students to list different ways thermal energy is transferred to food in cooking. These methods of thermal energy transfer can be categorized as either conduction, convection, or radiation. Some forms of cooking (for example, browning a steak or freezing ice cream) rely on high rates of thermal energy transfer; others depend only on total amount of thermal energy transferred (for example, making soup in a slow cooker).

Chapter Overview

Students have learned how energy can be transformed, as well as how it can be transferred between objects. This chapter will investigate the transfer of energy among the particles that compose a solid, liquid, or gas. The kinetic energy and potential energy of the particles that make up matter is called thermal energy, and this energy can be transferred to other objects. These concepts are useful in explaining how engines and refrigerators work, and they lead to the second law of thermodynamics, one of the most fundamental laws of nature.

Before students study the material in this chapter, they should study:

- Conservation of energy
- Kinetic energy

To solve problems in this chapter, students will need a strong grasp of:

- Graphing data
- Scientific notation
- Significant figures
- Slope
- Solving linear equations

Introduce the BIG IDEA

Thermal Energy Hold an inflated balloon in front of students, and ask them if anything is moving inside the balloon. Students might answer that molecules are moving inside the balloon. Then ask students if these molecules have energy. Students should recognize that moving particles have kinetic energy. Inform students that the energy associated with an object's atoms and molecules is called thermal energy.

Go online!
connected.mcgraw-hill.com

Your online experience includes:

- **Video**
- **Concepts In Motion**
- **Virtual Investigations**
- **All Lab, Applying Practices activities, and PBL worksheets**
- **Online Practice**
- **Personal Tutor**
- **Multilingual eGlossary**
- **Presentation**

Use the LaunchLAB

In *Thermal Energy Transfer,* students will investigate a common transfer of thermal energy.

Use Physics T.V.

Video

WARM-BLOODED V. COLD-BLOODED You might remember from biology that some organisms generate body heat internally, while others modify their behavior to regulate their body heat. What is the physics of body heat?

1 Introduce

Bellringer

A Rubber-Band Thermometer As a demonstration or a student investigation, make a simple thermometer using a ruler, a rubber band, a round wooden pencil, and a pin (see diagram). Place the pencil across the ruler and stretch the rubber band across the ends of the ruler to hold the pencil in place. The pin is stuck into the side of the pencil to indicate how the pencil rotates. Hold a desk lamp near one part of the rubber band. Because rubber shrinks when warmed (unlike most substances), the pencil will rotate the pin to point toward the warmed side of the band. When the heat source is removed, the pin should rotate back to its original position. BL Visual-Spatial

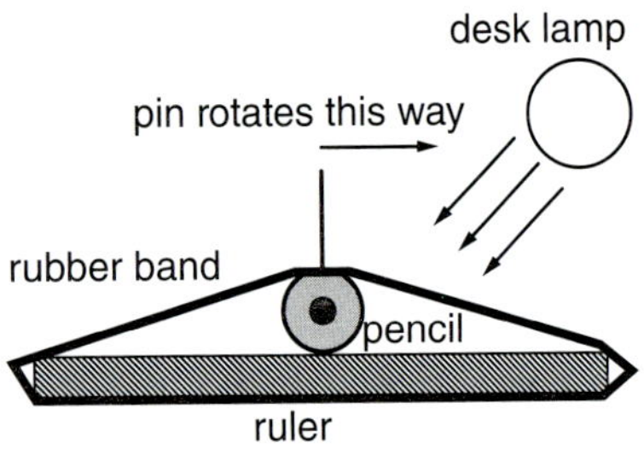

Tie to Prior Knowledge

Energy We have already investigated energy in several forms: linear kinetic energy, rotational kinetic energy, and elastic potential energy. Each atom or molecule in a material can have energy in one or more of these forms. These particles transfer energy among themselves through elastic and inelastic collisions. This transfer of thermal energy is called heat.

2 Teach

Thermal Energy and Temperature

student text pp. 320–321

Concept Development

Thermal Energy and Temperature If the classroom arrangement permits, make a large open area in the center of the room. Ask the students to gather in the center of the space with their arms straight down at their sides so their shoulders are touching (crowd them, but do not arrange them all facing any particular direction). Inform them they are now molecules of a solid at a low temperature. Tell them to vibrate, move up and down on their toes and bend their knees a bit, but not to lift their feet or arms. They are now a solid at a higher temperature (vibrating in a fixed position). It's now time to melt! Tell them to vibrate as before but to also move their feet slowly while the heel of one foot maintains contact with the toes of another student. Additionally, they may now raise their arms as much as needed but they must maintain hand-to-hand contact with at least two other students at all times (a good place for molecular high-fives). The solid has now melted, and they are a liquid. BL Kinesthetic EL

Equilibrium and Thermometers

student text pp. 322–324

Use Figure 4

For the hot object (A) and the cold object (B) to come to thermal equilibrium, the thermal energies of the particles must be exchanged through particle collisions. Ask students how the two bodies reach thermal equilibrium. Because thermal energy is transferred from a hot body to a cold body, there is a gradual transfer of thermal energy from left to right as the collisions occur. The rate of thermal energy transfer is affected by the initial temperature difference of the two bodies and by their area of contact. OL

Quick DEMO

Thermal Expansion

Estimated Time 10 minutes

Materials two beakers, ice water, hot tap water, large alcohol thermometer

Procedure Put the thermometer in a beaker of ice water. After the thermometer has adjusted to the low temperature, remove it and put it in a beaker of hot tap water. Ask students what causes the alcohol to rise in the thermometer. Thermal energy from the molecules in the hot water transfers to the alcohol particles in the thermometer. As the thermal energy of the alcohol increases, the liquid expands and rises up the fixed-volume capillary tube inside the thermometer.

Then ask students to explain the liquid's thermal expansion on the molecular level. According to the kinetic molecular theory, as the temperature of the liquid increases, its molecules have higher average kinetic energy. Because the molecules move faster, they collide with each other more often and more violently, which causes them to occupy more space.

PHYSICS PROJECT CHALLENGE

Calibrate a Thermometer Explain to students that a simple gas thermometer can be made with a small straw or plastic tube, a piece of modeling clay, and a small amount of colored water in a clear plastic bottle. Have students put one end of the straw into the colored water and use the clay to seal the neck of the bottle while holding the straw in place. Blow into the straw so that the water rises in the straw above the neck of the bottle. Warm the thermometer by placing it in the sun, or cool it by resting it in a bucket of ice water. As the temperature changes, the air trapped in the bottle will expand or contract, and the water will move up and down the straw. Ask students how they can calibrate and mark the tube so that it is a useful thermometer. They can mark the water level once the thermometer has reached thermal equilibrium in a place of known temperature. Then they should repeat the procedure in several other areas with known temperatures. AL Kinesthetic

Identify Misconceptions

Temperature Scales Students might be uncertain about which temperature scale to use in calculations. The Kelvin scale is the only thermodynamically correct temperature scale, and it can be used in most calculations. Because the temperature interval for the Celsius scale is the same as the temperature interval for the Kelvin scale, it is permissible to use the Celsius scale if only differences in temperatures are important. The Fahrenheit scale is never used in problems. Fahrenheit temperatures must be converted to temperatures in the Celsius or Kelvin scales before doing calculations.

Content Background

Other Common Thermometers Glass thermometers containing liquids are fragile and are useful only over a limited range of temperatures. In many furnaces, temperatures are measured by a thermocouple. A thermocouple is a very simple and rugged device made by twisting together the ends of two wires made of different metals, such as a copper wire and an iron wire. This twisted connection is called a thermocouple junction. If the free ends of the wires are connected to a voltmeter, one can measure a small voltage that changes with the temperature of the copper-iron junction.

Heat and Thermal Energy Transfer

student text pp. 324–325

Critical Thinking

Colors of Clothing Ask students if the fashion trend from lighter-colored clothes in the summer to darker ones in the winter might have any scientific basis. The darker clothes absorb more radiant energy, thus transferring more thermal energy to the wearer during the colder weather. BL

Use the PhysicsLAB

In *Solar Collectors,* students can investigate how efficiently different materials collect radiant energy from the Sun.

Reinforcement

Heat, Thermal Energy, and Temperature Have students, in groups of two, draw a concept map that relates the following key concepts or quantities: thermal energy, temperature, molecular motion, heat, mass, and specific heat. Students should use short explanatory phrases to explain the links on the concept map. BL Visual-Spatial EL

Specific Heat

student text pp. 325–326

Content Background

Specific Heat For most practical purposes, specific heat can be considered to be constant over a reasonably wide range of temperatures. It is a property of most materials that their specific heat decreases at very low temperatures. The specific heat of all materials must go to zero as the temperature is decreased to absolute zero.

ADDITIONAL IN-CLASS EXAMPLE

Use with Example Problem 1.

Problem Sometimes a short circuit in an electrical wiring system can produce enough thermal energy to melt a wire. How much thermal energy must be transferred to a 20.0-g piece of copper wire to raise it from room temperature (25.0°C) to its melting temperature (1085.0°C)?

Response

$Q = mC(T_f - T_i)$

$= (0.0200\ kg)(385\ J/(kg{\cdot}°C))(1085.0°C - 25.0°C)$

$= 8162\ J$

Use the PhysicsLAB

In *Heating and Cooling,* students can observe thermal energy transfers, temperature changes, and the specific heat of water.

Measuring Specific Heat

student text pp. 327–328

Quick DEMO

Calorimetry

Estimated Time 5 minutes

Materials thermometer, 250-mL beaker, water, calcium chloride (sold at hardware stores for drying humid basements), measuring spoon

Procedure Pour 200 mL of water into the beaker and measure its temperature. Add one teaspoon (5 mL) of calcium chloride to the water and mix briefly. Measure the solution temperature until it stabilizes. Add another teaspoon of calcium chloride and repeat. Measure the temperature again. Repeat once or twice more. Graph the measured temperatures versus the number of teaspoons added. From the temperature differences and the specific heat of the water, the students can calculate the energy released when the calcium chloride dissolves in the water. This technique is one way to measure the energy released or absorbed by a reaction.

Using an Analogy

Calorimeters It might help students understand how a calorimeter works if they compare its chamber to a vacuum flask that is used to store hot or cold beverages. The vacuum flask does not allow thermal energy to readily enter or leave the system. Similarly, the walls of a calorimeter must be well insulated. If thermal energy can enter or leave the system, then the change in water temperature will not accurately reflect the thermal energy absorbed or given off by the sample in the chamber.

DIFFERENTIATED INSTRUCTION

Struggling Students Students often have problems in calculations of the final temperatures in calorimetry because they misuse minus signs or fail to account for all the thermal energy exchanged. Remind them that exchanging thermal energy is like exchanging money: the money goes from one person to another and is not lost. Also, they should expect that the final temperature in a calorimeter will be somewhere between the initial temperatures of the coldest and hottest objects. Finally, when calculating the ΔT in a formula, the correct sign can always be obtained by subtracting the initial temperature from the final temperature. **BL**

Concept Development

MAINIDEA Thermal energy and temperature are concepts that deal with collections of large numbers of very small particles. Thermal energy is the total amount of kinetic and potential energy that these particles possess aside from the overall energy of an object. Temperature is the average kinetic energy of these particles. A small object at a high temperature, when brought into contact with a large object at a low temperature, will transfer some of its thermal energy to the large object until the two objects are at the same temperature. On a microscopic level, the particles of the small object have transferred some of their kinetic energy to the particles of the large object. Eventually the particles of both objects will have the same average kinetic energy.

Use the PhysicsLAB

In *How many calories are there?,* students can experience a real-life application of the use of specific heat.

ADDITIONAL IN-CLASS EXAMPLE

Use with Example Problem 2.

Problem A 0.025-kg block of an unknown substance is at a temperature of 82°C. It is placed in a calorimeter with 0.025 kg of water at 22°C. The system reaches an equilibrium temperature of 27°C. What is the unknown substance?

Response

$$C_{unknown} = \frac{-m_W C_W \Delta T_W}{m_{unknown} \Delta T_{unknown}}$$

$$= \frac{-(0.025\ kg)(4180\ J/(kg\cdot°C))(27°C - 22°C)}{(0.025\ kg)(27°C - 82°C)}$$

$= 380\ J/(kg\cdot°C)$, which corresponds to copper

Content Background

Specific Heat and Atomic Mass In 1819, Alexis-Thérèse Petit and Pierre-Louis Dulong demonstrated that the product of specific heat and atomic mass is very nearly constant for a broad range of solid elements. Thus, the specific heats of these elements are inversely proportional to their atomic masses. Therefore, with the specific heat of a new element determined, one could quickly approximate its atomic mass. Jöns Jacob Berzelius found this discovery useful in improving his table of atomic mass, a precursor to the periodic table of the elements.

Animals and Thermal Energy

student text p. 329

Discussion

Question Water plays a unique role in living things. Our bodies are made mostly of water. What would be the possible effects on humans if the specific heat of water and the heat of vaporization did not have unusually large values and instead were quite low?

Answer Our body temperature would be more easily influenced by external conditions. The high value of the specific heat of water means that a lot of thermal energy must be added or removed to change the temperature of the human body. For example, when a person drinks a cold beverage, his or her body temperature does not decrease. Water's high heat of vaporization also helps cool our bodies when we sweat. With a lower heat of vaporization, we might have to sweat a great deal more to remove excess thermal energy. OL

Logical-Mathematical

3 Assess

Assess the MAIN IDEA

Thermal Energy Fill two clear glasses with water, one with hot water and one with cold water. Add a drop of food coloring to each glass (do not stir the water). Ask students to observe what happens to the food coloring as it disperses through the water. Why does the food coloring disperse at different rates in the two glasses? The hot water has more thermal energy, so the molecules of the hot water have a higher average kinetic energy and disperse the food coloring more quickly.

Check for Understanding

Freezing Water on Bridges Ask students why signs near bridges in the northern parts of the United States often say, "CAUTION: Bridge Freezes Before Road Surface." The underside of the bridge is exposed to cold air, which can cool the bridge very rapidly and cause any water on its upper surface to freeze. The ground under a road insulates the road's surface and provides additional mass that must be cooled to 0°C before water will freeze on its surface. OL

Extension

Clothing for the Outdoors Suppliers have developed sophisticated clothing materials for hikers, campers, and mountaineers to use in cold weather and at high altitudes. Ask the students to research the useful properties of items such as fiberfill insulation, goose down, polymers for underwear and shirts, fleece and synthetic fleece, and reflective blankets. OL

Reading Checks and Caption Checks

Reading Check, *student text p. 322*
Answers will vary. Possible answer: You have just come into the house on a snowy day and kick off your snow boots just inside the door. When you first enter the house, your boots and the snow on them are not in thermal equilibrium with the air in the house. As the snow melts and your boots warm up, the water and the boots reach thermal equilibrium with the air inside.

Caption Check, *student text p. 322*
Because thermal energy is transferred from a warmer object to a colder object, thermal energy is transferred from your forehead to the thermometer. Once thermal equilibrium is reached, the thermometer will accurately measure your temperature.

Reading Check, *student text p. 323*
Absolute zero is the lowest possible temperature that can be obtained. At absolute zero, all thermal energy would be removed from an ideal gas.

Caption Check, *student text p. 324*
Possible answers: If you step on a hot sidewalk in the summer with bare feet, the thermal energy will be conducted from the sidewalk to your foot. When you open the door to an air-conditioned building on a warm day, thermal energy is transferred from the outdoors into the building by convection. Thermal energy is radiated from an electric heater.

Reading Check, *student text p. 325*
The specific heat of a material is a measure of the amount of energy that must be added to a unit mass of the material to raise its temperature one temperature unit.

Caption Check, *student text p. 325*
Copper has a low specific heat, so t is a good choice for the bottom; it will conduct heat from the stove to the inside of the pan. Stainless steel has a somewhat higher specific heat, so it will transfer thermal energy to the food, but isn't prone to temperature fluctuations. Plastic has a very high specific heat, making it a good insulator for the handle.

Practice Problems
student text p. 326

1. 5.3×10^4 J

2. \$0.36

3. a. 10.0 K

b. 21 K

c. For temperatures above 0°C, water is the better coolant because it can absorb thermal energy without changing its temperature as much as methanol does.

Practice Problems
student text p. 328

4. 9.04×10^2 J/(kg•°C)

5. 2.53×10^2 J/(kg•°C)

6. 45.0°C

7. 28.5°C

8. 42.1°C

Section 1 Review
student text p. 329

9. No; your sense of the floor temperature is actually the transfer of thermal energy to or from your feet. Tile conducts thermal energy from your feet more efficiently than a carpet or the air in the room.

10. a. 278 K

b. −239°C

c. 485 K

d. 43°C

11. No; heat is measured in joules (J), and specific heat is measured in joules per kilogram-kelvin (J/(kg•K)).

12. The basketball has kinetic energy when it is in motion and gravitational potential energy when it is above the ground. Some kinetic energy is transformed to thermal energy and sound energy when the basketball hits the ground. The total thermal energy of the basketball depends on the total energy of its particles.

13. Yes; if the mass of the cold water exceeds the mass of the hot water, the cold water could have the same thermal energy because of the greater number of particles in the bowl.

14. A potato has a high specific heat and conducts thermal energy poorly, so it loses its thermal energy slowly.

15. Potatoes do not conduct thermal energy well. Increasing surface area by cutting a potato into small parts increases the rate of thermal energy transfer into the potato.

16. They are good thermal conductors and have relatively low specific heats.

17. The plastic spoon has a low thermal conductivity, so it does not transmit thermal energy to your tongue very well.

18. Some of the water molecules are evaporating at the surface of the liquid. When the warm particles contact the cool air above the pot, the molecules condense into liquid water again. This condensed water forms a mist above the pot.

student text pp. 330–339

1 Introduce

Bellringer

Work and Internal Energy Have each student take a clean steel paper clip and reshape it into a "U." Tell students to hold the bend in the middle of the clip to their upper lip. Students should notice that the clip feels slightly cool. Then ask them to bend their clips vigorously several times and have them again touch the clip to their upper lip. They should notice that the clip feels warmer because of the work done on the metal. **BL** **Kinesthetic** **EL**

Tie to Prior Knowledge

Friction Students have seen that mechanical energy is transformed into thermal energy due to friction. This energy transformation increases the temperatures of objects. Have the students rub their hands, creating friction, and notice that their hands get warmer. Someone camping might rub his or her hands to warm them or hold them near a campfire. The camper is performing work by rubbing his or her hands. The fire warms the hands primarily though radiant energy. The first law of thermodynamics relates work, heat, internal energy, and temperature.

2 Teach

Changes of State

student text pp. 330–333

Content Development

MAINIDEA One way to demonstrate the energy required to change a solid to a liquid and then to a gas would be the melting of an ice sculpture. Begin with small ice sculptures (colored water frozen in a candy mold). Place two in dishes: one at room temperature and one under a heat lamp. Put a third one in a zipper storage bag and ask students to hold it in their hands and roll it around. The more energy that is applied, the quicker the ice will melt. Under the heat lamp, it will eventually evaporate.

Use the MiniLAB

In *Melting,* students can observe water change state and observe water's melting point.

Use the PhysicsLAB

In *Heat of Fusion,* students can investigate calorimetry.

ADDITIONAL IN-CLASS EXAMPLE

Use with Example Problem 3.

Problem 1.0 kg of water is warmed from room temperature (25.0°C) to boiling, then half of its initial volume is vaporized. How many joules of thermal energy must the stove provide to do this?

Response

$$Q = mC\Delta T + (1/2)mH_V$$

$$= (1.0\ \text{kg})(4180\ \text{J/(kg·°C)})(100.0°\text{C} - 25.0°\text{C}) + \frac{1}{2}(1.0\ \text{kg})(2.26\times10^6\ \text{J/kg})$$

$$Q = 1.4\times10^6\ \text{J}$$

PHYSICS CHALLENGE ACTIVITY

Melting Temperatures in Solutions The addition of a substance to water to form a solution changes the water's melting and boiling temperatures. For example, salt is sprinkled on icy sidewalks to lower the freezing point of the water. Antifreeze added to an automobile cooling system increases the boiling temperature of the water. Ask students to answer the following questions and report their findings to the class. How do the melting and boiling temperatures change with the number of particles dissolved in water? Do all solutes have the same effects on melting and boiling? How can these changes in temperature be explained in terms of what students have learned about melting and boiling? **AL** **Logical-Mathematical**

The First Law of Thermodynamics

student text pp. 334–336

Use the MiniLAB

In *Convert Energy,* students can investigate the relationship between work and thermal energy.

Identify Misconceptions

Engines and Energy Students might think that a car engine converts all the chemical energy of gasoline into the work that drives the car. Instead, the engine uses only part of the available chemical energy to produce useful work. The rest is expelled as waste heat. The engine also requires a lower-temperature reservoir into which the waste heat can be expelled. A car engine would not work on a planet where the atmospheric temperature is higher than the combustion temperature.

Critical Thinking

Thermal Energy Released by Condensation Ask students why exposing the skin to steam can cause a much more severe burn than exposing the skin to boiling water. At normal pressures, the temperature of steam can exceed 100°C, but the temperature of liquid water cannot. Steam at 100°C has much more thermal energy than water at 100°C. Also, steam that touches the skin can condense as water at 100°C, releasing its large heat of vaporization in the process. OL

DIFFERENTIATED INSTRUCTION

Visually Impaired Put several drops of peppermint oil in one evaporating dish and several drops of orange oil in another. Put the dish with the peppermint oil on a hot plate set on low, and put the dish with orange oil in a beaker of crushed ice. Put the two dishes equidistant from the class. The odor of peppermint oil reaches students first and is much stronger. Ask students to explain their observation. Because the hot plate is constantly adding thermal energy to the peppermint oil, more of the peppermint oil's molecules have sufficient kinetic energy to escape the attractive forces of the liquid. BL Kinesthetic EL

Use Figure 17 and Figure 20

Energy Diagrams These figures diagram the transfers of thermal and mechanical energies in engines and refrigerators. Show the two figures together and ask the students to compare them. The differences have to do with the directions of the arrows for W, Q_H, and Q_C. Would it be possible to use a heat engine to provide the work W to operate a refrigerator? Could that refrigerator then be used to provide the heat Q_H to run another heat engine that could produce more work, W_2? This could be done, but it is not useful. Some of the useful energy is lost in each of these steps due to the inherent waste heat lost in the engine and the refrigerator. OL

Concept Development

Waste Heat Ask the students to cite examples of waste heat produced in our homes and around our cities and to explain the effect that this waste heat has on the environment near cities. Examples might include the thermal energy removed from a building by air conditioning and the thermal energy from automobile engines. This waste heat can cause local temperatures to rise, especially in urban areas. OL Logical-Mathematical EL

REAL-WORLD PHYSICS

Heating Homes The Chinese were the first to use central heating for their homes. Instead of building a fire inside the house, they would build one outside. The fire heated air that circulated by convection currents through hollow spaces under the flooring. The floor tiles heated up, heating each room efficiently. The Inuit reversed the concept in their igloos. By digging the entryway lower than the inside of the igloo, they created a natural pressure barrier to keep the hot air in. Because cold air is more dense and sinks to the lowest point, the entryway fills with cold air from the outside, stopping the flow of air. The warmer air inside the igloo stays inside because it is less dense than the colder air in the entry.

Reinforcement

Engines and Refrigerators Have the students work in pairs to create a concept map that relates the following concepts or quantities: heat engine, refrigerator, heat, work, efficiency, and temperature reservoir. Students should use short explanatory phrases to explain the links on the concept map. BL Visual-Spatial EL

Content Background

Work on Gases How do you perform work on an object? If an external force F stretches a wire in length by an amount ΔL, then the amount of work done on the wire is $W_{external} = F\Delta L$. The work done by the wire would be $W_{wire} = -F\Delta L$. According to the first law of thermodynamics, it follows that $\Delta U_{wire} = Q - W_{wire} = Q - (-F\Delta L) = Q + F\Delta L$. Students will learn, in the next chapter, that pressure = force/area is a useful formula to use in working with gases. The work done by an external pressure in compressing a gas is $W_{external} = -P\Delta V$, where ΔV is the change in volume of the gas. The minus sign applies because the volume decreases as the external pressure increases.

The Second Law of Thermodynamics

student text pp. 337–339

Discussion

Question If you want to increase the entropy of a pot of water by a given amount ΔS, would it require more thermal energy or less thermal energy if the water were already warm than if the water had just melted?

Answer Because $\Delta S = Q/T$, the amount of thermal energy required to increase the entropy of the water by ΔS depends on the initial temperature of the water. The colder water would need less thermal energy applied in order to increase the entropy by a given amount.

OL Logical-Mathematical

REAL-WORLD PHYSICS

Entropy in Other Fields The concept of entropy as a measure of the dispersal of energy has become useful in other fields of study. Computer scientists employ mathematical recipes that use randomness to seek speedy solutions to complex problems. These algorithms sometimes include maximizing the entropy of the mathematical system. The entropy concept is useful in describing certain types of mathematical codes, including those used by spies and the genetic codes stored in DNA. Ideas related to entropy have also been employed to predict the behavior of financial markets.

PHYSICS PROJECT ACTIVITY

Solar Collectors The collection of solar energy can be used to produce electrical power. Ask students to design and build simple solar-energy collectors using materials found around the house, such as cardboard, umbrellas, aluminum foil, plastic cups, and tape. The goal should be to reflect sunlight to a collection point that includes a small beaker of water. The amount of energy collected can be calculated by measuring the increase in temperature of the water. Student teams should present their designs with the reasoning behind their design decisions, their collection devices, and their measurements of the devices' performance.

3 Assess

Assess the MAIN IDEA

Efficiency and Entropy Ask students what would happen to the temperature in a room if an air conditioner were left running on a table in the middle of the room. The temperature of the room would increase. The thermal energy removed from the room by the cool air is reintroduced to the room through the heat exhaust coils on the other side of the air conditioner. Even if the same amount of thermal energy was removed by the cool air that was reintroduced to the room by the exhaust coils, the heat produced by the compressor motor would still warm the room.

Check for Understanding

Heat, Work, and Entropy Ask students what makes heat different from work and which of the two affects entropy. Ask them to draw arrows on a diagram of a heat engine that show how entropy increases in the operation of a heat engine. Heat is the spontaneous transfer of thermal energy from a warmer object to a colder object. The addition of thermal energy increases the entropy of an object. Work can be done on an object without increasing its entropy. In a heat engine, the entropy of the hot reservoir is decreased, and the entropy of the cold reservoir is increased. For the engine to operate continuously, the entropy of the engine must remain constant. OL

Extension

Entropy and Computing Ask students to discuss the following: As a computer performs calculations, it is organizing information. Therefore, the computer's entropy is decreasing. Discuss whether the processes of computation violate the second law of thermodynamics. The computer can be viewed as a closed system in which entropy is decreased by work done on the system. However, as it does this the computer releases thermal energy, thereby increasing the entropy of the rest of the universe. So, in keeping with the second law of thermodynamics, the entropy of the universe always increases. AL

Reading Checks and Caption Checks

Reading Check, *student text p. 336*

A more efficient engine would use less fuel for a given amount of work because less of the energy released would be converted to waste heat, and more energy would be used to perform useful work.

Reading Check, *student text p. 338*

In any closed system (for example, the entire universe), the total amount of entropy can never decrease.

Practice Problems

student text p. 333

19. 3.75×10^4 J

20. 502 kJ

21. $H_f = 3.34\times10^5$ J/kg; $H_v = 2.259\times10^6$ J/kg

22. 9.5×10^7 J

23. 9.40×10^2 kJ

Practice Problems

student text p. 334

24. 75 J

25. 1.8×10^3 J

26. 9 times

27. 2.6×10^4 stirs

28. 200 kJ

Physics Challenge

student text p. 338

1. $\Delta S = Q/T = \frac{mC\Delta T}{T} = \frac{(1.0\text{ kg})(4180\text{ J/(kg·K)})(274\text{ K} - 273\text{ K})}{273\text{ K}}$

$= 15$ J/K

2. $\Delta S = Q/T = \frac{mC\Delta T}{T} = \frac{(1.0\text{ kg})(4180\text{ J/(kg·K)})(354\text{ K} - 353\text{ K})}{353\text{ K}}$

$= 12$ J/K

3. $\Delta S = Q/T = \frac{mC\Delta T}{T} = \frac{(1.0\text{ kg})(130\text{ J/(kg·K)})(274\text{ K} - 273\text{ K})}{273\text{ K}}$

$= 0.48$ J/K

4. $\Delta S = Q/T = \frac{mH_f}{T} = \frac{(1.0\text{ kg})(3.34\times10^5\text{ J/kg})}{273\text{ K}}$

$= 1.2\times10^3$ J/K

In general, applying a given amount of heat will increase entropy by a greater amount when the starting temperature is lower. This is because entropy measures dispersal of energy, and a given amount of heat will not disperse as much in a substance that is already at a higher temperature. Also, a change of state generally results in a larger change in entropy than heating by 1°C does.

Section 2 Review

student text p. 339

29. Some of the thermal energy from the hot reservoir is used to do work (transformation to mechanical energy), and some is transferred to the cold reservoir, raising its temperature. Entropy is increased whenever thermal energy flows from a warmer object to a colder object and raises its temperature.

30. The condensing steam released its heat of vaporization into the room and was then circulated back to the boiler where it was reheated.

31. 20.8 kJ

32. 3.5×10^5 J

33. Part of the kinetic energy of the hammer is absorbed as thermal energy by the lead block. The hammer's energy is 4.0 kJ, and the change in thermal energy of the block is 2.0 kJ. About half of the hammer's energy went to the lead block.

34. The water at the top has gravitational potential energy that is dissipated into thermal energy when the water splashes at the bottom.

35. 0.293°C

36. The gas releases thermal energy at its combustion temperature. The natural gas molecules break up and combine with oxygen. The thermal energy is distributed in many new ways, and the natural gas molecules cannot be readily reassembled.

37. When the water evaporates, it absorbs thermal energy from the air.

ON THE **JOB**

2's Company, 3's a Crowd . . .

Pedestrian Dynamics

Purpose

Students will learn how physics models are used to explain and quantify pedestrian dynamics (the flow of people in a certain space) to help design better systems for crowd flow and dispersal.

Background

In 1971 L.F. Henderson published a paper titled "The Statistics of Crowd Fluids" in which he showed that equations from the Maxwell-Boltzmann theory of a homogeneous gas composed of statistically independent particles in equilibrium on a two-dimensional surface could be applied to crowd movement. He noted that at low densities a crowd resembled the particles in a gas and at higher densities, a liquid. In the 1990s Dirk Helbing improved and mathematically founded this phenomenological fluid-dynamic approach on the basis of a pedestrian-specific gas kinetic model. Recent attention has focused on more microscopic (individual pedestrians) models, using theories regarding social force and attempts to predict pedestrian behavior statistically.

Teaching Strategies

- Ask students to remember a time when they were part of an extremely dense crowd, how quickly the forces can change direction, and whether they felt helpless to move in any particular direction.
- Ask students to study their building's evacuation plan and suggest any ways it can be improved.
- Ask students to research or think of any novel architectural ideas to help with crowd flow. One suggestion to improve evacuations in building is to put columns in front of exits, which should help to reduce friction effects and prevent "arching," seen when bottlenecks are created at narrow exits. Other suggestions include dispersal of entrances and exits rather than having centralized ones, avoiding obscured doorways and stairs, and eliminating confusing or ambiguous paths.

GOING**FURTHER** >>>

Expected Outcomes The width of an exit is perhaps the most important factor in the flow of a crowd. Studies of plane evacuations with narrow exits have found that passengers can leave faster when they cooperate rather than compete with one another to get out faster; however, for wider exits the opposite is the case.

CHAPTER 12 ANSWERS

SECTION 1

Mastering Concepts

38. The mechanical energy is the sum of the potential and kinetic energies of the ball considered as one mass. The thermal energy is the sum of the potential and kinetic energies of the individual particles that make up the mass of the ball. The temperature is a measure of the average kinetic energy of the particles of the ball.

39. No; there are no particles that have energy in a vacuum.

40. No; there is a distribution of velocities of the atoms or molecules.

41. Your skin measures thermal energy flow to or from itself. The metal doorknob absorbs thermal energy from your skin faster than the wooden door, so it feels colder, even if the door and the knob are at the same temperature.

42. The two objects will change temperatures depending on their masses and specific heats and will eventually reach the same temperature. The temperature changes are not necessarily the same for each.

Mastering Problems

43. 1.64×10^4 J

44. 2.02×10^4 J

45. 127 J/(kg•K)

46. 63°C

47. 1.00×10^3 J/(kg•K)

48. 0.87 L

49. 15 kg

50. 370 s (6.2 min)

51. D < E < C < A < B

52. 12.7°C

SECTION 2

Mastering Concepts

53. Yes; when you melt a solid or boil a liquid, you add thermal energy without changing the temperature.

54. In freezing, wax releases thermal energy.

55. When the water in the cover evaporates into the dry air, it must absorb an amount of thermal energy proportional to its heat of fusion. In doing this, it cools off the canteen.

56. Inside the house, the coolant is vaporizing in the coils to absorb energy from the rooms.

Mastering Problems

57. 6.68×10^6 J

58. 2.47×10^5 J/kg

59. 29°C

60. 3.09×10^4 J

61. 290°C

62. 1.0×10^1 m

Applying Concepts

63. It should make no difference. Either way, the water is at the same temperature.

64. Methanol, because it has a lower specific heat; for a given mass and heat transfer, it generates a bigger ΔT, because $Q = mC\Delta T$.

65. The specific heat of aluminum is much greater than that of lead; therefore, it melts more ice.

66. As they evaporate, they absorb their heat of vaporization from the skin.

67. The water on the leaves will not freeze until it can release its heat of fusion. This process keeps the leaves warmer longer. The heat capacity of the water slows down the cooling below 0°C.

68. No; the cup with block A will be hotter because block A contains more thermal energy.

69. In the northern hemisphere, the sunlight comes from the south. In most places it is easier to heat a house in winter than to cool a house in summer, so north-facing windows make it easier to keep a house cool in the summer while still letting indirect light in and warming the house somewhat during the winter.

Mixed Review

70. efficiency = 34%; waste heat = 3500 J/s

71. 0.016°C

72. **a.** Answers will vary. Sample answer: ". . . There is 1.0 kg of water in the beaker. A 5.0-g piece of 0°C zinc is placed in it. What is the final temperature of the zinc and water?"

b. Answers will vary. Sample answer: ". . . If 75 J of thermal energy are added to it, how much does its entropy increase?"

73. 12°C

74. 1.1 kg; thus, you need slightly more ice than tea, but this ratio would make watery tea. Let the tea cool to room temperature before adding the ice.

75. The copper block has 2.33 times as much mass as the aluminum block.

76. 12 m/s

77. 2.0×10^{-5} kg

Thinking Critically

78. 4.8×10^{-19} J/molecule

79. Answers will vary. Sample answer: "A certain amount of water is warmed from 250 K to 260 K and its entropy increases by 75 J/K. What is the mass of the water?"

80. **a.** 0.0313 J/K

b. 0.103 J/K

81. 0.0478 kg

82. Answers will vary. Answers should reflect changing average temperatures on Earth, different weather patterns, plant and animal species dying out, etc.

Writing in Physics

83. In 1799, heat was thought to be a fluid that flowed from one object to another. However, Count Rumford thought that heat was caused by the motion of particles. He did not do any quantitative measurements, and his ideas were not widely accepted. In 1843, Joule, doing careful measurements, measured the change in temperature caused by adding heat or doing work on a quantity of water. He proved that heat is a type of energy, and that energy is conserved. Joule deserves the credit and the eponymous unit.

84. The large specific heat and large heats of fusion and vaporization mean that water, ice, and water vapor can store a lot of thermal energy without changing their temperatures too much. The implications are many. The oceans and large lakes moderate the temperature changes in nearby regions on a daily and seasonal basis. The day-to-night temperature variation near a lake is much smaller than the day-to-night temperature variation in the desert. The large heat of fusion of water controls the change of seasons in the far north and south. The release of energy by freezing water in the fall and its absorption in the spring slows the temperature changes in the atmosphere. Water absorbs and stores a lot of energy as it vaporizes. This energy can be used to drive meteorological events such as thunderstorms and hurricanes.

Cumulative Review

85. 0.98 m/s^2; 3.9 rad/s^2; 35.3 N; 12.5 rad/s

86. 3.4×10^{3} J

87. 10 times in 1 h

Standardized Test Practice

student text p. 345

Multiple Choice

1. C
2. A
3. D
4. A
5. C
6. C
7. A
8. C
9. B

Free Response

10. To melt: 152 kJ; to boil: 1030 kJ; the conversion to steam requires 878 kJ more energy; to form steam: 190×10^2 kJ; the difference in energy between the phase changes is much greater than the energy required to warm water in the liquid state.

Rubric

The following rubric is a sample scoring device for free response questions.

Points	Description
4	The student demonstrates a thorough understanding of the physics involved. The response may contain minor flaws that do not detract from the demonstration of a thorough understanding.
3	The student demonstrates an understanding of the physics involved. The response is essentially correct and demonstrates an essential but less-than-thorough understanding of the physics.
2	The student demonstrates only a partial understanding of the physics involved. Although the student may have used the correct approach to a solution or may have provided a correct solution, the work lacks an essential understanding of the underlying physical concepts.
1	The student demonstrates a very limited understanding of the physics involved. The response is incomplete and exhibits many flaws.
0	The student provides a completely incorrect solution or no response at all.

CHAPTER 13

States of Matter

About the Photo

Merced River The Merced River originates in Yosemite National Park and was responsible for the formation of Yosemite Valley. Ask students if they have ever floated down a river. The concept of pressure can explain how a boat can be made of material that is denser than water if it is the right shape. Did they notice variations in speed along the course of the river? The speed of a river depends in part on its cross-sectional area and, therefore, also on the depth of the water at a given point.

Chapter Overview

This chapter discusses various properties of matter that depend on its state and, therefore, on its thermal energy. Pressure, volume, and temperature are explored via the gas laws. Forces in liquids include cohesive and adhesive forces. Concepts of pressure and buoyant force are quantified by using Pascal's law and Archimedes' principle, and Bernoulli's principle relates pressure to velocity of a fluid in motion. The distinction between crystalline solids and amorphous solids is explored. Most solids expand and contract depending on their temperature.

Before students study the material in this chapter, they should study:

- Kinetic energy
- Mass and weight
- Temperature
- Thermal energy
- Vectors versus scalars

To solve the problems in this chapter, students will need a strong grasp of:

- Scientific notation
- Significant Figures
- Solving linear equations

Introduce the BIG IDEA

Ask students if they have ever noticed dust particles randomly jittering in mid air. This is an example of Brownian motion. Brownian motion is the result of the kinetic energy that all atoms and molecules have, and is directly related to the phenomena of pressure, buoyancy, and temperature.

Go online!
connected.mcgraw-hill.com

Your online experience includes:

- **Video**
- **Concepts In Motion**
- **Virtual Investigations**
- **All Lab, Applying Practices activities, and PBL worksheets**
- **Online Practice**
- **Personal Tutor**
- **Multilingual eGlossary**
- **Presentation**

Use the LaunchLAB

In *Measuring Buoyancy,* students can observe and take measurements of a buoyant object.

Use Physics T.V.

Video

PLASMA You've probably explored the three main states of matter, but how much experience do you have with plasma? You might be surprised!

1 Introduce

Bellringer

A Pneumatic Lift You can demonstrate the force of a fluid by making an inflatable pillow. Seal a small trash bag with masking tape. Insert a straw into a hole in the bag, and seal where the straw enters the bag. Ask a student to inflate the bag slightly. Place two textbooks on the bag, and ask the student to raise them. Ask students to describe the forces the bag exerts on the books and to describe the forces and work that the student does on the book by blowing into the straw. Forces on the books: gravity and the upward force from the air in the bag. The force the bag exerts on the books equals the area of the bottom book times the pressure in the bag. The student does work on the books by moving air into the bag. **BL** **Kinesthetic** **EL**

Tie to Prior Knowledge

Forces Remind students about forces and drawing free-body diagrams. When an object is not accelerating, the sum of all forces acting on that object must be zero. Explain that pressure is a force divided by an area. Pressure is a scalar, not a vector.

2 Teach

Liquids and Gases

student text p. 348

Critical Thinking

MAIN IDEA The two fluids that most affect our lives are air and water. Ask students to write down as many properties as they can of each. List the properties in two columns for easy comparison. Ask students what happens when they move through these fluids. Answers will vary. Water v. Air: heavy/light, drink/breathe, river flow/breeze, sits in glass/stays in balloon, meniscus/no meniscus, wets objects/dries objects, visible/not visible, viscous/not viscous, definite volume/no definite volume, not compressible/compressible. Students may compare the effort needed for swimming or paddling to waving their hands in the air. **BL** **Linguistic**

Pressure

student text pp. 349–350

Concept Development

Pressure and Force Some students might confuse pressure with force; emphasize that pressure is the quotient of force divided by the area over which that force is exerted. In other words, pressure is force per unit area.

Identify Misconceptions

Direction of Pressure Students might think that air or water pressure is only in a downward direction when they learn that a column of air or water exerts pressure on the region beneath it. Explain that the column exerts pressure on the region beneath it in all directions, even to the side and upward. For example, an underwater diver experiences water pressure from all sides, not just from above.

ADDITIONAL IN-CLASS EXAMPLE

Use with Example Problem 1.

Problem How does the average pressure of an elephant standing on the ground compare to that exerted by a person standing on the ground? The person weighs 640 N. Each shoe has an area of about 0.016 m^2 in contact with the floor. The elephant weighs 4.13×10^4 N and has four feet, each with an area of about 0.14 m^2 in contact with the ground.

Response Find the average pressures for each, using $P = F/A$.

$$P_{\text{person on floor}} = \frac{640\ \text{N}}{(2)(0.016\ \text{m}^2)} = 2.0\times10^4\ \text{Pa}$$

$$P_{\text{elephant on ground}} = \frac{4.13\times10^4\ \text{N}}{(4)(0.14\ \text{m}^2)} = 7.4\times10^4\ \text{Pa}$$

Though the elephant's weight is almost 65 times that of the person, the pressure exerted by the elephant on the ground is not even four times greater than the pressure exerted by the person's soles.

Discussion

Question At high pressures, solids can crack. Ask students to discuss why the largest land animals on Earth are smaller than the largest animals in the ocean.

Answer If an animal on land were to become large enough, the pressure on the bones could cause them to crack (because stress due to force of gravity causes strain). The water surrounding a whale, however, supports it uniformly, while all the weight of an elephant is concentrated on four feet. **OL** **Logical-Mathematical**

PHYSICS PROJECT ACTIVITY

Teacher to Teacher

Pressure v. Area Demo For a fixed force, the larger the area, the less pressure exerted.

Materials paper cups, 6-in by 48-in (about 15 cm by 120 cm) board

Procedure Have a student place one foot on one cup and push down on the cup with his or her body's weight. The pressure should crush the cup. Then place 10–16 cups upside down on the floor. Lay the board on top and have the student carefully step onto it. The cups will not crush. Have two more students step on each side of the board without getting too close to the edge. The board and the cups should support the weight.

The Gas Laws

student text pp. 351–353

Use the MiniLAB

In *Pressure* students can investigate how much pressure they exert when they stand on one foot.

Quick DEMO

Charles's Law

Estimated Time 5 minutes

Materials hair dryer, partially inflated polyester-film helium balloon, string, small object

Procedure Do this on a cool or cold day or use a refrigerator to cool the balloon. Tie a small object to a partially inflated polyester-film helium balloon so that the balloon floats but cannot take off. Take the balloon outside or place it in the refrigerator. When exposed to the cooler temperature, the balloon will shrink and become less buoyant so that it nearly rests on the ground. Bring the balloon back inside. Ask students to predict what will happen to the balloon when you heat it up with the hair dryer. The balloon will rise up toward the ceiling and possibly lift the small object.

Use Figure 4

Gas Laws The gas laws present a good opportunity to reinforce graph interpretation skills. For instance, Boyle's law indicates an inverse relation between pressure and volume if temperature is held constant. Charles's law indicates a direct relationship between volume and temperature if pressure is held constant. **OL** **Visual-Spatial** **EL**

Content Background

The Liquid Barometer In 1643 Evangelista Torricelli demonstrated that mercury in a tube rises to the same level even when the tube is tilted. Because Torricelli allowed no air to enter the tube, the pressure at the bottom of the column of mercury equaled the pressure exerted by the surrounding air. Torricelli realized he could measure the pressure exerted by the atmosphere by measuring how high the mercury column rose. The height of the column of mercury in the tube is proportional to the atmospheric pressure. Barometers, which measure atmospheric pressure using this principle, are used to indicate changes in weather.

Reinforcement

Ideal Gas Law Activity Have students confirm the units in the ideal gas equation by first solving the equations for k ($Pa{\cdot}m^3/K$) and for R ($Pa{\cdot}m^3/(mol{\cdot}K)$). Then have them substitute the appropriate units for *P, V, T, N,* and *n* in the equations. **OL** **Logical-Mathematical** **EL**

Careers

Commercial Gases The production, storage, transportation, and delivery of gases is an important industry. Engineers in the gas-supply business must understand the operation of pressurized systems as well as the chemistry of the gaseous materials. Breathing gases, which are mainly mixtures of N_2 and O_2, are sold in high-pressure tanks to hospitals, physicians, and scuba divers. Giant pipelines transport the natural gas used in home heating and in the chemical industries. Insulated trucks transport natural gas that has been liquefied at low temperatures. Carbon dioxide, which puts the fizz in sodas and is used for many chemical processes, can be transported as a liquid or a solid. Even helium is often transported in the liquid form.

ADDITIONAL IN-CLASS EXAMPLE

Use with Example Problem 2.

Problem An oxygen storage bottle used by a medical patient with poor breathing has a volume of 4.0 L and contains gas at a pressure of 1.3×10^7 Pa at 300.0 K. The bottle releases oxygen for the patient to breathe at atmospheric pressure. What is the volume of oxygen that is delivered to the patient at atmospheric pressure? How many moles of oxygen are in the bottle?

Response To find the volume, use the combined gas law with $T_1 = T_2$ and solve for V_2. $\frac{P_1V_1}{T_1} = \frac{P_2V_2}{T_2}$ and $T_1 = T_2$

Then $V_2 = \frac{P_1V_1}{P_2}$

$= \frac{(1.3\times10^7 \text{ Pa})(4.0 \text{ L})}{1.0\times10^5 \text{ Pa}} = 520 \text{ L}$

To find the moles, use the ideal gas law, $PV = nRT$, and solve for n.

$n = \frac{P_1V_1}{RT_1}$

$n = \frac{(1.3\times10^7 \text{ Pa})(4.0\times10^{-3} \text{ m}^3)}{(8.31 \text{ Pa·m}^3/(\text{mol·K}))(300.0 \text{ K})}$

$n = 21$ mol

Thermal Expansion and Plasma

student text p. 354–355

Content Background

Candles, Where the Apparent Weight Is Zero
Combustion in space takes a different form compared to combustion on Earth. The teardrop shape of a candle flame is a result of Earth's gravity and convection currents. Where the apparent weight is zero (for example, in Earth's orbit), instead of seeing a tiny dense core of blue close to the wick topped with a blaze of yellow-white shooting upward in the candle flame, you see a small dome of intense blue. Think of it as a purer flame—without the influence of gravity, there are no convection currents to carry dust to the flame, and such impurities are what make a candle burn yellow.

Using Models

Ice Chips Ask students why they can pass their hands through water, but not through ice. Particles in a liquid can move past each other, but particles in a solid cannot. Point out that water has some unusual properties. For example, most solids expand when they melt. Water, however, expands when it freezes. Ice floats for the same reason it chips: its molecular structure. Take a box of small, round, plastic foam balls, each one representing a water molecule. Pour them out onto a table; they seem to flow. Show students a diagram of the crystalline structure of ice. Have them work in pairs to connect the small balls (water molecules) with toothpicks so that they form a similar lattice structure to model a block of ice. It is impossible now to pour the water molecules. See whether they can break a block of "ice" into chips. BL Kinesthetic EL

Use an Analogy

Solid and Liquid Forms of Water Ask students the following question: Does ice have greater or smaller volume than water? greater Have students explain their answer in terms of particles. The following analogy can be used to explain this: Ask students how many people fit in an elevator. When students have agreed on an answer, ask them how many people would fit if each of those people had his or her arms in casts, out at the sides of their bodies. Explain the solid (people with casts) and liquid (people without casts) forms of water in terms of this analogy. OL

3 Assess

Assess the MAIN IDEA

Pressure in a Tire Sketch an automobile on the board, with its four tires resting on a flat road. If each tire is inflated to a pressure of 2.0×10^5 Pa and each tire has an area of 0.024 m^2 in contact with the road, what is the weight of the automobile? 1.9×10^4 N

Check for Understanding

Putting on the Pressure Have each student estimate the surface area of his or her body. It will probably be between 1 and 2 m^2. Have students compute the total force due to air pressure on the surface of his or her body. Then ask why a person's body does not collapse under such a large force. An equally large outward force acts within each person's body. OL EL

Extension

Living in Space Astronauts wore space suits on the Moon because there is no atmosphere there. Ask students whether an astronaut could walk on the Moon using only the air-breathing apparatus that scuba divers use. Ask them to explain their answers. No; such an apparatus would only supply oxygen. Because being on the surface of the Moon is like being in a vacuum, the astronaut also needs to be surrounded by a pressurized environment.

Reading Checks and Caption Checks

Caption Check, *student text p. 348*
gases in the atmosphere, including water vapor, oxygen, and carbon dioxide

Caption Check, *student text p. 349*
3.3×10^4 Pa; find the astronaut's weight on the Moon (about one-sixth the weight on Earth) and the area of his or her shoes; use these values to calculate the pressure exerted by the astronaut.

Caption Check, *student text p. 352*
If volume is constant, the equation becomes $\frac{P_1}{T_1} = \frac{P_2}{T_2}$.

Reading Check, *student text p. 354*
As fluids are heated, they expand. As a result, heated air becomes less dense so it rises. When liquid near the bottom of a heated pan increases in temperature, it becomes less dense and rises.

Practice Problems
student text p. 351

1. 1.2×10^5 N
2. 1.0×10^2 kPa
3. 23 kPa
4. 2.7×10^4 N directed from the inside of the house outward
5. 8.9×10^{-2} m^2

Practice Problems
student text p. 354

6. 3.4 m^3
7. 5.1×10^2 g
8. 2.40×10^2 kPa
9. 0.83 m^3

Section 1 Review
student text p. 355

10. Liquids and gases are fluids that flow and do not have a definite shape. Exerting forces on liquids and gases can change the volume and temperature of them. The combined gas law represents the relationship between pressure and gases. Plasma is a gaslike state of negatively charged electrons and positively charged ions. Plasma is also a fluid.
11. **a.** The pressure of the air is the same.
    b. Because $F = PA$, the total force of the air is greater on the box with the greater area. The second box has twice the surface area, so it has twice the total force of the first box.
12. 3.1×10^1 m^3
13. As the water is heated from 0°C, the density will increase until it reaches a maximum at 4°C. On further heating to 8°C, the density of the water will decrease.
14. 9×10^2 K
15. 0.0224 m^3
16. 28.1 mol; 8.1×10^2 g
17. There are an equal number of particles in the two samples. In an ideal gas, the size of the particles is not relevant to the volume of the gas or the pressure exerted by the gas.

SECTION 2 Forces Within Liquids

1 Introduce

Bellringer

Molasses in January Slowly pour a thick fluid, such as shampoo or molasses, into an empty beaker. Explain that one factor that affects how liquids move is viscosity, or the resistance to flow. Attractive forces between molecules determine how easily molecules move past each other. These intermolecular forces also produce the liquid's surface tension. Have students describe how the fluid moves as you pour it. It moves slowly; it also adheres to the surface of the beaker. Point out that temperature can also affect viscosity. The folk expression "moving as slowly as molasses in January" refers to the fact that in cold weather the viscosity of molasses increases. BL Visual-Spatial EL

Tie to Prior Knowledge

Models of Materials Thermal energy is related to the kinetic energy of the particles that make up a substance. Review the description of the freely moving particles that make up a gas and the model of a solid as particles connected by springs.

2 Teach

Cohesive Forces

student text pp. 356–357

Critical Thinking

Cohesive Forces and Viscosity The cohesive forces in a fluid cause that fluid to be viscous. Ask students whether it is possible to have cohesive forces in a gas. yes The gas we are most familiar with, air, has a very low viscosity—the cohesive force among its particles is small. Ask students how you might observe the viscosity of air. It is evident, for example, in a wind tunnel where smoke makes the motion of air visible.
OL Logical-Mathematical

Use Figure 7

Surface Tension Point out to students how the water strider's feet depress the water as if the surface were a soft, stretched material, such as the material on a trampoline. Certain insects that are predators of the water strider spray a chemical that destroys the surface tension so the water strider cannot glide across the surface.

PHYSICS PROJECT ACTIVITY

Applications of Viscosity Viscosity is an important property of fluids, from motor oils to paints to cosmetics. In motor oil, viscosity controls the flow of oil as it lubricates between two sliding surfaces. If the oil is too viscous, it causes the surfaces to stick. If the oil is not viscous enough, it flows away from between the surfaces. Modern motor oils are designed to maintain proper viscosity on cold days and hot days alike. Have students investigate how viscosity is measured and devise a method to compare the viscosities of fluids such as motor oil, vegetable oil, shampoo, and water.
OL Kinesthetic EL

Concept Development

Surface Tension Challenge students to float a paper clip on water. Ask them what happens when they touch the water with a piece of soap. Touching the water with a piece of soap destroys the surface tension, causing the paper clip to sink. BL Kinesthetic EL

Adhesive Forces

student text p. 357

REAL-WORLD PHYSICS

Capillary Forces In Plants The capillary force that helps pull water up inside a tree is the same force that allows paper towels to soak up water. Place one end of a long strip of paper towel in a beaker of water, and observe how the water migrates upward.

Discussion

Question Every day we use ink to write on different types of surfaces. Ask students the following questions: Would you expect all inks have the same properties? Why or why not? What are desirable properties for inks in ballpoint and felt-tip pens and in inkjet printers? Are the same inks useful for all surfaces? Explain.

Answer Answers will vary. Ballpoint pen inks are viscous so they can roll off the ball. Inkjet inks must be less viscous, but their surface tension must allow them to form small drops. On some papers, an ink might spread too much through capillary action.
OL Logical-Mathematical

DIFFERENTIATED INSTRUCTION

Visually Impaired Help students understand how the surface of water can seem curved by providing a hands-on demonstration. Place modeling clay in a plastic cup. Form the surface of the clay into a concave meniscus. Have students feel the curvature. Explain that water adhering to the sides of a test tube forms this type of rounded surface. Add more clay to the cup to form a convex surface. Have students feel this curvature. Explain that water's surface tension when a cup is full forms this type of rounded surface. OL Kinesthetic

Evaporation and Condensation

student text pp. 357–358

Identify Misconceptions

Evaporation and Temperature Demo Some students believe that water must be boiled in order to change phase from a liquid to a gas. Ask them whether water can change phase from a liquid to a gas below its boiling point, 100°C. yes Put a flat dish containing water on a desk overnight and have students see the evidence of evaporation the next day.
OL Logical-Mathematical EL

Concept Development

MAINIDEA Helium is one of the most abundant elements in the universe and is a gas at a wide range of temperatures. When cooled to within a few degrees of absolute zero, helium can become a liquid with zero viscosity. This "superfluid" can quickly climb the walls of its container and flow away. What is the relationship between cohesive forces and adhesive forces in such a fluid? The adhesive forces between the fluid and the walls of the container must be much greater than the cohesive forces between the molecules of the fluid. This also accounts for its lack of viscosity.

Reinforcement

Summarize Properties Ask students to compare lists of the properties of gases and liquids and determine which properties are most typical of gases and which are most typical of liquids. Have them identify properties based on kinetic energy as well as properties based on the forces between the particles that make up the gas or liquid. OL

Use the PhysicsLAB

In *Evaporative Cooling,* students can collect and organize data about the evaporation of alcohols.

3 Assess

Assess the MAINIDEA

Condensation and Evaporation Students are familiar with the processes of condensation and evaporation of water. Ask them to explain these processes in terms of cohesive forces, temperature, and pressure. What does air pressure have to do with evaporation? Cohesive forces between water molecules slow down the process of evaporation of water. If the air pressure is greater, this will also inhibit evaporation because the particles of liquid water have to overcome air pressure as well as cohesive forces. Temperature measures average kinetic energy, so a higher temperature indicates there is more energy to overcome these forces.

Check for Understanding

Fog at Night Mysteries that take place in London, England, often start with a foggy night scene. Ask students why fog forms at night. Without heat from the Sun, the air cools and condenses moisture that was absorbed into the air during the day. This is the same reason dew appears on grass in the morning. BL

Extension

Capillary Action and Clothing Some clothes designed for sports or the outdoors can wick away moisture from the body and help retain body heat. Ask students how this would work. The material uses capillary action to move sweat from the skin to outside the clothing; moisture evaporates outside the clothes, not on the skin. So, the clothing insulates the skin from the cooling effect of evaporation. OL

Reading Checks and Caption Checks

Reading Check, *student text p. 358*

As particles with higher-than-average kinetic energy escape, the average kinetic energy of the remaining particles is lowered.

Section 2 Review

18. Yes; adhesive tape is sticking to something different from tape. A cohesive group is a collection of people working together.

19. The paper clip should be placed carefully and flatly on the surface of the water. This will reduce the weight per unit area of water surface on which it will rest.

20. If the paper clip breaks through the surface of the water, it will sink.

21. Alcohol has a greater adhesive attraction to glass than mercury does. The cohesive forces of mercury are strong enough to overcome its adhesive forces with glass.

22. Because alcohol evaporates easily, there is a very noticeable evaporative cooling effect. This procedure can lower the body temperature of the baby.

23. Beth could weigh the glass before cooling it in the refrigerator. Then she could remove it from the refrigerator and allow moisture to collect on the outside. Finally, she would weigh the glass a second time.

SECTION 3 Fluids at Rest and in Motion

student text pp. 359–366

1 Introduce

Bellringer

Fluids and Pressure Half-fill a large beaker with water. Place a smaller, inverted beaker into the water. Ask students to observe the water levels in both beakers before and after you insert the inverted beaker. Ask them to explain the changes in water levels. Why does the water not go up into the smaller beaker? The water level in the large beaker rises because the small beaker and the trapped air displace water. The water does not rise in the small beaker because the air pressure is forcing it downward. **BL** **Visual-Spatial**

Tie to Prior Knowledge

Floating Forces This section further explores pressure as the quotient of force divided by the area over which it is exerted. Students should have plenty of life experience with buoyant forces: swimming, floating on rafts, and observing helium balloons.

2 Teach

Fluids at Rest

student text pp. 359–360

Quick **DEMO**

Pressure and Depth in a Coffee Can

Estimated Time 5 minutes

Materials coffee can, nail or punch, hammer, water, sink or trough, tape

Procedure Punch three small holes at different heights in the side of a large coffee can. Cover the holes with strong, waterproof tape. Fill the can with colored water and set it in a large trough or washbasin. Remove the pieces of tape. Ask students to explain why the streams of water that emerge from the holes travel different distances from the can. The stream of water emerging from the bottom hole has the greatest horizontal velocity (v) because the pressure inside the can is greatest at the bottom of the can. Conversely, the stream emerging from the top hole has the lowest horizontal velocity because the pressure inside the can is least at the top of the can.

Swimming Under Pressure

student text pp. 360–364

Use the PhysicsLAB

In *Under Pressure,* students can analyze the forces from a fluid.

Concept Development

The Pressure Is Off Ask students to draw a diagram of the forces acting on a swimmer. The diagram should show gravity pulling downward but may be unclear about the source of the upward push. Ask students how putting on a life jacket changes those forces. The life vest adds some weight but increases the buoyant force even more. Have students make a similar force diagram for a hot-air balloon carrying a basket. What would be the effect of increasing the size of the balloon? It might add weight, but the increase in the buoyant force is greater. **OL** **Visual-Spatial** **EL**

Use Figure 12

This figure shows how the buoyant force arises from the differences in pressure on the top and the bottom of an object. Ask students to resketch this figure using shapes of different lengths and height; have them draw in the force vectors. Suppose the submerged object were a rectangular box with sides of unequal length. Discuss how the buoyant force on such a box would be the same no matter how the box is oriented. **OL** **Visual-Spatial** **EL**

Critical Thinking

Buoyancy of Air The first balloons to carry humans into the air were filled with heated air, not helium. Have students consider how a hot-air balloon provides enough lift to carry a gondola with passengers. Ask them whether such a balloon would work better or worse on a cold day. The balloon should work better on a cold day when the surrounding air is denser. Ask students to explain this in terms of what they know about gases and thermal motion. Students should also be able to explain this in terms of the ideal gas law as it leads to the kinetic-molecular concept of temperature. The hot air in the balloon is less dense than the air around it because the air molecules have more thermal energy and move farther apart. The density difference between the hot air and the colder surrounding air is what produces the buoyant force that lifts the balloon. **OL**

Content Background

Archimedes' Principle According to legend, King Hiero of Syracuse asked Archimedes (287–212 B.C.) to determine whether his new crown contained all the gold given to the artisan. As legend has it, when Archimedes lowered himself into his bath water, he realized that the water in his bathtub was displaced, and his body seemed to weigh less. He leaped from his bath and ran through the streets shouting, *"Eureka!"* ("I have found it.") Archimedes reasoned that a denser object, such as gold, would displace a much lower volume of water. He found that the crown was indeed not pure gold. The new crown displaced more water than the same weight of gold because it occupied more volume.

Use the PhysicsLAB

In *The Buoyant Force of Water,* students can demonstrate the relationship between the buoyant force and the weight of an object.

Identify Misconceptions

Apparent Weight and Mass Knowing that they can lift objects more easily under water, students might think that the objects have less weight or mass. Explain that the mass of the submerged object is unchanged and that the force of gravity on the object ($F_g = mg$) is the same as before. A buoyant force pushes upward on the object, however, so the apparent weight is less.

PHYSICS PROJECT ACTIVITY

Teacher to Teacher

The Effect of Atmospheric Pressure

Purpose to demonstrate the effect of atmospheric pressure on a solid

Materials empty aluminum soda can, tongs, Bunsen burner and ring stand or hot plate, goggles, protective gloves, metal pie pan filled with ice water

Procedure Wear goggles and gloves. Put 5 mL of water in the soda can. Heat the water until steam comes out the top of the soda can. Using tongs, remove the can from the heat and immediately invert it into the pan of ice water so that the top of the can is on the bottom of the pie pan. The reduced temperature will cause the pressure inside the can to decrease, and the external atmospheric pressure will collapse the can.

DIFFERENTIATED INSTRUCTION

Visually Impaired Buoyancy, pressure, and viscosity can be investigated by feeling. A student should be given the opportunity to feel the effect of buoyancy by pushing or lifting objects floating or submerged in an aquarium or sink filled with water. A student can investigate viscosity by stirring water or a thick shampoo with a finger. Have students describe the sensations they feel. Ask students the following questions: What do they experience when they try to submerge a foam block or an ice cube in water? What does this tell them about buoyancy? How does stirring hot cocoa compare to stirring pudding? What does this tell them about viscosity of fluids? BL EL

ADDITIONAL IN-CLASS EXAMPLE

Use with Example Problem 3.

Problem A child folds a piece of aluminum foil that has a mass of 0.012 kg into a boat that floats on the water. What volume of water does the foil boat displace?

Response Because the boat floats, the buoyant force is $F_{buoyant} = F_{g,\,aluminum} = m_{aluminum}g = \rho_{water}V_{water}g$. Then $V_{water} = \frac{m_{aluminum}}{\rho_{water}} = \frac{0.012\ kg}{1.0\times10^3\ kg/m^3} = 1.2\times10^{-5}\ m^3$.

Problem The child crumples the same boat into a tight ball that sinks in the water. How much volume does the foil displace, assuming that no air is trapped? What is the apparent weight of the foil ball under water?

Response The volume of the aluminum foil ball is $V = \frac{m_{aluminum}}{\rho_{aluminum}} = \frac{0.012\ kg}{2.70\times10^3\ kg/m^3} = 4.4\times10^{-6}\ m^3$.
This is the volume of water displaced by the submerged ball. The apparent weight is its weight minus the buoyant force on the submerged ball:
$F_{apparent} = m_{aluminum}g - \rho_{water}V_{water}g = (0.012\ kg)(9.8\ N/kg) - (1.0\times10^3\ kg/m^3)(4.4\times10^{-6}\ m^3)(9.8\ N/kg) = 0.074\ N$

Use the PhysicsLAB

In *The First Forensic Scientist,* students can reinforce the concepts of mass, volume, and density.

Bernoulli's Principle

student text pp. 364–366

Discussion

MAINIDEA Hold two pages of writing paper vertically in front of your face and blow between the flat surfaces. Ask students why the two sheets are attracted to each other. The higher airflow—the greater velocity of air molecules—between the pages causes a lower air pressure. The higher pressure outside the pages pushes them together. OL Visual-Spatial EL

Reinforcement

Bernoulli's Principle Remind students that pressure, buoyancy, and Bernoulli's principle are properties of liquids and gases alike. Ask them why buoyancy is not as obvious in the air and why we mostly use gas examples for Bernoulli's principle. The lower density of air makes its buoyant force less. Most objects move through liquids slowly, so the effects of Bernoulli's principle are often small in liquids. OL

PHYSICS CHALLENGE ACTIVITY

Designing Boats and Personal Watercraft Designing powerboats and recreational items such as personal watercraft involves an understanding of buoyant forces, the volume of water displaced, the center of gravity for balance, and Bernoulli's principle. For instance, engineers design many recreational boats and personal watercraft to float in shallow water. When these watercraft move at high enough velocities, they begin to plane on the surface, displacing less water and meeting less resistance to movement. Have students research how such machines are designed so that they are stable and so that they plane at certain speeds. OL Linguistic

3 Assess

Assess the MAINIDEA

Archimedes' Principle and Density Ask students if Archimedes' principle applies to objects that are less dense than water. How would you design an experiment that uses Archimedes' principle to determine the weight of such an object? Objects that float still displace a volume of water that is equal to their weight. This is why less dense objects float "higher up" in water the less dense they are. A graduated cylinder could be used to measure the rise in water level to determine how much water is displaced.

Check for Understanding

Bernoulli's Principle and Driving on the Highway When a large truck passes a car on the highway, the car may be pulled sideways toward the truck. Ask students why that occurs. The air between the car and the truck has to move at a higher velocity than the surrounding air, causing the air pressure between the car and the truck to decrease. This results in a net force toward the truck. OL

Reading Checks and Caption Checks

Caption Check, *student text p. 359*
F_2 would increase because the increased force is transmitted throughout the fluid.

Caption Check, *student text p. 362*
An object will float if the weight of the object is less than the buoyant force.

Reading Check, *student text p. 365*
As the velocity of a fluid increases, the pressure exerted by the fluid decreases.

Practice Problems
student text p. 360

24. 8.0×10^{1} N

25. 8.8×10^{3} N

26. 0.4:1

Practice Problems
student text p. 363

27. 1.6×10^{3} N

28. 6.2×10^{-2} m^3

29. 1.09×10^{3} N

30. 8.8×10^{2} N

31. 4.9×10^{-2} m^3

Section 3 Review
student text p. 366

32. About 1/4 cup of sugar is dissolved in the regular drink, making it denser than water. The diet drink has a small amount of artificial sweetener. The diet soda pop is less dense than the sugar-laden regular soda pop.

33. 24 N

34. 0.9 m^3

35. The cork is about one-tenth as dense as water.

36. a. 1.5×10^{5} Pa
b. 1.3×10^{3} N

37. a. The block of aluminum will displace more water.
b. Each will displace the same volume of water.

38. The fast-moving air of the tornado has a lower pressure than the still air inside the house. Therefore, the air inside the house is at a higher pressure and produces an enormous force on the windows, doors, and walls of the house. This pressure difference is reduced by opening doors and windows to let the air flow freely to the outside.

SECTION 4 Solids

student text pp. 367–371

1 Introduce

Bellringer

Slime The difference between a solid and a liquid is sometimes subtle. A good example is the elementary school example of slime made with borax and white glue or the commercially available polyvinyl alcohol. Pass around small amounts of slime for the students to investigate. Ask students the following questions: Is slime a solid or a liquid? Does it retain its shape or flow? Does it bounce? Can it be cut? When put in a refrigerator, does it solidify and then melt when worked at hand temperature? The answers depend on the slime used (many recipes are available), but all are a type of polymer that has a high viscosity, strong cohesion, and no well-defined melting temperature. Slime shares properties with liquids and solids. BL Kinesthetic EL

Tie to Prior Knowledge

Thermal Expansion and Kinetic Energy Review the model of a solid as a collection of particles connected by springs. Discuss how the addition of energy causes those particles to move more and more vigorously until the bonds between them break. Discuss how this would affect the size of the solid in one, two, and three dimensions. OL

2 Teach

Solid Bodies

student text pp. 367–368

Concept Development

Solid Intermolecular Interactions Changes in the thermal motion of atoms in a solid cause changes in the structure and in the dimensions of that solid. Unlike fluids, the intermolecular interactions in a solid hold it together in such a way that a solid tends to keep its shape. Therefore, solids are elastic and expand or contract in a well-defined way. Some of the most interesting solids are made of molecules that can rotate, such as water, or large polymeric molecules that can rotate and bend, such as rubber or plastics or the slime discussed in the Bellringer.

Careers

Materials Science and Engineering A materials scientist or materials engineer investigates and designs materials for specific uses in machines, vehicles, or electronic devices. These applications may require materials that have a certain strength, hardness, ductility, or thermal expansion characteristic. The engineer works with the chemical composition and the structure of the material to produce the desired properties. For instance, materials engineers design special metal alloys for the light, strong, and heat resistant parts needed for high-speed aircraft. A ceramics scientist may design heat-resistant, low-expansion, high-thermal-conductivity parts for auto engines or special electronic devices.

Thermal Expansion of Solids

student text pp. 368–371

Content Development

MAIN IDEA Effects of thermal expansion and contraction can often be observed on the surface of glazed ceramics such as teacups and coffee mugs. If the glaze used does not expand at the same rate with temperature changes as the underlying ceramic or is too thin, the glaze can separate from the ceramic surface. This results in a spidery network of fine lines where the ceramic is no longer protected by the glaze. Shivering is the opposite phenomenon. Glazing flakes off where it is too thick and becomes compressed when the underlying ceramic contracts as it cools.

Critical Thinking

Expansion and Density Ask students what happens to the mass and density of an object when it expands. The number of atoms and the total mass of the object do not change, but the volume increases. Therefore the density—the mass per volume—decreases when the material expands. OL

Identify Misconceptions

Relative Expansion Students might not understand that the amount of thermal expansion of an object is a relative effect. That is, when heated 10°C, a metal meterstick will expand in length ten times more than a 0.1-m ruler under the same conditions. Because of this effect, an object made of a single material will retain its shape when heated or cooled.

REAL-WORLD PHYSICS

Thermostats Some students might think that a chilly room warms faster by setting the thermostat higher and that a hot room cools faster by setting the thermostat lower. In fact, a thermostat is an on-off switch and does not operate like the accelerator in a vehicle. For example, moving the thermostat to 75°F in a 65°F room will not make the temperature rise to 70°F more quickly than it would have if the thermostat had simply been set at 70°F.

Reinforcement

Concept Building Temperature is a measure of the average kinetic energy of the particles in an object, and thermal expansion depends on the motion of these particles. Ask students to construct a concept map using temperature, thermal expansion, the coefficient of expansion, solids, gases, and the combined gas law. Ask them to discuss their maps with other students and then lead the class in a discussion about the maps. One possible map would look like this: OL EL

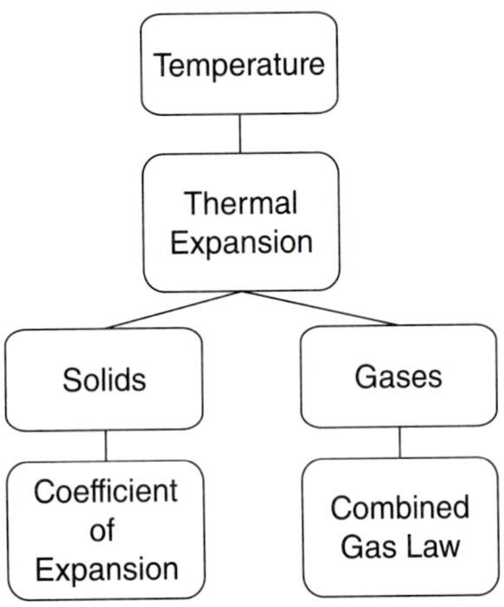

PHYSICS PROJECT ACTIVITY

The Materials World Ask students to choose a material to investigate and report on the physical properties that make it useful. Ask students the following questions: In what states of matter is the material used? What are the values of its properties such as its coefficients of thermal expansion and its density? What other materials fill same or similar niches? OL

ADDITIONAL IN-CLASS EXAMPLE

Use with Example Problem 4.

Problem At 30.0°C, a rectangular bar of aluminum has dimensions 0.0150 m × 0.125 m × 1.45 m. The bar is heated uniformly to a temperature of 310.0°C. What are the new dimensions of the bar?

Response $L_2 = L_1 + \alpha L_1(T_2 - T_1)$ or,
$L_2 = L_1(1 + \alpha(T_2 - T_1))$ where $(1 + \alpha(T_2 - T_1))$
$= 1 + (23\times10^{-6}/°C)(310.0°C - 30.0°C)$

The new dimensions are:

$(0.0150\text{ m})(1 + (23\times10^{-6}/°C)(310.0°C - 30.0°C))$
$= 0.015\text{ m}$

$(0.125\text{ m})(1 + (23\times10^{-6}/°C)(310.0°C - 30.0°C))$
$= 0.13\text{ m}$

$(1.45\text{ m})(1 + (23\times10^{-6}/°C)(310.0°C - 30.0°C))$
$= 1.5\text{ m}$

Use the MiniLAB

In *Jumpers,* students can demonstrate the different rates of expansion for metals.

3 Assess

Assess the MAINIDEA

Volume Expansion v. Linear Expansion Ask students to compare the coefficients of linear expansion and volume expansion for the solids listed in **Table 2.** What regularities can students find in these values? How might one account for this? For the solids, the coefficient of volume expansion is about three times that of linear expansion. This makes sense because volume expansion is essentially linear expansion in three dimensions.

Check for Understanding

Thermal Expansion Ask students what it would mean if they obtained a negative answer for the change in length in a thermal expansion problem. The negative result would indicate contraction, not expansion. Ask them how they would check if their answer to such a problem made sense. Most objects expand when heated, but the change in length is usually very small. If you calculate that the object doubles in length when cooled, something is not right! OL

Reteach

Open a Jar Ask students why heating a tight, screw-on metal lid on a glass jar by running hot water on the lid often causes the lid to loosen. As the temperature of the metal rises, the lid expands, enlarging its opening. The metal heats faster and expands more than the glass. AL

Reading Checks and Caption Checks

Caption Check, *student text p. 368*

The pavement would crack.

Practice Problems

student text p. 370

39. 6.1 mm

40. 12 cm

41. 2 mL

42. a. 4.4×10^4 L

b. The gasoline volume decreased because the temperature decreased. The mass of the gasoline remained the same.

43. 1.7×10^{-4} cm

44. −0.072%

Physics Challenge

student text p. 371

$L_{copper} = L_{aluminum} + L_{iron}$

and

$\alpha_{copper} L_{copper} \Delta T = (\alpha_{aluminum} L_{aluminum} + \alpha_{iron} L_{iron})\, \Delta T$

Substituting $L_{aluminum} = L_{copper} - L_{iron}$, this gives

$$L_{iron} = \frac{(\alpha_{copper} - \alpha_{aluminum}) L_{copper}}{\alpha_{iron} - \alpha_{aluminum}}$$

$$L_{iron} = \frac{(1.00\ \text{m})(17 \times 10^{-6}\text{°C}^{-1} - 23 \times 10^{-6}\text{°C}^{-1})}{12 \times 10^{-6}\text{°C}^{-1} - 23 \times 10^{-6}\text{°C}^{-1}}$$

$= 0.54$ m

$L_{aluminum} = L_{copper} - L_{iron} = 1.00\ \text{m} - 0.54\ \text{m} = 0.46\ \text{m}$

Section 4 Review

student text p. 371

45. Fit the door tightly.

46. The wax could be considered a solid because it has a definite volume and shape. It could be considered a viscous liquid because the particles do not form a fixed crystalline pattern.

47. To double its length, the temperature change would need to be 6.3×10^4°C. The copper would be vaporized at that temperature.

48. The coefficients of volume expansion are much greater for liquids than for solids.

49. Particles in a liquid are farther apart and less tightly bound. Because the particles are free to flow past one another, a liquid cannot be bent.

50. Let x be the width of the gap (measured along the arc of the ring). Let θ be the measure of the arc (degrees). Let L be the length of the ring (i.e., circumference less the gap width x).

Then: $\frac{\theta}{360} = \frac{x}{L + x}$ from which $x = L\frac{\theta}{360 - \theta}$.

Since an object retains its geometry and shape while it is heated, the above relationship between the gap width and the ring length continues to be true:

$$x + \Delta x = (L + \Delta L)\left(\frac{\theta}{360 - \theta}\right).$$

According to the definition of the coefficient of thermal expansion: $\Delta L = \alpha L \Delta T$. Substitute into the above equation and subtract the original equation to get:

$$\Delta x = \alpha L \Delta T\left(\frac{\theta}{360 - \theta}\right).$$

The gap width increases proportionately to the ring length increase.

HOW IT **WORKS**

Bending the Rules!

Background

Football, as it's called outside the United States, is the most popular sport on the planet. The World Cup occurs every four years and includes 32 teams from different qualifying nations. The motion of a ball in flight depends on the initial power of the kick (generating lift and/or spin, depending on where the ball is struck), where the ball contacts the foot, and, conversely, where the foot contacts the ball. Factors such as wind and atmospheric pressure (at different altitudes) can also have an effect on a ball's flight. Consider a ball that is spinning about an axis perpendicular to the flow of air across it. When the ball spins, the pressure on one side of the ball is greater than the pressure on the other side.

Teaching Strategies

- Show students the actual video of Roberto Carlos's kick on the internet. Make sure you view the video before showing your class.
- Bring students out to a soccer field and have them try bending shots into a goal. Make sure they wear proper attire—tennis shoes or sneakers are not optimal compared to soccer-specific shoes but will suffice.
- Compare kicking a curving soccer ball with throwing a curveball in baseball.

GOING**FURTHER** >>>

Expected Outcomes Student responses should indicate how the physical properties of each projectile are suited to the games in which they are used to play.

ANSWERS

SECTION 1

Mastering Concepts

51. Force depends only on the push or the pull on an object. Pressure depends on the force as well as the area over which the force is applied.

52. The liquid's volume will remain unchanged and may not fill the container. The gas will expand to the volume of the container and may fall in pressure.

53. Both have no definite volume or shape. The particles of plasma are more energetic, and plasmas can conduct electricity.

54. The Sun's plasma is extremely hot, but more importantly it is very dense—denser than most solids on Earth.

Mastering Problems

55. a. 8.3 N

b. 1.7×10^{2} Pa

56. B < A < E < D < C

57. 3×10^{1} cm

58. a. 0.32 moles

b. 14 g

59. 0.14 m^3

60. a. 3.2×10^{5} Pa

b. 31 psi

SECTION 2

Mastering Concepts

61. To melt, the ice must absorb an amount of energy from the air and the water that is equal to its heat of fusion. It will cool the air above it.

62. The water evaporates into the air, absorbing energy from the canteen and the water inside.

SECTION 3

Mastering Concepts

63. The equilibrium tubes illustrate that pressure is independent of the shape of the container.

64. Changes in pressure are distributed equally to all parts of the container. The pressure at the top increases.

65. Size or shape of the body of water does not matter, only the depth. The pressure is the same in each case.

66. No, it does not. The apparent weight of the displaced fluid is zero because the fluid is in free fall. Thus, there is no buoyant force.

67. The water pressure decreases because of Bernoulli's principle.

Mastering Problems

68. a. 1.7×10^{5} Pa

b. 3.9×10^{4} Pa

69. 8.4×10^{2} Pa

70. a. 6.33×10^{-5} m^3

b. 19.0×10^{3} kg/m^3; it is made of gold.

71. a. 203 N

b. 197 N

72. 0.70 m

73. 26.0 N

74. 24 N

75. −10.0 N

76. 5.0×10^{1} N

SECTION 4

Mastering Concepts

77. The atoms in a crystalline substance are arranged in an ordered pattern. In an amorphous substance, the atoms are randomly arranged.

78. No. The coefficient is a measure of the expansion of an object relative to its total length. Units and total length do not change α.

Mastering Problems

79. 1.4×10^{-4}°C^{-1}

80. 1.3×10^{-5} m

81. 0.1 m

82. 3.1×10^{-2} m

83. 1.6×10^{-3} m^3

84. -1.2×10^{2}°C

85. 0.55 m^3

86. 7.6×10^{-2} cm^3

87. 2.4 cm^3

88. **a.** 3.8×10^{-4} m
b. 2.3×10^{-3}

89. **a.** 0.348 m^3
b. 0.703 m

90. Parts 1 and 2 experienced a greater reduction in length than parts 3 and 4; therefore, parts 1 and 2 must have been made of aluminum, which has a larger coefficient of expansion than steel.

Applying Concepts

91. The pressure increased. The weight stayed the same, but the weight per area increased.

92. $1\ \text{Pa} = 1\ \text{N/m}^2 = 1\ (\text{kg}\cdot\text{m/s}^2)/\text{m}^2 = 1\ \text{kg/m}\cdot\text{s}^2$

93. It would sink deeper into the water because it would have a greater weight.

94. 136 cm

95. The cohesive forces are strongest in mercury and weakest in acetone. The stronger the cohesive force, the more spherical the drop will be.

96. The cohesive forces of water are greater than those of alcohol.

97. a-1, b-2, c-6, d-6, e-6

98. The ovenproof glass expands less than the soft glass when heated. The water does not rise as high in A because the soft glass tube is expanding in volume.

99. Platinum's coefficient of thermal expansion is similar to that of glass, so it expands and contracts as the glass does. Copper has a much larger coefficient than glass.

Mixed Review

100. 7.4×10^{5} Pa

101. 6.2×10^{-6} m^3

102. Answers will vary, but a correct form of the answer is, "A sample of gas initially has a pressure of 77.0 kPa and a volume of 25.0 L. When allowed to cool to room temperature, the pressure drops to 61.2 kPa and the volume expands to 28.0 L. What was the initial temperature of the gas?"

103. The volume coefficient of expansion of mercury is greater than that of aluminum. Therefore, as they are heated, the aluminum becomes denser relative to the mercury and would sink deeper into the mercury.

104. The water expands: 0.735 mL. The beaker expands: 0.76 mL. The water level will go down slightly but not enough to notice.

105. **a.** 2.5×10^{3} N
b. 75 cm

106. Atmospheric pressure is lower at higher altitudes. Therefore, the mass of the volume of fluid displaced by a balloon of the same volume is less at higher altitudes. To obtain the same buoyant force at higher altitudes, a balloon must expel more gas, requiring higher temperatures.

Thinking Critically

107. Answers will vary. A possible form of the correct answer would be "...if it is submerged in a tank of water and released, what will be its acceleration?"

108. The physician weighs the person normally, and then weighs the person totally submerged. Weight has to be added to the weighing device because the density of a human is normally less than the density of water. The volume of water displaced by the person should also be measured. The average density of the person can be calculated from the balance of forces that hold the person in equilibrium under water.

109. **a.** 9.5 percent of the floating sphere would be submerged.
b. 7×10^{1} N
c. 8×10^{-2} m^3

110. The density of the water in the bag, the fish, and the plastic are all near the density of water in the aquarium. Therefore, the bag should float with the water level in the bag at the same height as the water level in the aquarium.

Writing in Physics

111. Rubber bands are made of long rubber molecules called polymers that act like chains with many long links. The properties of rubber come from the ability of the chain links to twist and turn. When the rubber is colder, the polymer links are stretched out in a straight line, like the links in a steel chain that you hold at one end and let hang freely. Because the links are ordered that way, the polymers have relatively little disorder, or entropy. Adding heat to the polymers increases their thermal motion. The links begin to shake about, and their disorder increases. If you shake a chain like this, you will see that its average length becomes less than if the chain were hanging motionless.

112. Gay-Lussac was a French scientist who was also interested in high-altitude balloon ascents. He discovered that when gases are at the same temperature and pressure, their volumes react in ratios of small, whole numbers. Gay-Lussac's work contributed to the discovery of water's formula by showing that two volumes of hydrogen gas reacted with one volume of oxygen gas. Avogadro built on Gay-Lussac's work to develop the relationship between moles of a gas and volume.

Cumulative Review

113. a. $2.6\ m/s^2$

b. 22 N

114. a.

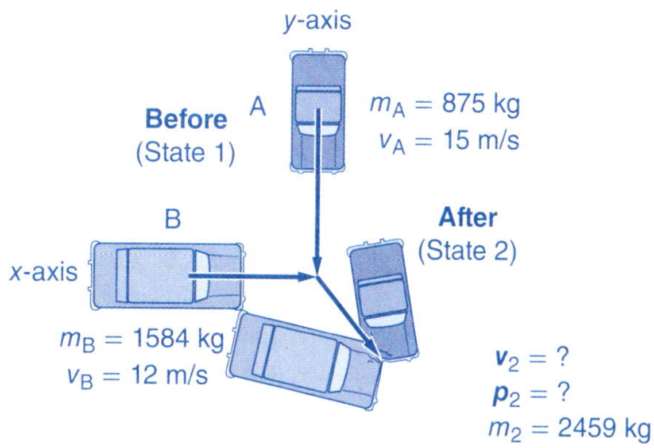

b. 9.4 m/s, 55° east of south

c. 8.2 m

115. 2.89×10^3 N

Standardized Test Practice

student text p. 379

Multiple Choice

1. B
2. B
3. D
4. C
5. A
6. B
7. D
8. B

Free Response

9. $V_2 = \frac{P_1V_1}{P_2}$

$= \frac{P_1V_1}{P_{\text{atmosphere}} + \rho hg}$

$= \frac{(101.3\times10^3 \text{ Pa})(125 \text{ mL})}{(101.3\times10^3 \text{ Pa} + (1.00\times10^3 \text{ kg/m}^3)(1.27 \text{ m})(9.8 \text{ N/kg}))}$

$= 111 \text{ mL}$

Rubric

The following rubric is a sample scoring device for free response questions.

Points	Description
4	The student demonstrates a thorough understanding of the physics involved. The response may contain minor flaws that do not detract from the demonstration of a thorough understanding.
3	The student demonstrates an understanding of the physics involved. The response is essentially correct and demonstrates an essential but less-than-thorough understanding of the physics.
2	The student demonstrates only a partial understanding of the physics involved. Although the student may have used the correct approach to a solution or may have provided a correct solution, the work lacks an essential understanding of the underlying physical concepts.
1	The student demonstrates a very limited understanding of the physics involved. The response is incomplete and exhibits many flaws.
0	The student provides a completely incorrect solution or no response at all.

CHAPTER 14

Vibrations and Waves

About the Photo

Wave Motion Have students look at the photo and tell what is moving in the photograph. water waves Have students describe how the waves are moving. in a pattern because they look about the same space apart Ask students what the pattern of motion might mean for a floating object. An object might bob up and down or the waves might carry the object from its position.

Chapter Overview

Periodic motion can be observed in the motion of pendulums and water waves. The chapter introduces the concept of periodic motion and the properties and behavior of mechanical waves. Wave properties are illustrated by waves in coiled springs and waves in water. The chapter discusses several types of wave behavior—interference, reflection, and refraction.

Before students study the material in this chapter, they should study:

- Adding vectors
- Elastic potential energy
- Gravitational potential energy
- Kinetic energy
- Tension
- Work, energy, and power

To solve the problems in this chapter, students will need a strong grasp of:

- Graphing Data
- Scientific Notation
- Significant Figures
- Sine, Cosine, and Tangent
- Solving linear equations
- Solving quadratic equations

Introduce the BIG IDEA

Periodic Motion Have students recall the motion of a playground swing. Ask students to describe the motion. back and forth on the same path Explain that this type of motion is called *periodic motion.*

Go online!
connected.mcgraw-hill.com

Your online experience includes:

- **Video**
- **Concepts In Motion**
- **Virtual Investigations**
- **All Lab, Applying Practices activities, and PBL worksheets**
- **Online Practice**
- **Personal Tutor**
- **Multilingual eGlossary**
- **Presentation**

Use the LaunchLAB

In *Wave Interaction,* students will observe different kinds of waves, along with their associated kinds of oscillations.

Use Physics T.V.

Video

PLAYGROUND PHYSICS? You probably remember playing on swings when you were younger. What did you have to do in order to keep going? How do swings compare to the periodic motion of a pendulum?

1 Introduce

Bellringer

Periodic Motions Have students observe the periodic motions of various objects, such as swinging pendulums, a plucked string, as well as a metal ruler or meterstick with one end clamped to a tabletop and set in motion by releasing the opposite end from an up or down displacement. Help students generate and discuss questions such as: What is necessary to start the motion? What maintains the motion? What eventually happens to the motion? Is the motion regular in displacement and/or time? **BL** **Visual-Spatial** **EL**

Tie to Prior Knowledge

x*, *v*, *a*, and *F Have students review concepts of displacement, velocity, acceleration, and force introduced in previous chapters.

2 Teach

Mass on a Spring and Hooke's Law

student text pp. 382–385

Use Figure 2

Force Magnitude-Stretch Distance Graphs Have students look at **Figure 2.** Have students recall that the area under the curve of a force versus displacement (stretch distance on the *x*-axis) graph represents work. Review the connection for students:

The area of a triangle is $A = \frac{1}{2}bh$.

Thus, $W = \frac{1}{2}x_1F_1 = \frac{1}{2}x_1kx_1$

$W = \frac{1}{2}kx_1{}^2$

Use Figure 3

MAIN IDEA Have students look at the bar graphs in **Figure 3.** Be sure students recognize that the force exerted on the object by the spring is not the same magnitude (bars) and direction (vectors) throughout the compression-stretching cycle. Point out that if this were so, the bar graphs in **Figure 3** would show *PE* and *KE* the same in each phase. For example, point out that the bar graphs for $t = 0.2$ s and $t = 0.4$ s show how *PE* and *KE* change as the spring is compressed from its equilibrium position (the vertical dotted line) to the $v = 0$ position. Also, show that the force acts in the opposite direction of the object's motion (as shown by the $\boldsymbol{F}_{sp}$ vector) as the coils of the spring are compressed.

ADDITIONAL IN-CLASS EXAMPLE

Use with Example Problem 1.

Problem A 560-N cyclist sits on a bicycle seat and compresses the two springs that support it. The spring constant equals 2.2×10^4 N/m for each spring.

a. How much is each spring compressed?

b. By how much does the compression increase each spring's elastic *PE*?

Response

a. $F = kx$

$x = F/k = \left(\frac{1}{2}F_w\right)/k$

$$x = \frac{\frac{1}{2}(560\text{ N})}{2.2\times10^4\text{ N/m}} = 1.3\times10^{-2}\text{ m}$$

b. $PE_{sp} = \frac{1}{2}kx^2 = \frac{1}{2}(2.2\times10^4\text{ N/m})(1.3\times10^{-2}\text{ m})^2$

$= 1.9$ J

Pendulums and Resonance

student text pp. 386–387

Concept Development

Force Direction For the simple pendulum shown in **Figure 4,** $\boldsymbol{F}_g + \boldsymbol{F}_t = \boldsymbol{F}_{net}$. Model this by drawing a vector-addition diagram for when the pendulum bob is at the position on the right, as seen in **Figure 4.**

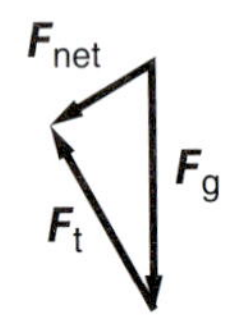

Ask students why F_T is greater than F_g. The bob travels an arc of a circle, so the net force at the bottom of the path must be centripetal (toward the center of the circle). **OL** **Visual-Spatial** **EL**

Content Development

Pendulum Bob Have students look again at **Figure 4** and consider the factors upon which the restoring force of a pendulum, $\mathbf{F}_{net}$, depends. Ask students to explain why the period of a pendulum is independent of the mass of the bob. The restoring force depends upon *mg,* and therefore, because $a_{net} = F_{net}/m$, a_{net} is independent of *m.* Because motion is described by a_{net} and the initial conditions, the motion is independent of mass.

OL Visual-Spatial EL

Use the PhysicsLAB

In *Pendulum Vibrations,* students will investigate how a pendulum can be used to study the properties of waves.

DIFFERENTIATED INSTRUCTION

Struggling Students Represent F_{net}, a_{net}, and v for the motion of a simple harmonic oscillator by the following graphs.

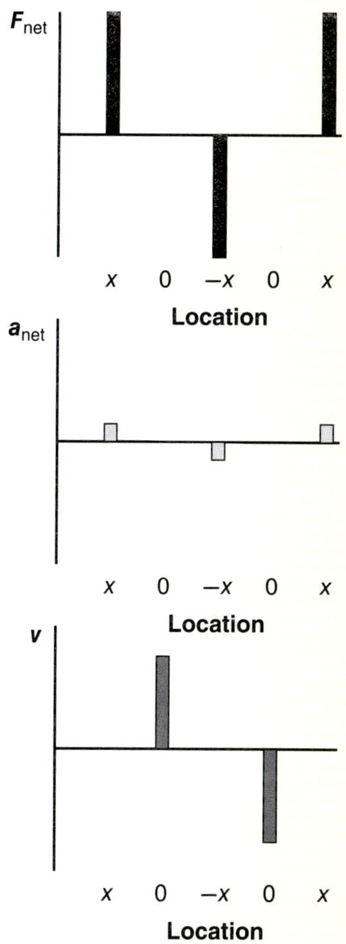

BL Visual-Spatial EL

Applying Physics

Foucault Pendulums Foucault pendulums are set up in a number of places around the country, such as the lobby of the General Assembly building at the United Nations and at the California Academy of Sciences. Because friction from air currents on the wire, vibrations in the cable, and other factors could stop the swing of a Foucault pendulum in a few hours, the energy lost with each swing must be replaced to avoid the damped motion and keep the swing going. To counteract the resistances, an iron collar is installed on the wire and surrounded by a doughnut-shaped electromagnet. As the weight swings out, an electronic device turns on the electromagnet and then turns off as the weight swings back. This interaction keeps the weight swinging but does not interfere with the direction of the swing.

Content Background

Small-Angle Oscillations The restoring force acting on the bob, as shown in **Figure 4,** is $-mg \sin \theta$, where θ is the angle between the pendulum and the vertical. This expression is valid for a pendulum's motion between $-90° \leq \theta \leq 90°$. A pendulum does not, however, move with simple harmonic motion unless θ is less than approximately 15°, in which case $\sin \theta \approx \theta$ (when θ is measured in radians) and the horizontal displacement of the bob equals approximately θ. Thus,

$$k = \frac{mg}{x} : \frac{mass \frac{length}{time^2}}{length} = \frac{mass}{time^2}.$$

For this small angle approximation, the restoring force is directly proportional to the displacement, which is required for simple harmonic motion.

ADDITIONAL IN-CLASS EXAMPLE

Use with Example Problem 2.

Problem What is the period of a simple pendulum with a length of 0.25 m?

Response

$$T = 2\pi\sqrt{\frac{l}{g}} = 2\pi\sqrt{\frac{0.25\ \text{m}}{9.8\ \text{N/kg}}}$$

$$T = 1.0\ \text{s}$$

3 Assess

Assess the MAINIDEA

Periodic Motion On the board, draw a cutaway view of a cylindrical silo with a conical top. From the apex of the conical top, draw a pendulum down nearly to the floor of the silo. Tell students that the period of the pendulum is 12 seconds. Have students find the height of the silo.

$T = 12\ \text{s} \qquad g = 9.8\ \text{N/kg}$

$$T = 2\pi\sqrt{\frac{l}{g}}$$

$$L = \frac{T^2 g}{4\pi^2}$$

$$L = \frac{(12\ \text{s})^2(9.8\ \text{N/kg})}{4\pi^2}$$

$L = 36\ \text{m}$

Check for Understanding

Equilibrium Position Ask what is true about the magnitude of the restoring force acting on an object in simple harmonic motion, the magnitude of its acceleration, and its speed as it passes through its equilibrium position. $F_{restoring} = 0$, $a = 0$, $v = v_{max}$ OL

Extension

Pendulum Motion Construct a pendulum that has a length of about 1.0 m and set it in motion. Have students trace the bob's path on a sheet of paper with a pencil that is moving back and forth across the sheet in tandem with the bob. Then, have students repeat, this time asking them to keep their tracing hand in place and with their free hand, pull the sheet at a constant speed perpendicular to the pencil's motion. Have students discuss their sketches and explain that the sketches represent a graph of amplitude versus time. Ask students what on the graph represents the pendulum's period. the interval between two points that have the same amplitude and direction of motion

AL Kinesthetic EL

Reading Checks and Caption Checks

Reading Check, *student text p.382*

The requirement for an object to be in simple harmonic motion is that the magnitude of the restoring force on the object is proportional to the magnitude of the displacement of the object from equilibrium. (Remember that both force and displacement are vector quantities, which have both magnitude and direction.)

Caption Check, *student text p. 382*

The displacement would be 0.5x m.

Reading Check, *student text p. 383*

The importance of the negative sign in Hooke's law is that it indicates that the restoring force on the object (the restoring force being a vector pointing from the object toward the equilibrium point) is opposite in direction to that of the displacement of the object from equilibrium (the displacement being a vector pointing from the equilibrium point to the object).

Caption Check, *student text p. 384*

The mass-and-spring system has the greatest potential energy at the positions of maximum displacement, at $t = 0$ s, $t = 0.4$ s, and $t = 0.8$ s.

Reading Check, *student text p. 386*

The period of a simple pendulum is independent of the mass of its bob.

Practice Problems

student text p. 385

1. 2.0×10^2 N/m
2. 1.96 J
3. 0.32 m
4. 0.61 m

Practice Problems

student text p. 387

5. 2.0 s
6. 0.16 m
7. 9.1 N/kg

Physics Challenge

student text p. 385

1. Conservation of energy implies that the car's gravitational potential energy at the top of the hill will be equal to the spring's elastic potential energy when it has brought the car to rest. The equations for these energies can be set equal and solved for x.

   $PE_g = PE_{sp}$, so $mgh = \frac{1}{2}kx^2$

   $x = \sqrt{\frac{2mgh}{k}}$
2. The height is doubled and x is proportional to the square root of the height, so x will increase by $\sqrt{2}$.
3. In the case of an ideal spring, the spring will propel the car back to the top of the hill.

Section 1 Review

student text p. 387

8. The pendulum swings back and forth, following the same path each cycle and requiring the same amount of time to complete each cycle.
9. The energy of the second spring is 4.0 times greater than the energy of the first spring.
10. Hang the same object from both springs. The one that stretches less has the greater spring constant.
11. If the graph is a straight line, the rubber band obeys Hooke's law. If the graph is curved, it does not.
12. To double the period, the length must be quadrupled; to halve the period, the length is reduced to one-fourth its original length.
13. At that speed, the tire's rotation frequency matches the resonant frequency of the car.
14. Both are periodic motions. In uniform circular motion, the accelerating force is not proportional to the displacement. Also, simple harmonic motion is one-dimensional and uniform circular motion is two-dimensional.

SECTION 2 Wave Properties

student text pp. 388–393

1 Introduce

Bellringer

Wave Motion ***CAUTION: All should wear goggles.*** Use a coiled spring toy (or a long coiled spring) with a ribbon or piece of yarn tied to the center of the spring. Stretch the spring slightly between several floor tiles or long enough to be able to see waves in the spring. Ask the students how they could move the ribbon perpendicular to the length of the spring without touching the ribbon. Moving one end of the spring from side to side will move the ribbon perpendicularly to the spring length. A similar investigation could be used to identify other wave motion features. Lay the spring along a seam between two tiles. Then, moving one end of the spring one tile width (or enough so a wave can be observed) and back quickly will create a sine wave crest and trough in the spring. Creating a standing wave will make this measureable. OL Visual-Spatial EL

Tie to Prior Knowledge

Simple Harmonic Motion Have students review concepts developed in their study of Newton's laws of motion, the law of conservation of energy, and simple harmonic motion.

2 Teach

Mechanical Waves

student text pp. 388–389

Use Models

MAIN IDEA Have students demonstrate "the wave," which is often performed by fans in stadiums. Point out that the students are actually modeling a transverse pulse that propagates the wave without changing the positions of the people. Thus, no matter is transferred in the process. BL Kinesthetic EL

Content Development

Surface Waves Have students work with a wave tank or large basin to create water waves. Add enough water to the container to float a small cork. Put a small cork in the water and make it bob up and down to form waves. Have students note that the medium does not move in the direction of wave travel for transverse waves. The cork bobs as would a floater on the end of fishing line. To make the effect more visible, carefully add a drop of food coloring to the water, then make more waves with the cork to show the color does not spread in the direction of wave travel. If possible, have students observe changes in wavelength as a result of higher or lower wave frequencies in the basin or tank. BL Visual-Kinesthetic EL

Quick DEMO

Longitudinal Waves

Estimated Time 10 minutes

Materials coiled spring toy

Procedure Use a coiled spring toy as described in this section's Bellringer activity or conduct the demonstration in a long hallway by extending the spring and having a student securely hold one end. Have students observe the spring as you take several coils on one end and alternately compress and extend them to send longitudinal pulses down the spring. Have students describe the motion. Students will observe a series of alternating compressions and rarefactions moving along the spring. OL Visual-Spatial EL

REAL-WORLD PHYSICS

Inventor and Physicist Hertha Marks Ayrton was a British physicist who lived from 1854 to 1923. She studied the role of wave action in the formation of rippled beaches and did research into carbon arc lamps. Because of her research interest in fluids, she invented the Ayrton fan that could safely remove gases from underground chambers. It became a deterrent to wartime gas attacks. She also patented a device called a line divider. This drafting tool could divide a line segment into any given number of equal parts. She was the first woman to read a paper to the British Royal Society and was also awarded the Hughes Medal for her original research.

Wave Properties

student text pp. 389–393

Identify Misconceptions

Amplitude and Pulse Speed Some students might think that waves with greater amplitude have higher speed. Ask students whether a wave's amplitude affects its speed, and have them write a paragraph supporting their statement. Then, do the following Quick Demo. OL Linguistic

Quick DEMO

Amplitude and Pulse Speed

Estimated Time 10 minutes

Materials spring, stopwatches

Procedure Use a long spring as described in this section's Bellringer or conduct the demonstration in a long hallway by extending the spring and having a student hold one end. Have several students use stopwatches to measure the time for a small-amplitude pulse, and then for a large-amplitude pulse, to travel the length of the spring. Because the time measurements and distances are similar, students might infer that the pulse speeds are similar.

PHYSICS PROJECT ACTIVITY

The Perfect Wave Have students research information on the characteristics of ocean waves with regard to wave speed, amplitude, wavelengths, and other wave characteristics. Research might include descriptions of wave trains, fully-developed seas, tidal waves, wave interference, why and how waves break, the different types of breakers, and how a surfer identifies the "perfect" wave. Information on the instruments used to measure waves, how someone on a ship measures a wave, and the biggest wave ever recorded could also be of interest to students. Encourage students to develop comparative graphics to depict these concepts. OL Linguistic-Visual

Content Background

Dispersive Media In some materials, the velocity of a wave depends not only on the medium but also on the frequency of the wave. Such a medium is called a dispersive medium. Shallow water is a dispersive medium for surface waves. Similarly, light waves undergo dispersion as they travel though glass and diamond. Even though little dispersion takes place as the sound waves of audible sounds travel through air, the speed of higher frequency ultrasonic sound waves in air is measurably greater than the speed of sound waves in the audible range.

Content Development

Amplitude, Energy, and Medium Ask students if a wave of small amplitude in a heavy spring carries more energy than a wave with a larger amplitude in a lighter spring. Have them explain their answers. It is possible, but not necessarily so. The factors that influence the energy of a wave and thus its amplitude include such characteristics as the force exerted on the spring, the resistance of the spring, and the elasticity or springiness of the material. Exerting the same force on springs of different materials would result in different amplitudes, with the more massive, heavier spring exhibiting a smaller amplitude. If the two springs exhibited equal amplitudes, you could conclude that the force exerted on the larger spring was greater than the force exerted on the smaller spring because more work was needed to overcome the resistive forces of the larger spring. AL

Use Figure 11

Wave Properties Have students look at **Figure 11.** List wave properties on the board, and have students identify which ones are shown in, or can be determined from, each graph. **Figure 11** left side: *amplitude, crest, trough, and wavelength;* **Figure 11** right side: *amplitude, crest, frequency, period, trough* OL Visual-Spatial EL

Reinforcement

Simple Harmonic Motion Have students who performed the Extension activity in Section 1 compare the graphs with **Figure 11**'s right side. Point out that the displacement of most mechanical waves can be modeled as simple harmonic motion plotted over time.

Activity

Feel the Waves To prepare tactile representations of waves, photocopy **Figure 11** and tape it to a piece of oak tag. Represent the displacement axis and the curve on each graph by carefully tracing the vertical axes and the curves using a needle-nosed glue dispenser. Sprinkle the glue with sand, let dry for a few minutes, and then shake off any excess sand. To represent the *x*-axis, glue a length of uncooked spaghetti along the horizontal axis of **Figure 11.** Repeat using a length of pipe cleaner for the *t*-axis in **Figure 11.** Have the students investigate the graphs as you explain what each axis and the curve on each graph represent. OL Kinesthetic EL

ADDITIONAL IN-CLASS EXAMPLE

Use with Example Problem 3.

Problem An 855-Hz disturbance (wave) moves through an iron rail at a speed of 5130 m/s.

a. What is the wavelength of the disturbance?

b. What is the period of the disturbance (wave)?

Response

a. $v = \lambda f$

$$\lambda = \frac{v}{f} = \frac{5130\ \text{m/s}}{855\ \text{Hz}} = 6.00\ \text{m}$$

b. $T = \frac{1}{f} = \frac{1}{855\ \text{Hz}} = 0.00117\ \text{s}$

Concept Development

$v = \lambda f$ Explain to students that they also will use the equation $v = \lambda f$ when they study sound, light, and quantum theory.

Discussion

Problem Have students recall that transverse waves can be represented graphically, as in **Figures 10** and **11.** Ask students how they might represent longitudinal waves graphically.

Response Answers will vary. Students might plot compressions and extensions visually as varying densities of shading or numerically as the number of coils per unit length as functions of time and location. OL

3 Assess

Assess the MAIN IDEA

Wave Properties Have students draw and label the parts of a transverse wave. Tell students to include at least 3 places where wavelength can be measured. Students should recognize that wavelength does not have to be measured from crest to crest or from trough to trough. BL Linguistic-Visual EL

Check for Understanding

Wave Characteristics Ask students to draw a *y*-displacement-location graph and a *y*-displacement-time graph of a transverse wave. Label the parts of the wave that each graph shows. Have students show how the wave's period or its wavelength can be determined. OL Visual-Spatial EL

Reinforcement

Wave Properties Review basic wave properties. Label three columns: property dependent on source, property dependent on the medium, and property dependent on both source and medium. List the properties under each category: period, frequency and amplitude, speed, and wavelength, respectively. Have students discuss how to verify these listings. BL Interpersonal

Reading Checks and Caption Checks

Caption Check, *student text p. 388*
In transverse waves, the displacement of the medium is perpendicular to the direction the wave travels. In longitudinal waves, the displacement is parallel to the direction the wave travels.

Reading Check, *student text p. 389*
If the amplitude of the wave is tripled, then the energy per second is increased by a factor of 3^2 or 9.

Reading Check, *student text p. 390*
Changing a wave's amplitude, frequency, or wavelength does not affect a wave's speed.

Caption Check, *student text p. 391*
The period is 0.4 s.

Practice Problems

student text p. 393

15. a. 343 m/s

b. 2.29×10^{-3} s

c. 0.787 m

16. You should shake it at a lower frequency because wavelength varies inversely with frequency.

17. 2.45 m/s

18. The frequency decreases to two-thirds of its original value.

19. 2.50 m

20. 0.600 m/s

21. 12.0 m/s

22. The frequency is one-half of its original value.

23. The wavelength is one-half of its original value.

24. The wavelength increases to 1.5 times its original length.

25. a. 338 m/s

b. 451 Hz

c. 2.22×10^{-3} s

Section 2 Review

student text p. 393

26. Loosely tie a piece of yarn somewhere near the middle of a rope. With your partner holding one end of the rope, shake the other end up and down to create a transverse wave. Note that while the wave moves down the rope, the yarn moves up and down but stays in the same place on the rope.

27. The amplitude and velocity remain unchanged, but the frequency increases while the period and wavelength decrease.

28. In longitudinal waves, the particles of the medium vibrate in a direction parallel to the motion of the wave. Nearly all mediums—solids, liquids, and gases—transmit longitudinal waves.

29. It takes time for the pulse to reach the other end in each case. It travels faster on the rope than on the spring, and fastest in the metal rod.

30. The swimmer's energy is transferred to the water in a small space over a short time, whereas the raindrops' energy is spread out in area and time.

1 Introduce

Bellringer

Wave Crests and Troughs Allow students to investigate the images projected from a ripple tank. Ask them to relate the alternating bright lines and dark bands projected on a screen with physical characteristics of a wave. The thicker water of the crests acts like a lens to refract and focus the light so the bright lines are produced by crests and the darker bands by troughs. OL Visual-Spatial EL

Tie to Prior Knowledge

Wave Characteristics Students apply the concepts of crest, trough, wavelength, and the equation $v = \lambda f$ to analyze reflection and refraction.

2 Teach

Waves at Boundaries

student text pp. 394–395

Identify Misconceptions

Wave Reflection Ask students to write a paragraph defending their answer to the following question: Does a wave slow down after reflection? OL Linguistic

PHYSICS PROJECT ACTIVITY

Wave Behavior Have groups of students analyze wave behavior using video cameras. Allow each group time to discuss their ideas and then narrow the analysis to a specific wave behavior (propagation, reflection, refraction, and constructive or destructive interference as a result of superposition), a specific medium (spring, rope, or water), and a method of video recording the behavior. Urge the groups to use the stop-action capabilities of a camcorder to analyze their data and present the analysis to the class. If students do not have access to personal video recording devices, have the media department of your school schedule time so that students can use the school's equipment and facilities. OL Interpersonal

Use Figure 14

Reflection from Rigid Boundaries Have students look at **Figure 14.** Point out that the reflected pulse shown in **Figure 14** is inverted relative to the incident pulse. Newton's third law explains the inversion. The pulse arriving at the fixed boundary exerts a force in the direction of its displacement on the boundary. The boundary exerts an equal and opposite force on the medium. Because the medium is elastic, it deforms and reflects as an inverted pulse.

Use the MiniLAB

In *Wave Reflection,* students will observe whether reflection changes the speed of a water wave.

Reinforcement

Wave Speed Have students note that a wave's speed is unaltered by reflection or interference, and ask them to speculate on why this is so. They might suggest that the wave continues to move through the same medium after it has reflected off a surface. Discussion should emphasize that only a change in the medium will change the wave speed. BL

Superposition of Waves

student text pp. 395–397

Use the PhysicsLAB

In *Interference and Diffraction,* students will investigate superposition of water waves.

Use the MiniLAB

In *Wave Interaction,* students can observe what happens when waves interact with each other.

Use Figure 16

Incident and Reflected Waves Ask students to explain if the middle panel in **Figure 16** indicates that the reflected wave is inverted. The wave is not inverted because the figure shows a crest being reflected as a subsequent crest. Next, ask students why the node in the left panel of **Figure 16** does not move. The waves are of the same amplitude, so they undergo destructive interference. OL

Use Models

MAIN IDEA Have students recall that people sometimes use their hands to do "the wave" at sporting events. Ask students what kind of wave this models. a transverse wave Have students show how "the wave" could be altered to demonstrate constructive and destructive interference of single pulses of 2, 1, 0, −1, and −2. These pulse heights might be represented by standing with hands upraised, standing, sitting, crouching, and crouching with hands lowered, respectively. Allow students to demonstrate their interference models. OL Kinesthetic EL

DIFFERENTIATED INSTRUCTION

Struggling Students Help students recognize standing waves by using a vibrating string with which nodes and antinodes can be seen. Have students identify the nodes and antinodes. Vary the tension in the string to produce different numbers of nodes and antinodes. Help students understand the connection between the numbers of nodes, antinodes, and frequency. BL Visual EL

Waves in Two Dimensions

student text pp. 397–399

Content Background

Waves in Two Dimensions Have students imagine a far away point disturbance. Tell students that at great distances from a point disturbance, the rays indicating the direction of the wave front can be considered parallel. At such distances, the wave front would appear as a linear crest. A bobbing linear object can also act as a source of a circular wave front. The adjacent wave fronts interfere constructively. Ask students to describe how the wave front moves in this case. The resulting wave front is parallel to the source and moves perpendicularly away from it.

REAL-WORLD PHYSICS

Classes of Seismic Waves Earthquakes produce four kinds of seismic waves. Primary waves, or P-waves, and secondary waves, or S-waves, travel outward from the quake's focus beneath Earth's surface. P-waves are longitudinal waves, and S-waves are transverse waves. Energy that reaches the surface creates Love waves and Rayleigh waves. Love waves cause the surface to vibrate in a back-and-forth motion. Rayleigh waves create an undulating, elliptical motion that causes the ground to rise and fall. Surface waves are responsible for most of the damage produced by earthquakes.

Reinforcement

Refraction Have students observe refraction of water waves. Place a glass plate on the bottom of a wave tank with edge angled to the approaching wave front. It helps to have minimal water at the shallow end of the tank. Supplement with short videos of refraction of waves so students get a clearer picture of this type of wave refraction.

Use the PhysicsLAB

In *Reflection and Refraction,* students will explore the change of wave direction in reflection and refraction.

3 Assess

Assess the MAIN IDEA

Wave Behavior Have students create three diagrams: 1) two different wave pulses coming toward each other from opposite ends of a coiled spring toy or piece of string or rope; 2) the superposition of the two pulses as they pass through each other; 3) the waves after passing through each other. OL Visual EL

Check for Understanding

Changes in Wave Properties Ask students what two wave properties change in refraction. speed and wavelength Ask what third property often changes also. direction BL

Extension

Two Point Sources Draw the following figure on the board.

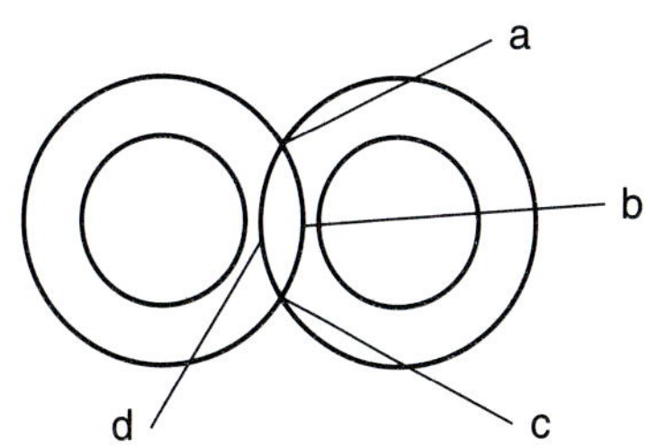

Explain to students that the figure shows circular waves from two point sources. Ask students to label two locations at which the medium is undergoing maximum constructive interference and two locations at which the medium is undergoing maximum destructive interference. a and c: maximum constructive interference; b and d: maximum destructive interference Visual-Spatial EL

Reading Checks and Caption Checks

Caption Check, *student text p. 394*

The amplitude of the incident wave is approximately twice the amplitude of the reflected wave. Since energy is proportional the square of the amplitude, the energy of the incident wave is approximately 4 times the energy of the reflected wave.

Reading Check, *student text p. 396*

The medium's displacement is zero at a node. Displacement is at a maximum at the antinode.

Caption Check, *student text p. 397*

The wavelength would be $(1/2)L$.

Reading Check, *student text p. 397*

Rays are always perpendicular to wavefronts.

Reading Check, *student text p. 398*

Angle of incidence and angle of reflection are measured from the normal, which is a line drawn perpendicular to the barrier.

Caption Check, *student text p. 399*

The wavelength decreases as the wave travels into the shallow water.

Reading Check, *student text p. 399*

The equation $\lambda = v/f$ shows that the wavelength λ of a wave is proportional to its speed v, if its frequency, f, is constant, as it is when a wave is refracted. If the speed of a refracted wave is half that of the incident wave, then the wavelength of the refracted wave must be half of that of the incident wave.

Section 3 Review

student text p. 399

31. Frequency remains unchanged. In general, amplitude, wavelength, and velocity will change when a wave enters a new medium. Direction may or may not change, depending on the original direction of the wave.

32. Answers will vary, but both pulses should have displacements in the same direction.

33. yes, if they strike the boundary while traveling normal to its surface, or if they have the same speed in both media

34. The number of nodes is always one greater than the number of antinodes.

35. The left panel of **Figure 16** behaves like a rigid wall because the reflected wave is inverted; the wave in the center panel behaves like a reflection from a non-rigid boundary because the boundary is an antinode and the reflected wave is not inverted.

HOW IT **WORKS**

Events of Magnitude

Background

Major earthquakes transfer energy to buildings and other structures through surface waves. Without careful design, buildings can be significantly damaged or even collapse due to oscillations from these waves. In Japan, which is prone to frequent earthquakes, builders are especially concerned with how earthquakes might damage structures. This concern helped drastically reduce damage and loss of life during the March 2011 Tohoku earthquake in Japan.

Teaching Strategies

- Have students discuss how they think a computer-controlled pendulum might control the sway of buildings. Inform students that inertia causes the large masses of the pendulum to move more slowly than the building around them. In effect, this means that the motion of the pendulum is sometimes in the opposite direction of the swaying building, preventing the building from swaying too far and exceeding the flexibility of the materials from which it is built.
- The Landmark Tower in Yokohama, Japan is not the only building specifically designed to withstand earthquakes. Have students research the technologies in some of these other buildings. Possibilities include the Taipei Financial Center in Taiwan and the Transamerica Pyramid in California.
- Have students brainstorm other ways of preventing damage from earthquakes. Discuss how different building materials—brick, concrete, steel, etc.—respond to the stresses that earthquakes present. Interested students might wish to investigate recent advances in materials science.

GOING**FURTHER** >>>

Expected Outcomes Students should report methods that dissipate or absorb seismic energy, such as mass dampers, flexible construction, thick foundations, reinforced or sliding walls, and base or seismic isolation systems using roller bearings, rubber pads, and fluid-filled shock absorbers. Japanese building codes apply these techniques to rules for short, medium, and tall buildings. Short buildings (less than 3 stories) must have reinforced walls and thick foundations. Medium buildings (up to 100 feet) often rest on pegs or shock absorbers, which turn lateral shaking into thermal energy by sliding side to side. Tall buildings have flexible construction. Students may compare the Landmark Tower with Japanese temples that have withstood many earthquakes because of their flexible construction. Another earthquake-resistant building is the Roppongi Hills Mori Tower in Tokyo, which uses an oil mass damper system.

CHAPTER 14 ANSWERS

SECTION 1

Mastering Concepts

36. Periodic motion is motion that repeats in a regular cycle. Examples include the oscillation of a spring, the swing of a simple pendulum, and uniform circular motion.

37. Frequency is the number of cycles or repetitions per second, and period is the time required for one cycle. Frequency is the inverse of the period.

38. Simple harmonic motion is periodic motion that results when the restoring force on an object is directly proportional to its displacement. A block bouncing on the end of a spring is one example.

39. The spring stretches a distance that is directly proportional to the force applied to it.

40. The spring constant is the slope of the graph of *F* versus *x*.

41. The potential energy is the area under the graph of *F* versus *x*.

42. no; yes; no, as long as the amplitude is less than about 15°; the gravitational field, *g*

43. Resonance will occur when a force is applied to an oscillating system at the same frequency as the natural frequency of the system.

Mastering Problems

44. 27 N/m

45. 0.12 m

46. 0.35 J

47. 0.29 m

48. a. 20 N/m

b. 2.5 J

49. 0.21 m

50. A = B < C = D

SECTION 2

Mastering Concepts

51. When throwing a ball, matter (in the ball) is transferred from one place to another. In a mechanical wave, energy transfers without matter traveling from place to place.

52. A transverse wave causes the particles of the medium to vibrate in a direction that is perpendicular to the direction in which the wave is moving. A longitudinal wave causes the particles of the medium to vibrate in a direction parallel to the direction of the wave. Surface waves have characteristics of both.

53. a. Speed of the waves depends only on the medium and cannot be changed.

b. Frequency can be changed by changing the rate at which the waves are generated.

54. Wavelength is the distance between two adjacent points on a wave that are in phase.

55. Once the pulse has passed, the point is exactly as it was prior to the pulse reaching the point.

56. A pulse is a single disturbance in a medium, whereas a periodic wave consists of several adjacent disturbances.

57. Frequency is the number of vibrations per second of a part of the medium. Velocity describes the motion of the wave through the medium.

58. They are the same.

59. Points are in phase when they have the same displacement and the same velocity. Otherwise, the points are out of phase. Two crests are in phase with each other. A crest and a trough are out of phase with each other.

60. The energy carried by a wave is proportional to the square of its amplitude.

Mastering Problems

61. 8.3 s

62. 4.0 m/s

63. a. 0.29 m/s

b. 0.21 s

64. a. 1.9 m/s

b. 2.1 m

65. a. 1.50×10^{3} m/s

b. 1.00×10^{-6} s

66. 1350 m

67. a. 550 Hz

b. 280 complete waves

c. 1.7×10^{2} m

68. 3.6 m/s

69. 8.1×10^{2} km

SECTION 3

Mastering Concepts

70. The frequency depends only on the rate the thin rope is shaken and the thin rope causes the vibrations in the thick rope.

71. The reflected pulse will be inverted.

72. Nothing, the medium does not move.

73. Bare areas are antinodal regions where there is maximum vibration. Sugar-covered areas are nodal regions where there is no vibration.

74. A standing wave exists and the string can be touched at any of its five nodal points.

75. The wavelength and direction of the wave fronts change. The frequency does not change.

Mastering Problems

76. 1. The amplitude is doubled.

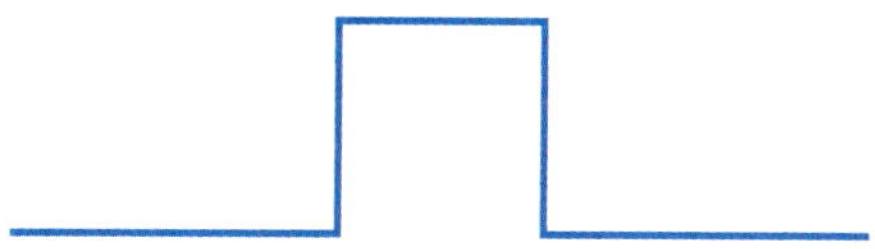

2. The amplitudes cancel each other.

3. If the amplitude of the first pulse is one-half of the second, the resultant pulse is one-half the amplitude of the second.

77. a. 2.4×10^{-3} s

b. Pulses are inverted when reflected from a more dense medium, so the returning pulse is inverted (below).

c. 15 cm from the other end, where the distances traveled are the same

78. Because wave speeds are the same, the largest wavelength has the smallest frequency. Wavelengths are A: 18 cm, B: 15 cm, C: 20 cm, D: 12 cm, so the frequency rankings are D > B > A > C.

Applying Concepts

79. At the bottom of the motion, the elastic potential energy is at a maximum, while gravitational potential energy is at a minimum and the kinetic energy is zero. At the equilibrium position, the *KE* is at a maximum and the elastic potential energy is zero. At the top of the bounce, the *KE* is zero, the gravitational potential energy is at a maximum, and the elastic potential energy is at a maximum. The total mechanical energy is conserved.

80. No; the space station is in free fall, and therefore the apparent value of g is zero. The pendulum will not swing.

81. in the first case, longitudinal waves; in the second case, transverse waves

82. The frequency of the waves will increase; the speed will remain the same; the wavelength will decrease.

83. As the frequency increases, the period decreases.

84. As the frequency increases, the wavelength decreases.

85. approximately two squared, or four times the energy

86. The period of the vibration must equal the time for the wave to go back and forth across the pan to create constructive interference.

Mixed Review

87. 2.4 s

88. a. Range is 190 m to 550 m

b. Range is 2.8 m to 3.4 m

89. a. 1/4 wavelength

b. 0.72 s

c. 1.4 Hz

90. $k = 23$ N/m

91. a. surface waves

b. The displacement is perpendicular to the direction of the wave—in this case, up and down.

c. 1.5 s

d. 0.67 Hz

e. 2 m/s

f. 2.7 m

92. 0.57 N/m

93. a. 2.4×10^{2} N/m

b. 1.1 kg

94. a. 6.3 s

b. There should be no difference. *T* is not affected by mass.

c. At the bottom of the swing, *KE* is at a maximum.

d. At the top of the swing, *PE* is at a maximum.

e. At the top of the swing, *KE* is at a minimum.

f. At the bottom of the swing, *PE* is at a minimum.

95. a. 22,000 N/m

b. 1.1 J

96. a. 7.5 s

b. 1.0×10^{1} s

c. 0.10 Hz

d. The arm on the new structure is 11 m longer.

e. Because of the square relationship, there would need to be a 4 times increase in the length of the pendulum, or a 300% increase.

97.

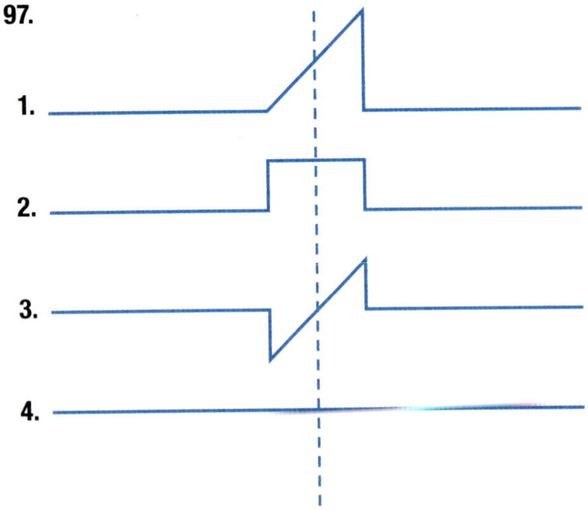

98. a. The clock must be made to run faster. The period of the pendulum can be shortened, thus increasing the speed of the clock, by shortening the length of the pendulum.

b. The length would need to shorten by $l_1 - l_2 = 0.150\ \text{m} - 0.135\ \text{m} = 0.015\ \text{m}$.

c. need to shorten by $l_1 - l_2 = 0.775\ \text{m} - 0.770\ \text{m} = 0.005\ \text{m} = 5\ \text{mm}$

99. 63.7 N

Thinking Critically

100. Answers will vary, but a correct form of the answer is, "A 1.65-kg watermelon is placed in the basket of a produce scale at the supermarket. If this stretches the spring of the scale by 1.5 cm, what is the scale's spring constant?"

101. a. "… If the block is initially pulled to a position 10 cm below the equilibrium position and then released, how fast will it be moving when it passes through equilibrium?"

b. "… If the block has a mass of 0. 85 kg, how much will the spring be stretched at the equilibrium position?"

c. "… The block oscillates such that it goes through 7 complete cycles in 6 s. What is the frequency of the oscillations?"

102. a. Graph should show an upward slope.

b. 21 N/m

c. 2.5 J

103. a. 40 N/m

b. 5 J

c. The force is not constant as the spring is stretched. The average force, 10 N, multiplied by the distance does give the correct work.

104. 3 Hz

Writing in Physics

105. Huygens proposed the wave theory of light, and Newton proposed the particle theory of light. The law of reflection can be explained using both theories. Huygen's principle and Newton's particle theory are opposed, however, in their explanations of the law of refraction.

Cumulative Review

106. a. 8.8×10^{6} J

b. The minimum amount of work must equal *KE*, or 8.8×10^{6} J. The engine had to do more work than was dissipated in work done against friction.

c. 11 m/s^2

107. 2×10^{-3} kg/s

Standardized Test Practice

student text p. 407

Multiple Choice

1. D

2. C

3. B

4. B

5. A

6. A

7. C

8. C

9. D

10. C

Free Response

11. $k(\text{m}) = \text{kg}\cdot\text{m/s}^2$

$$k = \frac{\text{kg}\cdot\text{m/s}^2}{\text{m}}$$

$$k = \frac{\text{N}}{\text{m}}$$

CHAPTER 15

Sound

About the Photo

Sound Intensity The relationship between human perception of sound and the physics of sound is complicated and indirect. Even so, the differences between music and noise, amplitude and loudness, and frequency and pitch are real and can be studied scientifically.

Chapter Overview

Sound is a pressure variation that travels as a longitudinal wave. This chapter examines sound characteristics, such as direction, pitch, loudness, and speed. The chapter then links physics and music by discussing resonance in air columns and on strings. The chapter closes with a discussion of sound quality, sound reproduction, and noise.

Before students study the material in this chapter, they should study:

- Fundamentals of waves
- The law of reflection
- Periodic motion
- Temperature
- Uniform motion in one dimension

To solve the problems in this chapter, students will need a strong grasp of:

- Graphing data
- Scientific notation
- Significant figures
- Solving linear equations

Introduce the BIG IDEA

Sound Waves Sound is a wave that can be described using all of the concepts that apply to mechanical waves, including reflection, refraction, interference, and superposition. Most sounds encountered in daily life are complex superpositions of many different frequencies.

Go online!
connected.mcgraw-hill.com

Your online experience includes:

- **Video**
- **Concepts In Motion**
- **Virtual Investigations**
- **All Lab, Applying Practices activities, and PBL worksheets**
- **Online Practice**
- **Personal Tutor**
- **Multilingual eGlossary**
- **Presentation**

Use the LaunchLAB

In *Producing Musical Notes,* students can use a stemmed drinking glass to produce a musical note.

Use Physics T.V.

Video

MAKING MUSIC From woodwinds to brass to percussion, every instrument makes a different kind of sound. What is the physics of making music?

student text pp. 410–417

1 Introduce

Bellringer

Sound and Energy Use a tuning fork and a beaker containing a small amount of water to show that the tines of a sounding tuning fork are transmitting energy to the matter around them. Strike a tuning fork and have students observe its sound. Show students the motion of the tines by plunging the tines into the water in the beaker. Have students infer that a vibrating tine transfers kinetic energy to the material surrounding it in a repetitive pattern. BL Auditory-Musical EL

Tie to Prior Knowledge

Properties of Waves Students extend their understanding of the properties of transverse waves (frequency, wavelength, and amplitude) to the similar properties that describe longitudinal waves, including sound waves.

2 Teach

Sound Waves

student text pp. 410–412

Use Models

Three-Dimensional Sound Wave Model the propagation of a sound wave in three dimensions by inflating a spherical balloon. Point out that the motion of the balloon's surface models the motion of a sound wave moving away from its source in all directions.

Use an Analogy

Sound Intensity Have students recall that as a spherical balloon expands, its color fades; the intensity of its color decreases because the pigment is being spread over a greater surface area. A sound also fades as it moves from its source because the intensity of the sound energy (energy/unit area) is less at a greater distance. The decrease in intensity occurs because the sound energy passes through an increasingly larger surface as the wave expands.

Use the PhysicsLAB

In *Sound Travels Through Air,* students can measure how long it takes sound to travel down and back in a long tube.

Identify Misconceptions

Pitch and Intensity Demo Possible misconceptions about pitch are revealed by asking students how they think the pitch of a tuning fork may change if its loudness decreases. Some students might incorrectly respond that the pitch will decrease. Address this misconception by first matching the frequency of a struck tuning fork with that of a tone generator. Next, strike the tuning fork and decrease the intensity of the tone generator so that it matches the tuning fork as the fork's intensity fades. Students should notice that the pitch does not decrease. BL Auditory-Musical EL

Reinforcement

Properties of Longitudinal Waves Have students diagram how the motion of the particles of the medium through which a sound wave moves indicates that it is a longitudinal wave. The particles move back and forth, parallel to the direction of the wave. BL EL

Detection of Pressure Waves

student text p. 412

DIFFERENTIATED INSTRUCTION

Struggling Students Point out that a sound wave is a pressure variation that is transmitted through matter, and that the definition implies that pressure varies with time. Illustrate this implication by drawing a pressure-time graph representing the wave as a sine curve. Emphasize that the graph is like a movie of how the pressure varies at a single point along the path of the sound wave. Ask students what term describes the time interval for the movie to repeat itself. period Show that a pressure-time graph yields the period of a sound wave. OL

REAL-WORLD PHYSICS

Early Warning System The Chinese used pottery jars to detect the sounds of an advancing army. Leather membranes were stretched over the mouths of empty, 80-L pottery jars. The jars were lowered into deep shafts that were dug a few paces apart, and soldiers with good hearing were stationed near these shafts. Not only could the soldiers hear the sounds of an approaching army, but by listening to the different sounds from the shafts, they could judge from which direction the enemy was coming and from how far away. They could do this because the vibrations from the feet of the moving soldiers transferred to the earth and radiated in all directions, including to the pottery jars and their leather caps, which in turn vibrated and created sound.

Perceiving Sound

student text pp. 413–414

Critical Thinking

Negative Sound Levels Ask students to explain what a negative sound level, such as −10 dB, is. Start with 0 dB. Because decibels are measured relative to another sound, a measurement of 0 dB occurs when sound pressure is equal to the reference level. A negative sound level has a pressure that is smaller than the reference pressure. OL

Use the MiniLAB

In *Sound Characteristics,* students can investigate physical factors that affect the sounds of dropped coins.

DIFFERENTIATED INSTRUCTION

Visually Impaired For activities that require discerning differences in pitch, students can use two flashlights—one with a red filter and the other with a blue filter. (The filters can be made from cellophane and attached to the flashlights with rubber bands.) Pair a hearing-impaired student with a hearing student facilitator. During demonstrations or activities involving pitch changes, the facilitator can turn on the blue light to indicate an increase in pitch or the red light to indicate a decrease in pitch. Alternatively, students can use a lamp with a dimmer switch to indicate changes in intensity.
COOP LEARN Visual-Spatial

Concept Development

MAIN IDEA Point out that the definition of a sound wave also implies that pressure varies with location. Illustrate this implication by drawing a pressure-position graph representing the wave as a sine curve, perhaps starting with an amplitude other than zero. Point out that this graph is a "snapshot" of the pressure along the path of the wave at one instant. Ask students what term describes the distance between locations that have identical snapshots. wavelength Show that a pressure-position graph yields the wavelength of a sound wave. OL

Use the PhysicsLAB

In *What is a decibel?,* students can use a sound-level meter to measure the sound levels in various situations.

The Doppler Effect

student text pp. 414–417

Use Figure 6

Ask students how the wavelength of the sound wave detected by observer A compares to the wavelength of the sound wave detected by observer B. It is shorter. Have students relate the wavelength to the frequency of the sound wave striking each observer's ear. Observer A detects a higher frequency than observer B does. How does the pitch perceived by each observer compare to the pitch that would be heard if the same sound source were at rest? If the source were at rest, the pitch would be lower for observer A and higher for observer B. They would both perceive the same pitch. OL

Content Background

Sonic Booms As can be seen in **Figure 6,** the motion of a sound source distorts the sound wave leading it. When the sound source is moving faster than the speed of sound (supersonic speeds), a conical wavefront called a shock wave is produced. The energy of the shock wave is concentrated at the surface of the cone and is the cause of a jet plane's sonic boom. At a much smaller scale, the crack of a whip is also a sonic boom because the tip of the whip is moving faster than the speed of sound.

Concept Development

Range of the Doppler Shift Emphasize that the Doppler shift is used to measure the velocity of a source of either mechanical waves, such as sound, or electromagnetic waves, such as light, over a wide range of scales, from atoms to autos to galaxies.

Quick **DEMO**

Doppler Effect

Estimated Time 15 minutes

Materials foam ball, electronic buzzer (oscillator), batteries

Procedure Before the demonstration, assemble the buzzer and batteries. Hollow out the foam ball so that the buzzer and batteries fit snugly within. Turn on the buzzer and place it inside the ball. Toss the ball to a student, and have the student note any changes in pitch as the ball is caught. Repeat with other students. Students should note a slight decrease in the pitch as they catch the ball. Point out that the ball is at rest once it is caught. Because the pitch dropped, the pitch of the ball as it was traveling toward the students must have been higher. Explain that the difference in pitch caused by motion of the sound source is an example of the Doppler effect.

ADDITIONAL IN-CLASS EXAMPLE

Use with Example Problem 1.

Problem A local fire station has a roof-mounted siren that has a frequency of 975 Hz. If you are on your bike moving away from the station at a speed of 6.00 m/s, what will be the frequency of the sound waves reaching your ear? Assume that the air temperature is 20°C.

Response $f_d = f_s(1 - v_d/v)$

$$f_d = 975\ \text{Hz}\left(1 - \frac{6.00\ \text{m/s}}{343\ \text{m/s}}\right)$$

$$f_d = 958\ \text{Hz}$$

Discussion

Seeing a Doppler Shift

Question How do astronomers use the Doppler shift?

Answer Because the Doppler shift occurs with light as well as sound, astronomers can note the change in the wavelength of light emitted by a distant, moving source. As a galaxy moves away from Earth, the apparent frequency of the light it emits decreases, and the wavelength increases. This is referred to as a redshift. The equivalent phenomenon in sound is more easily observed as a decrease in pitch. OL

3 Assess

Assess the MAIN IDEA

Doppler Effect and Hearing If a stationary source of a 20-Hz tone at 80 dB begins moving away from you at 50 m/s, what frequency would you hear? The frequency of the sound wave that reaches you would be about 17 Hz, but this is out of the range of normal human hearing so you may not be able to hear it at all.

Check for Understanding

Modeling Sound Waves Represent a sound wave by a sine curve on a pressure-position and a pressure-time graph. Ask students to label the segments of the graphs indicating the wavelength, period, and amplitude. Have students relate the labels to properties of sound such as pitch and intensity. BL Visual-Spatial

Extension

Doppler Shift Ask students for examples of the Doppler shift they may have experienced. Have them analyze the situations. Make sure students note that the shift is a change in apparent pitch caused by motion of the observer and/or the sound source. OL

Reading Checks and Caption Checks

Reading Check, *student text p. 411*
346 m/s

Caption Check, *student text p. 413*
about twice as loud

Caption Check, *student text p. 414*
$\Delta f_A = -v_s/\lambda$; $\Delta f_B = +v_s/\lambda$

Practice Problems

student text p. 416

1. 488 Hz
2. 392 Hz
3. 548 Hz
4. 3.52 MHz
5. 19 m/s

Section 1 Review

student text p. 417

6. frequency; amplitude
7. The student's sketch should resemble a sine wave, with appropriate labels and with time increasing continuously and displacement varying between minimum and maximum values.

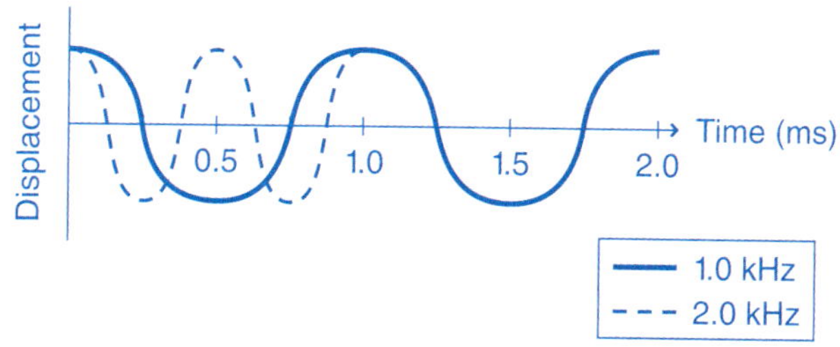

8. affected: speed and wavelength; unaffected: period and frequency
9. The sound pressure level increases by a factor of 10 for every 20-dB increase in sound level. Therefore, 60 dB corresponds to a 1000-fold increase in the sound pressure level.
10. The velocity of sound is greater in solids than in gases. Therefore, sound travels faster in steel rails than in air, and the rails help focus the sound so it does not die out as quickly as in air.
11. **a.** They would differ in intensity. Larger insects would reflect more of the sound energy back to the bat.

    b. An insect flying toward the bat would return an echo of higher frequency. An insect flying away from the bat would return an echo of lower frequency.
12. No; the car must be approaching or receding from the detector for the Doppler effect to be observed. Transverse motion produces no Doppler effect.

1 Introduce

Bellringer

Straw Flutes Distribute a plastic soda straw to each student. Have students cut one end of the straw to a point and flatten this end by chewing on it gently. (This will result in a reedlike end on the straw.) With a little practice, students can blow on the reed and produce a musical tone. Have several students cut their straws shorter and then compare the tones they make with others in the class. Have students discuss how the straw might produce a sound and how the pitch of the sound relates to the length of the straw. **BL** **Auditory-Musical**

Tie to Prior Knowledge

Sound in Music Students apply their understanding of sound waves to the concept of resonance and to the properties of standing waves in air columns and on strings.

2 Teach

Sources of Sound and Resonance in Air Columns

student text pp. 418–422

Discussion

Question Ask students to classify musical instruments as closed-pipe or open-pipe resonators and to explain why.

Answer Although many instruments can be difficult to classify, the trumpet and flute are examples of open-pipe resonators because both ends of these instruments are open, and the sound waves reflect off an open end and are inverted. A lip reed instrument, such as the clarinet, is a closed-pipe resonator because only one end of this instrument is open to air, while the reed end is (almost) closed to air. In these instruments, the closed end has a much smaller cross-section than the open end, which is enough to cause a reflection like that from a completely closed end. The pan pipe and didjeridu are other closed-pipe resonators. One of the oldest closed-pipe instruments is the conch shell. Playing a closed-pipe resonator is like blowing air into a bottle at the correct angle. **BL**

Identify Misconceptions

Pipes and Air Columns Remind students that the standing waves in pipes occur in the column of air within the pipe and not in the material of which the pipe is made. Explain that the sound moves through the air because the vibrating source, such as a reed of an instrument, produces the regular oscillations in the air pressure. With every cycle of the wave, air rushes in and out of the ends of the pipe to create the compression and rarefaction for the wave.

Use the PhysicsLAB

In *Speed of Sound,* students can investigate the speed of sound in a closed-pipe resonator.

PHYSICS CHALLENGE ACTIVITY

Graphing Sound Waves Suggest that students use sound display equipment to show pressure-time representations of the sound waves of pure tones. The equipment might include a graphing calculator with CBL, oscilloscope tracings, or computer interfacing. Sources of pure tones might include tuning forks, tone generators, or certain toys. Have students analyze the graphs for period and frequency. Urge students to post the analyzed graphs in the classroom for classmates to inspect. **AL** **Visual-Spatial**

Concept Development

Representing Standing Waves It is possible to graphically represent a standing wave on a string or in an air column. For a string, it makes sense to display the displacement of the string versus distance along the string. In an air column, both pressure and displacement of the air molecules can be displayed as functions of distance along the column.

REAL-WORLD PHYSICS

Hearing and Frequency The human auditory canal serves as the air column of a closed-pipe resonator. When sound enters the outer ear, it is in the form of a pressure wave. When the pressure wave reaches the middle ear, or the closed end of the pipe, it is converted to a mechanical wave, as the bones and eardrum of the middle ear begin to vibrate. In addition to the vibration reflecting the sound waves to the outer part of the closed pipe, it also reflects the wave to the fluid of the inner ear behind the eardrum. For us to perceive sound, the motion of this fluid activates the hair cells located in the inner ear, each of which is sensitive to a particular frequency of vibration. These cells initiate nerve impulses to the brain, which interprets the sound.

Content Background

Pressure and Displacement Graphs As can be seen in **Figure 12,** displacement nodes correspond to pressure antinodes. This relationship occurs because the molecules on either side of a displacement node are moving away from and then toward each other. As the molecules approach, the pressure increases to a maximum and then falls to a minimum as they recede. Therefore, this location is a pressure antinode because the pressure oscillates through it at maximum and minimum values. At displacement antinodes, the molecules on either side are moving in phase. Here the pressure change is zero, and thus the location of a displacement antinode corresponds to a pressure node.

Use the PhysicsLAB

In *How fast does sound travel?* students can investigate the relationship between the wavelength of sound waves in an open tube to the speed of sound.

Resonance on Strings

student text pp. 422–423

Activity

MAIN IDEA Allow students access to a piano with an open lid. Arrange students into small groups. While a student in the group depresses the damper (right) pedal of the piano, have another student loudly hum a tone for a short time. When the student has stopped humming, have the others listen intently and they will discern that a string in the piano is producing a similar pitch. Let other groups repeat the activity. Have students relate the effect to resonance between the frequency of the sound wave they produce by humming and the fundamental frequency of the string. BL Auditory-Musical

Critical Thinking

Differences in Standing Waves Point out that a 1.00-m-long brass rod held in the middle can be made to resonate with a longitudinal standing wave that has a wavelength of 2.00 m and a fundamental frequency of 1750 Hz. The rod acts the way an open pipe or a string does, with resonant frequencies at whole-number multiples of f, so that $f = \frac{v}{2L}$ and $\lambda = 2L$. On the other hand, for a closed pipe, $f = \frac{v}{4L}$, and a 1.00-m-long closed pipe produces a standing wave with an equal wavelength but a fundamental frequency of only 85.8 Hz. Ask students what other quantity differs between the two "pipes." The speed of sound differs. The fundamental frequency of a standing wave on a brass rod is greater because the speed of longitudinal waves in brass is greater than the speed of the longitudinal waves in air. OL Logical-Mathematical

PHYSICS PROJECT ACTIVITY

Pitch It Have students work in small groups to construct a pair of instruments out of common items, such as soda bottles, rods, and rubber bands. Together, the instruments have to sound some standard musical interval, which must be identified by the use of a tuned instrument, such as a piano, or a tone generator. Have students sketch pressure-time graphs of the waves produced by each instrument and the superposition of the waves when both instruments are played simultaneously. That is, they must show the total amplitude of the vibrations from both instruments. Allow the groups to demonstrate their instruments and explain their graphs to other groups. OL Auditory-Musical

Use the MiniLAB

In *Sounds Good,* students can use a frequency generator to investigate whether a specific wind instrument is an open-pipe resonator or a closed-pipe resonator.

Reinforcement

Standing Pressure Waves Have students explain why **Figures 13** and **16** are amplitude-position graphs. Both show the wave amplitude variation from one location to another along a medium. OL

Quick **DEMO**

Resonance in Rods

Estimated Time 10 minutes

Materials 50- to 200-cm length of 10- to 15-mm-diameter aluminum rod, rosin

Procedure Hold the rod securely at its center with the thumb and forefinger of one hand. Place some rosin on the thumb and forefinger of the other hand. Run your thumb and forefinger briskly along the rod until it starts to vibrate. Adjust the pressure of the fingers until the rod starts to vibrate loudly. Have students note the pitch. Repeat the demonstration. When the rod is sounding, grab the midpoint of the vibrating portion with your free hand. A rod can produce a longitudinal standing wave. Because you supported the rod at its midpoint, the center of the rod was a node and $\lambda_1 = 2L$. When you grabbed the midpoint of the vibrating portion of the rod, you eliminated an antinode, and the rod stopped vibrating.

Sound Quality

student text pp. 423–426

DIFFERENTIATED INSTRUCTION

Visually Impaired Allow students with visual impairment to examine musical instruments in the band room or those that students bring in. Have students determine the location of the source of the sound waves (such as lips, string, reed) and a means of controlling the fundamental frequency (such as length of air column, tension and length of a string). Have students also investigate how pitch, loudness, and tone quality are controlled with each instrument. **BL** **Auditory-Visual**

Critical Thinking

Pitch and Temperature Have students explain why woodwind instruments tend to "go sharp" (pitch of notes rises), but stringed instruments "go flat" as temperature rises. As temperature increases, the velocity of sound increases, but the length of the woodwind instrument, which determines the wavelength, changes little. Thus, the frequency varies only with the velocity of sound. On the other hand, a string expands, causing the tension to decrease and, in turn, lower the frequency. **AL**

ADDITIONAL IN-CLASS EXAMPLE

Use with Example Problem 2.

Problem A pipe organ has an open pipe with a resonating height of 9.75 m. Ignoring end corrections, what is the frequency of the longest sound wave produced by this pipe? Assume the speed of sound is 343 m/s.

Response Obtain the resonance length of the lowest note using $L = \lambda/2$, then use $v = \lambda f$ to solve for f, 17.6 Hz.

REAL-WORLD PHYSICS

Hearing Spectrum Audiograms are graphical representations of one's hearing. They are similar to sound spectra because they are plots of intensity versus frequency. Audiograms are used in audiometry, which is a medical testing procedure for hearing loss. A patient is exposed randomly to sounds at frequencies of 250 Hz, 500 Hz, 750 Hz, 1000 Hz, 2000 Hz, 4000 Hz, 6000 Hz, and 8000 Hz, one ear at a time, through earphones. The minimum intensity at which the patient perceives the sound is recorded. The resulting audiogram is analyzed against norms to determine the extent of hearing loss.

Use Figure 18

Ask students how many harmonics are shown in the sound spectrum of the violin in the figure. 5 Ask students to rank the instruments in order of increasing pitch of the fundamental being played by each. guitar, violin, steel drum

PHYSICS CHALLENGE ACTIVITY

Piano Strings Ask students to imagine a piano whose strings are all the same thickness and are at the same tension, so that the wave velocity of each string is 343 m/s. What is the length of the shortest wire if it produces a tone of 4190 Hz? 0.0409 m What is the length of the longest wire if it produces a tone of 27.5 Hz? 12.5 m Do actual pianos use this range of string lengths? No; real pianos use a variety of string thickness and tension to produce the full range of frequencies within a reasonable range of string lengths. **AL** **Kinesthetic**

Beats
student text p. 426

Use Models
Beat Period Demo Construct two similar pendulums with lengths of 0.4 m and 0.6 m. Arrange the pendulums so that they align vertically and the bobs can be released directly behind each other. Release the bobs and have students observe that both bobs reach their starting point in tandem about every 8 s. This time interval is the period of the beat, T_{beat}. Have students determine the beat frequency. $f_{beat} = 1/T_{beat} = 0.1$ Hz
OL Logical-Mathematical

Sound Reproduction and Noise
student text p. 427

Activity
Noise Waves Students can use an oscilloscope to observe differences in sound waves produced by noise and music. First, have them watch the waves generated on the oscilloscope as they produce musical sounds. They could make the sounds using either simple or real musical instruments. Next, have students observe the waves produced when they use the instruments to make sounds they consider to be noise.
BL OL Auditory-Musical EL

3 Assess

Assess the MAIN IDEA
Singing in the Shower Unusually rich, loud tones can be generated when singing in the shower. How can you explain this phenomenon in terms of the physics of music? A shower stall acts as a closed-pipe resonator. Middle C is 262 Hz, so a shower stall is about the ideal size to resonate with tones around this frequency. The hard surfaces of most showers reflect sound and contribute to the resonance as well.

Check for Understanding
Pipe Harmonics Have students draw pressure-position representations of standing waves in closed- and open-pipe resonators. Ask them how the wavelengths of standing waves are related to the length of each pipe. closed-pipe resonator: $\lambda = 4L$, $(4/3)L$, $(4/5)L$; open-pipe resonator: $\lambda = 2L$, L, $(2/3)L$ OL Visual-Spatial

Extension
So That's Timbre Urge students to listen to *The Young Person's Guide to the Orchestra,* Opus 34, by Benjamin Britten. Have them respond by writing a brief paragraph describing differences in timbre among the woodwind, brass, string, and percussion families of instruments.
BL Auditory-Musical

Reading Checks and Caption Checks

Reading Check, *student text p. 420*

A node is the stationary point where two equal wave pulses meet and are in the same location. An antinode is the place of largest displacement when two wave pulses meet. In the pressure graphs, the nodes are regions of mean atmospheric pressure. At the antinodes, the pressure oscillates between its maximum and minimum values. In the case of the displacement graph, the antinodes are regions of high displacement and the nodes are regions of low displacement.

Caption Check, *student text p. 420*

Areas of mean atmospheric pressure are represented by nodes in the pressure graphs.

Caption Check, *student text p. 421*

If open and closed pipes of the same length are used as resonators, the wavelength of the resonant sound for the open pipe will be half as long as that for the closed pipe.

Reading Check, *student text p. 423*

The greater the tension in the string, the faster the wave moves along it.

Practice Problems

student text p. 424

13. 970 m/s

14. 859 Hz

15. 0.39 m

16. a. 64.7 Hz

b. 129 Hz and 194 Hz

Physics Challenge

student text p. 425

1. $F_T = \frac{mv^2}{4L}$

2. $F_T = \frac{mv^2}{4L}$

$$F_T = \frac{(0.0010\ \text{kg})(343\ \text{m/s})^2}{4(0.400\ \text{m})}$$

$$F_T = 74\ \text{N}$$

Section 2 Review

student text p. 427

17. a. vocal cords

b. a reed

c. the player's lips

d. a string

18. The longer the tube, the lower the resonant frequency it will produce.

19. The length of the tube should be one-half the wavelength.

20. 740 Hz, 1100 Hz, 1500 Hz

21. a. 35.7 Hz

b. 0.10 m

22. Each instrument produces its own set of fundamental and harmonic frequencies, so they have different timbres.

23. It is either 389 Hz or 395 Hz. You can't tell which without more information.

24. The tuning fork's sound is amplified greatly when it is pressed against other objects because they resonate like a sounding board. They sound different because they resonate with different harmonics; therefore, they have different timbres.

PHYSICS THAT'S ENTERTAINMENT!

Sounds Good!

Theater Acoustics

Background

Engineers renovating modern theaters consider the acoustic properties of not only the walls and ceilings, but also the floors, the stage, and the seats. Carpeted floors often absorb sound better than hard floors. The acoustic properties of the stage can be customized for spoken-word theater—where actors' voices need to be projected—or for ballet—where the sounds of dancers' movements need to be dampened. Even audience seats can differ in their acoustic properties, and some theater renovations that focus on sound management must include seat replacement.

Teaching Strategies

- Explain that when building other theaters later on, the Greeks attempted to reconstruct the special acoustics of the theater at Epidaurus. They could not recreate it in other venues, however, because they did not know what caused the effect.
- Show students a diagram of how interference can create loud and soft spots in the audience of a theater. Review with them the conditions that cause constructive and destructive interference.
- Have students find out about the acoustic requirements for certifying a movie theater for IMAX films. Suggest that they present their findings in a brochure aimed at theater owners.

GOINGFURTHER >>>

Expected Outcomes Students should address placement of speakers of different types and the material, design, and placement of acoustic treatments, including panels designed to absorb, reflect, or diffuse sound. They might also discuss solutions to acoustic problems presented by flooring, windows, and overly large or small rooms.

CHAPTER 15 ANSWERS

SECTION 1

Mastering Concepts

25. Sound waves can be described by frequency, wavelength, amplitude, and speed.

26. Light travels at 3.00×10^8 m/s, while sound travels at 343 m/s. Officials would see the smoke before they would hear the pistol fire. The times would be less than actual if sound were used.

27. Pitch is the human perception of frequency. Loudness is the human perception of amplitude.

28. all types of waves

29. Doctors can measure the Doppler shift from sound reflected by the moving blood cells. Because the blood is moving, sound gets Doppler shifted, and the compressions either get piled up or spaced apart. This alters the frequency of the wave.

Mastering Problems

30. 1.7 km

31. 5.1×10^2 m

32. 5200 m/s

33. 9.8×10^4 Hz

34. 5.707 m

35. 1.45×10^3 m/s

36. 350 Hz

37. 0.0175 s

38. 11.3 m

39. a. A typical rock concert is 110 dB, so a 40-dB reduction is needed.

b. A barely audible whisper is 10 dB, so the actual level would be 50 dB, or that of an average classroom.

40. a. Each 20 dB increases pressure by a factor of 10, so 10 times greater pressure.

b. 100 times greater pressure

41. 2.0 m/s

42. a. 349 m/s

b. 0.436 s

43. 3.4×10^2 m

44. 0.353 mm

45. a. 1300 m

b. 580 m

46. 1000 times

47. a. 335 Hz

b. 356 Hz

48. a. 2.80×10^2 Hz

b. 2.63×10^2 Hz

SECTION 2

Mastering Concepts

49. a vibrating object and a material medium

50. The frequency of the note is the same as the natural resonance of the crystal, causing its molecules to increase their amplitude of vibration as energy from the sound is accepted.

51. While marching in step, a certain frequency is established that could resonate the bridge into destructive oscillation. No single frequency is maintained under "route step."

52. Tuning forks produce simple, single-frequency waves. Musical instruments produce complex waves containing many different frequencies. This gives them their timbres.

53. the sound quality or timbre

54. The slide of a trombone varies pitch by changing the length of the resonating column of vibrating air.

Mastering Problems

55. 442 Hz and 448 Hz

56. 540 Hz

57. 2.9 kHz

58. 2.7 kHz

59. $E < D < C < B < A$

60. 1100 Hz, 1800 Hz, 2600 Hz

61. a. 255 m/s

b. 392 Hz and 588 Hz

62. a. 10.5 m

b. 8.20 Hz

63. 442.5 Hz or 437.5 Hz

64. 6.70×10^2 mm

65. 2.0×10^2 Hz

66. 4.0×10^2 Hz

67. a. 1.0×10^{-6} N

b. 1.5×10^{-6} N

c. 0.58 Pa

68. 1.7×10^2 m/s

Applying Concepts

69. a. There is no change in frequency.

b. The wavelength increases.

70. 100 times greater

71. The speed of sound = 343 m/s = 0.343 km/s = (1/2.92) km/s; or, sound travels approximately 1 km in 3 s. Therefore, divide the number of seconds by three. For miles, sound travels approximately 1 mi in 5 s. Therefore, divide the number of seconds by five.

72. First, if you had heard a sound, you would have heard it after you saw the explosion. Sound waves travel much more slowly than electromagnetic waves. Second, in space the density of matter is so small that the sound waves do not propagate. Consequently, no sound should have been heard.

73. Red light has a longer wavelength and, therefore, a lower frequency than other colors. The Doppler shift of their light to lower frequencies indicates that distant galaxies are moving away from us.

74. Answers will vary, but a correct form of the answer is, "A tuning fork is defective, playing an A at 442 Hz instead of 440 Hz. At what speed must you move away from the tuning fork so that you hear the correct pitch?"

75. a. Frequency will increase.

b. Wavelength will decrease.

c. Wave velocity will remain the same.

d. Amplitude will remain the same.

76. $\lambda = 4L$ and $v = f\lambda$ so $v = 4fL$. If v increases and L remains unchanged, f increases and pitch increases.

77. The frequencies are getting farther apart.

78. The pitch is increased; the frequency is twice as high for an open pipe as for a closed pipe.

79. The strings have different tensions and masses per unit length. Thinner, tighter strings produce higher notes than do thicker, looser strings.

Mixed Review

80. Answers will vary. A possible form of the correct answer is, "If a stationary police officer hears a pitch of 525 Hz as the car approaches in a 30 km/h zone, is the car speeding?"

81. 5.4 s

82. 365 m/s; 294 m/s

83. 22.3 kHz

84. 2.3 m/s

85. a. 68.6 m/s

b. 153 mph. The car would be moving dangerously fast. Do not try the experiment.

86. 180 N

Thinking Critically

87. a. The graph shows an inverse relationship between frequency and wavelength.

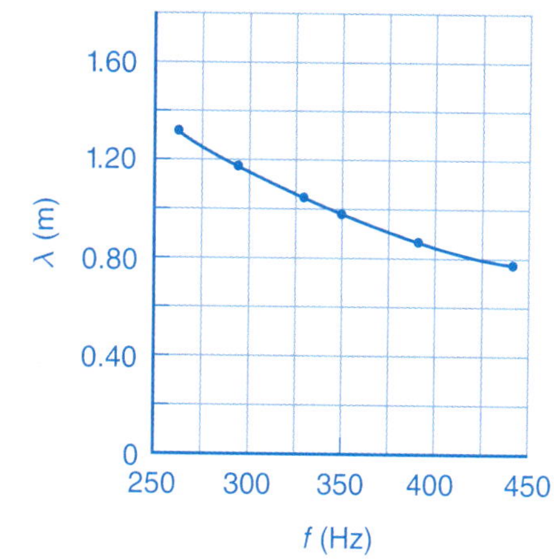

b. The graph shows a direct relationship between period ($1/f$) and wavelength. The speed of sound is represented by the slope, ~343 m/s.

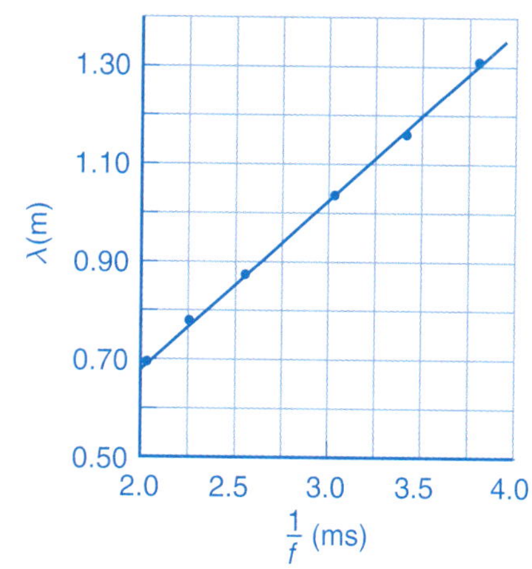

88. The graph should show a fairly steady frequency above 300 Hz as it approaches and a fairly steady frequency below 300 Hz as it moves away.

89. You could start the watch when you saw the hit and stop the watch when the sound reached you. The speed would be calculated by dividing the distance, 200 m, by the measured time. The measured time would be too large because you could anticipate the impact by sight, but you could not anticipate the sound. The calculated speed would be too small.

90. The Sun must be rotating on its axis in the same manner as Earth. The Doppler shift indicates that the left side of the Sun is coming toward us, while the right side is moving away.

91. Measure the mass and length of the string to determine μ. Then clamp the string to a table, hang one end over the table edge, and stretch the string by hanging weights on its end to obtain F_T. Calculate the speed of the wave using the formula. Next, pluck the string in its middle and determine the frequency by matching it to a frequency generator, using beats to tune the generator. Multiply the frequency by twice the string length, which is equal to the wavelength, to obtain the speed from the wave equation. Compare the results. Repeat for different tensions and other strings with different mass per unit length. Consider possible causes of error.

Writing in Physics

92. Answers will vary. A report on violin construction might include information about the bridge as a link between the strings and body and information about the role of the body in causing air molecules around the violin to vibrate. Students might also discuss the ways in which the woods and finishes used in making violins affect the quality of the sound produced by the instruments.

93. Students should discuss the work of Edwin Hubble, the redshift and the expanding universe, spectroscopy, and the detection of wobbles in the motion of planet-star systems.

Cumulative Review

94. a. Westward and southward are positive. See Online Solutions Manual for sketch of system.

b. 3.6 kg•m/s at 34° north of west; 1.8 m/s at 34° north of west

95. no work, because the force and the displacement are perpendicular

96. a. 11 kJ

b. 150 N

Standardized Test Practice

student text p. 435

Multiple Choice

1. A

2. D

3. C

4. C

5. B

6. B

7. B

Free Response

8. 328 m/s

Rubric

The following rubric is a sample scoring device for free response questions.

Points	Description
4	The student demonstrates a thorough understanding of the physics involved. The response may contain minor flaws that do not detract from the demonstration of a thorough understanding.
3	The student demonstrates an understanding of the physics involved. The response is essentially correct and demonstrates an essential but less-than-thorough understanding of the physics.
2	The student demonstrates only a partial understanding of the physics involved. Although the student may have used the correct approach to a solution or may have provided a correct solution, the work lacks an essential understanding of the underlying physical concepts.
1	The student demonstrates a very limited understanding of the physics involved. The response is incomplete and exhibits many flaws.
0	The student provides a completely incorrect solution or no response at all.

CHAPTER 16

Fundamentals of Light

About the Photo

Colors Have students examine the picture. Ask students to describe the colors they observe in the picture. Possible answers: The pictures have more bright colors than the environment shown in the background, even though it, too, is an illustration rather than a photograph. Then have students discuss how they can see the colors. Students might describe what happens in the eye, applying their knowledge of biology.

Chapter Overview

This chapter presents the basic properties of light and explains them via two different models—the ray model and the wave model. The first section describes the ray model and illumination based on the inverse-square law and describes experiments to determine the speed of light. In the second section, the wave model is used to explain diffraction, color, polarization, and the Doppler effect.

Before students study the material in this chapter, they should study:

- Energy
- Properties of waves
- Sound
- States of matter
- Velocity

To solve the problems in this chapter, students will need a strong grasp of:

- Graphing data
- Scientific notation
- Significant figures
- Solving linear equations

Introduce the BIG IDEA

Turn on a flashlight. Ask students what happens to the light from the bulb. It goes through the glass or plastic at the front of the flashlight to the eye. Ask students what this suggests about light: is it made of waves or particles? Possible answer: The glass or plastic might stop particles, so light might be wave-like, and, like sound, sometimes travel through solid materials.

Go online!
connected.mcgraw-hill.com

Your online experience includes:

- **Video**
- **Concepts In Motion**
- **Virtual Investigations**
- **All Lab, Applying Practices activities, and PBL worksheets**
- **Online Practice**
- **Personal Tutor**
- **Multilingual eGlossary**
- **Presentation**

Use the LaunchLAB

The lab *Light's Path* allows students to observe and start thinking about what path light takes as it travels.

Use Physics T.V.

Video

MIXING COLORS Everyone knows that yellow and blue make green, right? But is that true of mixing paint colors or combining colored light? Or both? Discover the physics of how colors combine.

SECTION 1 Illumination

1 Introduce

Bellringer

Illumination Hold up two blocks of paraffin wax (about half an inch thick) horizontally, one on top of the other. Ask students what they observe. The block on the bottom is darker in color than the one on top. Now flip the blocks over, and ask students what they see. The block that is now on the bottom appears darker in color. Repeat several times. The one on the bottom will always appear darker. Ask for possible explanations. Less light strikes the bottom block because the top block reflects and absorbs some light that would have reached the bottom block. OL Visual-Spatial EL

Tie to Prior Knowledge

Inverse-Square Relation Students will have learned about inverse-square laws in the context of universal gravitation. In this chapter, students will learn how an inverse-square law applies to the illuminance of a point source of light.

2 Teach

Light's Path and the Ray Model of Light

student text, pp. 438–440

Discussion

Problem How do you see an object?

Response A luminous source must create light, which travels to the object, is reflected by the object, and then travels to your eye where nerve tissues absorb it and signal the brain for interpretation or identification.

Use Figure 1 and Figure 2

MAIN IDEA Point out that the light rays in **Figure 1** travel in straight lines to objects on the ground. Have students think about what happens when reaching for something in water, such as a rock at the bottom of a clear pond or stream or a decoration in an aquarium. The object is not always where we thought it was, and we often have to make adjustments to figure out precisely where to reach for it. When you see a solid object in the water, light rays traveling from the object must pass through the air-water interface. When the light reaches the interface, first when entering the water and second when leaving the water, the light is refracted; that is, the light rays are deflected from the path they had before they were refracted. We perceive the position of the object based on the direction of the refracted ray leaving the water, not on where the object actually is. BL

Quick DEMO

Light Sources

Estimated Time 5 minutes

Materials small lightbulb

Procedure Turn on the lightbulb in a darkened room. Ask students if they can see the lightbulb from any position in the room. Ask students what this means about the way light travels from the source to their eyes. The source sends light in all directions away from it at the same time.

Critical Thinking

Source of Light Rays Ancient Greek philosophers used to think that our eyes are the sources of the light rays that enable us to see. What observations are inconsistent with, and thus refute, this theory? First, we are unable to see in a dark room. Second, when we look at a bright light, the iris of the eye will contract. Students may think of other observations as well. OL

Quick DEMO

Seeing

Estimated Time 5 minutes

Materials laser, chalky erasers

Procedure Shine the laser across the front of the room. Ask students to describe what they see. They should observe a spot on the wall from the laser, but no laser beam across the front of the room. Now, clap the chalky erasers together above the path of the laser. Ask students what they see. They should observe a beam where the laser's light reflects off the chalk dust. Ask students what they can conclude from these observations. Light must travel into the eye to be seen. The spot on the wall is visible because the laser light reflects from the wall and into their eyes. The laser beam is visible when there is chalk dust in the air because, in that case, some of the laser light scatters from the chalk dust into the eyes. When there is no chalk dust in the air, there is nothing to appreciably scatter the light and so the laser light from the side cannot be seen.

Reinforcement

Opaque, Translucent, Transparent Have students list various materials and whether they are opaque, transparent, or translucent. Discuss what happens when light reaches each type of material. Transparent objects transmit light so that objects can be seen clearly through them; translucent materials scatter and transmit light so that objects cannot be seen clearly through them; and opaque materials prevent light from passing through them. Also discuss the ways in which objects with different transparencies are used: for example, transparent glass at a salad bar, opaque screening in a dressing room, and translucent window shades. OL Visual-Spatial EL

Quantity of Light

student text, pp. 440–442

Use an Analogy

Inverse-Square Law To help students understand the dependence of illuminance on distance from the source, draw an analogy to the law of universal gravitation. This law states that as the distance between a mass and an object increases, the force decreases. Likewise, as the distance between an illuminated object and the light source increases, the illuminance of the object decreases.

Careers

Photographers When a photographer uses a flash to illuminate a subject, the distance between the flash and the subject must be taken into account to get the correct exposure. If the distance to the subject is doubled, the illuminance decreases by a factor of four. To compensate for the reduced illuminance, the photographer increases the size of the aperture, the opening through which light enters the camera. If the photographer moves twice as far from the subject, the diameter of the aperture must be doubled. Doubling the diameter of the aperture quadruples the area of the aperture and therefore quadruples the amount of light entering the camera.

Concept Development

Luminous Flux and Luminous Intensity Ask students which light source would be more effective: a 1000-lm lightbulb or a 100-cd lightbulb. By converting either candelas to lumens or lumens to candelas, you find that the 100-cd lightbulb produces more light. OL

Applying Physics

Architecture An energy-saving trend in architecture is to enhance the use of daylighting. This approach aims for buildings to capture as much natural light as possible to minimize the amount of electric light needed for areas occupied during the day. Daylighting must take place, however, without overheating an area, creating too much glare, or creating uneven light distribution. To design these areas properly, architects must account for the angle at which sunlight hits the interior of a building during various times of the day and year and incorporate various types of windows, among other measures. To control brightness from the Sun, clouds, sky, and reflected light, outside overhangs and inside window blinds are used to reflect light. To limit light transmission from too much sunlight, darker glass is used, and limits are set on visible light transmittance.

Reinforcement

Inverse-Square Law The inverse-square law for illumination describes an ideal situation involving a point source of light. Discuss with students situations that are close to the ideal and those that depart from it. For example, the longer the distance is from the source compared to the size of the source, the more agreement there is between actual measurements and the law. Additionally, the law assumes that particles in the air between the source and the point of interest do not absorb any light. OL

PHYSICS CHALLENGE ACTIVITY

Law of Illumination Set up a grease-spot Bunsen photometer. This should have an unknown light source on one end and a known intensity lightbulb on the other end. Adjust the photometer so the illuminance is the same from both sides. Have students determine the intensity of the unknown source. The students should place a white piece of paper with a grease spot between two light sources. One light source should be of known intensity, and the other should be an unknown light source. Students should adjust the position of the paper until the grease spot disappears. They should measure the distance from each light source to the grease spot. If the lightbulb is specified in lumens, they need to convert to candela by dividing by 4π. Then, using the inverse-square law, they can equate the ratios of luminous intensity over distance squared for each side, and solve for the unknown intensity. AL Visual-Spatial EL

Use the PhysicsLAB

In *Measuring Ultraviolet Light,* students will explore how different materials affect the transmission of light.

DIFFERENTIATED INSTRUCTION

Struggling Students Use a small pegboard and a small, intense flashlight. Remove the bulb cover. Have students hold the light and pegboard so that light passes through the holes in the pegboard onto a solid board. Position the pegboard so that it is halfway between the light and the other board. Darken the room and have the students draw circles around the illuminated areas on the pegboard and on the other board. Have students compare the areas of the two circles. The illuminated area on the solid board is four times larger than the illuminated area on the pegboard. Repeat the procedure with the pegboard positioned one-third the distance between the light and the other board. The illuminated area on the solid board is nine times the area on the pegboard.
BL Kinesthetic

Surface Illumination

student text p. 443

Use Models

Point-Source Illuminance To better visualize the measurement of point-source illuminance, have students place two square flat boards upright and facing each other with a source of light in between the two boards at an equal distance from each board. The light can be a small lamp without a shade. Because the bulb emits light in every direction, the two boards receive the same amount of light. Now have students place four or five boards at different distances and angles from the lamp. Each of these objects is illuminated differently, and the boards that are closer are more illuminated.
OL Kinesthetic EL

Reinforcement

Ray Model of Light Ask students to draw a concept map of the ray model of light relating the concepts of luminous and illuminated sources, types of media, quantity of light, speed of light, and inverse-square law.
OL Visual-Spatial EL

Use the PhysicsLAB

In *Light Intensity and Distance,* students will measure, calculate, and graph the effect of distance on light intensity.

REAL-WORLD PHYSICS

Beyond the Incandescent Lightbulb Note the growing availability and use of compact fluorescent lamps (CFL) for home illumination, the proliferation of high-visibility light-emitting diode (LED) billboards, and flat-screen liquid crystal display (LCD) and plasma technology for computers and televisions. Have students discuss where they have seen these products used.

ADDITIONAL IN-CLASS EXAMPLE

Use with Example Problem 1.

Problem You shine an 8.0×10^2-lm lightbulb onto a person who is standing 2.0 m away. What is the illuminance of the light on the person?

Response

Assume the lightbulb acts as a point source.

$$E = \frac{P}{4\pi r^2} = \frac{8.0\times10^2\ \text{lm}}{4\pi(2.0\ \text{m})^2} = 16\ \text{lx}$$

The Speed of Light

student text, pp. 445–446

Reinforcement

Speed of Light Have students investigate how the speed of light is used to measure great distances in space. They each should research an astronomical object that is well-known or currently under study and report back to the class on its distance from Earth in light-years and on how that measurement was obtained.
OL Linguistic

Use the MiniLAB

In *The Speed of Light,* students will investigate ways to measure how fast light travels.

PHYSICS PROJECT ACTIVITY

Measuring the Speed of Light Scientists other than those mentioned in this chapter also attempted to measure the speed of light. For students who are interested in how measuring the speed of light developed and the techniques that were used to investigate it, have them research and write a brief paper on the history of these investigations. Students who are interested in precise experimental procedures may want to research further Michelson's experiments to measure the speed of light and write a brief report. *The Master of Light—A Biography of Albert A. Michelson,* by Michelson's daughter, Dorothy Livingston, contains many details about his experiments and some amusing personal anecdotes. **OL** **Linguistic**

3 Assess

Assess the MAIN IDEA

Light at a Boundary Use a laser pointer demonstration to re-emphasize the reflection and refraction of light. Use the laser pointer with a mirror and chalky erasers to demonstrate the reflection of a light beam. Then use a prism and chalky erasers to demonstrate the refraction of light. Have students describe what happens to the light in each case. The light rays travel in straight lines. The light reflects and bounces off the mirror. The light ray passes through the material of the prism but changes direction at the boundaries of the prism. Next, stand at different distances from a wall of the classroom, and shine the laser (or a flashlight) onto the wall. Have students describe what happens to the spot. The spot size changes depending on the distance of the light source from the wall. Ask students how the total light energy in the light spot changes as the spot size changes. The total amount of luminous flux in the light spot stays the same (if the absorption and scattering of light in air is ignored).

Check for Understanding

Illuminance and Luminous Intensity Have students distinguish the meaning and units of luminous intensity and illuminance and how the two are related by the inverse-square law. Luminous intensity is the luminous flux (rate at which a light source emits light) divided by 4π. It is measured in candela. Illuminance, the rate at which light strikes a surface, is measured in lux, or lm/m^2. The illuminance at a given point is proportional to the luminous intensity divided by the square of the distance from the point source. **BL**

Extension

Light Rays Ask students to imagine how the way people see things would be affected if light traveled in a curve instead of a straight line. Have students write a paragraph describing the change that they envision. **OL** **Linguistic**

Answers to Reading Checks and Caption Checks

Reading Check, *student text p. 438*
When light from a flashlight or the Sun streaming through a window is made visible by dust particles in the air, the light rays are in a straight line. You can see this when light streams through trees in a forest as well.

Caption Check, *student text p. 439*
If the room had no light, you would not be able to see anything.

Reading Check, *student text p. 440*
You can see the fish through the fishbowl because the glass is transparent. You can see the fishbowl itself because the glass reflects light.

Reading Check, *student text p. 442*
Luminous intensity is a measure of the luminous flux that fall on 1 m^2 of the inside of a 1-m-radius sphere. The relationship of luminous intensity to illuminance is an inverse-square relationship.

Reading Check, *student text p. 445*
He made his measurements as part of a project to improve maps by calculating the longitude of locations on Earth.

Practice Problems
student text p. 444

1. After the lamp is moved, the illumination is one-ninth of the original illumination.
2. 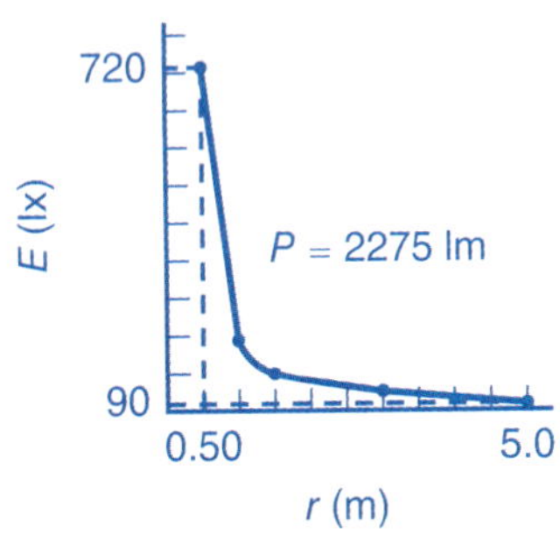

3. 7.1 lx
4. 3.2 m
5. 2.0×10^1 lx
6. 8.0×10^3 lm
7. If hung at 1.5 m, the illuminance would be 61.91 lx, which is far too dim. To get 175 lx from 1.5 m, you need 4950-lm bulbs.

Section 1 Review
student text p. 446

8. Possible answers may include sunlight through a window made visible by dust particles and the fact that your body blocks light, creating a shadow.
9. You would use a translucent window shade to keep people from looking in or out, while still allowing daylight in. You would use an opaque window shade to keep the daylight out.
10. One lightbulb provides an illuminance that is twice as large as two of the lightbulbs at twice the distance. $E_2 = \left(\frac{1}{2}\right)E_1$.
11. 27 cd
12. 0.71 m
13. 9 km
14. 3.840×10^8 m
15. 3.01×10^8 m/s. This method appears to be accurate because it results in a value close to the accepted value for the speed of light.

1 Introduce

Bellringer

Diffraction Ask students why it is possible to hear someone talking in the next room, even though you cannot see the person. Answers will vary. Students might mention how light and sound move differently through a doorway if one is open. Some might mention that sound can go through a wall, while light cannot. If students say that sound travels through the wall, but light does not, point out that you would be able to hear them if the door was open even if sound did not travel through the wall ("soundproof"). Ask students to explain how this could work. Be ready to explain that a person cannot see around a corner, even though light waves may be diffracted like sound waves. Both sound and light have wave properties, so they can diffract through open doorways. Discuss how the wavelengths of sound waves are typically comparable to or greater than the size of an open doorway, so diffracted sound waves (especially large wavelengths) can be detected by the human ear. The wavelengths of light are so much smaller than a doorway (note that they are roughly a billionth of the size of a doorway) that light diffracted around an open doorway is not nearly strong enough to be detected by the eye. **OL** **Logical-Mathematical**

Tie to Prior Knowledge

Transverse Waves Students will apply some of their knowledge from their study of transverse waves. They should remember the relation $\lambda = \frac{v}{f}$.

Doppler Effect Students also will be familiar with the Doppler effect from their previous study of sound.

2 Teach

Diffraction and the Wave Model

student text p. 447

Content Development

Diffraction Project a red laser pointer across a darkened classroom onto a painted wall or a sheet of white paper. Use the edge of a metal ruler to create a shadow on the wall, and have students observe the diffraction of the beam. **OL** **Visual** **EL**

Color

student text pp. 448–450

Concept Development

Speed of Light and Color Ask students to imagine that a supernova occurred somewhere in space and that the light from this stellar explosion was just arriving at Earth. Ask them what they would expect to see if the speed of light in empty space was different for different colors of light. Light from one part of the spectrum would arrive first, then gradually sweep through the rest of the spectrum. This is not what is observed, because the speed of light in empty space is the same for all colors of light. **OL** **Logical-Mathematical**

Content Development

Colors Show students objects of several different colors, then make the room very dark. Have students look at these objects when illuminated by an ordinary flashlight. Then, place various color filters (such as red, blue, and green) over the flashlight and observe the same objects. Ask students to reflect on how this experience may be different from ones they have had when they used pigments. **OL** **Visual-Spatial** **EL**

Use the PhysicsLAB

In *A Whole Spectrum of Possibilities,* students will analyze light spectral data.

Quick DEMO

Colors of Light

Estimated Time 10 minutes

Materials white screen, three primary light sources (red, blue, and green)

Procedure Shine the light from the three sources onto the screen, so that there is a region where all three overlap, regions where each pair of colors overlap, and regions where each primary color is visible by itself. Ask students to observe and describe what they see. The three primary colors together form white light; blue and green light make cyan; red and green light make yellow; and blue and red light make magenta.

Identify Misconceptions

Primary Colors Students might have learned in art class that the primary colors are blue, yellow, and red. Color is detected in human eyes by cone cells that are sensitive to red, green, and blue. Paints and dyes use different primary colors which often are red, yellow, and green. Light mixing has three basic colors, cyan, yellow, and magenta, from which other colors can be created.

Content Background

Color Vision Students might have already learned that there are two types of cells in the back of the eye, rods and cones. When light hits one of these cells, it causes a chemical reaction that produces a nerve impulse. The nerves in the retina combine the information from several rods and cones, do some "data processing," and pass the signal on to the brain, where the image is created.

About 95 percent of the cells are rods, which are sensitive to faint light and are therefore responsible for night vision. There are three types of cones, which are concentrated in the center of the retina. Each type of cone is sensitive to a range of wavelengths in the light spectrum. One type has its greatest sensitivity in the short-wavelength (blue) part of the spectrum. A second type has its greatest sensitivity around the middle (green) part of the spectrum, and the third type has its greatest sensitivity a bit more toward the long-wavelength (red) part. There is overlap between the sensitivities of the three types of cones, especially between the last two mentioned. That means that much of the light we see simultaneously excites more than one type of cone.

Critical Thinking

Light Reflection Ask students what color of light is likely to be least healthful for a plant and why. Chloroplasts in plants cannot use green light for photosynthesis. Most plants look green because they reflect this green light. So, green would be the least healthful light for a typical plant. Discuss that there are several other wavelengths of light that plants do not absorb that fall in the yellow, orange, red, and green colors. They combine to generally average out in the green range. Each plant varies in its green color as a result. OL

Reinforcement

The Colors of Color Television Have students use a powerful magnifying glass or hand-held microscope to look at a picture on a color television set. They should see that the picture is actually made up of very small dots that cannot be seen with the naked eye. The dots are called pixels, and each one contains a red light, a green light, and a blue light. The varying brightness of each pixel creates a range of additional colors and shades. When all the color pixels are dimmed, black is created; when all the pixels become brighter, white is created. The mixing of color lights is called color by addition, or additive color. OL Visual-Spatial EL

REAL-WORLD PHYSICS

Why Leaves Change Color in Autumn Throughout spring and summer, photosynthesis takes place in the leaves of deciduous trees. Chlorophyll absorbs sunlight, creates food for the tree, and determines the green color. In addition to chlorophyll, the cells contain other pigments, including *carotenoids* (yellow, orange, and brown) and *anthocyanins* (red and purple). The vast amount of green pigment in the leaves masks these other colors during the growing season. When the days grow shorter, the chlorophyll breaks down and the other pigments become visible. The mixing together of pigments and the interaction between sunlight and other substances in the leaves, such as glucose and waste products, bring out additional colors.

Extension

Color by Addition on a Computer Many computer programs feature a "color-chooser" dialog, which allows the user to set arbitrary combinations of red, green, and blue intensities. Students can experiment with the "color by addition" concepts discussed in this chapter.

Use the MiniLAB

In *Color by Temperature,* students will shine a dimmer-controlled light through a prism and note any changes in the observed colors emerging from the prism as the brightness of the light and temperature of the bulb is increased.

Use the MiniLAB

In *Polarization of Reflected Light,* students will compare the effects of polarizing filters on both direct and reflected light.

Polarization of Light

student text pp. 451–452

Use Figure 19

To help students understand how a polarizing filter works, place a piece of rope through a long slit in a piece of heavy cardboard. Send transverse waves along the rope. When the waves are polarized in the same direction as the slit, they will pass through. When the waves are polarized perpendicularly to the slit, they do not pass through. Explain that although this model helps visualize polarization, it is not completely accurate; in reality, the material of the filter is absorbing the energy from the light wave and re-emitting it in a particular way depending upon the orientation between the two. OL Visual-Spatial EL

DIFFERENTIATED INSTRUCTION

Physically Challenged Give students polarizing filters and ask them to walk around and observe various objects through the filters, rotating the filters and observing what happens to what they see. To accommodate students with limited mobility, provide a variety of portable objects that can be hand viewed, such as a calculator with an LCD display or a liquid crystal thermometer. Students should find that the filter does not change the color of most objects and that most objects do not change their appearance as the filter is rotated. Smooth reflecting surfaces, such as windows, will change noticeably upon rotation of the filter. Liquid crystal displays will disappear when the filter is correctly rotated. BL Kinesthetic EL

Critical Thinking

Polarization Have students imagine an unpolarized light source and two polarizing filters that are arranged so that no light emerges from the second one. Have students imagine that a confident student walks up and asserts that he or she can place a third polarizing filter between the first two such that light will then pass through all three polarizing filters. Ask students whether he or she can in fact do this. Yes, it can be done. Although the intensity of the light coming from the first filter is reduced by the middle filter, the middle filter also shifts the angle of polarization of the light, thereby aligning some of the remaining light with the polarizing axis of the last filter and allowing some light to pass through it. OL

Reinforcement

Polarized LCD Displays Like LCD computer screens, digital watches and calculators also use polarizing plates, liquid crystals, and pixels to create displays. However, these displays show only two types of areas—those that are lit up (usually the background) and those that are not lit (usually the numbers). Ask students to explain how these polarized devices operate. To create the lighted background areas, the LCD cells are rotated and arranged so that the light's polarization is exactly 90 degrees between the two perpendicular plates, which lets the light through the second plate. To make the darkened areas, or the areas where the numbers can be seen, electric fields create a signal. The signal arranges the liquid crystals so that they are no longer rotated 90 degrees. When the crystals are in line with the filters, the light is blocked from going through the pixels and the second plate. OL

PHYSICS CHALLENGE ACTIVITY

Photons Light can behave like, and can be thought of as, either waves or particles (called photons) depending on the conditions under which light is observed. Both views are useful for understanding the wide range of behavior of light. The wave picture, for example, can help explain polarization. For a different perspective, have interested students investigate light as a collection of photons. They can discuss the production of photons by light sources, how the photons interact with objects, how human eyes absorb the photons and perceive images, and related topics. Light sources, such as sodium vapor lights, that generate photons, as well as other practical applications, could be included in the discussion. Have students write a brief essay on the topic of their choice. AL Linguistic

Use the PhysicsLAB

In *Polarization,* students will explore the polarization state of various light sources using a polarizing filter.

Use the PhysicsLAB

In *Reducing Glare,* students will explore the effects of polarizing filters on light and also determine whether certain light waves are polarized. One use of polarizing filters is reducing glare.

Speed, Wavelength, and Frequency of Light

student text pp. 453–455

Discussion

Problem Ask students to imagine a duck floating in a pond. It is not swimming, but it is flapping its wings at a constant frequency and causing concentric circular waves to spread out from its position. Now suppose that the duck decides to paddle to the shore, thereby changing its position. What will happen to the circular waves? How does this phenomenon demonstrate the Doppler effect?

Response The duck will continue to produce circular waves; however, the waves will no longer be concentric. Presuming that the duck's speed is slower than the speed of the water waves, the waves in front of the duck will have a shorter wavelength than the waves behind the duck. OL

Reinforcement

MAINIDEA Repeat one or more of the demonstrations used to introduce the Doppler effect for sound waves to show the similarities between the two manifestations of the effect. Also, explicitly discuss the major differences—the speed of light is always the same in a given medium, and wind will not affect such a shift for light.

Content Background

Interferometers Michelson's interferometer is an optical instrument that has many applications in optics. Michelson originally invented the interferometer for experiments looking for the "ether," the medium light was believed to move through at the time. Since the planet Earth is speeding around the Sun, it was thought that the interferometer in the laboratory was traveling through the ether. Hence, there should be a Doppler shift when the direction of motion of the interferometer was changed relative to the ether. In the light velocity experiments Michelson did with Morley in the late 19th century, there was no such Doppler shift. This showed that no such ether existed. Interferometers are now widely used in the testing of precision optical instruments, such as telescopes.

3 Assess

Assess the MAINIDEA

Light in Space Have students imagine that an astronaut shines a green laser light out the window of a space ship while travelling from Earth to the Moon. Ask students how the Doppler effect will change the light, as observed by a person standing on Earth who sees the light from the ship. Since the source of light is moving away from the observer, the light will be shifted to a longer wavelength. Have students explain how the wavelength shift will be different if the space ship is travelling in the opposite direction. Since the source of light is moving toward the observer, the light will be shifted to shorter wavelengths.

Check for Understanding

Colors Have students work in pairs. Ask one of the pair of students to explain to the other student about color formation by both additive and subtractive processes. Then have the other student in each pair refine the first student's explanation. Ask for different pairs to share their explanations with the class using diagrams or props such as colored gels or filters.

OL Visual-Spatial EL

Extension

Have students further explore the nature of vision by learning about the different types of color-blindness. AL

Answers to Reading Checks and Caption Checks

Reading Check, *student text p. 449*

Color by the addition of light occurs when colors of light are combined to form a new color, such as red, blue, and green light combining to form white light. Some substances, such as dyes and pigments, absorb certain wavelengths of light, while reflecting others. The color that an object appears to be is a result of the material absorbing specific light wavelengths and reflecting the rest.

Caption Check, *p. 449*

In white light, the yellow die absorbs blue light and reflects red and green. In red light, the die reflects all of the red light, which makes it appear red in color.

Caption Check, *p. 450*

Leaves reflect green light.

Practice Problems

student text p. 455

16. 5.85×10^{14} Hz
17. 6.03×10^{14} Hz
18. 4.97×10^{-7} m
19. 7.02×10^{6} m/s

Physics Challenge

student text p. 452

1. The analyzer filter allows some light to pass through, since its polarizing axis is not perpendicular to the polarizing axis of the first filter. The last polarizing filter can now pass light from the analyzer filter, since the polarizing axis of the analyzer filter is not perpendicular to the polarizing axis of the last polarizing filter.
2. I_1 is the intensity of light from the first filter; $I_{analyzer}$ is the intensity of light from the analyzer filter; and I_2 is the intensity of light from the last filter.

   $I_{analyzer} = I_1 \cos^2 \theta$

   $I_2 = I_{analyzer} \cos^2(90° - \theta)$

   $I_2 = I_1 \cos^2(\theta) \cos^2(90° - \theta)$

Section 2 Review

student text p. 455

20. Red-shifted light is shifted to longer wavelengths. When a light source and an observer are moving away from each other, the observed light is red-shifted. Blue-shifted light is shifted to shorter wavelengths. When a light source and an observer are moving toward each other, the observed light is blue-shifted. This shift in the apparent wavelength is called the Doppler effect.
21. yellow (a mixture of the other two primaries, red and green)
22. **a.** yellow

    b. yellow

    c. black
23. The secondary pigment colors are red, green, and blue. Objects appear these colors because the pigments they contain reflect those colors of light while absorbing others. For example, red pigments absorb green and blue light, so only red is reflected.
24. Yellow and magenta pigments are used to produce red. Yellow pigment absorbs blue, magenta pigment absorbs green, and neither absorbs red so the mixture would reflect red.
25. See whether the glasses reduce glare from reflective surfaces, such as windows or roadways.
26. The polarizing axis should be oriented vertically, since the light reflecting off the road will be partially polarized in the horizontal direction. A vertical polarizing axis will filter horizontal waves.
27. Yes; because $v = \lambda f$ and $\lambda = v/f$, when v decreases, so does λ.
28. The spectral lines of the emissions of known atoms are blueshifted in the light we see coming from Andromeda. We would be moving toward Andromeda due to gravitational attraction. This gravitational attraction could be due to the mass of the Milky Way or other objects located near the Milky Way.

PHYSICS THAT'S ENTERTAINMENT!

3-D Movies

Journey into the Third Dimension

Background

The use of 3-D for entertainment dates back to the mid-1800s when Dr. Charles Wheatstone created the stereoscope using two photographs taken from two different perspectives, viewed through special lenses that direct one image to each eye. Students might be familiar with 3-D viewer toys that are essentially a modern application of the same technology.

The illusion of 3-D can also be created by superimposing two images on the same surface. Students might have seen this effect used in illustrations or movies that required red-blue or red-green glasses. Modern 3-D movies use polarizing filters to create the same effect.

Teaching Strategies

- Have students close one eye and focus on an object in the distance. Direct students to cover the object with a thumb, holding their arms outstretched in front of them. When students switch which eye is open and which is closed, they should notice that the distant object stays in basically the same position while their thumb appears to "jump" to the side. Have students repeat the experiment with their thumbs closer to their faces (instead of with their arms outstretched); they should notice that the effect from before is even more pronounced. Students should recognize that this demonstrates how binocular vision can be used to judge depth; the differences in perspective between the views from each eye are more pronounced in nearby objects.
- If possible, bring in several pairs of polarizing-filter 3-D glasses for students to investigate. If there are enough for students to try them on, have them look at each other while wearing the glasses. Students should note that while looking at another student wearing 3-D glasses, one of the lenses will appear to be opaque black. Have students close one eye while looking at another student wearing 3-D glasses. As they switch which eye is closed, the lens on the other student that looks opaque will appear to switch.
- Students might be under the impression that older 3-D movies used red-blue or red-green glasses (called anaglyphic 3-D). Inform students that very few movies utilized anaglyphic lenses. The vast majority of 3-D films—even during the "3-D craze" of the 1950s—utilize polarizing-filter 3-D glasses.

GOINGFURTHER >>>

Expected Outcomes The technology used to create glasses-free 3-D in handheld devices is sometimes called parallax 3-D. It uses angled pixels or reflective surfaces to send images to each eye. The viewer of such a 3-D illusion would need to be centered in front of the screen and situated at an ideal distance—usually 45–60 cm away. For this reason, this type of projection is usually limited to small screens, such as handheld video game consoles or small televisions.

SECTION 1

Mastering Concepts

29. Answers may include: desired brightness in buildings and parks can be achieved and distances can be calculated by using light's travel time.

30. A luminous source emits light. An illuminated object is one on which light falls and is reflected.

31. It is mainly illuminated. The filament is luminous; the frosted glass is illuminated. You barely can see the hot filament through the frosted glass.

32. Ordinary nonluminous objects are illuminated by reflected light, allowing them to be seen.

33. A transparent object is a material through which light can pass without distortion. A translucent object allows light to pass but distorts the light to the point where images are not discernable. An opaque object does not allow light to pass through.

34. The illumination on a surface is directly proportional to the intensity of the source and inversely proportional to the square of the distance of the surface from the source.

35. Roemer was the first to measure the time that it took for light to travel between two points.

Mastering Problems

36. 2.0 lx

37. 3.84×10^{8} m

38. The 100-W (1620-lm) setting is needed.

39. **a.** 3.9×10^{9} m

b. 25 km/s

c. 3.0×10^{1} km/s, so fairly accurate.

40. 1.31×10^{3} lm

41. Possible answer: How far from a 1100-lm lightbulb should you place a screen so that it has an illuminance of 96 lx?

42. 3×10^{4} km; the mirror would be half this distance, or 15,000 km away. Earth is 40,000 km in circumference, so this is three-eighths of the way around Earth.

SECTION 2

Mastering Concepts

43. Diffraction is most pronounced around obstacles approximately the same size as the wavelength of the wave. We are more accustomed to obstacles of the size that diffract the much larger wavelengths of sound.

44. Violet light has the shortest wavelength.

45. red, yellow, green, blue, violet

46. 400 nm to 700 nm

47. Longer wavelengths will bend the most and so the order from bending the most to bending the least is red > yellow > green > blue > violet.

48. White light is a combination of all the colors, or at least the primary colors.

49. No; they have no vertical or horizontal components.

50. Because the galaxy is distant, it is most likely moving away from Earth. The wavelength actually would shift away from the wavelength of green light toward a longer red wavelength. If it shifted toward the blue wavelength, the wavelength would be shorter, not longer. This would indicate the galaxy is getting closer to Earth, and no galaxy outside the Local Group has been discovered moving toward us.

Mastering Problems

51. 7×10^{-7} m

52. 3.09×10^{6} m/s; the original assumption was valid.

53. The reflected light is partially polarized in the direction parallel to the surface of the lake and perpendicular to the path of travel of the light from the lake to your eyes.

54. 1.95×10^{7} m/s; the original assumption was valid.

55. Possible answer: "… and is mounted on a spaceship moving away from Earth at 30 percent the speed of light. What wavelength would an observer on Earth measure for the laser?"

56. An unrealistic value would make the galaxy seem to be moving away from us at a speed close to or greater than the speed of light ($v \approx c$). If this were the case, use of the low-speed Doppler shift equation would give a wavelength difference of $(\lambda_{app} - \lambda) = +\frac{c}{c}\lambda$. When solved, this would give an apparent wavelength of $\lambda_{app} = 2\lambda$. It would be twice as large as the actual wavelength. So any apparent wavelength close to or greater than twice the actual wavelength would be unrealistic.

Applying Concepts

57. Illumination $E \propto 1/r^2$. The illumination at screen B will be one-fourth that at screen A because it is twice as far from the source.

58. The insides are painted black because black does not reflect any light, and thus there is no interference while observing or photographing objects.

59. **a.** no

b. Distance is doubled, so the illumination of the page is one-fourth as great.

60. Fire trucks should be painted yellow-green, 550 nm, because less light has to be reflected to the eye for the fire truck to be seen.

61. Blue pigment of a police car will absorb the red and yellow light. Dark blue police cars would not be very visible. If a community wants its police cars to be visible, they should buy yellow cars.

62. Illuminance decreases, as described by the inverse-square law.

63. No change; distance does not affect the luminous intensity of the lamp.

64. Light scattered from the atmosphere is polarized, but light scattered from the clouds is not. By rotating the filter, the photographer can reduce the amount of polarized light reaching the film.

65. **a.** Red cellophane reflects red light and absorbs blue and green light.

b. Cellophane transmits red light.

c. Blue and green light are absorbed.

66. Mix cyan and magenta.

67. yellow

68. Black; almost no light would get through because the light transmitted through the first filter would be absorbed by the second.

69. 4.65×10^7 m/s. That is over 100 million mph. If he does not get a ticket for running a red light, he will get a ticket for speeding.

Mixed Review

70. 2.3 m

71. Humans hear over a range of about nine or ten octaves (20 Hz to 10,240 or 20,480 Hz); however, human vision is less than one "octave."

72. The time for light to travel 1.6 km is a small fraction of a second (5.3 μs). The sound travels about 340 m/s, which is about one-fifth of the 1.6 km every second, and takes about 4.7 s to travel 1.6 km.

73. $\pm 3.3 \times 10^{-12}$ m/s

Thinking Critically

74.

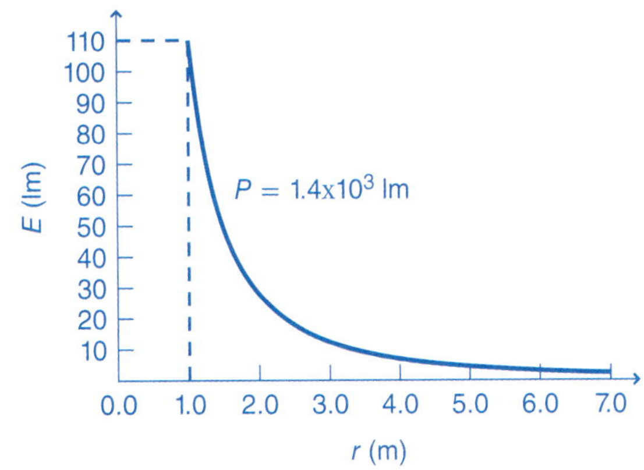

a. hyperbola

b. inverse square

75. It was not precise enough. He was not able to measure the small time intervals involved in a terrestrial measurement.

Writing in Physics

76. Answers will vary. Much information can be found in this chapter. Students should put this information in their own words. Students can also do further research to add to the information provided.

77. Answers will vary. Begin with the element thorium. Heat it to the melting point of platinum. At this temperature, the thorium will glow. Surround the thorium with an opaque material that can take the high temperature. Leave an opening that is one-sixtieth of a square centimeter in size. The candela is defined as the amount of steady flow of light energy that is emitted by the thorium through the opening under these conditions.

Cumulative Review

78. a. 2.1×10^{2} N

b. 1.7×10^{2} N

79. 171 kg

80. $V_f = 252$ mL, $V_f = 249$ mL.

81. 530 m/s

82. 8.8 cm

Standardized Test Practice

student text p. 461

Multiple Choice

1. B
2. C
3. D
4. C
5. D
6. D
7. B
8. B

Free Response

9. They are approaching each other.

   $v = 2.97 \times 10^7$ m/s

10. $I_2 = 0.25I_0$

Rubric

The following rubric is a sample scoring device for free response questions.

Points	Description
4	The student demonstrates a thorough understanding of the physics involved. The response may contain minor flaws that do not detract from the demonstration of a thorough understanding.
3	The student demonstrates an understanding of the physics involved. The response is essentially correct and demonstrates an essential but less than thorough understanding of the physics.
2	The student demonstrates only a partial understanding of the physics involved. Although the student may have used the correct approach to a solution or may have provided a correct solution, the work lacks an essential understanding of the underlying physical concepts.
1	The student demonstrates a very limited understanding of the physics involved. The response is incomplete and exhibits many flaws.
0	The student provides a completely incorrect solution or no response at all.

CHAPTER 17
Reflection and Mirrors

About the Photo

Reflection of Light Have the students examine the photograph. Ask them to describe what they see. They see a reflection of buildings from the surface of a sculpture. Ask them to imagine that they are standing next to the sculpture and looking at the building reflection, and to describe the path that the observed light takes. The light that illuminates the building travels to the sculpture surface and reflects into the eyes of the observer. Ask the students why the image of the building is distorted. The curvature of the mirror causes the distortion of the image.

Chapter Overview

The chapter expands on the law of reflection and applies it to analyze image formation by plane mirrors. Virtual images and their properties are discussed. The law of reflection is applied to image formation by concave and convex mirrors. Real images and their properties are discussed. Finally, the mirror equation is used to quantify relationships between object and image locations and sizes.

Before student study the material in this chapter, they should study:

- Fundamentals of waves
- The law of reflection (Ch. 14)

To solve the problems in this chapter, students will need a strong grasp of:

- Significant figures
- Solving linear equations

Introduce the BIG IDEA

Hold up a mirror to the class and ask them what they see. They see their reflection or the reflection of other objects. Then, breathe onto the mirror or use a spray bottle to "fog up" the mirror surface. Point out that they can no longer see their reflection. Ask the students what changed about the mirror surface. Answers will vary. The water droplets caused surface roughness that scattered the light.

(l) Alan Copson/Photographer's Choice/Getty Images, (r) Comstock/PictureQuest

Go online!
connected.mcgraw-hill.com

Your online experience includes:

- **Video**
- **Concepts In Motion**
- **Virtual Investigations**
- **All Lab, Applying Practices activities, and PBL worksheets**
- **Online Practice**
- **Personal Tutor**
- **Multilingual eGlossary**
- **Presentation**

Use the LaunchLAB

In *Images from Mirrors,* students will study the conditions that produce a clear image on a screen.

Use Physics T.V.

Video

SEEING AROUND CORNERS Have you ever seen a round mirror in a store, on a building, or even on a car's side mirror? What's the point of a mirror like that? Explore the physics of reflection.

1 Introduce

Bellringer

Reflections Place a variety of objects on a desk and have students observe whether they can see their reflections from each object. Include some objects that produce spectral reflections, such as a metal can, a flat mirror, and a spoon. Also include some objects that do not produce spectral reflections, such as cardboard, a clear plastic cup, and a piece of crumpled aluminum foil. Students will see clear images in the smooth, mirrored surfaces, unclear images in the nonsmooth, mirrored surfaces, faint images in the shiny, opaque surfaces, and no images in the rough, opaque surfaces. Ask students whether they think light is reflected off the rougher surfaces and why. BL Kinesthetic EL

Tie to Prior Knowledge

Angles of Reflection Students have previously learned about the reflection of mechanical waves at a boundary. In this section, students will apply this concept to determine the angles of reflection of light waves from smooth and rough surfaces.

2 Teach

Reflected Images and The Law of Reflection

student text pp. 464–466

Identify Misconceptions

Light Demonstration Students may believe that they are able to see light that is not directed toward their eyes. Explain to students that when light reflects off a surface, they see only the rays that reflect directly toward their eyes. Demonstrate this by having students try to observe the beam of light from a laser pointer that you shine across the room. ***CAUTION: Warn students never to look directly at a laser beam.*** If you shine the light toward an open door, students cannot see the beam. If you shine the light toward a wall, they can see the rays that reflect off the wall toward their eyes. Sprinkle some chalk or talcum powder in the path of the laser beam so that students can now see it. Ask them why they can now see the beam. Some of the laser light is reflected off the dust particles toward their eyes. BL Visual-Spatial EL

Use Models

MAIN IDEA Help students model the law of reflection with a ball bouncing off a wall. First, draw a line normal to the wall. Then draw two other lines that model an incident ray and a reflecting ray—they should be at equal angles to the normal and touch the same point on the wall. Have a student roll the ball, such as a golf ball, on the floor along one angled line. The class should notice that the ball bounces along the other angled line, in the same way that a light ray reflects off a surface. Although the analogy between the reflecting of light and the bouncing of a ball has some use, make sure that students do not draw too much from this analogy. A light beam and a ball are made of different "stuff" and different theories explain their behaviors (at least at the level of classical physics, on which most of this book focuses). BL Kinesthetic Visual-Spatial EL

PHYSICS PROJECT ACTIVITY

Illusions Magicians often use mirrors to create illusions for audiences. Directors of older movies sometimes used mirrors to produce special effects. Ask students to investigate the ways in which mirrors can be used to create these effects. Have them work in groups or as individuals to create and perform a skit or trick using a mirror illusion for the class. Interested students may even try to get pointers by visiting a local magic shop or interviewing a magician who performs in their area. OL Kinesthetic EL

Quick DEMO

Reflecting Surfaces

Estimated Time 5 minutes

Materials laser pointer, white marker board, sheet of white paper, plane mirror, laser pointer

Procedure

1. In a dimly lit room, shine the laser pointer directly toward a white board. Some light may reflect along the board because of scratches and imperfections in the board.
2. ***CAUTION: Warn students never to look directly at a laser pointer.*** Lay the paper on a desk next to a wall. Shine the laser pointer at about a 45° angle toward the paper. Ask students why they see diffuse, reflected light on the wall. The surface of the paper is rough, and the reflected light is scattered.
3. Lay the mirror on the desk next to a wall. Then, shine the laser pointer at about a 45° angle toward the mirror. Ask students why they see a dot of light reflected onto the wall. The smooth surface of the mirror reflects the light in a narrow beam.

Concept Development

Silvered Mirrors Most of the reflection from an ordinary mirror occurs at the silver coating on the underside of the glass. A little light does reflect from the front surface, which sometimes produces a faint image. Front-silvered or front-surfaced mirrors produce higher-quality images because light reflects directly off the silver coating without passing through glass. Therefore, they produce only one reflection, which is a major reason for their use.

Content Background

Mirror Coatings The first mirrors, dating back to antiquity, were made of polished metal such as bronze, tin, or silver. Glass mirrors were introduced in fourteenth-century Venice. By the sixteenth and seventeenth centuries, mirrors were made by pressing an amalgam of mercury and tin onto a sheet of glass and later draining off the excess mercury. The method on which modern mirror coatings are based was developed by German chemist Justus von Liebig in 1835. A compound of silver and ammonia was poured onto the glass surface. Then, a reducing agent, such as formaldehyde, was added to reduce the compound to metallic silver. Present-day mirrors are made by spraying molten aluminum or silver onto plate glass in a vacuum.

ADDITIONAL IN-CLASS EXAMPLE

Use with Example Problem 1.

Problem A light ray strikes a plane mirror at an angle of 15° to the normal. The mirror then rotates 20° around the point where the beam strikes the mirror so that the ray's angle of incidence increases. The axis of rotation is perpendicular to the plane of the incident and to the reflected rays. What is the final angle of reflection of the light ray?

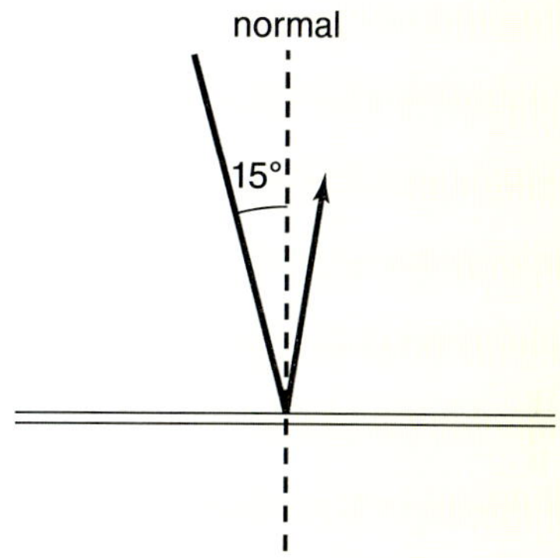

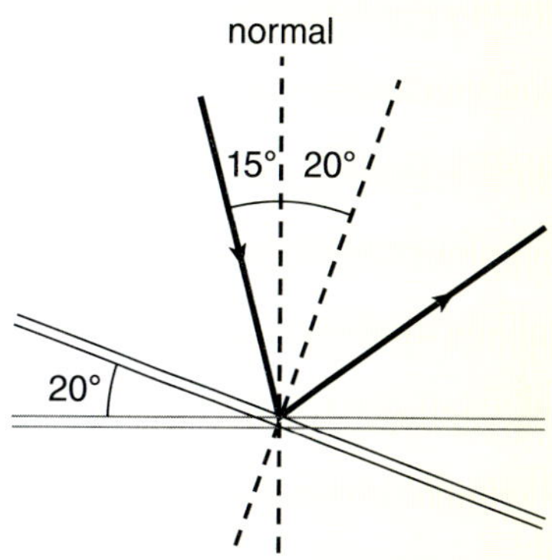

Response $\theta_{r,\,final} = 35°$

Discussion

Problem Ask students the following questions: A good mirror must be smooth, and it must reflect most of the light that strikes it, but is this enough? Why doesn't a smooth, white surface make a good mirror?

Response A typical mirror reflects well because of its silvered backing. Although a white surface reflects well, it appears white because it is mostly diffuse reflection. It is not as smooth as a silvered surface. OL

Objects and Plane-Mirror Images and Properties of Plane-Mirror Images

student text, pp. 468–470

Discussion

Problem Ask students why you can see your reflection if you look out a window at night, but not during the day. Why is it easier to see outside at night if the inside lights are off?

Response Window glass transmits most light, but some light is reflected. This reflected light is not as visible during the day because it is much dimmer than the sunlight that is transmitted in from outside. At night, the reflected light is visible because there is very little light outside. OL

REAL-WORLD PHYSICS

Multiple Plane Mirrors Using information covered in this section, have students study how plane mirrors can be used in combination to make multiple images. For example, they can study a three-plane mirror in a clothing store and draw a ray diagram, using ray tracing to determine the location of each image in the three-plane mirror. Students also can research other uses of combinations of plane mirrors and draw ray diagrams for them. OL Visual-Spatial EL

Critical Thinking

Reflection Demo Set up two small mirrors about 10 cm apart so that they face each other. Scrape a small hole in the paint on the back of one mirror about one-third of the way down from the top. Place a soda can or other object halfway between the mirrors. Have students look through the hole and describe the reflections. It appears that reflections continue to infinity, with the sizes of the images getting smaller. Have students explain this effect. The first image in each mirror is of the real soda can. The second image in each mirror is a reflection of the first image of the other mirror. Each of the first images are three times farther away from the opposite mirrors than the real can, so the second images appear to be three times farther behind the mirrors than the first images. This repeats, with each image becoming smaller because it appears farther away from and behind the mirror. OL Visual-Spatial EL

Careers

Optical Engineers Students who enjoy astronomy and do well in optical physics may be interested in a career as an optical engineer at a space observatory. Optical engineers are responsible for the design, alignment, and maintenance of optical systems in the telescopes. In addition to understanding mirror and lens systems, optical engineers must be able to use and maintain the instrumentation that supports the system. Students interested in this field should obtain at least a bachelor's degree in engineering or physics. They also should learn as much mathematics as possible.

Reinforcement

Perspective Demo Be sure students understand that being able to see an image behind a plane mirror is dependent on the viewer being in a position from which the image is visible. To emphasize this, place a mirror against a wall in the middle of the classroom. Have one student stand in front of the mirror and another student stand far to the right of the mirror. Place an object far to left of the mirror so that the first student cannot see the image of it, but the second student can. Let each student in the class stand at each position to reinforce how perspective of the viewer affects what can be seen using a mirror. BL Kinesthetic Visual-Spatial EL

Use Figure 7

Ask students how tall a mirror must be so that a person can see his or her entire reflection. To answer this question, have a student stand in front of a full-length mirror within reaching distance. Have the student point at his or her head and feet. Point out that the top of the mirror must be at the student's height and that the bottom of the mirror can be no more than half the student's height from the ground. Have students study the angles formed by the light rays in **Figure 7** to study this effect. By applying the law of reflection, they should see that the student observes a full-sized image in a mirror half of his or her height. OL Visual-Spatial EL

Use the PhysicsLAB

In *Position of Mirror Reflection,* students will measure the location of an image from a plane mirror.

Use the PhysicsLab

In *A Little Time to Reflect,* students will determine the path of a light ray incident to, and reflected from, a flat mirror. They will compare their measured values of incidence and reflection angles.

DIFFERENTIATED INSTRUCTION

Visually Impaired When performing the MiniLab, pair a visually impaired student with another student who can describe the distances and image properties. Visually impaired students may understand the concept of plane-mirror images better if you allow them to walk from the position of an object to the plane mirror. Explain to them that if they could continue walking into the mirror, the image would be the same distance behind the mirror as the distance they just walked. However, the image would not really be there when they arrived because it is a virtual image.
Kinesthetic

Use the MiniLAB

In *Virtual Image Position,* students will use a camera to determine the distance of an image from a plane mirror.

3 Assess

Assess the MAINIDEA

Place a flashlight on a table, aimed horizontally. Dim the lights to the room and use a small mirror to demonstrate that the light spot can be directed at the ceiling by holding the mirror at a 45° angle (from the horizontal direction). Draw a diagram of the mirror and light path and on the board, and ask the students to explain what is happening. Since the angle of incidence equals the angle of reflection (45° in this case), the light is reflected straight up. Now use the mirror to demonstrate how it can be used to aim the light spot at will.

Check for Understanding

Image Location Demo Ask students whether the point on the surface of a mirror at which light from an object reflects is the same as the location of the image formed by the reflected light. The image is behind the plane mirror, not on or in the surface of the mirror. One way to demonstrate this is to have students hold a small mirror and place their thumbs on the front of the mirror. By applying their thumbs directly to the surface, they can see clearly that the reflection is behind the surface of the mirror. **BL** **Visual-Spatial** **EL**

Reteach

Mirror Image Flat mirrors do not reverse up and down or left and right, but they do reverse front and back. To reinforce this, have students take turns standing and facing a mirror. Have the student who is in front of the mirror point to his or her left. Ask the student if the mirror image is pointing in the same direction as he or she is. yes Now have the student point to the right. Ask the same question as before. yes Finally, have the student point straight ahead, that is, in the direction he or she is facing. Ask the question once again. no This demonstrates straightforwardly what is meant by the statement that mirrors reverse only front and back. Why, then, does it *seem* that mirrors reverse left and right? The answer lies in psychology, not physics. Your mirror image looks likes you *would* look had you rotated yourself 180°, in which case left and right *would be* reversed. Have two students face each other and point to the left. They will point in opposite directions. Mentally you think that your image is like that person opposite you and is pointing in the reverse direction, when actually it is pointing the same direction.
BL **Kinesthetic** **Visual-Spatial** **EL**

Reading Checks and Caption Checks

Caption Check, *student text p. 468*

The boy is illuminated by light directly from the lamp and by light reflected from the mirror.

Caption Check, *student text p. 469*

In a plane mirror, the position of the image (x_i) is the same distance behind the mirror as the object (x_o) is in front of the mirror.

Practice Problems

student text p. 467

1. Water fills in the rough areas and makes the surface smoother. The surface normals are then parallel.
2. 35°
3. a. 42°
   b. 48°
   c. 84°
4. 51°
5. 30°
6. 90°

Section 1 Review

student text p. 470

7. 10°
8. $x_1 = -3$ m; $h_1 = 50$ cm; the image is virtual.
9. The law of reflection applies to individual rays of light. Rough surfaces make the light rays reflect in many different directions.
10. Specular: polished metal, window glass, smooth water. Diffuse: paper, rough metal, plastic milk jug, ground glass.
11. 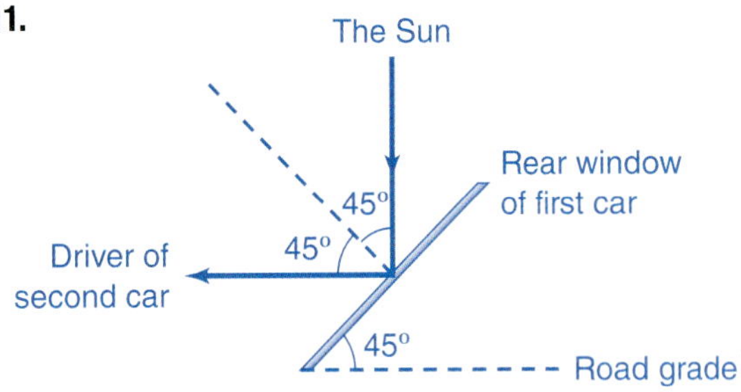

    The Sun's position directly overhead would likely reflect light into the driver's eyes, according to the law of reflection.
12. The incoming light reflects off the surface of the object in all directions. This enables you to view the object from any location.

SECTION 2 Curved Mirrors

student text pp. 471–481

1 Introduce

Bellringer

Different Types of Mirrors Have students look into a plane mirror and then look into the front and back sides of a shiny spoon. Point out that the shape of a reflecting surface affects the image it produces.
BL **Kinesthetic** **EL**

Tie to Prior Knowledge

Image Properties Students studied properties of images produced by flat reflecting surfaces in Section 1. In this section, they will extend these concepts to curved reflecting surfaces. They will learn how the focal length is related to object and image position.

2 Teach

Properties of Curved Mirrors

student text pp. 471–472

Reinforcement

Focal Point Emphasize the difference between the focal point, *F*, and the focal length, *f*. The focal length is the distance between the mirror and the focal point.

Use an Analogy

Focal Point Demonstrate the concept of a focal point by wrapping a rubber band loosely around the middle of about 20 pieces of thin skewers or uncooked spaghetti. Spread the skewers out on both sides of the rubber band. Point out that the skewers converge at the rubber band and then spread out, just as light reflects off a concave mirror, converges at the focal point, and then spreads out again.

Use Figure 10

Students may notice that the rays in this figure are drawn to the plane perpendicular to the principal axis, rather than to the face of the mirror. Explain that this is an approximation and they will study the reason for it later in this section. For now, tell students to be sure to draw this plane on the surface of the mirror at the point where the principal axis intersects the face of the mirror. **OL**

Critical Thinking

Image Dimming Ask students whether they think a mirror is a perfect reflector. Suggest that a mirror might reflect 90 percent of the light incident upon it. Have students imagine light bouncing off three such mirrors, one after the other. Ask students what percentage of the original light would bounce off the third mirror. Because about 90 percent of the incoming light reflects from each mirror surface, the total amount after three reflections is $0.90 \times 0.90 \times 0.90 = 0.73$, or 73 percent. Ask students how a curved mirror could be used to rebrighten an image. A large curved mirror can make the image appear brighter by putting the same amount of light into a smaller area. **OL**

Ray Diagrams for Concave Mirrors Real Images with Concave Mirrors Virtual Images with Concave Mirrors

student text pp. 472–477

PHYSICS PROJECT ACTIVITY

Uses of Mirrors Have students look through astronomy magazines or other science and technology periodicals to find an article about a modern technology that uses concave mirrors, such as the *Hubble Space Telescope*. Encourage them to read the article, learn as much as they can about how the technology uses mirrors, and prepare a presentation. **OL** **Linguistic**

Identify Misconceptions

Demonstrating Ray Diagrams When drawing ray diagrams, students will sometimes find that a ray does not intersect the picture of the mirror on the paper. Students might believe that the diagram is impossible to draw or that the mirror cannot form a full image of the object. Demonstrate how they can extend the plane perpendicular to the principal axis as far as necessary, up or down. The reflection of the ray occurs at this plane, just as if it had intersected the mirror. Using this model, the mirror is shown to form a full image, just as in reality. Demonstrate that even in the case of a spoon, a student will be able to see the entire upper part of his or her body, which is obviously taller than the bowl of the spoon.

DIFFERENTIATED INSTRUCTION

Struggling Students Drawing ray diagrams can be confusing to students when they first encounter it, but obtaining this skill is important to solving optical problems. Divide the class into small groups and provide each group with two or three problems involving concave mirrors. Have each group apply the above Problem-Solving Strategies step-by-step to solve the problems. Students should draw the ray diagrams independently, but they can benefit from discussing each step with the group.
BL COOP LEARN Interpersonal

PHYSICS PROJECT ACTIVITY

Spherical Aberration Have students investigate spherical aberration of real images using large demonstration-sized concave mirrors. First, have students mask the outer regions of the mirror with manila file folders. They should notice that the image is less intense, but that it suffers less spherical aberration. Next, have them mask the center of the mirror. Again the image is less intense, but the spherical aberration is more noticeable. It is interesting to use a hole with a diameter of about 5 cm as a mask and to move the hole around over the surface of the mirror. Students should compare the images produced by different regions of the mirror's surface. OL Kinesthetic EL

Applying Physics

A malfunction in a measuring device during polishing made the outer edge of the *Hubble Space Telescope's* primary mirror too flat. Although the imperfection was only about the depth of 1/50 of a human hair, it was enough to make *Hubble's* viewing only slightly better than that of a land-based telescope. COSTAR (Corrective Optics Space Telescope Axial Replacement) positioned five pairs of corrective mirrors at locations that successfully countered *Hubble's* spherical aberration. Have interested students research and prepare a class presentation about COSTAR and other optical improvements made on *Hubble Space Telescope.* OL

Content Background

Spherical Aberration Students should understand that spherical aberration is an intrinsic defect, not an imperfection in a mirror. It occurs even in perfectly shaped spherical mirrors. Spherical aberration can be avoided by producing aspherical mirrors. However, the production of aspherical surfaces for precise astronomical uses is difficult because the glass surfaces must be ground and polished. For applications in which less-precise images are acceptable, a molded, aspherical plastic base for a mirror can be easily produced.

Convex Mirrors

student text pp. 476–477

Use the PhysicsLAB

In *Concave Mirror Images,* students will use a concave mirror to observe the conditions needed to produce real and virtual images.

Use the MiniLAB

In *Finding the Focal Point,* students will observe if and how much a convex mirror focuses sunlight on a piece of paper. They will then do the same with a concave mirror.

Calculating Image Position

student text pp. 477–480

Concept Development

MAINIDEA Discuss what happens to the image produced by a concave mirror as the object position approaches the focal point from the direction of the mirror and from a large distance from the mirror. The virtual image height gets larger and larger, and then disappears when the object reaches the focal point. The light rays are reflected as parallel rays that never meet. The mirror and magnification equations suggest that the virtual image is infinitely far away and infinitely large. OL

Identify Misconceptions

Object Location Students may believe that it is necessary for an object to be on the principal axis of a concave mirror for the mirror to be able to form an image. For a concave mirror, the object can be to the side as long as any amount of mirror surface is visible from the location of the object. Have students use a concave magnifying mirror to investigate this concept. BL Visual-Spatial EL

Activity

Magnification Some students might need help understanding the concept of image magnification by curved mirrors. Cut a manila folder into several rectangles of different dimensions. Give one rectangle to each pair of students. Demonstrate how students can use a ruler to draw an object using a simple magnification ratio such as 2:5. Students should include the scales on their drawings. Pairs of students should then exchange diagrams and determine the height of the original objects from the scale drawings. OL Visual-Spatial EL

ADDITIONAL IN-CLASS EXAMPLE

Use with Example Problem 2.

Problem A concave mirror has a radius of curvature of 24.0 cm. A 6.4-cm-tall object is held 26.0 cm from the mirror. Where is the image and how tall is the image?

Response

$x_i = 22.3$ cm. The image is 22.3 cm in front of the mirror.

$h_i = -5.5$ cm. The image is 5.5 cm tall and inverted.

Concept Development

■ Benefits of Ray Diagrams Be sure to point out that the ray-drawing techniques serve as a check on algebraic methods and vice versa. The two methods should agree.

■ Image Height Equation The equation $h_i = -x_i h_o / x_o$ is used frequently for solving problems involving concave mirrors. Students will profit from deriving this equation from the magnification equation. To solve for image height, h_i, students simply need to multiply the ratios in the magnification equation by the object height, h_o.

$m = h_i / h_o = -x_i / x_o$

$(h_o)(h_i / h_o) = (-x_i / x_o)(h_o)$

$h_i = -x_i h_o / x_o$ OL

REAL-WORLD PHYSICS

Liquid Mirrors Have students imagine swirling milk in a cereal bowl until the milk has a parabolic shape. Explain that scientists use this concept to produce liquid telescopes. Mercury, a highly reflective metal that is liquid at room temperature, is placed in a large container. As the container is spun, the mercury becomes a parabolic concave mirror. A bowl of compressed air cushions the mercury from outside vibrations, and smooth plastic film on the surface of the mercury prevents air disturbances. Although liquid telescopes are inexpensive compared to those that employ solid mirrors, they have a limited viewing area because they must be positioned facing straight upward.

Quick DEMO

Image Position

Estimated Time 10 minutes

Materials concave mirror, battery-powered or chemically powered glowing object

Procedure

1. Dim the lights, and hold the object in one hand and the concave mirror in the other.
2. Face the mirror toward a white board or wall. Hold the object off the principal axis between C and F, but close to the focal point. A large inverted image of the object will appear on the wall.
3. Hold the object beyond C. You will have to angle the mirror face slightly away from the wall. A small inverted image will be formed on the wall between C and F.
4. Hold the object between F and the mirror. No real image will be formed.

Discussion

Problem Ask students if they would expect a difference in brightness between enlarged and reduced images.

Response Yes. As light rays diverge, the light is spread over a greater area. As light rays converge, the light is concentrated on a smaller area. As a result, enlarged images are dimmer and reduced images are brighter. This explains why the focal point of a concave mirror can be located by finding the brightest spot of reflected light. The focal point for a concave mirror is where rays that were parallel before striking the mirror come together. OL

ADDITIONAL IN-CLASS EXAMPLE

Use with Example Problem 3.

Problem A convex mirror has a radius of curvature of 28.4 cm. If you hold a 16-cm-long pencil 23.5 cm in front of the mirror, what will the image position and image height be?

Response

$x_i = -8.85$ cm; $h_i = 6.0$ cm

Mirror Comparison

student text, p. 481

Reinforcement

Uses of Mirrors Have students investigate the uses of mirrors, make a list of every mirror that they encounter in a one-day period, and describe each mirror's use. Students could make sketches to help explain how each mirror performs its intended function. If possible, allow students to bring some of the mirrors that they discover to class. BL **Interpersonal**

PHYSICS PROJECT ACTIVITY

Systems of Mirrors Many optical devices use systems of mirrors to obtain images with specific properties. Have students work in pairs to design a desktop optical system using two or more mirrors, including plane and curved mirrors. They should be able to explain their system, including information about focal lengths, object locations, and image properties. To further focus the activity, challenge students to devise a system using two plane mirrors that would allow them to undo the reversal of letters seen with a single plane mirror.
OL **Kinesthetic** EL

Reinforcement

Properties of Mirrors Divide the class into small groups. Have each group of students write true/false questions about the properties of mirrors on the front of index cards and write the answers with explanations on the back. The groups can exchange cards, and students can use them to quiz each other. BL **Interpersonal**

3 Assess

Assess the MAIN IDEA

Allow the students to use both convex and concave mirrored surfaces to examine the reflected images of their textbook. A large reflective spoon can be used for this if better mirrors are not available. Ask the students to describe how the image of the text appears. Answers will vary. The convex surface will shrink the appearance of the text. The concave surface may invert and shrink the text, or enlarge the text, depending on the exact relative position of the text and mirror surface.

Check for Understanding

Real and Virtual Images Have students each write and illustrate a paragraph comparing real and virtual images formed by a concave mirror. BL **Linguistic**

Extension

Convex Mirrors in Art Students can research famous paintings that include convex-mirror images. Some examples are *The Arnolfini Marriage* (1434) by Jan van Eyck and *Self-Portrait in a Convex Mirror* (1524) by Parmigianino. Have students each write a paragraph describing how the artist used the mirror in the painting. They also should describe the visual effects produced by the convex mirror. OL **Linguistic**

Reading Checks and Caption Checks

Caption Check, *student text p. 473*
Use the same procedure, but have the rays originate from a different part of the object.

Caption Check, *student text p. 475*
The flag and the desk appear inverted because they are beyond the focal point of the mirror.

Reading Check, *student text p. 475*
A virtual image will appear to be located behind the mirror surface.

Caption Check, *student text p. 477*
We are used to seeing smaller objects farther away. The convex mirror, not distance, decreases the size of an image. The object is not as far away as we usually perceive it to be.

Practice Problems

student text p. 479

13.

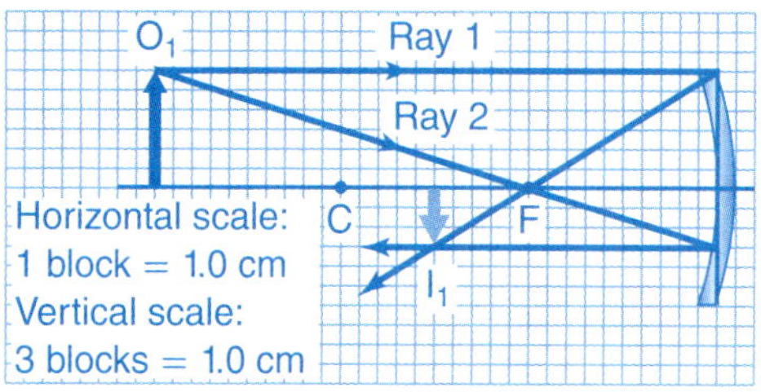

14. 28.8 cm

15. $x_i = 13.3$ cm; $h_i = -2.0$ cm

16. -1.9 cm

17. $x_o = 26.7$ cm; $h_o = 5.0$ cm

student text p. 480

18. $x_i = -8.57$ cm

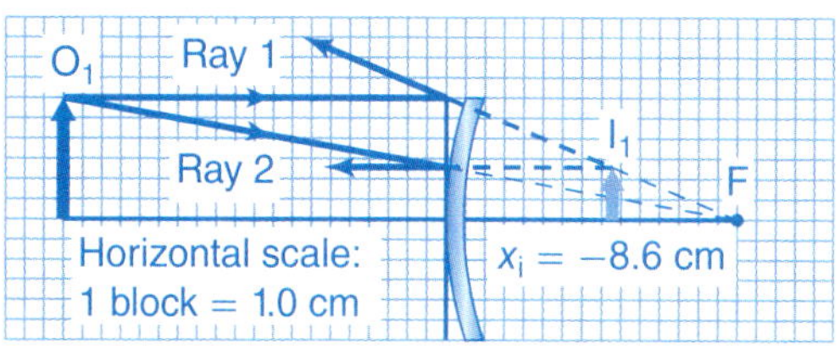

19. $x_i = -10.7$ cm, $h_i = 1.1$ cm

20. $x_i = -12.7$ cm, $h_i = 4.4$ cm

21. a. $x_i = -0.48$ m, **b.** $f = -0.60$ m

22. a. $x_o = 32$ cm, **b.** $f = -96$ cm

Physics Challenge

student text p. 475

1.

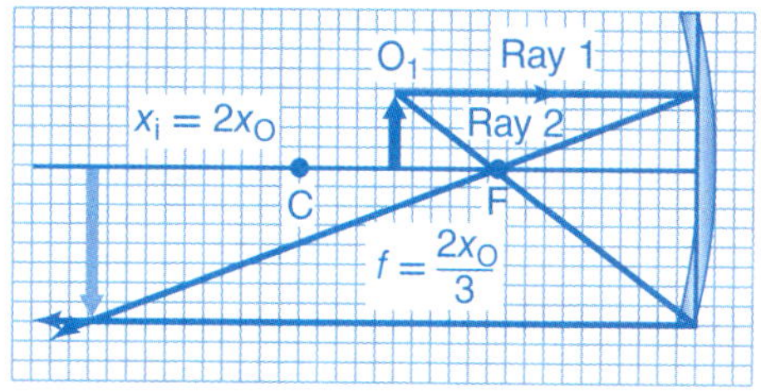

Distances will vary. A correct form of the answer would be: $x_o = 6$ cm, $x_i = 12$ cm, and $f = 4$ cm

2. $1/f = 1/x_o + 1/x_i$

$x_o = fx_i/(x_i - f)$

$x_o = f(2f)/(2f - f)$

$x_o = 2f$

$m = h_i/h_o = -x_i/x_o$

$h_i = -x_i h_o/x_o$

$h_i = -(2f)h_o/2f$

$h_i = -h_o$

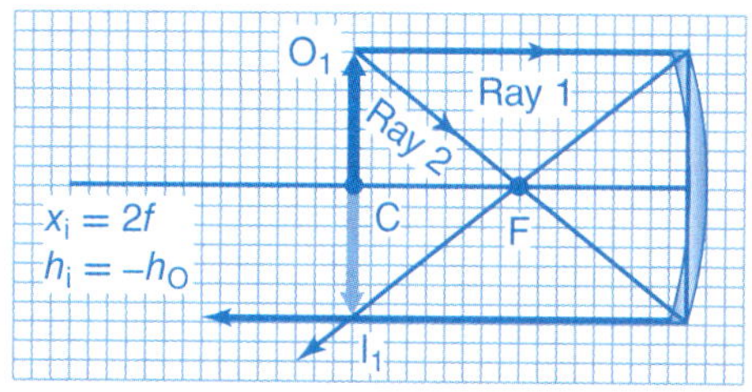

Distances will vary. A correct form of the answer would be: $x_o = 10$ cm, $x_i = 10$ cm, and $f = 5$ cm.

3. The object should be placed at the focal point.

Section 2 Review

student text p. 481

23. You should place the object between the mirror and the focal point. The image will be virtual.

24. $m = -0.82$

25. $x_o = 26.0$ cm

26. See Online Solutions Manual for ray diagram; $x_i = 26.4$ cm, $h_i = -3.6$ cm

27. See Online Solutions Manual for ray diagram; $x_i = -6.46$ cm, $h_i = 1.8$ cm

28. 29 cm

29. -36 cm

30. It would be less for a mirror whose height is relatively small compared to its radius of curvature; diverging light rays from an object that strike the mirror are more paraxial so they converge more closely to create an image that is not blurred. The curvature of a "short" mirror deviates less from parabolic.

FRONTIERS IN **PHYSICS**

Distant Earths

The Hunt for Exoplanets

Background

In astronomy, transition is when a planet passes in front of a star, blocking some of the star's light. The Kepler Mission searches for exoplanets by looking for evidence of transitions. It looks at the relative brightness of light coming from a star over time. A periodic dimming of the star's light indicates that a planet is likely passing in front of it during its regular orbit. Once astronomers have determined that the change in brightness is due to an exoplanet, the data can be used to determine the exoplanet's period, orbital size, size, and temperature.

Teaching Strategies

- As of September 2010, astronomers have discovered 490 exoplanets. Explain that astronomers expect to find many more. Have students research online to find the current number of known exoplanets.
- Model the habitable zone of a solar system by using a lamp to represent a star and a tennis ball to represent a planet. Explain that a planet's temperature is determined in part by its distance from the star. Show them the region around the lamp that represents the temperature range in which water could be in the liquid state on an orbiting planet.
- Have students find out about the field of astrobiology and the study of life in extreme environments on Earth. Have them make a poster that describes the conditions a planet must have to support life.

GOING**FURTHER** >>>

Expected Outcomes Student answers should include a description of the conditions on a known planet, such as temperature, pressure, presence or absence of oxygen, presence or absence of liquid water, amount of solar radiation, length of days and nights, etc. The description of an organism that could live on the planet should account for survival under each condition mentioned.

SECTION 1

Mastering Concepts

31. When parallel light is reflected from a smooth surface, the rays are reflected parallel to each other. The result is an image of the surface from which the rays originated. When light is reflected from a rough surface, it is reflected in many different directions. The rays are diffused. No image of the source results.

32. The *normal to the surface* is a line perpendicular to the surface.

33. The image is in a plane that is perpendicular to the mirror and the same distance behind the mirror as the object is in front of the mirror.

34. A plane mirror is a flat, smooth surface from which light is reflected by specular reflection. The images created by plane mirrors are virtual, upright, and as far behind the mirror as the object is in front of it.

35. No, the rays do not converge at a virtual image. No image forms and the student would not get a picture. Some virtual images are behind the mirror.

36. Place a sheet of plain paper or photographic film at the image location and you should be able to find the image.

37. The image is three-dimensional. You can tell because different parts of the image are visible from different vantage points. Objects closer to the mirror have images that are closer to the mirror. Objects far away have images that are also far away.

Mastering Problems

38. 108°

39. 38°

40. a. 60°

b. See Online Solutions Manual for diagram.

41. $\theta_r = 53°$; 106°

42. The ray from the top of the head hits the mirror halfway between the eyes and the top of the head. The ray from the feet hits the mirror halfway between the eyes and feet. The distance between the point the two rays hit the mirror is half the total height.

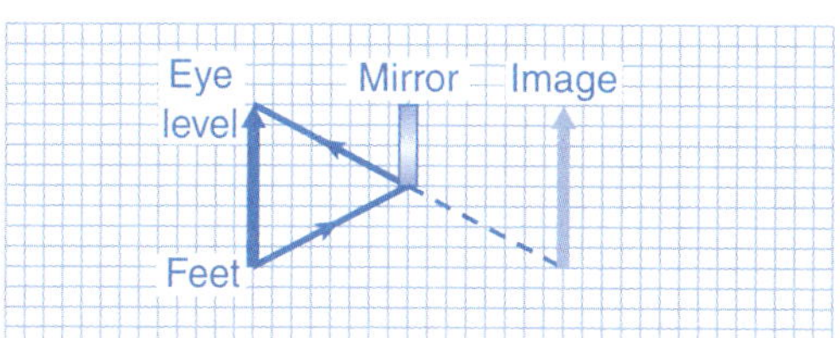

43. To see your knee, the mirror needs to be at a spot midway between your knee and eye. To see any part of your body, the mirror needs to be midway between your eye and that part.

44. The image is 1.2 m behind the mirror, so the camera lens should be set to 2.4 m.

45. 48°

46. Reflection from the first mirror is $\theta_{r,1} = \theta_{i,1} = 30°$. The angle the ray forms with the mirror is thus $90° - 30° = 60°$. Because the two mirrors form a 45° angle, the angle the ray reflecting off the first mirror forms with the second mirror is $180° - 60° - 45° = 75°$. The angle the ray forms with the second mirror is thus $\theta_{i,2} = 90 - 75° = 15°$. The angle of reflection from the second mirror is $\theta_{r,2} = \theta_{i,2} = 15°$.

SECTION 2

Mastering Concepts

47. Object must be located between F and the mirror.

48. Rays parallel to the axis that strike the edges of a concave spherical mirror are not reflected through the focal point. This effect is called spherical aberration.

49. $\frac{1}{f} = \frac{1}{x_i} + \frac{1}{x_o}$

50. $r = 2f$

51. The magnification is equal to the negative of the image distance divided by the object distance.

52. Convex mirrors are used as rearview mirrors because they allow for a wide range of view, allowing the driver to see a much larger area than is afforded by ordinary mirrors.

53. The light rays always diverge.

Mastering Problems

54. 1.8 m

55. $x_1 = 30.0$ cm; $h_1 = -1.8$ cm

56. 20.0 cm

57. −7.2 cm

58. 75 cm

59. $m = 0.5$

60. real; inverted; larger

61. Possible answer: An image is formed 0.75 m from a concave mirror with $f = 0.40$ m. How far from the mirror is the object?

62. D > E > C = A > B

63. $m = 5$

64. $x_i = 4.0$ cm; $h_i = -8.0$ mm

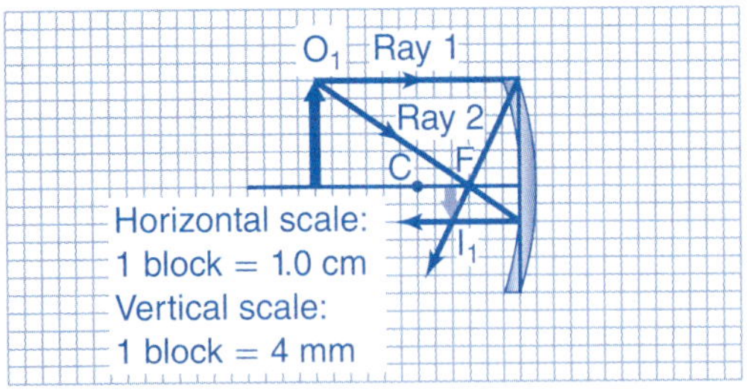

65. $x_i = 70.5$ cm; $h_i = -9.4$ cm

66. $x_i = -9.4$ cm; $h_i = 0.75$ cm

67. $x_i = 33$ cm; $h_i = -4.0$ cm

68. $x_i = -24$ cm; $h_i = 9.0$ cm

Applying Concepts

69. Less light is reflected back to the car from a wet road.

70. The smoother and glossier the pages are, the lesser the diffuse reflection of light and the greater the glare from the pages.

71. The image will be at C, the center of curvature, inverted, real, and the same size of the object.

72. The image will be between C and F, and will be inverted, real, and smaller than the object.

73. You should order a parabolic mirror to eliminate spherical aberrations.

74. You can use only a concave mirror with the object beyond the focal point. A convex mirror will not form a real image.

75. Convex mirror; it provides a wider field of view.

76. The image in a single convex mirror is always virtual, erect, smaller than the object, and located closer to the mirror than the object.

Mixed Review

77. Possible answer: "... that has a focal length of +35 cm. Where will the image be?"

78. −6.9 cm

79. 62°

80. $x_i = 22.9$ cm; $h_i = -1.8$ cm

81. −72 cm

82. 58 cm

83. The focal length is positive, so the spherical mirror is a concave mirror.

84. The image is −2.3 m tall; the negative means it is inverted. It is 14 m from the mirror.

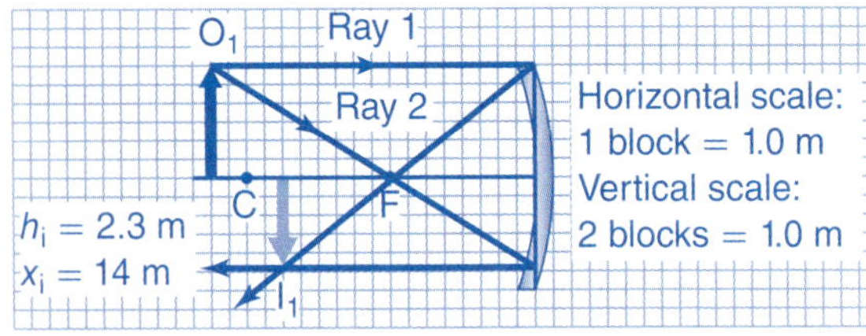

85. The image height is 1.0 cm, and its location is 2.7 cm from the mirror.

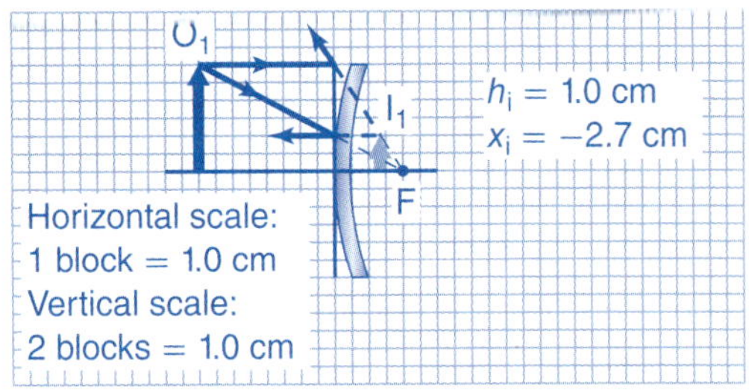

86. $f = -12$ cm

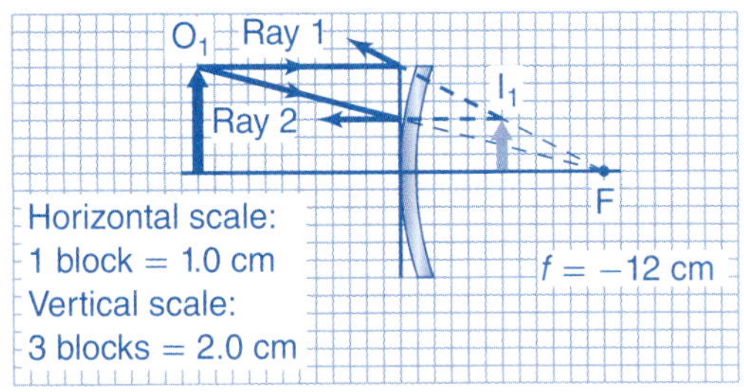

87. a. −1.5 m

b. 0.39 m

88. a. An enlarged, upright image results only from a concave mirror, with the object inside the focal length.

b. 32 mm

89. −0.68 m

90.

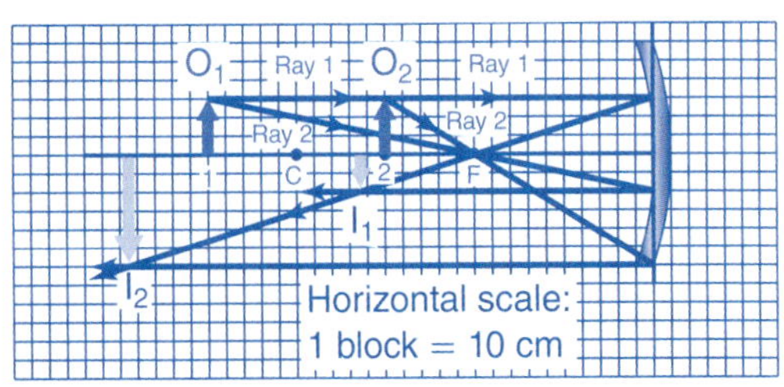

Thinking Critically

91. 11 cm

92. Beyond C, the image is smaller than the ball. As the ball rolls toward the mirror, the image size increases. The image is the same size as the ball when the ball is at C. The image size continues to increase until there is no image when the ball is at F. Past F, the size of the image decreases until it equals the ball's size when the ball touches the mirror.

93. The smaller mirror is concave to produce a real image at the eyepiece that is upright. The light rays are inverted by the first concave mirror and then inverted again by the secondary concave mirror.

94. -1.8 cm

95. For the primary mirror, $x_0 = \infty$. All light rays enter parallel. $x_i = f = 1.0$ m.

96. The convex mirror is placed to intercept the rays from a concave mirror before they converge. Rays incident on the convex mirror are already converging, not parallel. The convex mirror places the point of convergence in the opposite direction back toward the concave mirror, and lengthens the total distance the light travels before converging. This effectively increases the focal length compared to using the concave mirror by itself, thus increasing the total magnification.

Writing in Physics

97. Answers will vary depending on the mirrors and methods chosen by the students. Amateur methods usually involve rubbing two "blanks" against each other with varying grits between them. Methods used at national labs vary.

98. Answers will vary. Student answers might include information about deformation of a mirror from its own weight as size increases and how a mirror made of aluminum could impact this problem.

Cumulative Review

99. 0.18

100. 4.6×10^{-11} N; The force is provided by the frictional force between the bug and the flying disk.

101. 9.6 °C

102. 2.4 kPa

103. $T_{moon} = 7.0$ s; $T_{Earth} = 2.8$ s

104. a. The resonant frequency of an open pipe is twice that of a closed pipe of the same length. Therefore, the pipes of a closed-pipe organ need be only half as long as open pipes to produce the same range of fundamental frequencies.

b. No. While the two organs will have the same fundamental tones, closed pipes produce only the odd harmonics, so they will have different timbres than open pipes.

Standardized Test Practice

student text p. 489

Multiple Choice

1. D
2. D
3. A
4. D
5. C
6. C
7. D
8. C
9. A

Free Response

10. $h_i = 2.1$ cm

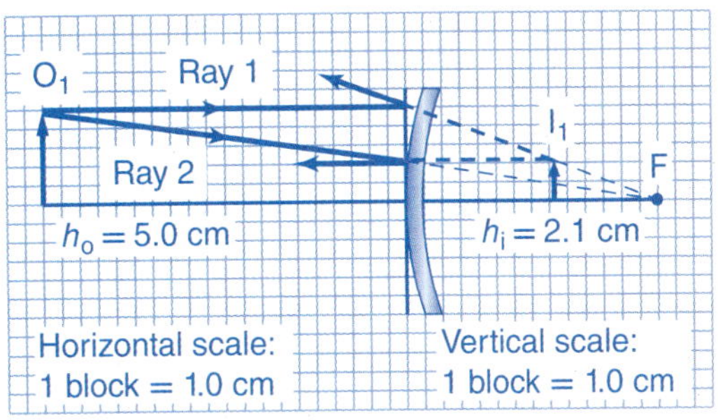

Rubric

The following rubric is a sample scoring device for free response questions.

Points	Description
4	The student demonstrates a thorough understanding of the physics involved. The response may contain minor flaws that do not detract from the demonstration of a thorough understanding.
3	The student demonstrates an understanding of the physics involved. The response is essentially correct and demonstrates an essential but less than thorough understanding of the physics.
2	The student demonstrates only a partial understanding of the physics involved. Although the student may have used the correct approach to a solution or may have provided a correct solution, the work lacks an essential understanding of the underlying physical concepts.
1	The student demonstrates a very limited understanding of the physics involved. The response is incomplete and exhibits many flaws.
0	The student provides a completely incorrect solution or no response at all.

CHAPTER 18

Refraction and Lenses

About the Photo

View Through a Lens Have students examine the photograph. Ask students to describe what is being viewed in the distance. the wooden pier or bridge Have students describe how the view through the lens differs from what is observed without using the lens. The lens shows the pier and the sky inverted. Ask students to speculate what is happening. The lens changes where the light goes as it passes through the lens.

Chapter Overview

Light traveling in a medium with a certain index of refraction changes speed when it passes into a medium with a different index of refraction. This change in speed alters the direction of the light if it strikes the boundary at an angle. Light that passes through a lens may produce an image with a size and orientation that are different from those of the original object. The eye and certain optical instruments are able to obtain clear images of small or distant objects because of the refraction of light.

Before students study the material in this chapter, they should study:

- Fundamentals of waves
- The law of reflection
- The mirror equation

To solve the problems in this chapter, students will need a strong grasp of:

- Scientific notation
- Significant figures
- Sine, cosine, and tangent
- Solving linear equations

Introduce the BIG IDEA

Light Strikes Surfaces Ask students what happens to light when it strikes a mirror. Light reflects an image back to the viewer. Have students speculate what might happen if mirrors did not have a silvery or solid backing. Light would pass through the glass. Discuss that the light could be affected by the mirror's glass.

Go online!
connected.mcgraw-hill.com

Your online experience includes:

- **Video**
- **Concepts In Motion**
- **Virtual Investigations**
- **All Lab, Applying Practices activities, and PBL worksheets**
- **Online Practice**
- **Personal Tutor**
- **Multilingual eGlossary**
- **Presentation**

Use the LaunchLAB

In *Broken Straw Appearance,* students will compare the degree to which light changes direction in three different liquids.

Use Physics T.V.

Video

LIGHTHOUSE LENSES How does a lighthouse produce a beam of bright, focused light? How are lighthouses similar to spotlights in a theater? It's all about the lens.

1 Introduce

Bellringer

Bent Pencil Place a pencil in a clear glass of water. Point out to students that the pencil appears to be broken. Move the pencil from side to side, and then move it closer and farther away from you. Students should notice how its width appears to change. Explain that this happens because light changes speed as it travels from one medium to another. Students will learn how this change in speed causes the light to change direction. BL Visual-Spatial EL

Tie to Prior Knowledge

Light at an Interface Remind students that they have learned that light can be absorbed, reflected, or transmitted at the interface between two mediums. Tell students that in this section they will learn that the transmitted light will change direction if it strikes the boundary at an angle. Students will also use their knowledge of sine and inverse sine.

2 Teach

Light and Boundaries and Snell's Law of Refraction

student text pp. 492–494

Concept Development

The Angle of Refraction Have students recall that in the law of reflection the angles are measured from the normal to the surface. Tell students that this is also true with refraction. The angle of refraction at a boundary is measured with respect to the normal on the opposite side of the surface from that of the incident ray.

Critical Thinking

3-D Refracted light rays travel in a plane. Ask students to describe the plane of travel of a refracted ray in a problem involving three dimensions. The plane of travel is affected by the incident ray and the normal to the surface. The refracted ray travels in the same plane. OL

Identify Misconceptions

Direction of Refraction Students may believe that light always bends toward the normal when it enters a material and away from the normal as it exits the material. Explain that the direction in which the light bends depends on the indices of refraction of the two materials. Light bends toward the normal only if the light travels into a medium with an index of refraction that is larger than that of the incident medium.

Reinforcement

Optical Illusion Some beverage glasses are made with extra-thick walls so that they appear to hold more beverage than they really do. Have students work in pairs to make drawings that show how it works.
OL Visual-Spatial EL

ADDITIONAL IN-CLASS EXAMPLE

Use with Example Problem 1 and Table 1.

Problem A ray of light is incident upon a layer of flint glass at an angle of 19.0°. What is the angle of refraction?

Response With the angle of incidence and both indices of refraction given, use Snell's law to find the angle of refraction.

$$n_1 \sin \theta_1 = n_2 \sin \theta_2$$

$$\sin \theta_2 = \left(\frac{n_1}{n_2}\right) \sin \theta_1$$

$$\sin \theta_2 = \left(\frac{1.00}{1.62}\right)(0.33)$$

$$\theta_2 = 11.6°$$

Understanding Concepts

Outlining As students work through this section, have them make an outline that includes the major topics as well as a short summary of each topic in their own words. They may also wish to draw diagrams to explain the topics. Allow class time for students to read and discuss their summaries. OL Linguistic

REAL-WORLD PHYSICS

Atmospheric Refraction Many fundamental principles of spherical and parabolic mirrors and lenses were discovered by Alhazen, a medieval mathematician (965–1039). Alhazen measured the reflection and refraction of light by mirrors and lenses, and determined that the curvature of a lens or mirror accounted for focusing. He developed geometric equations for image formation for spherical and parabolic mirrors. Using his knowledge of refraction, he was able to measure the refraction of the atmosphere. This led to two conclusions: first, that when twilight ends, the Sun is already 19° below the horizon, and second, that the height of the atmosphere is about 16 km, which is about the average height of the troposphere, the lowest layer of the approximately 100-km-high atmosphere.

The Meaning of the Index of Refraction

student text pp. 495–496

Use an Analogy

Bending Light Analogies can help students understand the concept of light changing direction when it passes from one material into another. Have students look carefully at **Figure 4.** Explain that the wave front approaching a region of a higher index of refraction can be compared to two wheels connected by an axle moving from a smooth surface onto a grassy area. When the first wheel touches the grass, it slows down. Because the other wheel is still moving quickly, the direction of the axle rotates toward the grassy area until the faster wheel reaches the grass, where it moves at the same speed as the other wheel. Emphasize to students the limits of this analogy: With light, there are no wheels and axle. Rather, considering the wave nature of light, refraction is a consequence of different parts of a wave front reaching the second medium at different times.

Concept Development

Light Interaction Students have learned that light slows down when it enters a region with a higher index of refraction, but they may not understand what property of a material causes it to have a higher or lower index of refraction. Explain that, as light passes through a material, atoms absorb and usually re-emit the light. Light moves more slowly through the material than through empty space because of the time it takes atoms to absorb and re-emit light. Different types of atoms need different amounts of time for this atom-light interaction. This means that different materials have different indices of refraction.

PHYSICS PROJECT ACTIVITY

Wheelchair Refraction A wheelchair can serve to model refraction. Have a student volunteer hold one wheel as stationary as possible as you move the other wheel forward. Point out the direction in which the chair turns. Repeat the demonstration with the other wheel stationary. Again, note the direction in which the chair turns. BL Visual-Spatial EL

Identify Misconceptions

Transmission and Reflection Ask students whether incident light can be both transmitted and reflected after reaching a boundary between two materials. Some students may mistakenly think that the answer is "no." Explain that, in general, light is both partially transmitted and partially reflected at a boundary. Have students look at a glass window or a pair of eyeglasses worn by someone in the classroom. They will see through the glass, but should also see a faint reflection. BL

Quick DEMO

Broken Pencil

Estimated Time 5 minutes

Materials pencil; large, rectangular, transparent, plastic block

Procedure

1. Hold the pencil behind the plastic block. Have students look at the block and the pencil from the front so that they see the pencil both through the block and over the block.
2. Next, have students move slightly from side to side so that they view the block and pencil from different angles.
3. Ask students to explain why the pencil appears to be broken and why the gap between the pencil seen through the block and over the block changes when you move from side to side. Encourage students to draw a ray diagram to help them understand these observations. The pencil appears to be broken because the perceived position of either the top or bottom part of the pencil depends on the angle at which the light ray from that part of the pencil enters your eye. Above the block, the light ray enters your eye at a different angle than the light ray that emerged from the block. The reason is the light above the block travels in a straight line from the pencil, but the light transmitted through the block is refracted twice. The gap changes as you move from side to side because, as your position changes, so too does the angle between the ray above the block and the ray transmitted from the block.
4. Ask students to explain why the pencil does not appear broken when viewed from a direction perpendicular to the block. If you view the pencil from a direction perpendicular to the block, the angle of incidence is zero degrees. By Snell's law, the angle of refraction must also be zero degrees.

Use the PhysicsLAB

The lab *How Does Light Bend?* directs students to measure the index of refraction of glass using several different angles of incidence. The investigation reinforces the fact that index of refraction depends on the material, not on the angle of incidence.

Total Internal Reflection

student text pp. 496–497

Challenge Demo

MAIN IDEA Have students prepare a demonstration for the class showing total internal reflection in a prism. Have students use a laser pointer to vary the angle of the incident light until the refracted light disappears. Have students make angular measurements of the incident, refracted, and reflected light beams and record their measurements in a table. After taking a few initial measurements, they would ideally show a progression, gradually increasing from a small angle of incidence to beyond a 90° angle. Encourage students to explore 180° deviation of the beam of light. ***CAUTION:*** *Do not look into the beam at any point during the activity.*

AL **Kinesthetic** **EL**

Quick DEMO

Total Internal Reflection in Water

Estimated Time 5 minutes

Materials 1-L plastic soda bottle, laser pointer, plastic tub, water

Procedure Punch a small hole in the side of the soda bottle. Place the bottle at the edge of a table so that the water will flow into the tub, then fill the bottle with water. Darken the room and shine the laser pointer through the bottle from the side opposite the hole, aiming at the hole. Students should be able to view total internal reflection of the red laser light through the water stream.

Critical Thinking

Water-Filled Pipes Have students think about the total internal reflection that they observed in the Quick Demo, *Total Internal Reflection in Water.* Explain that total internal reflection is impossible if the laser light is directed through a water-filled pipe instead. Ask them to explain why. A reasonable explanation is that the refractive index of the pipe material is higher than that of water, in which case total internal reflection would be impossible. Students might check this idea by looking up the index of refraction of the material of which the pipe is made, or even better, if possible, measure it with a clever experiment. **AL**

DIFFERENTIATED INSTRUCTION

Struggling Students Some students might have trouble understanding why light doesn't escape through the sides of optical fibers. Explain that the fiber cladding must have a lower index of refraction than that of the core. Draw a fiber on the board. Show students how only the light that strikes the cladding at an angle larger than the critical angle will be transmitted down the length of the fiber. Demonstrate this by shining a light at the side of the fiber. Students will observe that very little light propagates to the end. If you shine a light at the end so that much of the light strikes at an angle larger than the critical angle, much of the light will propagate to the other end.
BL Visual-Spatial EL

Dispersion of Light

student text pp. 498–499

Quick DEMO

Make a Spectrum

Estimated Time 10 minutes

Materials water, spherical flask, flashlight, cardboard with hole

Procedure Cover the end of the flashlight with the cardboard so that only a narrow beam of light emerges. Fill the flask with water. Shine light toward the flask so the light reflects onto a light-colored wall or other surface. The reflected light should form a spectrum on the wall.

Use Figure 8

Raindrop Refraction Point out that the incident ray of sunlight in **Figure 8** strikes the upper part of the raindrop. A ray that strikes just at the middle of the raindrop will pass through without being refracted or reflected. Have students draw incident rays striking the upper half, middle, and lower half of a raindrop.
OL Visual-Spatial EL

Content Background

Angle of a Rainbow Describe the angle of a rainbow in relation to the direction of the Sun's rays. To see the top of a rainbow, students would look 42° above the ground. Light that raindrops reflect and refract emerges at various degrees. There is a strong peak in intensity at about 42° for red light and 40° for blue light, producing an arc of colors.

Use Figure 9

Rainbow Lighting Point out that the sky is brighter inside the arc of a primary rainbow. This is because many light rays from the Sun reflect at angles of less than 42°. Light of different wavelengths creates the bright area within the rainbow. Little light is reflected beyond 42°. This creates a darkened area, known as Alexander's Dark Band, between the primary and secondary rainbows.

Use the MiniLAB

In *Personal Rainbow,* each student makes his or her own rainbow by spraying water from a garden hose.

Use the PhysicsLAB

In *A Whole Spectrum of Possibilities,* students will analyze refraction data.

3 Assess

Assess the MAINIDEA

Refraction Have students explain how pulling a child's wagon from a sidewalk to the grass is a model for refraction Students should describe how the sidewalk and the grass are two mediums, and as each wheel moves from one medium to the other, the path traveled by the wagon will change.

Check for Understanding

Light at a Boundary Students study reflection and refraction separately, but reinforce that these phenomena usually occur together. Provide students with a sketch of a light ray incident on the boundary of a second medium without showing the reflection and refraction. Have students complete the total diagram showing both reflection and refraction.
BL Visual-Spatial EL

Reteach

Degree of Bending Ask students how the degree of bending of light as it enters and exits a material is related to the index of refraction of the material and the average speed of light in that material. Snell's law describes the relationship: $n_1 \sin \theta_1 = n_2 \sin \theta_2$. According to the definition for index of refraction, the speed of light for the material is $v = \frac{c}{n_2}$.

Reading Checks and Caption Checks

Caption Check, *student text p. 495*
You can tell that the glass has a higher index of refractions, because the light is bent toward the normal as it enters the glass.

Reading Check, *student text p. 496*
There is an inverse relationship between the index of refraction and the speed of light in a medium.

Caption Check, *student text p. 497*
Index of refraction is smaller for warmer air. Due to the inverse relationship between speed of light and index of refraction, the speed of light will be greater near the road surface.

Caption Check, *student text p. 498*
A rainbow would not be seen if you were close enough to touch it. You can see a rainbow because the colors spread out with distance from the water droplets where they are dispersed.

Practice Problems

student text p. 494

1. 26.3°
2. 34.2°
3. 17.0°
4. 1.5
5. The medium is float glass.

Section 1 Review

student text p. 499

6. $n_{water} < n_{liquid} < n_{float\ glass}$, therefore, n_{liquid} must be between 1.33 and 1.52.
7. 1.46
8. 48.4°
9. 1.99×10^8 m/s
10. Yes, because $n_{glass} > n_{water}$. No.
11. float glass because it has a lower index of refraction and would produce total internal reflection
12. because of bending of light rays in the atmosphere; refraction
13. No, it would mean the speed of light in the medium is faster than it is in a vacuum.
14. In the east, because the Sun sets in the west and sunlight must shine from behind you in order for you to see a rainbow.

1 Introduce

Bellringer

Image-Producing Lenses This activity will require a small convex lens with a focal length between 100 and 300 mm, a 15-cm candle (with holder), a small white box, and a ruler.

Create a double-wall cardboard holder for the lens by folding a piece of thin cardboard in half. Create tabs or "feet" on one of the short sides so that the cardboard can stand upright like a book standing on end. Cut a circle in the walls of the top part of the cardboard slightly smaller than the lens. Staple the cardboard below and on both sides of the circle to create a holder for the lens, but leave the top open to slip the lens inside its cardboard holder. Arrange the lens in its holder between the candle and the box, so that an inverted image of the candle appears on the box. Measure the image of the object and screen distances. Diagram and discuss image formation. OL Visual-Spatial EL

Tie to Prior Knowledge

Refraction Through Lenses Review reflection and refraction, and have students think about how light passes through transparent objects. Remind students that light from a candle will pass through a lens. Tell students they will use the concept of refraction to study what happens when light passes through thin, spherical convex or concave lenses. Explain that the equation that relates focal length, image distance, and object distance for lenses is very similar to the mirror equation.

2 Teach

Types of Lenses, Convex Lenses, and Concave Lenses

student text pp. 500–503

Content Development

Focused Image Have students observe the formation of a real image with a convex lens. Pass around several convex lenses. Have students use them to focus the light from an overhead light source onto a piece of white paper. BL Kinesthetic EL

Identify Misconceptions

Complete Images Students might think that the size of a lens determines whether it will create a complete image of an object. They might also think that only a partial image is formed if part of the lens is covered. Explain that a lens of any size will create a full-sized image. Have students first look through a convex lens and observe the image.

On the board, diagram what students are seeing by starting with a single point on the drawing of the object. Draw rays leaving that point in all directions, with some rays reaching the lens. All the rays are bent or refracted and meet at a single point. Next, have students mask part of the lens and observe the object again. If the mask is placed at the right location, the rays will meet at a single point. Have students try different mask locations until they observe the effect.

Modify the board diagram accordingly. Next, you can show that a larger lens will collect more light rays, a smaller lens (or a portion of the lens) fewer rays. On the board diagram, draw rays from a second location on the object and repeat, noting that no matter how small the lens, both points on the object have related image points for the lens. Make sure students understand that masking the lens decreases the number of light rays; the image is still complete. OL Visual-Spatial EL

Use the MiniLAB

With *Lens Masking Effects,* students can explore what an image looks like when part of the lens is covered.

Connect to History

Camera Obscura The principle behind the camera obscura predates Aristotle, but the first significant improvement in the images it could produce came with the addition of a convex lens in the sixteenth century. First, explain how a camera obscura works or demonstrate using a pinhole in a darkened room. Then, have students research the history of a particular scientific or artistic application and prepare a report or presentation. Students may choose to report on people such as Johannes Kepler and Jan Vermeer, who made significant use of the device or to focus on how designs changed over time, up to and including the development of modern cameras. OL Linguistic

Critical Thinking

MAIN IDEA Have students describe how the image changes if an object, which is originally far from a convex lens, is moved slowly closer toward the lens. When the object is far from the convex lens, its image is inverted and reduced in size. When the object reaches a position equal to twice the focal length, the image is the same size as the object. The image size increases as the object moves toward the focal point. Just at the focal point, no image is produced. As the object moves from the focal point closer to the lens, the virtual image is upright and increases in size. Students can draw ray diagrams to help them visualize these relationships. OL

REAL-WORLD PHYSICS

Corrective Lenses Have students use eyeglasses with concave lenses to correct for nearsightedness or convex lenses to correct for farsightedness. Bring in several pairs of convex and concave eyeglasses. Have students compare and contrast what they observe looking through eyeglasses with lenses that are known to be concave or convex. Have students construct a chart correlating the changes observed in a single object with the thickness and shape of each lens. Have students consider factors other than magnification, such as qualitative changes in brightness.
BL Visual-Spatial EL

Careers

Optometrist A student who is interested in optical physics and who enjoys working with people may be interested in a career as an optometrist. An optometrist is a medical professional who diagnoses and treats visual disorders and diseases. Optometrists generally obtain a four-year undergraduate degree followed by graduation from a specialized school of optometry. Students who wish to become optometrists should study chemistry, physics, anatomy and molecular biology. Note that some students might wonder what the difference is between an optometrist and an ophthalmologist. An ophthalmologist is a medical doctor who specializes in treating diseases and defects of the eye. An optometrist specializes in correcting vision through use of glasses and contacts.

Use a Model

Prisms as Lenses Use equilateral prisms to model convex and concave lenses. Show students how to place two prisms upright on a table with the long rectangular faces touching and small triangular faces on the table. Place a piece of white cardboard about 10 cm from one side of the prisms. Have students shine a laser pointer through the prisms starting at the narrow end of one of the prisms, across the wider ends of the prisms and to the narrow end of the second prism. They will see how the pair of prisms bends the light inward like a convex lens.

Next, have students rearrange the prisms, so that they separate the rectangular faces, but leave one corner of each prism touching. Looking down at the prisms, they would see two tents standing next to each other. This arrangement resembles a concave lens. When the students move the laser pointer from left to right between the two prisms they will see how the light bends outward as with a concave lens.
OL Kinesthetic EL

Use the PhysicsLAB

In *Convex Lens Images,* students will explore the characteristics of convex lenses.

Use the PhysicsLAB

In *Convex Lenses and Focal Length,* students will use a convex lens to observe images for various object distances.

Use the MiniLAB

In *Water Lenses,* students will explore how a curved surface of water can act like a lens.

Lens Equations

student text pp. 503–505

Concept Development

Thin Lens Equation Help students to grasp connections among focal length, object distance, and image distance using the thin lens equation. Choose appropriate values for x_o and f, and solve the equation for x_i on the board. Have students predict how x_i would change if either x_o or f were increased or decreased. They should test their predictions by solving the equation for different values. OL Logical-Mathematical

ADDITIONAL IN-CLASS EXAMPLE

Use with Example Problem 2.

Problem A 5.0-cm-tall block is positioned 25.0 cm from a convex lens. The focal length of the lens is 14.0 cm. What are the position, height, and orientation of the block's image?

Response

$$x_i = \frac{fx_o}{x_o - f} = \frac{(14.0\text{ cm})(25.0\text{ cm})}{25.0\text{ cm} - 14.0\text{ cm}} = 31.8\text{ cm}$$

The image is 31.8 cm in front of the mirror.

$$h_i = \frac{-x_i h_o}{x_o} = \frac{-(31.8\text{ cm})(5.0\text{ cm})}{25.0\text{ cm}} = -6.4\text{ cm}$$

The image is 6.4 cm tall and inverted.

Reinforcement

Writing Problems Ensure that students understand how lenses converge or diverge light by having them write problems similar to the Example Problem and Practice Problems. Have partners write problems and discuss concepts. OL

Defects of Spherical Lenses

student text p. 506

Content Background

Spherical Aberration The amount of spherical aberration in a lens depends on the shape of both sides of the lens. A meniscus lens, which has one concave side and one convex side, has the most spherical aberration. Double convex lenses have the least aberration for spherical lenses. In any case, spherical aberration results in a range of positions at which light rays focus, instead of just one. Rays that pass through the center of the lens are focused at one point while those that pass through from the edge of the lens are focused at another point.

3 Assess

Assess the MAIN IDEA

Image Diagrams Have partners use different concave and convex lenses to view objects. Have students make drawings of how the objects look through each lens.

Check for Understanding

Distinguishing Lenses Mount a convex lens and a concave lens so students cannot see the sides or edges of the lenses. Have students, one at a time, look through the convex lens and then through the concave lens. Allow them to move the lenses, but don't tell them which lens is which. Ask students to identify the type of lens and describe how it affects the light that passes through it. BL Visual-Spatial EL

Extension

Lens Power Optometrists use the reciprocal of the focal length (in meters), $P = \frac{1}{f}$, to describe the power of a lens. The power is also known as the number of diopters of a lens. Have students start with the thin-lens equation and show that the image distance can be defined in terms of the lens power by the equation $x_i = \frac{x_o}{Px_o - 1}$, where i stands for image and o for object.

$$\frac{1}{f} = \frac{1}{x_i} + \frac{1}{x_o}$$

$$\frac{1}{x_i} = \frac{1}{f} - \frac{1}{x_o}$$

$$= P - \frac{1}{x_o}$$

$$= \frac{Px_o - 1}{x_o}$$

$$x_i = \frac{x_o}{Px_o - 1}$$

Then have students determine the number of diopters for a lens with a 2.0-m focal length and for a lens with a −0.5 m focal length. For a focal length of 2.0 m, $P = \frac{1}{f} = 0.5$. For a focal length of −0.5 m, $P = \frac{1}{f} = -2.0$.

Reading Checks and Caption Checks

Caption Check, *student text p. 502*

A virtual image is on the same side of the lens as the object. A real image is on the opposite side of the lens from the object.

Reading Check, *student text p. 503*

A concave lens always produces a virtual image because the light rays that pass through the lens diverge and cannot produce a real image.

Caption Check, *student text p. 506*

An achromatic lens is a system of lenses that uses differences in index of refraction to correct for chromatic aberration.

Practice Problems

student text pp. 505

15. position: 16 cm, height: −4.1 cm

16. position: 2.0×10^1 cm, height: 3.4 cm

17. $x_o = x_i = 50$ mm

18. position: 6.2 cm, height: −0.50 cm (inverted image); inverted

19. $x_i = 15$ cm

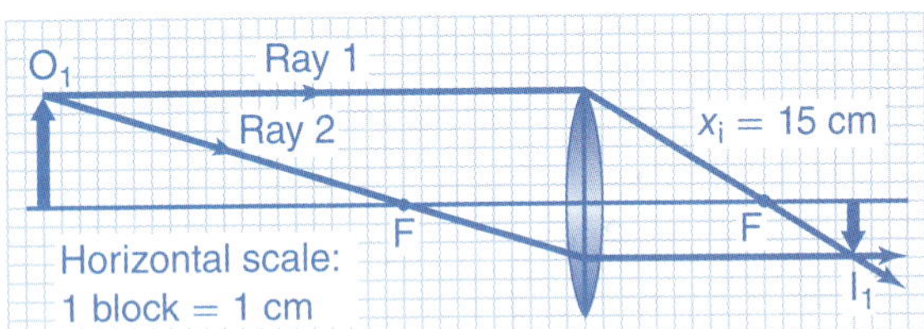

20. The location should be about 15 cm on the same side of the lens (−15 cm), and the image should be upright and about 1.5 cm tall.

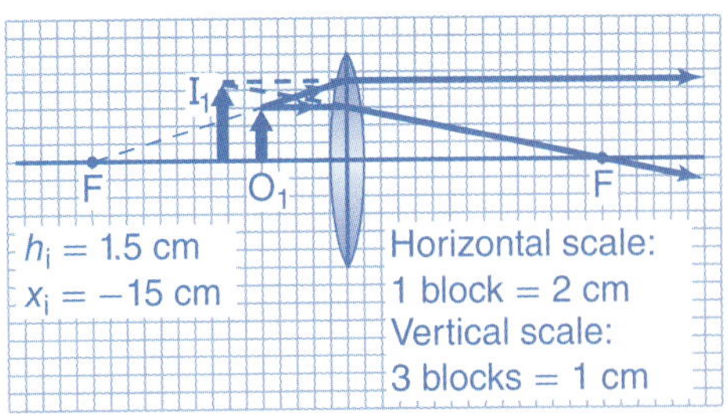

Section 2 Review

student text p. 507

21. If the object is located farther than twice the focal length from the lens, the size of the image is smaller than the size of the object.

22. a. The first and third lenses are converging.

b. The second and fourth lenses are diverging.

23. location: −3.0 cm, size: 4.5 cm

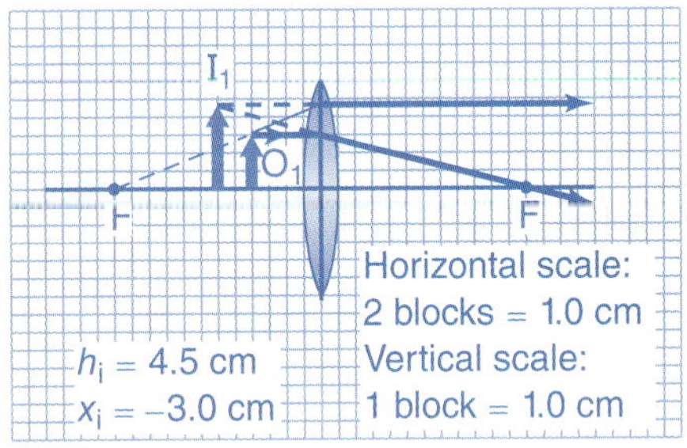

24. −4.0 m; the image is real ($x_i > 0$) and inverted ($h_i < 0$).

25. A diverging lens is a lens that spreads out the light rays that enter it. A diverging lens is typically concave. A converging lens is typically convex and focuses light rays together.

26. The thin lens approximation assumes that all refraction occurs at a plane, called the principal plane, which passes through the center of the lens. This approximation simplifies calculations significantly.

27. All precision optical instruments use a combination of lenses, called an achromatic lens, to minimize chromatic aberration.

28. A single concave lens will always produce a virtual, upright, reduced image.

29. See Online Solutions Manual for ray diagram. $x_i = 20$ cm; $h_i = -24$ cm

30. a. $x_i = 30$ cm, real image

b. $m = -5.0$ cm; The image is larger than the object.

c. $x_i = -20$ cm; virtual

$m = 5.0$; enlarged

31. closer to the lens

32. See Online Solutions Manual for sketch. The light rays will diverge.

1 Introduce

Bellringer

Focusing with Eyes Have students hold a pencil about 10 cm from their eyes and focus on it. Then, have them slowly look to a point at least a meter away and focus. If they do this several times, they will notice that their eyes become very tired. Explain that muscles in their eyes help them to focus on different distances.
BL Kinesthetic EL

Tie to Prior Knowledge

Using Lenses Review with students that they have learned how light refracts when it passes through convex and concave lenses. Tell students that they will learn ways in which lenses are used in daily life.

2 Teach

Lenses in Eyes

student text pp. 508–509

Use Models

Materials a large round flask filled with water, fluorescein dye, a slide projector, a piece of blank paper, eyeglass-type lenses (such as normal, long, long correcting, short, and short correcting)

The Eye Focuses Light Have students use a model of the human eye to observe how the eye can focus incoming light. Set up a model of the human eye using the flask filled with water and enough fluorescein dye to color the water so that a beam of light would be visible. Place the projector about two feet from the flask and turn it on.

Place a normal lens just in front of the flask in the path of light. Adjust the projector so that the cone of light in the flask converges at the back of the flask. Place the piece of paper behind the flask. Using the long lens, show students that the cone of light converges behind the flask. Placing the long correcting lens in front of the long lens should show how the lens makes the cone of light converge at the back of the flask. Repeat the demonstration using the short and short correcting lenses. Guide a discussion about how the eye focuses light and the causes of nearsightedness and farsightedness. OL Visual-Spatial EL

Use Figure 23

MAIN IDEA Point out that the left two images in **Figure 23** are drawn with parallel rays coming from a distant object. The right two images are drawn with nonparallel light rays coming from a nearby object. Ask students why the illustrations are drawn this way. To show what the lenses do: a nearsighted eye cannot focus on faraway objects; a farsighted eye cannot focus on nearby objects. OL

Applying Physics

Lenses Vary in Thickness Explain to students that small variations in thickness near the centers of contact lenses determine how those lenses refract light rays. Have them draw three sketches of the eye with light rays passing through a contact lens with each contact lens having a different thickness. Students should see that the thickness of the center of the lens determines the refraction of the light through that particular lens.
OL Visual-Spatial EL

REAL-WORLD PHYSICS

Correcting Nearsightedness Physicians have several methods for correcting nearsightedness without corrective lenses. Because most cases of nearsightedness are caused by the cornea bulging out too much, these methods are designed to flatten the cornea. In radial keratotomy, a surgeon makes small slits in the cornea in a pattern like the spokes of a wheel. In another technique, the top layer of the cornea is removed, as is a thin slice below it. The top layer is then sewn back on. Finally, high-powered lasers have been used to vaporize cells in the center of the cornea and to sculpt the cornea to correct its shape.

Refracting Telescopes, Cameras, Microscopes, and Binoculars

student text pp. 510–511

Identify Misconceptions

Sharp Images from Telescopes Students often think that the greatest benefit from using a telescope is the magnification of an image. Explain that astronomical objects are so far away that magnification has little effect on our ability to view them. The main benefit from using a telescope is that it increases the amount of light collected from the distant object, which results in a brighter image. Ask students to explain why inexpensive refracting telescopes sold in non-specialty stores do not produce clear images of distant objects, despite having high magnification lenses. Students may suggest that the objective lenses are too small to collect sufficient light to produce clear images, but it is more likely that the inexpensive lenses have spherical aberrations. OL

Concept Development

Telescope Differences Compare the optical properties of Keplerian and Galilean telescopes. In a Keplerian telescope, the object has a focus between the two lens elements, so the final image is inverted. In a Galilean telescope, the final image is not inverted because the incident rays from the top and bottom of the object being viewed do not cross at the focal point. The image, therefore, is upright. The Keplerian telescope has an advantage over the Galilean model because of its wider field of view. Ask students how the problem of inversion of the image can be corrected. Add a correcting lens.

PHYSICS CHALLENGE ACTIVITY

Early Telescopes The telescope was invented by Hans Lippershey in 1608. In 1609, Galileo built a telescope from information he received about Lippershey's work. Through careful modifications, Galileo was able to produce a magnification of 30x. Challenge capable students to make a model of a telescope like the one Galileo used. Provide them with a convex and a concave lens, two nesting tubes, and other materials to secure the tubes and lenses. A Galilean telescope employs a tube having a length equal to the difference between the focal lengths of the two lenses that are mounted inside. The resulting model should use the concave lens as the eyepiece and the convex lens as the objective. AL Kinesthetic EL

Critical Thinking

Microscope Objectives Microscopes typically have a rotating nosepiece that holds two or more objective lenses. Ask students to explain how the magnification of the microscope would change if you changed from a lens with a 16 mm focal length to a lens with a 4 mm focal length. The object would focus at a distance four times as close. Thus, magnification would increase by a factor of 4. OL

Reinforcement

Lens Applications Divide the class into small groups. Have students in each group compare the different applications of lenses presented in this section. BL Interpersonal

3 Assess

Assess the MAIN IDEA

Bifocals Have students look through a pair of bifocals. Students should look carefully through both parts of the bifocals and study the shape of the lenses. Have students explain how the images they see are produced by the top and bottom sections of the lenses. In bifocals, the upper part of the lens is more concave, letting you see things farther away that you might not be able to see; the lower part of the lens more convex, letting you see things closer to your eye that you might not be able to see.

Reteach

Microscope Lenses Bring in a microscope with 2 or more objective lenses, such as 10x, 50x, and 100x. Have students examine the optical system. Allow students to observe a small object using various objective lenses. Demonstrate the use of the iris and immersion oil when viewing objects with the 100x lens.

Extension

Applications of Lenses Have students research other applications of lenses, such as digital cameras, zoom lenses, wide-angle lenses, periscopes, or overhead projectors. Have students draw a diagram or write a brief description of each optical system they research. AL Linguistic

Reading Checks and Caption Checks

Reading Check, *student text p. 508*

The cornea is responsible for most of the light focusing in your eye. The lens fine-tunes the focus and allows you to see both near and far objects.

Caption Check, *student text p. 508*

The difference in the indexes of refraction is greater at the air-cornea boundary than at the boundary entering the lens, so most refraction occurs at the air-cornea boundary.

Caption Check, *student text p. 509*

Different parts of a bifocal lens have different shapes so that, depending on what part the user looks through, the lens can correct for both nearsightedness and farsightedness.

Reading Check, *student text p. 510*

These rays enter the objective convex lens, which focuses them as a real image at the focal point of the objective lens. The image is inverted compared to the object. This image then becomes the object for the convex lens of the eyepiece.

Caption Check, *student text p. 510*

Astronomical objects are observed in relation to other astronomical objects and the relative orientation is not changed if they are inverted.

Caption Check, *student text p. 510*

If the shutter is open longer, more light is allowed to reach the image sensor. This can make it possible to clearly see the image in dim light.

Physics Challenge

student text p. 509

1. 21°
2. 28°
3. Refraction is greater in air because the angle to the normal is smaller. Objects seem closer in water.
4. 22°

Section 3 Review

student text p. 511

33. A nearsighted person should use a concave lens. A farsighted person should use a convex lens.
34. The difference in index of refraction between the air and the cornea is greater than any other difference that light rays encounter when traveling toward the retina.
35. The lens will move closer to the sensor. Real images are always farther from the lens than the focal point. The farther away the object is, the closer the image is to the focal point.
36. You are using the light that strikes only a small area of the object. A brighter lamp could be used.

A CLOSER LOOK

Do you see what I see?

Gravitational Lenses

Purpose

Students will learn how gravitational fields can bend right to produce a lensing effect.

Background

Gravitational lensing gave the first evidence to support Einstein's general theory of relativity. Photographs taken during a total solar eclipse in 1919 showed that the Sun's gravity bent starlight in just the amount predicted by the theory. This discovery made headlines in newspapers all over the world and turned Einstein into a scientific celebrity.

In addition to an Einstein ring, gravitational lensing can produce a series of arcs or an image called an Einstein cross. These various images are the result of skewed alignment of the source, lens, and observer or two or more lenses acting together.

Teaching Strategies

- A gravitational lens can be simulated by using the base of a wine glass. Score an inexpensive wine glass around the narrowest part of the stem. Cover the glass with a cloth and break the stem parallel to the base. Wear gloves and goggles for protection. You can melt a rough break with an acetylene torch.
- Hold the wine glass lens in your hand and shine a penlight through the center of it onto a smooth surface. You will see a bright spot in the center and a ring of light around it.
- For another demonstration, draw a black dot about 5 mm in diameter on a piece of white paper. Place the "lens" flat side down on the paper and slowly move it over the dot. The dot represents a gravitationally lensed source. As the dot approaches the center of the lens, you will see arcs and perhaps the appearance of multiple dots around the center of the lens (Einstein cross). When the dot is in the center, you will see the Einstein ring appearance.

GOINGFURTHER >>>

Expected Outcomes One prediction of general relativity is that clocks run slower as gravity becomes stronger. This effect is tested every day with global positioning satellites. Another prediction of the theory was that the perihelion and aphelion of the planets (the places in their orbits where they are closest to and farthest from the Sun) will change over time. Mercury, the closest planet to the Sun, is the most affected, and this effect was measured for Mercury in 1859. Later it was shown that general relativity "predicted" this measurement. Gravitational red shift, when light tries to escape a large body, was another test, first performed at Harvard in 1960.

CHAPTER 18 ANSWERS

SECTION 1

Mastering Concepts

37. The angle of incidence is larger than the angle of refraction because air has a smaller index of refraction.

38. The angle of incidence is smaller than the angle of refraction because glass has a larger index of refraction.

39. The term *critical angle* refers to the incident angle that causes the refracted ray to lie right along the boundary of the substance when a ray is passing from a region of higher index of refraction to a region of lower index of refraction. If the incident angle exceeds the critical angle, total internal reflection will occur.

40. C > D > B > A

41. The speeds of the different colors of light traveling through air are the same.

42. During a lunar eclipse, Earth blocks the Sun's rays from the Moon. However, sunlight refracting off Earth's atmosphere is directed inward toward the Moon. Because blue wavelengths of light are dispersed more, red wavelengths of light reflect off the Moon toward Earth.

Mastering Problems

43. 1.24×10^8 m/s

44. 20.8°

45. 24.4°

46. **a.** 25.4°

b. 28.9°

47. 1.33; water

48. 49.7°

49. 53°; 1.1 m shallower

50. 1.96×10^8 m/s; 1.99×10^8 m/s

51. 60.8°

52. for red light: 12.0°; for blue light: 11.8°

53. $\sin\theta_2 = 1.09$ is not defined; therefore, total internal reflection occurs.

54. **a.** 28°

b. 32°

c. 53°

55. 13.7°

56. $\theta_g = 28°$; $\theta_c = 41.1°$
When the light ray in the glass strikes the surface at a 62° angle, total internal reflection occurs.

SECTION 2

Mastering Concepts

57. Convex lenses are thicker at the center than at the edges. Concave lenses are thinner at the center than at the edges.

58. The index of refraction of the material from which the lens is made also determines the focus.

59. All lenses have chromatic aberration, which means different wavelengths of light are bent at slightly different angles near their edges. An achromatic lens is a combination of two or more lenses with different indices of refraction that reduce this effect.

60. It is a real image that is located between F and 2F, and that is inverted and smaller compared to the object.

61. Another lens is included in the optics system of the projector to invert the image again. As a result, the image is upright compared to the original object.

Mastering Problems

62. $x_i = 34$ cm

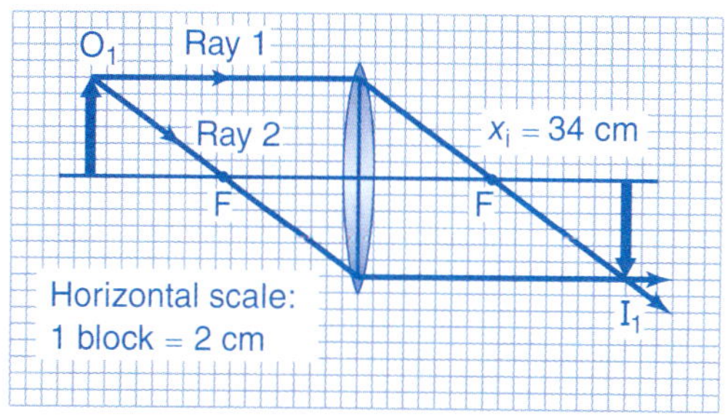

63. The image is 39.3 cm from the lens.

64. 14 cm

65. a. $h_i = -1.8$ cm; $x_i = 10.5$ cm, inverted

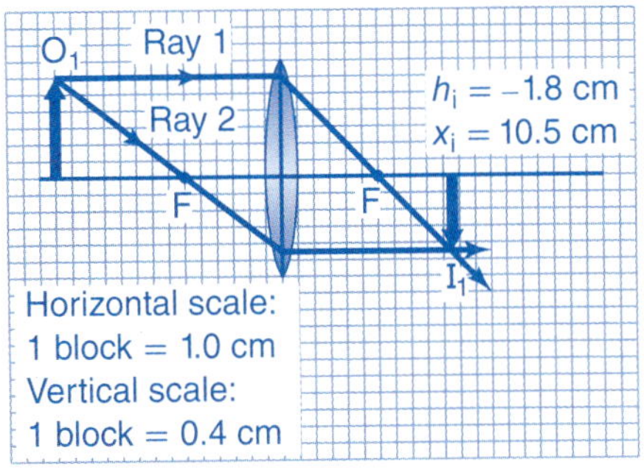

b. −1.8 cm; the image is inverted.

66. a. 6.00 cm

b. Since $x_{i,\,new}$ is negative, the image is virtual and on the same side of the lens as the object. $h_{i,\,new}$ is positive and the image is upright.

67. a. position: 7.5 cm, height: 3.0 cm

b. position: −15 cm, height: 6.0 cm; this is a virtual image that is upright compared to the object.

68. Answers will vary, but a correct form of the answer is, "An object is located 6 cm from a convex lens of focal length 4 cm. Where is the image located?"

SECTION 3

Mastering Concepts

69. Light entering the eye is primarily focused by the cornea. Fine focusing occurs when muscles change the shape of the lens, allowing the eye to focus on either near or far objects.

70. nearsightedness

71. real image, inverted

72. It improves the three-dimensional view.

73. The reflex mirror diverts the image onto a prism so that it can be viewed before taking a photograph. When the shutter release button is pressed, the reflex mirror moves out of the way so that the lens focuses the image onto the film or other photodetector.

Mastering Problems

74. a. 51 mm

b. 1.01×10^3 mm

75. 56 cm

76. 35 mm; for a distant object, x_o can be considered at ∞, thus $1/x_o$ is zero. According to the thin lens equation, $x_i = f$.

77. a. 66.7 cm

b. $-1.67\ h_o$; the copy is enlarged and inverted.

78. a. 6.0×10^1 mm

b. −5.0

c. −20.0 mm, or 20.0 mm beneath the eyepiece

d. -1.0×10^1

79. a. position: 21.0 cm, height: −0.49 cm; this is a real image that is inverted compared to the object.

b. position: -3.2×10^2 cm, height: -4.0×10^1 cm; this is a virtual image that is inverted compared to the object.

c. −4.0

Applying Concepts

80. The angle in substance A is smaller, so it has the larger index of refraction.

81. As the index of refraction of a material increases, the speed of light in that material decreases.

82. Even though Greenland is below the horizon, it is visible as a mirage due to the refraction of light.

83. This illustrates light reflected at angles larger than the critical angle, or total internal reflection.

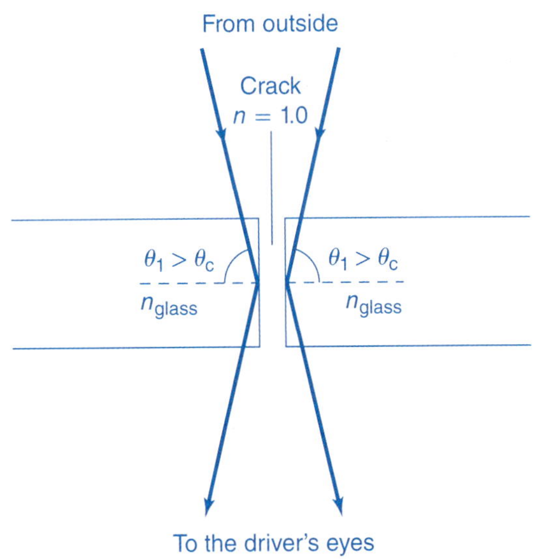

84. You can see a rainbow only when the Sun's rays come from behind you at an angle not greater than 42° with the horizon. In the northern hemisphere, sunlight nearly always originates from the south and is never behind you at an angle that would make it possible for you to see a rainbow.

In the southern hemisphere in the late afternoon, the sun would be setting to the west. You should look to the east to see a rainbow.

85. Blue light travels slower in a prism than red light does.

86. n_2 is the index of refraction for the medium the ray is traveling into. The critical angle will go up until $n_2 = n_1$.

87. Chromatic aberration for lenses is due to the dispersion of light (different wavelengths of light have different speeds in the lens and refract with slightly different angles). Mirrors reflect, and reflection is independent of wavelength.

88. Air and glass have the smaller critical angle of 41.1°. The critical angle for air and water is 48.8°.

89. The location of the image depends on the focal length of the lens and the distance of the object from the lens. Therefore, the location of the image doesn't change.

90. The magnification is much less in water than in air. The difference in the indices of refraction for water and glass is much less than the difference for air and glass.

91. Eyes can focus better in bright light because rays that are refracted into larger angles are cut off by the iris. Therefore, all rays converge at a narrow angle, so there is less spherical aberration.

92. Each side of the binoculars is like a refracting telescope. The objective lens image must, therefore, be between the eyepiece lens and its focal point to magnify the image.

Mixed Review

93. 7 cm

94. 1.07

95. 1.28×10^8 m/s

96. 2.7 min

97. See Online Solutions Manual for full solution. 8.3 mm

98. The angle of incidence must be in air. If we let substance 1 be air, then $n_1 = 1.000$. Let $n_2 = n$. Therefore,

$$n_1\sin\theta_1 = n_2\sin\theta_2;\ \sin\theta_1 = n\sin\theta_2;\ n = \frac{\sin\theta_1}{\sin\theta_2}$$

99. The critical angle for glass to air is $\theta_c = 42°$. Because $\theta_1 > \theta_c$, the light reflects back onto the glass and one cannot see out of an adjacent side. Total internal reflection occurs.

100. direction: 6.7; apparent depth: 8.9 cm; apparent depth/true depth = 0.75

101. 18 mm

Thinking Critically

102. for red light: 24.173°; for blue light: 23.543°; difference = 0.630°

103. 49.8°. In comparison, the critical angle for glass, $n = 1.54$, is 40.5°. The larger critical angle means that fewer rays would undergo total internal reflection in an ice core than in a glass core. Thus, they would not be able to transmit as much light. Fiber optic cables made of glass would work better.

104. Answers will vary, but a correct form would be "… which travels through air until it reaches an interface with a diamond. If the angle of incidence is 5°, what is the angle of refraction?"

105. position: −10 cm, height: 12 cm; the image orientation is not changed.

106. The light that passes through a lens near the edges of the lens is slightly dispersed, since the edges of a lens resemble a prism and refract different wavelengths of light at slightly different angles. The result is that white light is dispersed into its spectrum. The effect is called chromatic aberration.

107. It will get dimmer because fewer light rays will converge, but you will see a complete image.

Writing in Physics

108. Answers will vary. Students may find that it is necessary to simplify their chosen system for explanation purposes.

109. Answers will vary depending on the animals selected by the students.

Cumulative Review

110. 177 times

111. The pitch of the horn that is heard by the pedestrian will decrease as the car slows down.

112. **a.** 1×10^{-6} the value it was originally

b. 1×10^{-6} the value it was originally

c. They both follow the inverse square law of distance.

113. **a.**

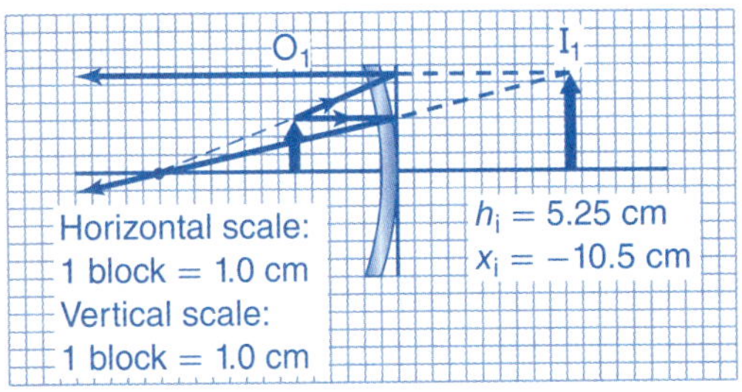

b. position: −10.5 cm, height: 5.25 cm

Standardized Test Practice

student text p. 519

Multiple Choice

1. C

2. D

3. A

4. C

5. B

6. C

7. B

8. D

9. B

Free Response

10. 55.9°

11. $m = -(-2.95\text{ cm})/(6.98\text{ cm}) = 0.423$; a negative image position with an image height that is reduced relative to the object means that the lens is concave.

Rubric

The following rubric is a sample scoring device for free response questions.

Points	Description
4	The student demonstrates a thorough understanding of the physics involved. The response may contain minor flaws that do not detract from the demonstration of a thorough understanding.
3	The student demonstrates an understanding of the physics involved. The response is essentially correct and demonstrates an essential but less than thorough understanding of the physics.
2	The student demonstrates only a partial understanding of the physics involved. Although the student may have used the correct approach to a solution or may have provided a correct solution, the work lacks an essential understanding of the underlying physical concepts.
1	The student demonstrates a very limited understanding of the physics involved. The response is incomplete and exhibits many flaws.
0	The student provides a completely incorrect solution or no response at all.

CHAPTER 19

Interference and Diffraction

About the Photo

Interference in nature Have students examine the photograph. Ask them why a peacock feather is so colorful. Possible answer: different parts of the feather are colored differently. Explain that a real peacock feather will change colors if you rotate it. Point out that simple coloration cannot explain this. Interference of light can explain this phenomenon.

Chapter Overview

This chapter explores the phenomena of interference and single-slit diffraction. Interference occurs when two or more coherent wave fronts combine to produce regions that are bright and other regions that are dark. Single-slit diffraction occurs when light is incident upon a single slit, and the Huygens' wavelets of the incident light interfere constructively and destructively. This produces regions of brightness and darkness.

Before students study the material in this chapter, they should study:

- Fundamentals of waves
- Periodic motion
- The law of reflection

To solve the problems in this chapter, students will need a strong grasp of:

- Scientific notation
- Significant figures
- Sine, cosine, and tangent
- Solving linear equations

Introduce the BIG IDEA

Have a student volunteer blow some large soap bubbles, and use a flashlight to illuminate them. Ask students to describe the colors of the bubbles. The bubbles are multicolored, and the colors on the bubble surfaces change and swirl. Point out that the soap solution is clear. Explain that light waves interact with the bubble surfaces to produce the observed coloration effects.

Go online!
connected.mcgraw-hill.com

Your online experience includes:

- **Video**
- **Concepts In Motion**
- **Virtual Investigations**
- **All Lab, Applying Practices activities, and PBL worksheets**
- **Online Practice**
- **Personal Tutor**
- **Multilingual eGlossary**
- **Presentation**

Use the LaunchLAB

In *Patterns of Light,* students will observe the diffraction of light produced by a compact disc.

Use Physics T.V.

Video

CD "RAINBOWS" Why do DVDs and CDs reflect rainbows? What about oil slicks and insect wings? Investigate the physics of iridescence.

1 Introduce

Bellringer

Interference of Two Coherent Sources Using an audio oscillator (sine-wave generator), place two speakers about 34 cm apart in the front of the room. Adjust the oscillator to 1 kHz and turn it on. Ask students to raise their hands if they can clearly hear the sound. Unless the walls are too reflective, the patterns of destructive interference will be obvious in the pattern of students with raised hands and those without. Repeat this activity with small changes in frequency or speaker separation. You also may allow students to walk around the room and listen to the interference pattern.
BL Auditory-Musical EL

Tie to Prior Knowledge

Waves, Geometry, and Trigonometry This chapter uses what students have learned about the behavior of waves to explain interference and diffraction patterns. Geometry and trigonometry will be used to analyze such patterns mathematically.

2 Teach

Incoherent and Coherent Light and Interference of Coherent Light

student text pp. 522–526

Quick DEMO

Coherence and Incoherence

Estimated Time 15 minutes

Materials ripple tank

Procedure

1. Demonstrate the waves produced by a single source.
2. Demonstrate coherent sources by adding a second source that is oscillating with the same frequency as the first.
3. Demonstrate incoherent sources by adding additional sources that are not at the same frequency.

In step 2, patterns of destructive and constructive interference will be observed. In step 3, rough water will occur.

Use Models

Double-Slit Experiment Use cardboard cutouts to model the effects of a double-slit experiment. First, draw a 2-m-long sine wave on a piece of stiff paper or cardboard. Make sure that the wave is composed of one complete wavelength. Cut along the wave to make two pieces of paper in the shape of waves. Tape the ends a short distance apart on a whiteboard so that the wave portion of each is on top and the waves are in phase. Extend them into the classroom, bringing them together at their ends, in phase. Slide the end of one paper wave down the length of the other paper wave until they are out of phase. The crest and trough locations model the pattern of light and dark bands.

DIFFERENTIATED INSTRUCTION

Visually Impaired While participating in the *Using Models* activity above, students can feel the curve of the transverse paper waves to understand how the two waves in a double-slit experiment can be in phase or out of phase at different locations where the waves intersect. Align the two paper waves so that they are in phase (crest over crest). Ask students what will happen if the two waves are added. The waves experience constructive interference, producing a wave with the sum of the amplitudes. Align the two paper waves so that they are out of phase (crest over trough). Ask students what will happen if the two waves are added. They experience destructive interference, and the waves cancel each other. BL Kinesthetic EL

Identify Misconceptions

The Nature of Light Waves Upon observing double-slit interference fringes, some students may think that the dark fringes are like shadows. Shadows are dark because no light is hitting the areas that are dark. Explain to students that light is landing on the dark fringes. Elicit from students the differences between a shadow and a dark fringe.

Use an Analogy

Bowl Interference The inside of a bowl represents a wave trough. The outside of an upside-down bowl represents a wave crest. When two bowls or upside-down bowls are stacked, they still have a bowl shape, just as two light-wave troughs or crests in superposition still form a trough or crest. When a bowl is filled to its rim, it is flat across the top. This is analogous to a wave in superposition with a wave trough. Students can think of it as the wave crest emptying its contents into the wave trough, filling it until it is a flat line with no wave shape.

Use Figure 6

MAINIDEA In **Figure 6,** the arcs represent wave crests moving outward from the slits. Midway between the crests are the troughs. Along the solid lines, where waves from the two slits overlap crest-to-crest and trough-to-trough along the solid lines, they interfere constructively, creating high light intensity. Along the dashed lines, where the waves from the two slits overlap crest-to-trough, they interfere destructively, creating low light intensity. The wavefronts are not sitting still but are moving from the slits toward the screen. The regions of high and low light intensity will remain fixed, even as wave fronts continue to pass through on their way to the screen. **OL** **Visual-Spatial**

Critical Thinking

Higher-Order Bands The equation for wavelength from a double-slit experiment was developed for the distance between the central bright band and the next bright band. Ask students the following questions: How can the model be extended to find the distance from the first to the second bright band? Is it possible to extend the model to any additional bands? The second bright band is formed when $S_1R = 2\lambda$. Using geometry, the length of P_0P_2 is twice the length of P_0P_1. Therefore, the difference between the first and second bands is again x. By extension, one can see that additional bands are equally spaced, but only as long as the angle remains small. **AL**

PHYSICS PROJECT ACTIVITY

Teacher to Teacher

Explore Differences and Similarities of Patterns Set up two lasers of different wavelengths and a double-slit mask for each of a different slit-separation distance. Aim and adjust the two patterns onto a screen, one above the other. The greater the distance between the screen and slits, the easier it is to see the patterns. Use this to explore how the variables in the setup of a double-slit experiment affect the pattern.

Use the PhysicsLAB

In *What is the Wavelength?* students will investigate the relationship between double-slit interference patterns and diffraction and wavelength.

Use the PhysicsLAB

In *Double-Slit Interference,* students will use a double-slit interference pattern to measure the wavelength of light.

ADDITIONAL IN-CLASS EXAMPLE

Use with Example Problem 1.

Problem When a different light source is used, the separation between the bands in the Example Problem is reduced to 19.5 mm. What is the wavelength of the new source?

Response $\lambda = \frac{xd}{L} = \frac{(19.5\times10^{-3}\text{ m})(1.90\times10^{-5}\text{ m})}{0.600\text{ m}}$

$= 618\text{ nm}$

Reinforcement

Effect of Increasing Slit Separation You can illustrate the effect of slit separation distance on double-slit interference using two identical overhead projector transparencies with concentric, alternating black and clear rings. This works best when the black rings are significantly narrower than the clear ones, and the rings are drawn with the same interval between them. Place the two transparencies on the projector. Use a sheet of paper to cover the bottom halves of the circles on both transparencies. The central dot (smallest circle) represents the origin at the slit. Slowly slide the transparencies away from each other to simulate increasing slit-separation distance. Have students discuss how the patterns of light and dark change. Stop at points where the rings are in phase and out of phase. BL Visual-Spatial

PHYSICS CHALLENGE ACTIVITY

Wavelength and Slit Separation How precisely can a wavelength be measured using two slits? What if the double-slit separation is not known? Suggest that students do a double-slit experiment using a HeNe laser ($\lambda = 632.8$ nm) to find the slit separation, and then use the same setup to determine the wavelength of light from a laser pointer. Point out that if they don't measure the separation between adjacent bands, but between, for example, the tenth band on one side and the tenth on the other, then they can achieve greater precision. Ask students what will limit the precision of the measurement. The ability to find and measure from the center of each band will limit the precision. AL Kinesthetic

Thin-Film Interference

student text pp. 527–530

REAL-WORLD PHYSICS

Nonreflective Eyeglasses Reflection creates glare that prevents television viewers from being able to see the eyes of a newscaster who is wearing eyeglasses without a nonreflective film. Nonreflective eyeglasses reduce the reflection of studio lights off the eyeglass lenses and into the television cameras.

REAL-WORLD PHYSICS

Thin Films How thick can a thin film be? There are two reasons why thin-film interference is not seen in window glass. The light reflected from the back surface must interfere with light reflected from the front surface. Because window glass is 2 or 3 mm thick, it is many times thicker than a thin film. Also, the two surfaces of a thin film are relatively parallel and flat, compared to the two surfaces of window glass. The variations in thickness of the glass and the lack of flatness means the two reflected waves do not undergo constructive or destructive interference.

Concept Development

Thin-Film Interference This interference phenomenon depends on the properties of reflected waves. Review vibrations and waves, describing when reflected waves are inverted relative to the incident wave. Then make the connection with light and the index of refraction. A good mnemonic to remember the rule for wave inversion is, "Low to high: phase change pi; high to low: phase change no."

Use the MiniLAB

In *Soap Film,* students will observe the inference pattern produced by soap film.

Quick DEMO

Thin Films of Air

Estimated Time 5 minutes

Materials plano-convex lens with a long focal length, thick flat glass, microscope slide with cover glass

Procedure

1. Place the convex lens on the flat glass on a dark, nonreflective surface. Next to it, place the microscope slide with the cover glass on top.

2. Illuminate both from the top, either with diffused white light or, preferably, a single color of light.

3. Have students observe both the convex lens and the microscope slide with the cover glass on the top for interference patterns. The lens will produce concentric rings called Newton's rings, which are caused by interference in the thin film of air between the two glass pieces. Pressing down on the lens changes the rings. The patterns from the microscope slide and cover glass will be less regular, but they will be changed even more by gentle pressure on the cover glass.

Content Background

Antireflection-Coated Camera Lenses Even if the surface of a lens is perfectly clean, only 96 percent of the light will be transmitted. In a fine camera, there can be up to 12 lens surfaces, which means that only 61 percent of the original light will reach the film. In the 1930s, Alexander Smakula at Zeiss Optics in Germany developed a practical antireflection coating. Multiple layers of materials such as MgF_2, TiO_2, and SiO_2 and metals such as Au and Cu are used to improve the transmission of light throughout the visible spectrum. As a result, more than 99.5 percent of the light can be transmitted through each surface. This method is often used in the manufacturing of eyeglasses, adding only a small amount to their cost.

ADDITIONAL IN-CLASS EXAMPLE

Use with Example Problem 2.

Problem If the oil film were on glass ($n = 1.55$) rather than water, what would be the minimum film thickness?

Response In this case, there are two phase changes. Therefore, for constructive interference, the minimum thickness is given by $2t = \frac{\lambda}{n_{\text{oil}}}$, which leads to $t = \frac{\lambda}{2n_{\text{oil}}}$. Thus,

$$t = \frac{555\text{ nm}}{2 \times 1.45} = 191\text{ nm}.$$

Extension

Nature's Thin Films As an out-of-class activity, have students research examples of thin-film interference in nature. Research findings will vary. Mussel shells and mother-of-pearl (nacre) have complex, multilayer thin films. Birds, beetles, and butterflies show iridescent colors resulting from both thin films and diffraction gratings, as presented in Section 2. Oil on the surface of tea shows colors. Minerals often have thin layers of transparent material; for example, mica produces interference. **OL** **Linguistic**

3 Assess

Assess the MAIN IDEA

Illumination Turn off the classroom lights and have a student volunteer blow large soap bubbles. Illuminate the soap bubbles using a flashlight with a green or red filter. Ask students to explain why parts of the bubble surfaces appear dark and some appear illuminated. The color bands are interference patterns. Regions that are bright exhibit constructive interference, and regions that are dark exhibit destructive interference.

Reteach

Antinodes and Bright Bands Demo Use a ripple tank (or simulations of a ripple tank available on the Web) to illustrate interference effects, such as double-slit interference. Ask students how the wavelength could be increased and what effect that would have on the interference pattern. Ask students what in the ripple tank is equivalent to the bright bands in a light-interference pattern. The wavelength is increased by decreasing the oscillation frequency. If this is done, the spacing of the nodal lines of destructive interference will be increased. The nodal lines of destructive interference correspond to the dark bands, and the nodal lines of constructive interference correspond to the bright bands.

Extension

Double-Slit Angle Demo Set up a laser and a double slit. Project the pattern on a screen. Ask students what will happen if the double slit is angled such that one slit is closer to the screen than the other. Angling the double slit in such a way will not affect L, because L is very large. However, it will reduce d. Although the actual separation will not change, explain that d is the slit-separation distance perpendicular to the light beam. From $x/L = \lambda/d$, reducing d increases the spacing of bright bands, represented by x.

Reading Checks and Caption Checks

Caption Check, *student text p. 524*

The width of the slits needs to be narrow enough to create a coherent light source.

Practice problems

student text p. 526

1. 418 nm

2. 1.88×10^{-2} m = 18.8 mm

3. 9.66×10^{-6} m = 9.66 μm

4. 0.755 m

Practice problems

student text. p. 529

5. 109 nm

6. 101 nm

7. 196 nm

8. 97.9 nm

9. 95.7 nm

Section 1 Review

student text p. 530

10. When a wave encounters a slit, the wave bends. Light is diffracted by the slits. Light from one slit interferes with light from the other. If interference is constructive, there is a bright band; if destructive, the region is dark.

11.

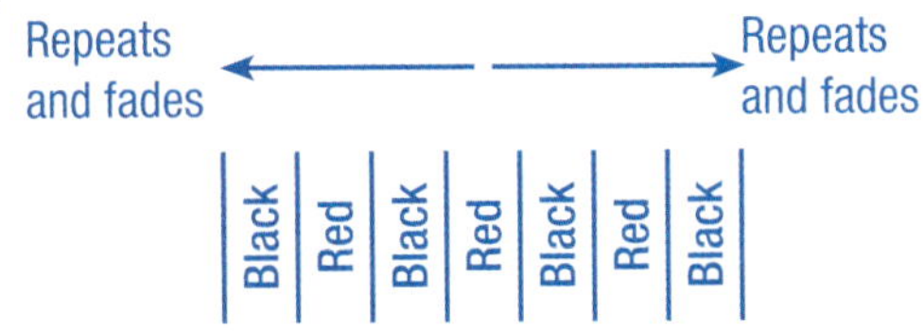

12. The light bands become more closely spaced than when using blue light.

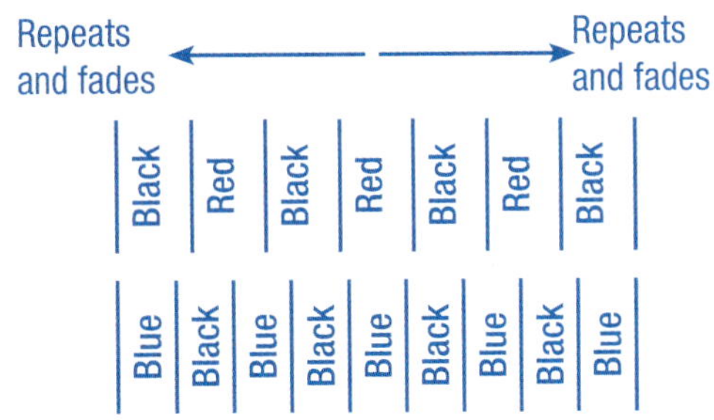

13. a. 324 nm

b. Any odd-integer multiple of 108 nm will produce a bright stripe.

14. 16.3 μm

15. $\sin \theta = \tan \theta$ to two significant digits up to 9.9°. An increase in the precision of the measurement reduces this angle to 2.99°.

1 Introduce

Bellringer

Diffraction Gratings Obtain a set of "rainbow glasses," sold at novelty stores, or a number of diffraction gratings. Have students look through them at sources such as gas-discharge lamps and compact fluorescent lamps. Ask students to describe what they see. Diffraction gratings will produce bright lines of specific colors from discharge tubes, depending on the gas used. **BL** **Visual-Spatial**

Tie to Prior Knowledge

Diffraction of Light Double-slit interference (Section 1) and light diffraction will be used to explain diffraction-pattern phenomena in this section. Lenses, which are considered to be apertures, produce ring diffraction patterns.

2 Teach

Single-Slit Diffraction

student text pp. 531–533

Concept Development

Interference of Huygens' Wavelets Modeling the single slit as a combination of double slits, explore with students the kind of model that might be used if the slit were divided first into four strips. The first and third strips would destructively interfere at the first minimum (dark band), as would the second and fourth. Try eight strips next. The first and fifth, second and sixth, third and seventh, and fourth and eighth would interfere destructively. Finally, explain the model if you divided the slit into 1000 or even 1 million tiny strips. **BL** **Kinesthetic**

Identify Misconceptions

Diffraction or Interference Students might think that the double-slit interference pattern consists only of the alternating bright and dark bands, and that there is no other pattern to it. They will be confused when looking through a double slit at a line filament. **Figure 6** represents the ideal case of double-slit interference, where each slit produces a line source of light. For this to be the case, the slit must have a width on the order of 1 μm. If slits are used that are 10 to 100 μm, then the double-slit interference pattern is still produced; however, the fringes get dimmer, then brighter, then dimmer again, relative to the bands of single-slit diffraction. With finite-width slits, single-slit diffraction is said to modulate double-slit interference.

Reinforcement

Diffraction Pattern Ask students how the width of the central maximum in the single-slit diffraction pattern ($2x_1$) depends on slit width, w, wavelength, λ, and distance to the screen, L. The width of the central crest increases as L or λ increases and decreases as w increases. **OL**

Quick DEMO

Diffraction Pattern

Estimated Time 10 minutes

Materials laser or laser pointer, adjustable slit or two index cards

Procedure Darken the room. Shine the laser light on a white wall or whiteboard. The size of the laser and the spread of light traveling in a straight line can account for the size of the spot on the wall. You can use a blocking card to demonstrate this. Insert the adjustable slit in the light beam and gradually close the slit. Students will first see the spot getting smaller as the slit cuts off the beam. However, just before disappearing, the beam will spread into a diffraction pattern.

Use Models

Connection Between Interference and Diffraction To help students understand single-slit diffraction, use the paper models of waves for two-slit interference patterns. Rather than drawing two lines on the board, draw one wide slit with an even number of lines within it. Have students fasten the wave models to the board at the positions of the two associated lines in the single slit. They should then adjust the angle until the ends of the paper waves destructively interfere. As they move the models to the next pair of strips, the point of destructive interference will shift slightly, but point out that given the tiny size of the slit and the large distance to the screen, the shift is not noticeable in the pattern.
OL **Kinesthetic** **EL**

PHYSICS PROJECT ACTIVITY

Wavelengths of LEDs Have students use either a laboratory spectrometer (if available) or a diffraction grating and metersticks to determine the wavelengths of a number of light-emitting diodes (LEDs). You can purchase various colors of LEDs (red, orange, yellow, green, and blue) at an electronics supply store. Place them in simple circuits to demonstrate their light. A camcorder can detect near-infrared wavelengths, so students can use IR LEDs, including those used in remote controls, in this activity. Students can investigate white LEDs to determine how they produce white light. **OL** **Kinesthetic**

Diffraction Gratings

student text pp. 534–537

Use the MiniLAB

In *Diffraction Gratings,* students will investigate how wavelength affects the diffraction pattern produced by a diffracting grating.

Discussion

Problem How many spectra does a diffraction grating produce?

Response Constructive interference occurs whether the light is diffracted to the left or right, so there is always one spectrum on each side of the central location ($x = 0$ or $\sin\theta = 0$). However, depending on the values of d and λ, there can be more than one on each side. The wavelength is given by $\lambda = (d/m)\sin\theta$, where m – 1, 2, 3, and so on. Spectroscopists often use the $m = 2$ or 3 spectra because the resolution is higher and measurements can be made more precisely. **OL**

Critical Thinking

Grating Selection The equation for the wavelength found by a diffraction grating spectrometer includes $\sin\theta$, which can never be larger than 1. Ask students the following question: When you are working with light containing wavelengths that are known to you, how would you specify the value of d to select the diffraction grating to use, considering that you want to use the $m =$ 2 spectrum? Because $\sin\theta \leq 1$, $\lambda \leq d$. Choose d to be larger than the largest wavelength to be investigated. When $m = 2$, $\lambda \leq d/2$, so d must be larger than twice the largest wavelength. **AL**

PHYSICS CHALLENGE ACTIVITY

MAIN IDEA Using a HeNe laser, have students measure the diameter of a hair. Shine the laser beam onto a hair positioned vertically across a slide holder or stiff card with a 2-cm-square cutout. Project the resulting interference pattern onto a screen and take measurements for the distance from the hair to the screen and the distance between the first dark band on either side of the central bright band. Ask how they can apply diffraction and interference concepts to measuring the width of the hair. Supply the wavelength of the HeNe beam to calculate $\lambda = d\sin\theta$. Also, explain that a narrow barrier, such as a hair, will create the same diffraction pattern as a slit of the same dimensions, so that they still can use the equation for a single bright band.

ADDITIONAL IN-CLASS EXAMPLE

Use with Example Problem 3.

Problem A red laser pointer, $\lambda = 670$ nm, is pointed at a CD. How far from the perpendicular would the spot be on the wall?

Response Use $d = 7.4{\times}10^{-7}$ m and $L = 1.25$ m. Then find $\sin\theta = \frac{\lambda}{d} = \frac{6.70{\times}10^{-7}\ \text{m}}{7.4{\times}10^{-1}\ \text{m}} = 0.92$. The angle that has a sine of 0.91 is 66°. To find the distance on the wall, use $x = L\tan\theta =$ (1.25 m)(2.25) = 2.81 m.

Discussion

Problem Why are the bright bands of a grating diffraction pattern narrower than the bands of a single-slit diffraction pattern?

Response The dark spaces between the bright bands of both the diffraction grating and the single slit are areas of destructive interference. Because the amount of destructive interference with a diffraction grating is greater than with a single slit, the dark areas are larger and hence the bright lines are narrower. **OL**

Quick DEMO

Shimmering Blue Light

Estimated Time 10 minutes

Materials white light source with slit, blue filter, diffraction grating

Procedure Darken the room. Place the filter between the light source and the slit, and direct the light at a wall with walking space in front of it. Have students walk along the wall while observing the slit. Students will see light mostly when they are in front of the slit. Then put a diffraction grating on the opposite side of the slit from the light source and have students walk along the wall while observing the diffraction grating. They will see shimmering blue light.

Resolving Power of Lenses

student text pp. 538–539

Use Figure 20

Light Sources In **Figure 20,** the triangles formed by light rays from the two light sources (objects) that cross in the center of the aperture and then separate to reach their image locations on the screen are similar triangles. Therefore, the ratio of the sides opposite the central angle is equal to the ratio of the altitudes of the triangles. This equality can be rearranged to give $x_{obj}/L_{obj} = x_1/L$. BL

Content Background

Resolution of Parallax Scientists since Copernicus (1473 to 1543) have been aware of the idea that the Sun is stationary and Earth rotates around the Sun once every year. However, telescopes manufactured before the middle of the eighteenth century were unable to detect this motion. This motion was detected in the eighteenth century by measuring the exact location of fixed stars at one point in the year and measuring the exact location again six months later. By this time, the resolution of telescopes improved to the point that the apparent shift in the location of the stars due to Earth's orbit around the Sun was observed.

Use the MiniLAB

In *Retinal Projection Screen,* students will investigate how the retina of the eye can be used as a screen to identify diffraction spectra.

3 Assess

Assess the MAINIDEA

Using a laser pointer, demonstrate the diffraction pattern using an adjustable slit. Ask students to describe what will happen to the width of the bands if the width of the slit is increased. Using the equation $2x_1 = 2\lambda L/W$, students can predict that the bands of light will decrease in width as the slit width is increased.

Check for Understanding

Diffraction Versus Interference Ask students to describe the similarities and differences between two-slit interference and single-slit diffraction. Similarities: Both are the result of the superposition of coherent light waves that have traveled different distances and interfere either constructively or destructively. Differences: With two slits, two sources interfere, and the pattern projected on a screen has evenly spaced regions of light and darkness. With one slit, numerous Huygens' wavelets from different parts of the slit interfere. The pattern on the screen has a wide, bright central crest and dim, unevenly spaced side crests. OL Linguistic

Extension

Resolution Interested students can explore the concept of resolution for invisible electromagnetic waves, which they will study in later physics chapters. For example, the wavelength used to bring satellite TV to homes is about 4 cm. If a satellite dish acts like an aperture with a 40-cm diameter, ask students how far apart two geosynchronous satellites would need to be so that the antenna's resolution could separate the signals. Geosynchronous satellites are about 3.6×10^8 m above Earth's surface, so $x_{obj} = 1.22\lambda L_{obj}/D = (1.22)(4\text{ cm})(3.6\times10^8\text{ m})/(40\text{ cm}) = 4.4\times10^7$ m, a long distance. OL

Reading Checks and Caption Checks

Caption Check, *student text p. 531*

Each color of light produces a central band of the same color as the source. A longer wavelength of light produces a wider band than a shorter wavelength of light.

Reading Check, *student text p. 532*

Destructive interference of Huygens' wavelets creates a dark band on the screen.

Reading Check, *student text p. 533*

The single-slit diffraction equation assumes that L is much greater than w.

Reading Check, *student text p. 535*

More information can be stored on the DVD that is read with a shorter wavelength of light.

Caption Check, *student text p. 536*

The small angle simplification used for the double-slit calculation does not apply for gratings.

Physics Challenge

student text p. 533

1. Use (1) $\lambda = x_1 w/L$,

(2) $v_{substance} = \lambda f$, and

(3) $n_{substance} = c/v$.

Combine (2) and (3).

(4) $n_{substance} = \lambda_{vacuum} f/(\lambda_{substance} f)$

$= \lambda_{vacuum}/\lambda_{substance}$

because the frequency remains constant as the light crosses a boundary. Rewrite (1) in terms of a substance in the space between the slits and the screen.

(5) $\lambda_{substance} = x_{min} w/L$ Combine (4) and (5) and solve for n. $n_{substance} = \lambda_{vacuum}/(x_{mi} nw/L)$

2. $x_1 = \dfrac{\lambda_{vacuum} L}{n_{substance} w}$

$= \dfrac{(6.34\times10^{-9}\text{ m})(1.15\text{ m})}{(1.33)(0.10\times10^{-3}\text{m})}$

$= 5.5\times10^{-3}$ m

$2x_1 = 1.1\times10^{-2}$ m

Practice Problems

student text p. 534

16. 8.6 mm

17. 2.43 m

18. 9.7×10^{-6} m $= 97\ \mu$m

19. 5.90×10^{2} nm

20. a. red, because the width of the central bright band is proportional to wavelength

b. for blue: 18 mm; for red: 25 mm

Practice Problems

student text p. 537

21. A full spectrum of color is seen. Because of the variety of wavelengths, dark fringes of one wavelength would be filled by bright fringes of another color.

22. 9.4×10^{-7} m

23. 0.449 m

24. 490 nm

25. 1.6×10^{3} slits/cm

Section 2 Review

student text p. 539

26. Band spacing is exactly the same as in the pattern produced by the two slits, but now light bands are much thinner and separated by wider dark bands.

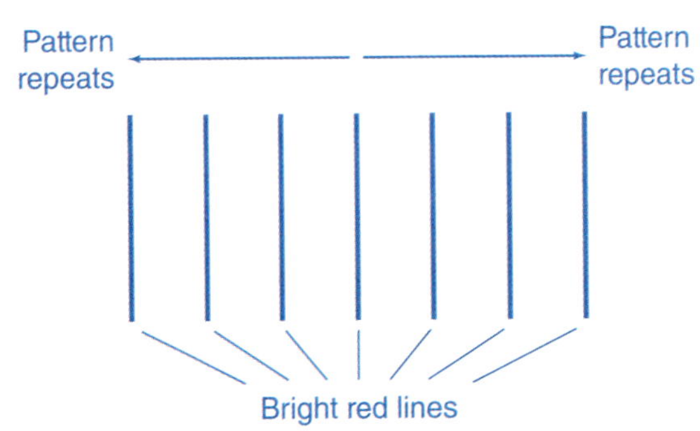

27. 2.2×10^{10} m

28. The first grating has more lines per millimeter. $\lambda \cong xd/L$, so the greater the dot spacings (x) the narrower the slit spacing (d) and thus more lines per millimeter

29. 9.3 mm

30. Determine if the violet or the red end of the spectrum makes the largest angle with the direction of the beam of incident white light. A prism bends the violet end of the spectrum the most, whereas a grating diffracts red wavelengths the most.

FRONTIERS IN **PHYSICS**

Nature's Crime Fighter

Iridescence in Butterflies and Currency

Background

Coloration that is due to optical effects rather than direct reflection of pigments is called structural color. Another example of structural color in nature is the Blue Morpho. These butterflies get their iridescent blue color from interference. Their wing scales have microscopic grooves that are separated from one another by a distance of half the wavelength of blue light. This structural pattern causes light of that particular wavelength to interfere constructively, while all other wavelengths interfere destructively.

Teaching Strategies

- Review with students that polarized light is light whose oscillations are all oriented in the same direction and that a polarizer is a filter that changes ordinary, unpolarized light to polarized light.
- Explain that the emerald color that the Peacock butterfly appears in ordinary light helps it blend in with the tropical foliage of its native Indonesia. Scientists hypothesize that its ability to appear a contrasting blue under different optical conditions could be an adaptation that allows it to hide from predators while being visible to other butterflies whose eyes perhaps detect light differently.
- Have students research other anti-counterfeiting methods and examples of paper currency with security strips or inks that can be detected as genuine or counterfeit under different optical conditions.

GOING**FURTHER** >>>

Expected Outcomes Students might include other examples iridescence on their Web sites, such as the optical effects produced by soap bubbles, thin films of oil, peacock feathers, fish scales, snake skins, mother-of-pearl, shiny beetles, opal, and other butterflies such as the Blue Morpho.

CHAPTER 19 ANSWERS

SECTION 1

Mastering Concepts

31. When monochromatic light is used, you get a sharp interference pattern; if you use white light, you get sets of colored bands.

32. All wavelengths produce the line in the same place.

33. Let the light fall on the double slit, and let the interference pattern fall on a sheet of paper. Measure the spacings between the bright bands, x, and use the equation $d = \lambda L/x$.

34. When the light strikes the front of the film, some reflects off this surface and some passes through the film and reflects off the back surface of the film. When light reflects off a medium with an index of refraction higher than that of the medium it is traveling in, it undergoes a phase shift of one-half wavelength; this happens to the light that initially reflects. In order for a colored band to be produced, two light rays must be in phase. If the thickness of the film is such that the ray reflecting off the back surface goes through a multiple of half of a cycle while passing through the film, the light rays arriving at your eye will be in phase and constructively interfere. Remember that the index of refraction of the film influences wavelength, so the thickness of the film must equal a multiple of half a wavelength of the light, divided by the film's index of refraction.

Mastering Problems

35. 451 nm

36. 94.0 nm

37. thinnest film: 75.8 nm; next thinnest film: 227 nm

38. 61.5 nm, 184 nm, 307 nm

39. A < D < C < B < E

SECTION 2

Mastering Concepts

40. The spacing is directly proportional to the wavelength, and because red light has a longer wavelength than violet, the red lines will be spaced farther apart than the violet lines.

41. The large number of grooves in diffraction gratings increases the intensity of the diffraction patterns. The grooves are close together, producing sharper images of light.

42. Small apertures have large diffraction patterns that limit resolution.

43. Answers will vary, but a correct form of the answer is, "Light of wavelength 530 nm is shone through a slit of width 0.20 mm. If a diffraction pattern is made on a screen 2.00 m away, what is the distance from the center of the pattern to the first minimum?"

44. Answers will vary. A possible form of the correct answer would be, ". . . If resolution were to be diffraction-limited, what would be the smallest angle that could be resolved if the incoming light had a wavelength of 550 nm?"

45. violet light, the color with the smallest wavelength

46. Horizontal scratches produce up and down spectra; vertical scratches produce the left and right. Two sets of scratches at an angle 45° to the horizontal produce the diagonal spectra.

Mastering Problems

47. 600 nm

48. 4.0×10^{-4} cm

49. 0.30 cm

50. 0.4 m

51. 1.1×10^{-2} cm

52. 2.6 cm

53. for red light: 49.3°; for blue light: 30.4°

Applying Concepts

54. a. complete destructive interference

b. complete constructive interference

c. complete destructive interference

55. The lines in diffraction pattern are narrower for the 10^5 lines/cm grating.

56. Shine each laser pointer at the grating, reflecting onto a nearby wall. The color with the longer wavelength will produce spots with a greater spacing on the wall because the spacing is directly proportional to the wavelength. (Mark is correct; red light has a longer wavelength than green light.)

57. A double-slit interference pattern consists of equally spaced lines of almost equal brightness. A single-slit diffraction pattern has a bright, broad central band and dimmer side bands.

58. Less diffraction results from the short wavelength of blue light.

59. a. interference

b. pigments

c. interference

d. refraction

60. The bands get wider and dimmer.

Mixed Review

61. a. 83 ridges/cm

b. 84 ridges/cm

62. a. 5.4×10^{-3} mm

b. 5.4 μm; the pixel is larger than the resolution of the lens.

63. 6.0×10^{2} nm, the light is reddish-orange. For other values of *m,* the wavelength is shorter than that of visible light.

Thinking Critically

64. A green spot at 0°, yellow spots at +30° and −30°, and two blue spots slightly closer in.

65. The angle of diffraction depends on the ratio of slit width to wavelength. Thus you would increase the width to 1.5*w*.

66. 22 km; Diffraction doesn't limit the sensing ability of your eyes, which are limited by aperture at a few hundred meters. More probable factors are the refractive effects of the atmosphere, like those that cause stars to twinkle, or the limitations of the retina and the optic area of the brain to separate two dim sources.

Writing in Physics

67. Student answers will vary. Answers should include Young's two-slit experiment that allowed him to measure the wavelength of light precisely.

68. Descriptions will vary but should include a description of thin-film interference.

69. Student answers will vary. Answers could include the correction of distorted wave fronts with a deformable mirror to improve resolution.

70. Student answers will vary. Answers could include diffraction in telescopes and microscopes, as well as spectroscopy.

Cumulative Review

71. 10 kJ

72. 0.12 m

73. $x_i = -24.0$ cm, $h_i = 4.0$ cm

74. $x_i = -11.7$ cm, $h_i = 3.11$ cm

Standardized Test Practice

student text p. 545

Multiple Choice

1. B
2. D
3. B
4. C
5. B
6. D
7. B
8. A

Free Response

9. 570 nm

Rubric

The following rubric is a sample scoring device for free response questions.

Points	Description
4	The student demonstrates a thorough understanding of the physics involved. The response may contain minor flaws that do not detract from the demonstration of a thorough understanding.
3	The student demonstrates an understanding of the physics involved. The response is essentially correct and demonstrates an essential but less-than-thorough understanding of the physics.
2	The student demonstrates only a partial understanding of the physics involved. Although the student may have used the correct approach to a solution or may have provided a correct solution, the work lacks an essential understanding of the underlying physical concepts.
1	The student demonstrates a very limited understanding of the physics involved. The response is incomplete and exhibits many flaws.
0	The student provides a completely incorrect solution or no response at all.

CHAPTER 20

Static Electricity

About the Photo

Lightning Bolts The effects of static electricity can be quite dramatic, as shown in this photo. Various effects in the atmosphere separate positive and negative charges, building up forces that, when released, create dramatic effects such as lightning. Ask the students in which direction the force on the electrons is acting in this photo. What other effects can affect the path of a lightning bolt? Questions such as this are addressed in this chapter.

Chapter Overview

This chapter introduces the electrostatic force that occurs between charged objects. This force is described by Coulomb's law, which states that the strength of the electrostatic force between point charges is proportional to the magnitudes of the charges and on the inverse square of the distance separating them.

Before students study the material in this chapter, they should study:

- Adding vectors in one dimension
- Adding vectors in two dimensions
- Newton's law of universal gravitation
- Newton's laws of motion
- Vectors v. scalars

To solve the problems in this chapter, students will need a strong grasp of:

- Scientific notation
- Sine, cosine, and tangent
- Solving linear equations

Introduce the BIG IDEA

Newton taught us that mass attracts mass through the force of gravity. In this chapter, you will learn that another property of matter, called its electric charge, can create force. Unlike mass, electric charge comes in two varieties—positive and negative charge. Objects with a like charge will experience a repulsive force and objects with unlike charge will experience an attractive force.

Go online!
connected.mcgraw-hill.com

Your online experience includes:

- **Video**
- **Concepts In Motion**
- **Virtual Investigations**
- **All Lab, Applying Practices activities, and PBL worksheets**
- **Online Practice**
- **Personal Tutor**
- **Multilingual eGlossary**
- **Presentation**

Use the LaunchLAB

In *Electrostatic Force: Another Field Force,* students can observe the forces of attraction and repulsion between static charges.

Use Physics T.V.

Video

LIGHTNING Does your hair ever stand up when you comb or brush it? Have you ever been shocked when petting your cat? Static electricity is everywhere—discover the physics behind the phenomenon.

(l)Comstock/Jupiterimages, (r)NOAA Photo Library, NOAA Central Library; OAR/ERL/National Severe Storms Laboratory (NSSL)

1 Introduce

Bellringer

Squirrelly Tape Before proceeding, please note that most static electric charge demonstrations work best when the indoor atmosphere is very dry. Obtain a dispenser that contains a roll of wide transparent tape. Cut a 25-cm strip from the roll. Dangle the strip with one end stuck to a finger in such a way that everyone can see how the strip moves. Ask students what happens when the strip comes near a surface. The strip bends toward an uncharged surface. This happens because the strip acquires charge as you remove it from the roll and induces an opposite charge when it comes near an uncharged surface.
BL **Visual-Spatial** **EL**

Tie to Prior Knowledge

Force The acceleration of the tape is evidence of the presence of a force. Students should have explored the effect of an applied force on the motion of an object earlier in their studies. The acceleration of a charged object indicates that static electric charge can produce a force.

2 Teach

Evidence of Charge

student text pp. 548–550

Identify Misconceptions

How Charging Occurs The idea that friction between objects is required to give them static charges is not true. What is required is that two different kinds of electric insulators come into contact and then separate. Some of the materials used in this chapter never need to be rubbed to become charged. For example, charge separates when the adhesive backing of transparent tape is peeled away from the nonadhesive surface on the roll. The details of how objects become charged are still an area of active inquiry; most answers are unknown.

Reinforcement

MAINIDEA Have students create a table summarizing their observations with charged strips of transparent tape. The table should relate the behavior of the bottom strip (*B*) and top strip (*T*) to any tests performed and with positive and negative charge conventions. Have students record their observations in the column under either *B* or *T* for each tested object depending on which the object repels. A plastic comb becomes negatively charged when rubbed. Use this example to establish the +/– convention. **OL** **Linguistic** **EL**

Use the PhysicsLAB

In *Static Charge,* students can observe the nature of static charge.

A Microscopic View of Charge

student text pp. 550–551

Critical Thinking

Forces on Protons In a neutral atom, the positively charged protons balance the negatively charged electrons. The protons are packed together inside the nucleus, and like charges have been shown to repel each other. Ask students what must be true of the interproton forces within a nucleus that prevents the protons from flying apart. An attractive force strong enough to overcome electrostatic repulsion must exist within the nucleus. This attractive force can operate only at very small distances. Scientists call this internucleon attraction the strong force. **OL**

Content Background

Elementary Charge Rutherford was the first to propose the concept of atomic structure that we use today—that atoms have a tiny central structure with positive charge surrounded by orbital electrons with negative charge. Robert Millikan, a contemporary of Rutherford's, showed in an experiment with oil drops that charge comes in integral units. In other words, charge is quantized. The electron and proton each carry a single quantum of opposing charge. By definition, the charge of the electron is negative and the charge of the proton is positive. Later in this chapter, students will learn that the value of the elementary charge is 1.602×10^{-19} C.

Quick **DEMO**

Separation of Painted Surfaces

Estimated Time 5 minutes

Materials two small pieces of cardboard, water-soluble paint

Procedure Place a thin coat of wet paint on one surface of a piece of cardboard. Place the unpainted piece of cardboard on the painted piece of cardboard so that the surfaces touch. Separate the two pieces of cardboard. Ask students how this is similar to a transfer of charge between a charged surface and an uncharged surface. Some of the paint would transfer from the original painted surface to the previously unpainted surface. When a charged object comes in contact with an uncharged object, some of the charges can transfer from the original charged surface to the previously uncharged surface. **OL** **EL**

Conductors and Insulators

student text, pp. 551, 552

Use the PhysicsLAB

In *Charged Objects,* students can compare and contrast the ability of materials to acquire and hold positive and negative charges.

Content Background

Conductors and Insulators There are many other examples of substances that act as either electric conductors or insulators, depending upon the conditions. Air is a poor conductor until it is ionized by a strong electric field. Then it becomes plasma—a gas-like state that easily conducts electric charge. In another example, dissolving salt in water increases the conductivity of the solution, even though neither substance is a good conductor on its own.

ACTIVITY

Real-World Insulators and Conductors Bring a variety of small electric parts, such as a piece of cable, a plug adaptor, computer connectors, and a circuit board with ceramic pieces, for the students to observe. Have them identify which aspects of the assemblies are the conductors and which are the insulators. OL Visual-Spatial EL

DIFFERENTIATED INSTRUCTION

Physically Challenged The manipulation of strips of transparent tape may be difficult for physically challenged students. It is possible to make a charge tester with two inflated balloons. Tie each balloon to one end of a 20- to 25-cm-long piece of string. Fasten the other end of the string to a stick. Keep the balloons near each other but not touching. The student charges each balloon by rubbing it all over with a piece of plastic wrap and observes that the balloons repel each other. Before proceeding, the student should touch the balloon all over to remove the charge. The student then rubs one balloon with plastic wrap and the other with a piece of wool and observes that the balloons attract. BL EL

Use Figure 5

Explain that the even distribution of charge on the conductor could be disturbed if someone brought another charged object nearby. Negative charges move relatively easily if an external charge approaches because electrons move very easily. On an insulating surface, although charges are much less free to move, molecules near the surface of the material can become polarized. OL

3 Assess

Assess the MAINIDEA

Imagine bringing a charged metal sphere into contact with an uncharged metal sphere. After separating them, would the force between them be attractive, repulsive, or nonexistent, and why? The force would be repulsive because the charge on the charged sphere would spread over both spheres, so that each sphere would have a net charge of the same type.

Check for Understanding

Type of Charge Ask students whether two uncharged objects can acquire the same type of charge when rubbed together. How could you do a simple experiment to verify your answer? The objects always acquire opposing charges. One object gains negative charge, the other loses it. You can test the charge on each object with a strip of transparent tape freshly removed from a roll. One object will attract the tape, but the other object will repel the tape. OL

Extension

Conductors and Insulators Under certain conditions, charges can move through a substance that is ordinarily an insulator. What is one such example? Lightning. During a lightning storm, what is it about staying inside a car that makes it safe? It is not the rubber tires on a car that protect you, but the fact that charges added to a conductor such as a car frame quickly spread over the outermost surface of the conductor. The reason for this is that like charges repel. By staying on the outer surface of a conductor, the distance between charges is maximized, as required by the repulsive force between them. Therefore, no excess charge comes inside the car, leaving you safe. OL

Reading Checks and Caption Checks

Caption Check, *student text p. 548*

Water molecules are polar and so can carry charge more easily than nonpolar molecules. On a humid day, many water molecules are in the air and can carry away excess charge. On a dry day, fewer water molecules are present and so charge is carried away more slowly.

Reading Check, *student text p. 550*

The bottom piece of tape is negatively charged and so is repelled by the negatively charged comb. The top piece of tape is positively charged, so it is attracted.

Caption Check, *student text p. 552*

Manufacturers coat wires with rubber because rubber is an insulator.

Reading Check, *student text p. 552*

Metals are good conductors because at least one electron on each atom is not tightly bound to the atom. Electrons can move freely throughout the metal.

Section 1 Review

student text p. 552

1. Bring a positively charged glass rod near the two strips of tape. The one that is repelled by the rod is positive.
2. The comb loses its negative charge to its surroundings and becomes neutral once again.
3. Bring an object of known charge, such as a negatively charged hard rubber rod, near the pith ball. If the pith ball is repelled, it has the same charge as the rod. If it is attracted, it may have the opposite charge or be neutral. To find out which, bring a positively charged glass rod near the pith ball. If they repel, the pith ball is positive; if they attract, the pith ball must be neutral.
4. The wool becomes positively charged because it gives up electrons to the rubber rod.
5. An apple contains equal numbers of positive and negative charges, so it is neutral.
6. The glass rod attracts electrons off the metal rod, so the metal becomes positively charged. The charge is distributed uniformly along the rod.
7. Because the copper is a conductor, it remains neutral as long as it is in contact with your hand.
8. The two-charge model can better explain the phenomena of attraction and repulsion. It also explains how objects can become charged when they are rubbed together.

1 Introduce

Bellringer

Electric Torque Suspend a wooden ruler by a string and make it still. Bring a neutral rod near the ruler. Then bring a charged rod near the ruler. Students should observe that the charged rod makes the ruler rotate. Have students discuss how a charged rod might cause the neutral ruler to turn. Assume the rod is negatively charged. Molecules in the ruler become polarized, creating a net positive charge on the near side of the ruler and a net negative charge on the far side of the ruler. This will create an attractive force between the positively charged part of the ruler and the rod, and a repulsive force between the negatively charged part of the ruler and the rod. However, because the positively charged part of the ruler is closer to the rod, the attractive force will be greater than the repulsive force, so the net force is attractive. BL Visual-Spatial EL

Tie to Prior Knowledge

Force and Gravitation You can introduce the electrostatic force by comparing it with the gravitational force. Both are long-range forces. Coulomb's law for the electrostatic force between point charges is similar in mathematical form to Newton's law of universal gravitation. For both, the magnitude of the force decreases with the square of the separation distance.

However, the electrostatic force and the gravitational force have important differences, too. The relative strength of the electrostatic force between common objects can be much larger than the gravitational force. In fact, only celestial-sized bodies, such as the Sun, the Moon, and Earth, are sufficiently large to produce easily observable gravitational force. The gravitational force between ordinary-sized objects is small, and only extremely sensitive equipment can detect such forces. Another difference is that the electrostatic force between two objects can be either attractive or repulsive, whereas the gravitational force between two objects is always attractive.

2 Teach

Forces on Charged Objects

student text pp. 553–556

Concept Development

Net Force Because forces are vectors, when more than one charge exerts a force on another charge, the net force on that charge is the vector sum of the individual forces.

Storing Charge A Leyden jar, the precursor of the capacitor (and a topic for student research), can store electrical energy. A Leyden jar may be used to transport electrical energy in the form of electric charges from one material to another.

REAL-WORLD PHYSICS

Controlling Electrostatic Discharge (ESD) Electronic components, such as those found in computer components, are vulnerable to damage from discharges of static electricity. People might not be aware that they often carry enough electric charge to damage such equipment. Technicians who work with sensitive components or explosives use antistatic mats and wear special metal wrist straps that provide a path for charge to flow quickly to ground. This prevents charge from accumulating on the body. To ensure that a person grounded with a wrist strap won't accidentally be electrocuted, a high-value resistor is placed between the wrist and the ground.

PHYSICS PROJECT ACTIVITY

Humidity and Electrostatic Discharge (ESD) Which atmospheric conditions are worst for electronic devices that are vulnerable to damage from electrostatic discharge (ESD)? low-humidity conditions First have students study the mechanisms by which equipment can be damaged by ESD. Then have them devise a procedure for evaluating the daily ESD threat to equipment. One approach may be to keep an electroscope in a certain location along with a consistent procedure for charging a particular object and bringing it near the uncharged electroscope. The degree of the electroscope's response each day could be categorized, along with the relative humidity at that time. Student reports could discuss any possible correlations. OL Kinesthetic

Discussion

Question Earth and its atmosphere act as a giant charge separator. The ground (Earth's surface) carries a negative charge whereas the conductive layer of the upper atmosphere is positive. What mechanism in the atmosphere maintains this global charge separation?

Answer The answer to this question is counterintuitive. It seems as if lightning should discharge Earth because its local effect is usually the discharge of particular clouds where charge imbalance has built up. But it is the worldwide action of lightning in thousands of daily thunderstorms that carries a net negative charge to the ground, leaving net positive charge in the atmosphere as a whole. Without lightning, this global charge imbalance would not be maintained. Air is not a perfect insulator, so it allows charge to drain off slowly. OL

Using Models

Grounding Activity In electric circuits, ground is a vast supply of charge, usually connected by a conductor to Earth, that remains essentially undisturbed regardless of the flow of charge. Any transferred charge will be tiny compared to the vast supply. Use a box full of foam packing peanuts and a charged rod to visualize the nature of ground. A box about 200 cm long by 100 cm wide filled to the top with a few hundred foam peanuts should be enough. When someone brings the charged object near the opening of the box, a few of the peanuts will jump up and stick to the object. But this will be a very small number compared to the total number of foam peanuts in the box. The box still looks full. Likewise, shaking a few peanuts (representing charges) back into the box causes a negligible change to the total in the box. BL Visual-Spatial EL

Content Background

Accumulated Charges Your body can become electrically charged from contact between your socks and the carpet. If you walk across the floor to greet someone, you might end up giving a small shock in addition to a handshake. Static electricity is worse in winter because the air tends to be drier. There are ways to reduce static charge. Avoid wearing the sort of fabrics that readily accumulate charges, such as wool and nylon. Before grasping a metal door handle with your bare hand, touch a key to the metal first to conduct the charge away from your body. You can also reduce the shock sensation by tapping the metal handle with your knuckles first. There may still be a spark, but it will be smaller.

Use the PhysicsLAB

In *Charge It Up,* students can observe the properties of charged objects. They can also observe charges produced by induction and conduction.

Use the MiniLAB

In *Investigating Induction and Conduction,* students can investigate inducing a charge on a neutral object and transferring charges by contact.

Coulomb's Law

student text pp. 557–560

Reinforcement

MAINIDEA How does the electrostatic force vary with distance? The electrostatic force is proportional to the inverse square of the distance between point charges. Coulomb's law provides a method to directly calculate electrostatic forces. However, provided the charge quantities remain constant, students can use ratios to compute the electrostatic force at new positions. Have students do the following series of quick calculations: If two charges separated by 4.0 cm experience a force of 90.0 N, what force will they experience if the distance between them is increased to 12.0 cm? The distance is tripled, so the force is reduced by a factor of $3^2 = 9$ to 10 N. What would the force be if the distance between the same charges decreased to 2.0 cm? The force would increase by a factor of $2^2 = 4$ to 360 N. OL

DIFFERENTIATED INSTRUCTION

Struggling Students To better visualize the force that two fixed point charges exert on a third point charge, students can make the scale drawing suggested by the Problem-Solving Strategies for Example Problem 1. Have them map out on graph paper the positions of each charge in the problem. Then they should sketch the lines of action due to each pair of forces by joining the charges with a straightedge. Finally, have them carefully add arrows of length proportional to the electrostatic force calculated for that pair. If needed, review how students can use the Pythagorean Theorem to find the resultant force. BL Visual-Spatial EL

Quick DEMO

Cup of Charge

Estimated Time 5 minutes

Materials one metal cup, one intact polystyrene cup, one polystyrene cup broken into small chips, a Van de Graaff generator

Procedure Place equal amounts of polystyrene chips in each cup. Ask students to predict what will happen when you place the cups on top of the Van de Graaff generator. Ask them to explain the differences observed. The chips remain in the metal cup but fly out of the polystyrene cup. Probe student explanations with questions. For example, students might think that charge is transferred through the polystyrene cup to the polystyrene scraps. How can this be if polystyrene is an insulator? The extra negative charges on the Van de Graaff generator repel negative charge in the chips, which become polarized. They then accumulate some extra negative charge from the inside of the cup because of their (local) excess positive charge. Once the chips accumulate sufficient negative charge, they are repelled by the excess negative charge on the Van de Graaff generator. Does this mean that polystyrene is actually a better conductor than metal? No. Students might also hypothesize that charge only accumulates on the outside of the metal cup. How could students further test this hypothesis? Place the metal cup on an insulating surface and connect the inside of the cup to the Van de Graaff generator with a wire. Now charge the inside of the cup with the Van de Graaff generator and observe the chips. Will charge eventually begin to accumulate on the inside of the metal cup, given enough time? No, charges will only accumulate on the outside of the cup because this is the lowest-energy state. **OL**

ADDITIONAL IN-CLASS EXAMPLE

Use with Example Problem 1.

Problem Use the configuration of charges shown in Example Problem 1 but move sphere C to a position that is 5.0 cm directly below sphere B and let the charge on C be +2.0 μC. Ask students to find the net force on sphere B.

Response Determine the force from sphere C on sphere B.

$$F_{\text{C on B}} = \frac{Kq_Bq_C}{r_{BC}^2}$$

$$= \frac{\left(9.0\times10^9\ \frac{\text{N}\cdot\text{m}^2}{\text{C}^2}\right)(3.0\times10^{-6}\ \text{C})(2.0\times10^{-6}\ \text{C})}{(5.0\times10^{-2}\ \text{m})^2}$$

$$= 2.2\times10^1\ \text{N}$$

Spheres C and B have opposite charges so they experience an attractive force. Therefore, the force of C on B is downward. The net force F_{net} on sphere B is the vector sum of $F_{\text{A on B}}$ and $F_{\text{C on B}}$, so the magnitude of F_{net} is

$$F_{\text{net}} = \sqrt{F_{\text{A on B}}^2 + F_{\text{C on B}}^2}$$

$$= \sqrt{(1.0\times10^2\ \text{N})^2 + (2.2\times10^1\ \text{N})^2}$$

$$= 1.0\times10^2\ \text{N}.$$

Determine the angle of the force:

$$\tan\theta = \frac{F_{\text{C on B}}}{F_{\text{A on B}}}$$

$$= \tan^{-1}\left(\frac{2.2\times10^1\ \text{N}}{1.0\times10^2\ \text{N}}\right)$$

$$= 12°$$

$F_{\text{net}} = 1.0\times10^2$ N, 12° below the *x*-axis.

PHYSICS CHALLENGE ACTIVITY

Calculate Charge You can have students suspend two balloons from strings and charge those balloons. Then have students make careful measurements of the balloons' separation distance r and separation angle θ. Ask students to calculate the charge q on each balloon. The force components F_T of the tension in each string are $F_T \cos\left(\frac{\theta}{2}\right) = mg$ and $F_T \sin\left(\frac{\theta}{2}\right) = \frac{Kq^2}{r^2}$, so $q = \sqrt{\frac{mgr^2}{K}\tan\left(\frac{\theta}{2}\right)}$. Use $K = 9.0\times10^9\ \text{N}\cdot\text{m}^2/\text{C}^2$.

The result may be order-of-magnitude accurate.

AL **Kinesthetic** **EL**

Identify Misconceptions

Coulomb's Law Ask students to describe the circumstances when Coulomb's law applies. Coulomb's law applies exactly only if the charges are concentrated at points. If the charge is distributed over a finite-sized object, Coulomb's law applies approximately if each object is very small compared to the distance between the objects.

Applications of Electrostatic Forces

student text pp. 560–561

Reinforcement

Electrostatic Actuator Ask the students to devise an electrostatic actuator, which is a device that converts an electrical signal into a physical action. One possible device consists of two interdigitated metallic combs. When charged with like charge, the combs will repel each other and separate. When charged with opposite charges, the combs will attract each other.

3 Assess

Assess the MAIN IDEA

Imagine a Van de Graaff generator is near a table on which 3 balls with equal negative charge and equal mass are placed. With respect to the Van de Graaff generator, the balls are positioned in a line with the first one a distance *r* from the center of the Van de Graaff generator, the second one a distance 2*r*, and the final one a distance 3*r*. The negative charge on the Van de Graaff generator will exert a repulsive force on the three balls. How will the initial acceleration of the balls be related? Newton's second law gives $F = ma$; Coulomb's law gives $F = Kq_1{}^2q_2{}^2/r^2$. Combining these equations and solving for the acceleration gives $a = Kq_1{}^2q_2{}^2/(mr^2)$. Thus, if the first ball has acceleration a_0, the second ball will have acceleration $a_0/4$, and the third ball will acceleration $a_0/9$.

Reteach

Test Charge Have students picture a $q_1 = -2.0\ \mu C$ charge and a $q_2 = -8.0\ \mu C$ charge that are 6.0 m apart. Ask students to use charge and distance ratios to determine the point where the net force on a $+1.0$-μC test charge would equal zero. The attractive force on the test charge due to the -2.0-μC charge must have the same magnitude as the force due the -8.0-μC charge. This happens on the line connecting the two charges. Because the charge ratio is $\frac{q_1}{q_2} = \frac{-2.0\ \mu C}{-8.0\ \mu C} = \frac{1}{4}$, the distance ratio must be $\left(\frac{x_{T\text{-}q_1}}{x_{T\text{-}q_2}}\right)^2 = \frac{1}{4}$ or $\frac{x_{T\text{-}q_1}}{x_{T\text{-}q_2}} = \frac{1}{2}$ (where $x_{T\text{-}q_1}$ is the distance from charge q_1 to the test charge), or $x_{T\text{-}q_1} = \frac{1}{2}x_{T\text{-}q_2}$. Thus, the distance from the test charge to the -2.0-μC charge will be $\frac{1}{2}$ that to the -8.0-μC charge, or $x_{T\text{-}q_1} = 2.0$ m and $x_{T\text{-}q_2} = 4.0$ m. OL

Extension

Electrostatic Force and Newton's Law A proton has approximately 2000 times more mass than an electron. Ask students what Coulomb's and Newton's laws suggest would happen if it were possible to release a free proton and a free electron in a region without the presence of other charges. The particles would experience attractive forces of equal magnitude (Newton's third law). According to Coulomb's law, this force is $-Ke^2/r^2$ where e is the elementary charge. Because the electron's mass is 2000 times less than that of the proton, the electron's acceleration is 2000 times greater than that of the proton (Newton's second law), so the particles collide much nearer to the proton's original position. AL

Reading Checks and Caption Check

Reading Check, *student text p. 554*
The leaves of the electroscope spread farther apart.

Caption Check, *student text p. 554*
If the magnitude of the positive charge on the rod is less than or equal to the magnitude of the negative charge on the electroscope, the leaves will move closer together as the electroscope is partially or fully discharged. If the magnitude of the charge on the rod is greater than that on the electroscope, the electroscope will become positively charged and the leaves will spread.

Practice Problems
student text p. 560

9. 1.6×10^{4} N attractive

10. 3.0×10^{-6} C

11. The force diagram is reflected about the y-axis with respect to the diagram shown in Example Problem 1. Magnitudes of all forces remain the same. The direction changes to 42° above the negative x-axis, or 138° counterclockwise from the positive x-axis.

12. The electrostatic force between the two charges decreases by a factor of $3^2 = 9$.

13. 0.068 N toward the right

14. 3.1 N toward the right

Physics Challenge
student text p. 560

1. $q = m\sqrt{G/K}$

2. The distance does not affect the expression for q because both forces are inversely related to the square of the distance, so the distance cancels out.

3. $q = (8.61\times10^{-11}\ \text{C/kg})(1.50\ \text{kg}) = 1.29\times10^{-10}\ \text{C}$

Section 2 Review
student text p. 561

15. The electrostatic force is proportional to the product of the two charges and inversely proportional to the square of the distance between them. The equation for this relationship is: $F_E = Kq_Aq_B/r^2$.

16. Electrostatic force is directly related to each charge. It is repulsive between like charges and attractive between opposite charges.

17. Electrostatic force is inversely related to the square of the distance between charges. If the distance is tripled, the force will be one-ninth as great.

18. The electroscope remains neutral.

19. As the leaves move farther apart, the electrostatic force between them decreases until it is balanced by the gravitational force pulling down on the leaves.

20. Charge separation, caused by the attraction of opposite charges and the repulsion of like charges, moves the opposite charges in the neutral body closer to the charged object and the like charges farther away. The inverse relation between force and distance means that the nearer, opposite charges will attract more than the more distant, like charges will repel. The overall effect is attraction.

21. To charge positively, touch the rod to the electroscope. To charge negatively, bring the rod near the electroscope. Ground the electroscope; remove the ground and then remove the rod.

22. The forces are equal in magnitude and opposite in direction.

23. Some charge on the metal sphere will be repelled to the opposite side from the plastic sphere, making the effective distance between the charges greater than the distance between the sphere's centers.

A CLOSER **LOOK**

When Sparks Fly

Gas Pump Explosions

Background

Gas-pump fires are more common in cool, dry weather. They are also more common when synthetic fibers, such as nylon seat covers, are involved. One solution to static buildup in car seats is to rub upholstery with dryer sheets or a similar product that prevents the buildup of charge due to friction between fabrics.

Teaching Strategies

- Compare the spark at a gas pump to lightning hitting a lightning rod. For both cases of static discharge, help students make the connection between the charged object (the storm cloud, the driver) and the electric conductor (the lightning rod, the gas-pump nozzle).
- Have students research the myth that cell-phone use can lead to gas-pump fires. Have them write a response to a friend that has emailed them about the danger of using a cell phone while pumping gas. Have them dispel the myth and include information about real hazards and safety tips.
- Discuss with students the use of antistatic ground straps for vehicles. These narrow straps hang down from the metal chassis of a car or truck, providing an electrical connection between the vehicle and the road in order to disperse static electricity that might otherwise build up in the car.

GOING**FURTHER** >>>

Expected Outcomes Students should include a description of how static buildup can produce a spark that will ignite gas fumes. They should also provide tips on how to discharge static electricity before refueling and how to avoid becoming recharged during refueling.

SECTION 1

Mastering Concepts

24. No; your hair must become positively charged in order to transfer a negative charge to the comb. Total charge (hair + comb) is conserved.

25. The paper is initially attracted to the comb because the comb causes separation of charge in the paper. The part of the paper pieces with positive charge is attracted to the comb. When the paper touches the comb, some of the excess negative charge in the comb is transferred to the paper. Because their charges are like, the paper is then repelled.

26. Student answers will vary but may include dry air, wood, plastic, glass, cloth, and deionized water as insulators and metals, tap water, and your body as conductors.

27. Metals contain free electrons; rubber has bound electrons.

SECTION 2

Mastering Concepts

28. They have been charged by contact as they rub against other clothes, and thus, are attracted to clothing that is neutral or has an opposite charge.

29. Rubbing the CD charges it. Neutral particles such as dust are then attracted.

30. No; net charge is the difference between positive and negative charges. The coin still can have a net charge of zero.

31. Electrostatic force is inversely proportional to the distance squared. As distance decreases and charges remain the same, the force increases as the square of the distance.

32. Bring the conductor close to, but not touching, the rod. Ground the conductor in the presence of the charged rod; then, remove the ground before removing the charged rod.

33. The $1/r^2$ dependence is only valid for point charges. The two disks could be viewed as a collection of point charges, but in order to calculate the r dependence, one would have to integrate over all point charges. This is only a problem for small separations. If the disks were much farther apart, they would behave like point charges.

Mastering Problems

34. 1.0×10^{-8} N, away from each other

35. 2.5×10^{2} N, toward the other charge

36. 3.2×10^{-19} C

37. 1.6×10^{20} electrons

38. 98 N, east

39. $q_A = 5.2\times10^{-7}$ C; $q_B = 1.5\times10^{-6}$ C

40. **a.** 18 N, right
b. 42 N, left

41. Answers will vary, but a correct form of the answer is, "A 3.0-μC charge is located between a 2.0-μC charge and a 5.0-μC charge, such that it is 0.25 m from the 2.0-μC charge and 0.45 m from the 5.0-μC charge. What is the net force on the -3.0-μC charge?"

42. 2×10^{5} C

43. Answers will vary. A possible form of the correct answer would be, "… and is placed 3.5 cm from another sphere with a charge of 2.1 μC. What is the magnitude of the electrostatic force they exert on each other?"

44. A > B = C > D > E

Applying Concepts

45. The distance would have to be decreased such that $r^2 = \frac{1}{3}$, or 0.58 times as far apart.

46. 2.32 N

47. Gravitational forces only can be attractive. Electrostatic forces can be either attractive or repulsive, and we can sense only their vector sums, which are generally small. The gravitational attraction to Earth is larger and more noticeable because of Earth's large mass.

48. The charge of the proton is exactly the same size as that of the electron but has the opposite sign.

49. Use a known insulator to hold one end of the object against the electroscope. Touch the other end with the charged rod. If the electroscope indicates a charge, the object is a conductor.

50. The natural spheres are initially attracted to the charged rod, but they acquire the same charge as the rod when they touch it. As a result, they are repelled from the rod.

51. The leaves will move farther apart when a positively charged rod is brought near the knob, but they will drop slightly when a negatively charged rod is brought near.

52. The charge in the cloud repels electrons on Earth, causing a charge separation by induction. The side of Earth closest to the cloud is positive, resulting in an attractive force.

53. After changing spheres A and B equally, sphere B is touched to two other equally sized balls that are touching each other. The charge on B will be divided equally among all three balls, leaving one-third the total charge on it.

54. The similar properties are the inverse-square dependence on distance and that the forces are both proportional to the product of two quantities (mass or charge). The difference is that mass has only one sign, so gravitational force is always attractive, whereas there are two signs of charge, so electrostatic forces can be either attractive or repulsive.

Mixed Review

55. 14 N, away from each other

56. 8.2×10^{-8} N, toward each other

57. 5.0×10^{-8} C

58. 6.7×10^{-7} C

59. 1.6×10^{-8} C

Thinking Critically

60. 2.3×10^{39}

61. a. $+2.00$ m on the x-axis

b. $+2.00$ m on the x-axis

62. $F_{net} = 3.7\times10^{2}$ N, 197° from the positive x-axis

63. a. 9.8×10^{-3} N

b. 5.7×10^{-3} N

c. 2.4×10^{-8} C on each ball

64. When the positive ion is exactly in the center of the rods, the force from the top rod exactly balances the force from the bottom rod. Likewise, the forces from the left and right rods exactly balance each other. If the ion moves up or down the closer rod exerts a stronger repulsive force, pushing the ion back toward the center. When the ion moves left or right the closer rod exerts a stronger attractive force, so the ion is pulled away from the center.

65. a. $F_A = 3.7\times10^{2}$ N, away (toward q_T); $F_B = 92$ N, toward (away from q_T)

b.

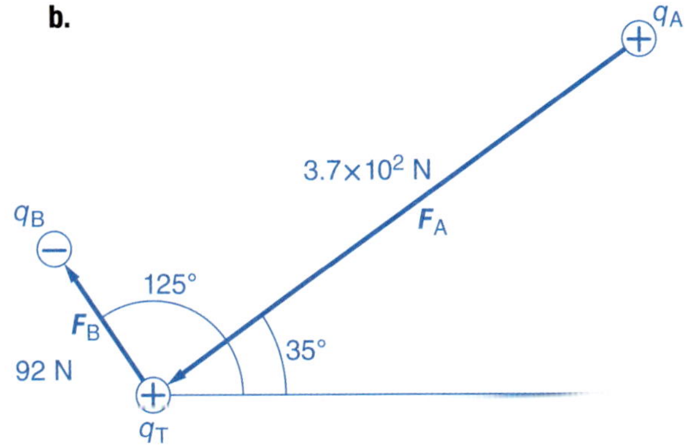

c.

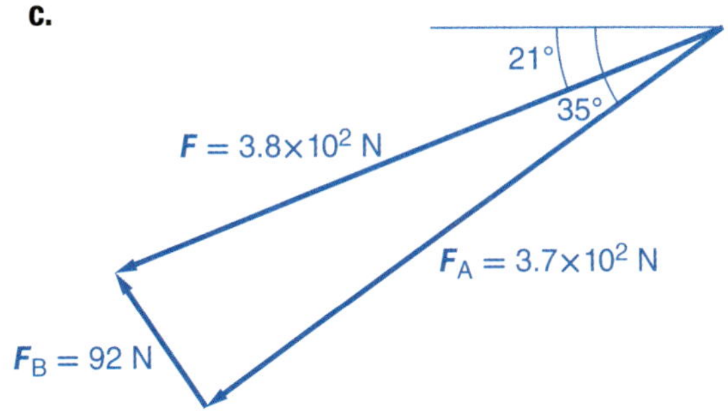

Writing in Physics

66. Student answers will vary but should include information such as the following. The Leyden jar, invented in the mid-1740s, was the earliest capacitor. It was used throughout the eighteenth and nineteenth centuries to store charges for electricity-related experiments and demonstrations. The Wimshurst machine was a device used in the nineteenth and early twentieth centuries to produce and discharge static charges. Wimshurst machines, which were replaced by the Van de Graaff generator in the twentieth century, used Leyden jars to store the charges prior to discharge.

67. Answers will vary, but students should describe the interactions between positive and negative charges at the molecular level. Students should note that the strength of these forces accounts for differences in melting and boiling points and for the unusual behavior of water between 0°C and 4°C.

Cumulative Review

68. Measure the length and period of the pendulum, and use the equation for the period of a pendulum to solve for *g*.

69. **a.** 1380 m

b. 1510 Hz

c. 1520 Hz

70. −48.0 cm

71. position: −10.9 cm; height: 1.09 cm; virtual and upright relative to the object

72. 0.00347°

Standardized Test Practice

student text, p. 567

Multiple Choice

1. D
2. C
3. A
4. B
5. C
6. C
7. C
8. D
9. A
10. B

Free Response

11. $\mathbf{F}_{net} = 0.46$ N in the positive x-direction

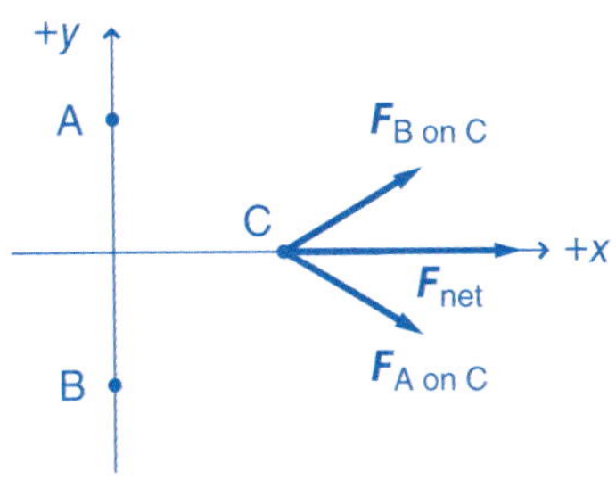

Rubric

The following rubric is a sample scoring device for free response questions.

Points	Description
4	The student demonstrates a thorough understanding of the physics involved. The response may contain minor flaws that do not detract from the demonstration of a thorough understanding.
3	The student demonstrates an understanding of the physics involved. The response is essentially correct and demonstrates an essential but less-than-thorough understanding of the physics.
2	The student demonstrates only a partial understanding of the physics involved. Although the student may have used the correct approach to a solution or may have provided a correct solution, the work lacks an essential understanding of the underlying physical concepts.
1	The student demonstrates a very limited understanding of the physics involved. The response is incomplete and exhibits many flaws.
0	The student provides a completely incorrect solution or no response at all.

CHAPTER 21

Electric Fields

About the Photo

If an electric field is large enough, it can tear electrons off their atoms and create a plasma discharge, as shown in the photo. Once freed of their oppositely charged partner, electrons and ions accelerate across the gap in opposite directions as a result of the force they experience in the electric field. The electric field is strongest near the sharp corners of the electrodes, where the discharge is white.

Chapter Overview

This chapter expands on the concept of electrostatic charge and investigates electric fields. In Section 1, Coulomb's law is applied to the concept of electric field, and students will read about how to model an electric field using electric field lines. In Section 2, students will read about applications of electric fields.

Before students study this chapter, they should study:

- Adding vectors in two dimensions
- Conservation of energy
- Electric charge
- Gravitational potential energy
- Kinetic energy
- Newton's law of universal gravitation

To solve the problems in this chapter, students will need a strong grasp of:

- Graphing data
- Sine, cosine, and tangent
- Slope
- Solving linear equations

Introduce the BIG IDEA

Like the force due to gravity, in which a gravitational field created by an object with mass exerts a force on another object with mass, the electric field created by an object with charge exerts a force on another object with charge. There is a difference, however: there is only one type of mass whereas charge can be either positive or negative. This means the force of gravity can be only attractive, whereas electrical forces can be either attractive or repulsive.

Go online!
connected.mcgraw-hill.com

Your online experience includes:

- **Video**
- **Concepts In Motion**
- **Virtual Investigations**
- **All Lab, Applying Practices activities, and PBL worksheets**
- **Online Practice**
- **Personal Tutor**
- **Multilingual eGlossary**
- **Presentation**

Use the LaunchLAB

In *Charged Objects and Distance,* students can observe how two charged objects interact at a distance.

Use Physics T.V.

Video

STATIC ELECTRICITY Have you ever felt a jolt of static electricity when touching a doorknob? Take a trip to your local science museum to discover the shocking truth about static electricity.

1 Introduce

Bellringer Activity

Forces at Work Charge a 1-m-long plastic tube such as the kind used to cover the shafts of golf clubs by rubbing it with a sheet of plastic kitchen wrap. Place an empty aluminum can on its side on a tabletop and observe what happens when you pass the charged plastic tube over it. Electrostatic forces will cause the can to roll in any direction without touching it with the tube. In this chapter, students will learn that electric fields exert forces that can do work.
BL Visual-Spatial EL

Tie to Prior Knowledge

Forces and Laws In a previous chapter, students read that objects can have a charge. They also learned that there are different kinds of charges—positive and negative. Coulomb's law explained the relationship between the magnitude of force of charges and the distance between the charges. This chapter expands on the idea of forces between charges and explores how they exert this force when they are not touching. Students will apply their knowledge of charges and Coulomb's law to the concept of the electric field.

2 Teach

Defining the Electric Field

student text pp. 570–573

Concept Development

Force Per Unit Emphasize the analogy between the gravitational field and the electric field. Write $g = F/m$ and $E = F/q'$ on the board and explain that the gravitational field is force per unit mass and that the electric field is force per unit charge.

Critical Thinking

Electric Field Strength Have students further consider their results from the LaunchLAB. Ask them if an isolated balloon could be charged by induction. no Ask them to explain what could happen to a charged balloon if they were to touch it. If students were to touch the balloon, the point on the balloon that they touch will discharge by sharing charges with the person touching it. Note that the entire balloon will not discharge because it's made from nonconducting material. OL

Content Background

Electric Fields and Health At the University of Bristol in England, The Human Radiation Effects Group has conducted experiments on a 400-kV 50-Hz power line where the maximum electric field strength 1 m above ground level was about 4 kV/m. They discovered that airborne pollutants get drawn into the electric field and end up concentrating beneath the power lines where they become polarized. The polarity sets up an oscillatory movement that makes the particles "stickier," so they are more likely to adhere to lung tissue. Human immune systems are accustomed to dealing with airborne pollutants, but when airborne pollutants are present in such concentrations and with increased adherence properties, they present a health risk.

Identify Misconceptions

Force v. Field Students might confuse the electric field around a charge with the force on a charge. Clarify that the electric field is a ratio measuring the electrostatic force per unit charge. ($E = F/q'$). Use Example Problem 1 and the In-Class Example to explore how the magnitude of the electric field must have changed if the force on a positive test charge is found to have doubled.

ADDITIONAL IN-CLASS EXAMPLE

Use with Example Problem 1.

Problem An electric field is measured using a positive test charge of 3.0×10^{-6} C. The test charge experiences a force of 0.24 N at an angle of 15° north of east. What are the magnitude and direction of the electric field at the location of the test charge?

Response Use $E = F/q'$, where $q' = 3.0\times10^{-6}$ C and $F = 0.24$ N. Solve for E:

$E = (0.24\text{ N})/(3.0\times10^{-6}\text{ C}) = 8.0\times10^{4}\text{ N/C}$

The test charge is positive, so the force on the test charge and the electric field are in the same direction; 15° north of east.

DIFFERENTIATED INSTRUCTION

Visually Impaired Visually impaired students often fail to realize that some phenomena occur in three-dimensional space. Help students grasp the fact that an electric field is actually three dimensional. Have students make a model of the electric field surrounding a positive charge. They can use modeling clay, toothpicks, pipe cleaners, or other suitable materials. Ask students to explain how the electric field lines are oriented when they present their model. **OL** **Kinesthetic** **EL**

ADDITIONAL IN-CLASS EXAMPLE

Use with Example Problem 2.

Problem What happens to the magnitude of the electric field when you halve the distance to the source charge? Compare the result here to Example Problem 2. Find the magnitude of the electric field at a point 0.15 m to the right of a sphere of charge of -4.0×10^{-6} C.

Response Because the electric field obeys an inverse square law, the electric field should increase by a factor of four if you halve the distance to the source charge. Use

$E=\frac{F}{q'}$ with Coulomb's law $F=\frac{Kqq'}{d^2}$, where $K=9.0\times10^{9}\ \text{N}\cdot\text{m}^2/\text{C}^2$, $q=-4.0\times10^{-6}$ C, and $d=0.15$ m. This gives

$$E=K\frac{qq'}{d^2q'}=K\frac{q}{d^2}=\frac{(9.0\times10^{9}\ \text{N}\cdot\text{m}^2/\text{C}^2)(-4.0\times10^{-6}\ \text{C})}{(0.15\ \text{m})^2}$$

$$=-1.6\times10^{6}\ \text{N/C}.$$

The negative sign indicates that the positive test charge q' experiences a force to the left (i.e., toward the negative point charge q).

Electric Field Demonstration

MAIN IDEA Blow up two balloons. Rub one of them on your hair to give it a negative charge (electrons from your hair are transferred to the balloon). Place the balloon on a nonconducting surface such as a table or the floor. Now charge up the second balloon in the same way and bring it close to, but not touching, the first balloon, so that the first balloon is repelled. Explain to the students that the electric field from the negatively charged balloon in your hand interacts with the negatively charged balloon on the floor or table to create a repulsive force. This demonstration can be repeated with a host of nonconducting objects of varying geometries. **OL** **Visual-Spatial**

PHYSICS CHALLENGE ACTIVITY

Electric Field Mills To avoid possible lightning strikes when launching a vehicle, engineers at NASA use a type of sensor called an electric field mill to assess the magnitude of atmospheric electric fields in and around clouds along the intended launch trajectory. Challenge students to research and build an electric field mill—it is possible using trash cans, cake pans, an electric motor, and other hardware. Make sure students explain how NASA uses electric field mills to determine the magnitude and polarity of the ambient electric field (by measuring the amplitude and phase of current to and from stators). Students can present their electric field mills to the class. **AL** **Kinesthetic** **EL**

Modeling the Electric Field

student text pp. 574, 575

Identify Misconceptions

Direction of Force Students might not understand that field lines, or lines of force, are directed from positive charges to negative charges. Emphasize that a positive test charge in an electric field experiences a force in the same direction as the local electric field. A negative test charge in an electric field, however, experiences a force in the direction opposite the local electric field.

Use an Analogy

Contour Lines Cartographers use latitude and longitude lines to show location, yet those lines do not actually exist; that is, one doesn't find latitude and longitude lines while out walking. Similarly, electric field lines do not exist in the real world. They are used to represent certain properties of an actual electric field.

DIFFERENTIATED INSTRUCTION

Struggling Students Have students draw diagrams to show electric field lines for various point charges. The field lines should lead away from positive charges and toward negative point charges. For two positive point charges, the field lines lead away from both charges, because like charges repel. For distances much greater than the separation between charges, the field lines should resemble those from a single double-positive point charge. For one positive and one negative point charge, the field lines go from the positive charge to the negative charge. For distances much greater than the separation between charges, the field lines should resemble that of neutral point (i.e., there should be no field lines). Field lines never cross one another.
BL Visual-Spatial EL

Van de Graaff Generator

student text p. 576

Use Figure 7

Most students associate the word *field* with a flat surface. Ask students to analyze the photograph of the person touching the Van de Graaff generator in **Figure 7.** They can see the front of the person's head, but what do they imagine the back of the head looks like? Hair is standing up. Ask students what they can deduce about electric fields from this photograph. Electric fields are three dimensional. In this example, they radiate out from the person's head.

Concept Development

Determine Charge Ask students to suggest another way to determine whether a Van de Graaff generator is charged positively or negatively. Answers may vary. For example, if they place the cathode (negative electrode) of a discharge tube near the charged sphere and the tube glows, then the generator is negatively charged. However, if the generator is charged positively, students will have to hold the anode (or positive electrode) toward the sphere to obtain the same glow.

Reinforcement

Earth's Charge At Earth's surface, there is an electric field of almost 150 N/C, directed downward. Ask what sign Earth's charge must have to create an electric field with this orientation. Earth must be charged negatively. OL

3 Assess

Assess the MAIN IDEA

Consider the demonstration with balloons that showed that electric fields can impart forces on charged (or polarized) objects. Ask the students how they could increase the force between the balloons. To increase the force, the balloons could be brought closer together, the charge on one or both balloons could be increased, or both. Ask them to draw the electric field lines emanating from one of the balloons when it is isolated, and from both balloons when they are brought close enough to repel each other. In the former case, the electric field lines will emanate radially from the balloon. In the latter case, the same will be true except that, between the balloons, the field lines will curve away from each other. OL

Check for Understanding

Electric Field Lines Have students draw the field lines for a large, positively charged plate. Consider only the area of the plate far from the edges of the plate. Away from the edges of the plate, the field lines are straight parallel lines that are perpendicular to the plates and point away from the plate on both sides.
BL Visual-Spatial EL

Extension

Electric Field Strength Ask students if there is a limit to how strong an electric field could be. Yes, there is a limit because it takes a collection of charges to produce an electric field. After reaching a certain density, these charges would begin to repel each other, making it impossible to add more charge. OL

Reading Checks and Caption Checks

Reading Check, *student text p. 571*

The variables in bold type are vectors, so it takes two quantities (magnitude and direction) to define them.

Caption Check, *student text p. 574*

The arrows show the direction of the force applied to a positive charge. Because like charges repel and unlike charges attract, the arrows must point away from a positive charge and toward a negative charge.

Caption Check, *student text p. 575*

The copper wires are where the charge is located.

Practice Problems

student text p. 572

1. 4.0×10^{1} N/C
2. 3.0×10^{6} N/C to the left
3. -3.2×10^{-8} C
4. See answers below.

Table 2 Sample Data		
Test Charge Strength (C)	Force Exerted on Test Charge (N)	Electric Field Intensity (N/C)
1.0×10^{-6}	0.30	3.0×10^{5}
2.0×10^{-6}	0.65	3.3×10^{5}
3.0×10^{-6}	0.45	1.5×10^{5}

5. 8.1×10^{-6} N, south
6. 1.6×10^{4} N/C toward q
7. **a.** No; the force on the 2.0-μC charge would be twice that on the 1.0-μC charge.

   b. Yes; you would divide the force by the strength of the test charge, so the results would be the same.

Practice Problems

student text p. 573

8. 2.6×10^{4} N/C
9. 6.5×10^{3} N/C
10. 2.5×10^{4} N/C east
11. -3.1×10^{-9} C
12. 7.7 m
13. Because the strength of the test charge q' and the force F are directly proportional, $F = (Kq/r^2)q'$. Therefore, the electric field, which is ratio of the force to the test charge, is independent of q': $E = F/q' = Kq/r^2$.
14. 7.5×10^{2} N/C
15. 6.4×10^{3} N/C

Section 1 Review

student text p. 576

16. To detect a field at a point, place a test charge at that point and determine if there is a force on it. To determine the magnitude of the field, divide the magnitude of the force on the test charge by the magnitude of the test charge. The magnitude of the test charge must be chosen so that it is very small compared to the magnitudes of the charges producing the field.
17. 6.25×10^{4} N/C east
18. The arrows around a positive charge point away from the charge; the arrows around a negative charge point toward the charge.
19. The field is a property of that region of space and does not depend on the test charge used to measure it. The force depends on the magnitude and sign of the test charge.
20. No; this charge is large enough to distort the field produced by the other charges with its own field.

Applications of Electric Fields

student text pp. 577–587

1 Introduce

Bellringer Activity

Carrying Charge Place an empty coffee can on an insulating platform, such as a polyfoam block, and charge it. Touch a conductor, such as a strip of aluminum foil, to the inside of the can and then to an electroscope. Make sure you hold the foil with an insulator so that you do not ground the foil. Point out that the electroscope leaves remain limp. Repeat; but this time touch the outside of the coffee can. Point out that the leaves now diverge. Ask the class what this tells them about how an object carries a charge. The charge is carried on the outside, not the inside, of an object. **BL** **Visual-Spatial** **EL**

Tie to Prior Knowledge

Potential Energy As in Section 1, students can continue to draw analogies from the force of gravity and gravitational potential energy. Reviewing potential energy will help students grasp the idea of electric potential energy.

2 Teach

Energy and Electric Potential

student text pp. 577–579

REAL-WORLD PHYSICS

Capacitor Safety Remind students that electronic devices are not only easily damaged, but they can do damage as well. Ask students if they have seen any sort of warning labels on computers, radios, televisions, and DVRs. Inform your class that they can still get a shock even when the power is turned off because the capacitors can remain charged. If they touch a charged electrode, they could get a significant shock because not all electrodes are grounded.

Use an Analogy

MAIN IDEA Electric potential difference is analogous to height in gravitational potential energy calculations. In the case of gravity, objects fall from a greater elevation to a lesser elevation, exchanging potential energy for kinetic energy. In the case of electric potential, positive charges move from a greater electric potential to a lesser electric potential (negative charges move in the opposite direction), also exchanging potential energy for kinetic energy. Extend this analogy by pointing out that positive charges move toward a region of lower electric potential the same way a boulder rolls downhill toward a region of lower gravitational potential. Similarly, just as a boulder will roll faster after rolling down a longer hill, charges will move faster after "falling" through a greater electric potential difference.

Concept Development

Electric Potential Difference Ask students whether they can measure electric potential difference at a single point. No; electric potential difference is the difference in electric potential between two points—a measure of the energy required to move one coulomb of charge from one point to the other point. **OL**

REAL-WORLD PHYSICS

Gel Electrophoresis This technique separates DNA fragments by size. Special enzymes cut the DNA where there is a specific sequence of amino acids. The result is a collection of electrically charged fragments of various lengths. DNA is placed at one end of a gelatin-like material and a potential difference is applied across the gel, causing DNA fragments to move to the other end. The larger a fragment is, the more slowly it moves through the gel. The DNA is dyed, and a photograph is taken to show how far each fragment migrates. Ask students, "Suppose you extract DNA from each of 15 snails. You find DNA fragments from 5 of the snails at the same location. What might you conclude?" Five of the snails have similar DNA and are likely related. **BL**

Electric Potential in a Uniform Field

student text pp. 580, 581

Use Figure 11

The electric field between two parallel plates is uniform only if the length and width of the plates are much larger than the distance separating them. Point out that the field is not uniform near the edges of the plates.

DIFFERENTIATED INSTRUCTION

Struggling Students On the board, write *Work = Charge × Electric Potential Difference.* Point out that this formula relates work to charge and potential difference. Ask students what units they would use in the formula. The unit for work is joule; coulomb is the unit of charge; the unit for potential difference is volts. Have students rewrite the formula replacing the variables with the correct units. 1 joule = (1 coulomb)(1 volt).
BL **Logical-Mathematical**

ADDITIONAL IN-CLASS EXAMPLES

Use with Example Problem 3.

Problem Two 30-cm-square charged parallel plates are 4.0 cm apart. The magnitude of the electric field between the plates is 2400 N/C. What is the electric potential difference between the two fields? (You might have to remind some students to convert 4.0 cm to 0.040 m.)

Response First solve for the unknowns. $\Delta V = Ed$

$E = 2400$ N/C

$d = 4.0$ cm $= 0.040$ m

$\Delta V = (2400 \text{ N/C})(0.040 \text{ m}) = 96 \text{ V}$

Problem How much work is required to move a proton from the negative plate to the positive plate?

Response $\Delta V = W/q$, so $W = q\Delta V$. Insert q and ΔV: $q = 1.602\times10^{-19}$ C, $\Delta V = 96$ V

$W = (1.602\times10^{-19} \text{ C})(96 \text{ V}) = 1.5\times10^{-17} \text{ J}$

Millikan's Oil-Drop Experiment

student text pp. 582,583

Quick DEMO

Charged Parallel Plates

Estimated Time 5 minutes

Materials plastic foam plate, aluminum pie plate, wool, foam or plastic cup, pith ball

Procedure Attach the cup to the pie plate; you will use the cup as a handle. Charge the foam plate by rubbing it with wool. Use the pith ball to demonstrate the field near the region you rubbed. Next lay the metal pie plate on the foam plate and touch it with your finger. This gives the pie plate the opposite charge. Use the pith ball to show that there is a field around it. Hold the pie plate parallel to the foam plate. The pith ball will show the field between the plates.

ADDITIONAL IN-CLASS EXAMPLES

Use with Example Problem 4.

Problem An oil drop suspended motionless between two 30-cm-square parallel plates weighs 1.5×10^{-14} N. The parallel plates are 2.4 cm apart, and the potential difference between them is 450 V. What is the charge on the oil drop?

Known:

$\Delta V = 450$ V

$F_g = 1.5\times10^{-14}$ N

$d = 2.4$ cm $= 0.024$ m

$F_e = F_g$

Unknown:

charge on drop, $q = ?$

number of electrons, $n = ?$

Response First use $F_e = qE$ and substitute $E = \Delta V/d$ to solve for q.

$$F_g = qE = \frac{q\Delta V}{d}$$

$$q = \frac{F_g d}{\Delta V}$$

$$q = \frac{(1.5\times10^{-14} \text{ N})(0.024 \text{ m})}{450 \text{ V}} = 8.0\times10^{-19} \text{ C}$$

Problem If the upper plate is positive, how many excess electrons are on the oil drop?

Response Solve using $n = \frac{q}{e}$.

$e = 1.602\times10^{-19}$ C

$q = 8.0\times10^{-19}$ C

$$n = \frac{8.0\times10^{-19} \text{ C}}{1.602\times10^{-19} \text{ C}}$$

$n = 5$ electrons

PHYSICS CHALLENGE ACTIVITY

Applications of Electric Potential Difference Interested students can learn about electrocardiography (EKG), electroretinography, electromyography (EMG), and electroencephalography (EEG). Have students research one of these medical procedures. Ask them to describe how electric potential difference applies to their chosen device or procedure when they present their findings to the class. AL Linguistic

Use the MiniLAB

In *Electric Fields,* students can observe an electric field.

Electric Fields Near Conductors

student text p. 584

Discussion

Question Why is it a good idea to touch a metal pole or a similar conductor before fueling up your car?

Answer It grounds any electric-charge build-up on your body, thus preventing a spark that could ignite the gasoline fumes and cause an explosion. Explain to students that they should not get into or out of a car while fueling the car because sliding on the seats can cause a charge build-up on their bodies. OL

Capacitors

student text pp. 585–587

Concept Development

Total Charge Clarify that the word *charge* in the definition of capacitance refers to the absolute value of charge on either plate. As a complete device, a charged capacitor has zero net charge because the charge on opposing plates is equal in magnitude but has the opposite sign. The opposite charges cancel each other out, leaving the overall device neutral.

Critical Thinking

Earth's Electric Field Remind students that Earth has an electric field. Propose that a lightning storm could be understood as a giant capacitor arrangement. Ask students to explain the parts of this capacitor. Earth acts as one of the charged plates; the clouds make up the other charged plate; the air between them is the insulator (also called the dielectric).

Use the MiniLAB

In *Construct a Capacitor,* students can learn more about capacitors.

PHYSICS PROJECT ACTIVITY

Electric Fields Around Objects Students can use pith balls to explore the electric field around objects that have different shapes. First students charge a piece of polyfoam sheet by rubbing it with a piece of wool. Then they suspend a metal sphere on an insulating thread and touch it to the piece of charged polyfoam. Students should find that the sphere repels the pith ball equally in all directions. Next have students suspend a metal rod from the insulating thread and touch it to the charged polyfoam. Ask students what they observe. The pith ball should be repelled equally along the entire length of the rod, from left and right as well as top to bottom. Near the end of the rod, there may be some nonuniformities in the electric field. BL Kinesthetic

Reinforcement

Capacitors Define a *capacitor* as "two conducting components separated from each other and charged with different amounts of charge (q_1 and q_2)." Ask students how a capacitor is able to store electrical energy. Because the two conducting components of the capacitor are charged with different amounts of charge q_1 and q_2, an electric field must exist between the conducting components. It is this electric field that stores the energy. Typically, q_1 and q_2 have the same absolute value but they are opposite in sign. OL

ADDITIONAL IN-CLASS EXAMPLE

Use with Example Problem 5.

Problem The electric potential difference between a charged sphere and Earth is 76.0 V when the sphere has been charged to 3.8×10^{-4} C. What is the capacitance of the sphere-Earth system?

Response Use $C = q/\Delta V$ and solve for C.

$\Delta V = 76.0\text{ V}$

$q = 3.8\times10^{-4}\text{ C}$

$$C = \frac{3.8\times10^{-4}\text{ C}}{76.0\text{ V}} = 5.0\times10^{-6}\text{ F} = 5.0\ \mu\text{F}$$

Quick DEMO

Potential Difference Across Capacitors

Estimated Time 10 min.

Materials three 1-F capacitors, hand-operated generator, connecting wires with clips, knife switch, battery-operated device such as a small flashlight or radio

Procedure Connect the three capacitors, knife switch, and hand-operated generator in a series circuit. **WARNING: Follow all safety precautions for use of capacitors. Make sure students do not touch bare wires or the capacitor. Make sure that all wires, switches, etc. are properly insulated.**
Turn the hand crank to energize the capacitors. When the capacitors are energized, open the knife switch and disconnect the generator. Connect the battery-operated device in place of the generator. Close the knife switch and observe. The battery-operated device should operate. Ask students why the device is operating. The device is getting electrical power from the energized capacitor. If time permits, connect the capacitors in parallel and observe the results. The battery-operated device should operate.

Use the PhysicsLAB

In *Storing Charge*, students can observe how a capacitor works.

Use the PhysicsLAB

In *Energizing Capacitors,* students can investigate the rate at which different capacitors are charged.

3 Assess

Assess the MAINIDEA

Imagine a test charge placed in a uniform electric field. When the charge is released, it accelerates toward a plate and strikes the plate with 1 J of kinetic energy. Ask the students to find the initial electric potential energy of the charge. Energy is conserved. All the initial electric potential energy is converted to kinetic energy (assigning the potential of the plates to be zero), so the initial electric potential energy of the charge must have been 1 J. Now imagine that the test charge is 1 C. What is the electric potential at the position where the test charge was released? Electric potential is electric potential energy per unit charge. In this case, it is (1 J)/(1 C) = 1 J/C. **OL**

Check for Understanding

Electric Field Have students explain how they can use a positive test charge to measure the magnitude of an electric field at a specific location. Place the positive test charge at the specific location and measure the force on the test charge from the electric field. Then ask what they would need to do to measure the electric potential difference between two points. They would have to measure the work done in moving the charge between the two points. **OL**

Extension

Breakage potential Tell students that a capacitor has a label with a value known as the breakage potential (also known as breakdown voltage or voltage rating). Ask students whether it would be possible to break a capacitor by placing too much charge on it. Yes; excess charge would result in an excessive potential difference (ΔV) between the capacitor's conducting components. If this potential difference is greater than the breakage potential, the resulting electric discharge will break down the dielectric in the capacitor, destroying its function. **AL**

Reading Checks and Caption Checks

Caption Check, *student text p. 577*

Bring the charges closer together. Because the charges are alike, they repel each other, so it takes work to bring them closer together. This work is stored as an increase in the electric potential energy of the system.

Caption Check, *student text p. 578*

Electric potential is defined as the work done moving a positive test charge between two points in an electric field divided by the magnitude of the test charge.

Reading Check, *student text p. 578*

Electric potential difference is the work that must be done on a charge to move it. Mathematically, this is expressed as $\Delta V = W_{on\ q'}/q'$, where $W_{on\ q'}$ is the work done on the charge, and q' is the magnitude of the charge.

Reading Check, *student text p. 582*

The smallest unit of charge is $e = 1.6\times10^{-19}$ C. The only possible charges are integer multiples of e.

Reading Check, *student text p. 584*

The metallic body of a car is an equipotential surface. Thus, the passengers will not feel any forces due to electric fields inside the car, even if a lightning strike causes a large change in the electric potential of the outer surface of the car.

Caption Check, *student text p. 584*

Negative charges repel each other. On the surface of a conductor, this repulsive force causes them to be evenly distributed over the conductor's surface.

Reading Check, *student text p. 584*

The lightning rod provides a low-resistance path by which the lightning strike can reach the ground.

Practice Problems

student text p. 580

21. 3×10^{2} V

22. 2×10^{4} N/C

23. 5.00×10^{2} V

24. 2.94×10^{-2} m

25. 7.9×10^{4} N/C

Practice Problems

student text p. 581

26. 4.5 J

27. 2.1×10^{4} N/C

28. 2.9×10^{-15} J

29. 1.8×10^{-14} J

30. 1.8×10^{7} J

Practice Problems

student text p. 583

31. gravitational force (weight) downward, friction force of air upward; equal in magnitude

32. 3.2×10^{-19} C, 2 electrons

33. 4.0×10^{4} N/C

34. 3.2×10^{-19} C, 2 electrons

Practice Problems

student text p. 586

35. 1.2×10^{23} C

36. the 6.8-μF capacitor, 1.6×10^{-4} C

37. the 3.3-μF capacitor, 1.1×10^{2} V

38. 3.3×10^{-5} C

39. 6.4×10^{-5} C

40. 1.0×10^{-5} F

Physics Challenge

student text p. 587

1. $F = q^2/Cd$

2. 2.6×10^{-4} C

Section 2 Review

student text p. 587

41. Answers will vary; Sample answer: Electric potential is potential energy per unit charge, and it equals the work required to move a test charge to a given location in an electric field.

42. Electric potential energy changes when work is done to move a charge in an electric field. Electric potential difference is the work done per unit charge to move a charge in an electric field.

43. V/m = J/C•m = N•m/C•m = N/C

44. The potential difference should be increased.

45. The drop is electrically neutral.

46. 5.6×10^{-6} C

47. **a.** The potential differences between the spheres is zero.

b. The charge per unit area on each sphere would be the same.

48. The charges on the metal dome produce no field inside the dome. The charges from the belt are transferred immediately to the outside of the dome.

HOW IT WORKS

He♥rt-Shaped Box

AEDs

Background

The heart pumps blood by first squeezing the top chambers and then, a moment later, the bottom chambers. The timing of these two squeezes results from the slight delay that occurs when the electric current generated in the sinoatrial node (the pacemaker) reaches the atrioventricular node, which slows for a moment before passing to heart muscle cells in the lower chambers of the heart. Cardiac arrest can result when the rhythm of these electric pulses becomes chaotic. The electric shock that an AED delivers might restore a normal rhythm, if delivered in time.

Teaching Strategies

- Describe the signs of cardiac arrest in a victim: sudden collapse, loss of consciousness, seizures or lack of movement, inability to respond when shaken, lack of breathing and lack of pulse, and bluish skin.
- Tell students it is important that the person suffering from cardiac arrest not be wet or laying in a puddle while the AED is operating. Ask students why this safety precaution is important. Lead them to conclude that, because water conducts an electric current, the electric shock delivered by the AED could travel to unintended places, causing injury to the victim and to others nearby.
- Have students find out about the Good Samaritan law and discuss how it could apply to AEDs.
- Encourage one or more students to attend an AED training course and present what they learned to the class.

GOINGFURTHER >>>

Expected Outcomes Students should discover through their research that AEDs are most useful where there are lots of people. Maps of AED locations might include schools, workplaces, airports, stadiums, and shopping malls.

CHAPTER 21 ANSWERS

SECTION 1

Mastering Concepts

49. The test charge must be small in magnitude relative to the magnitudes of the charges producing the field and be positive.

50. The direction of an electric field is the direction of the force on a positive charge placed in the field.

51. Electric field lines are used to represent the actual field in space around a charge. The direction of the electric field at any point is the tangent drawn to a field line at that point.

52. The closer together the electric field lines are, the stronger the electric field.

53. a.

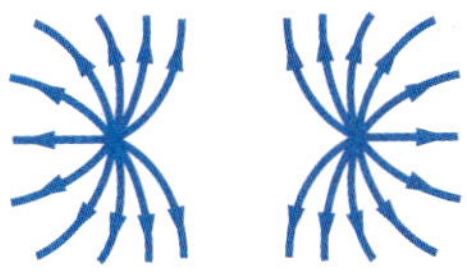

b.

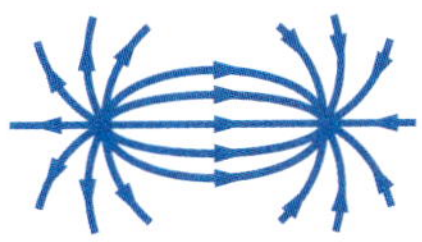

c.

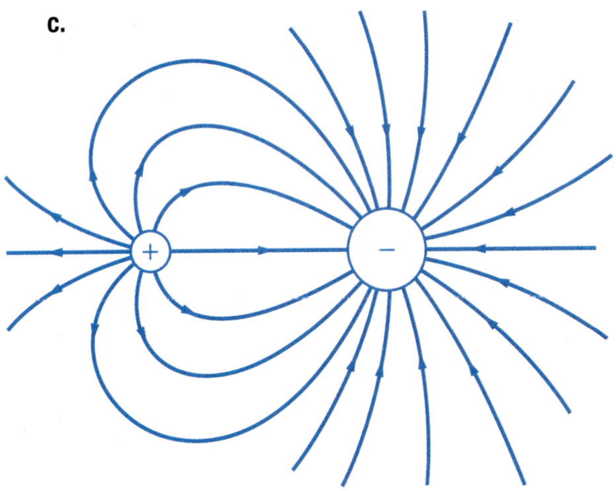

d.

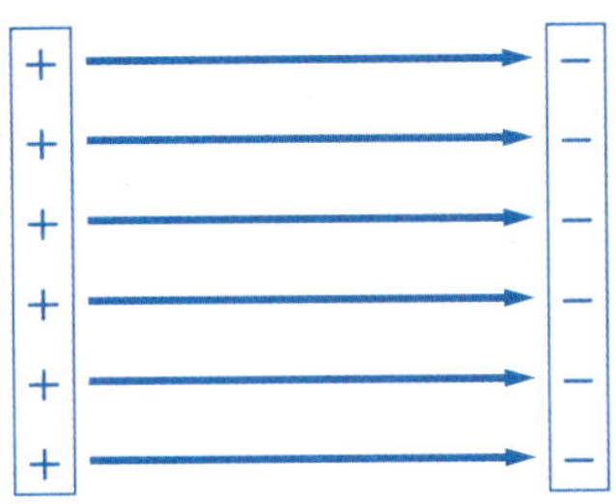

54. They end on distant negative charges somewhere beyond the edges of the diagram.

55. nothing

56. increases

Mastering Problems

57. 2.8×10^{-5} C

58. 6.7×10^{-7} C

59. 1.8×10^{5} N/C

60. 3.0×10^{4} N/C

61. a. upward

b. 2.4×10^{-17} N upward

c. 8.9×10^{-30} N, more than one trillion times smaller

62. a.

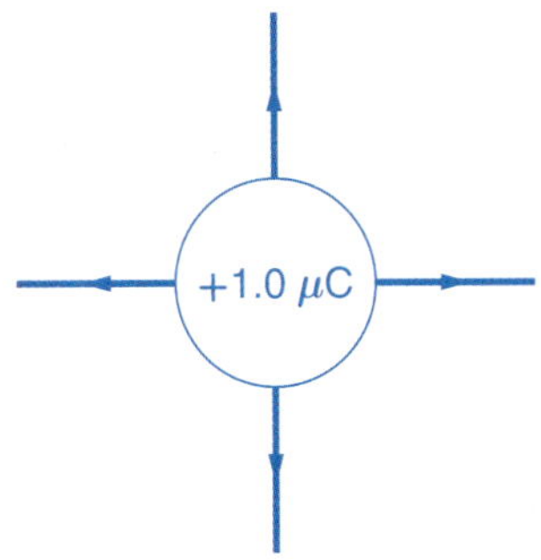

b.

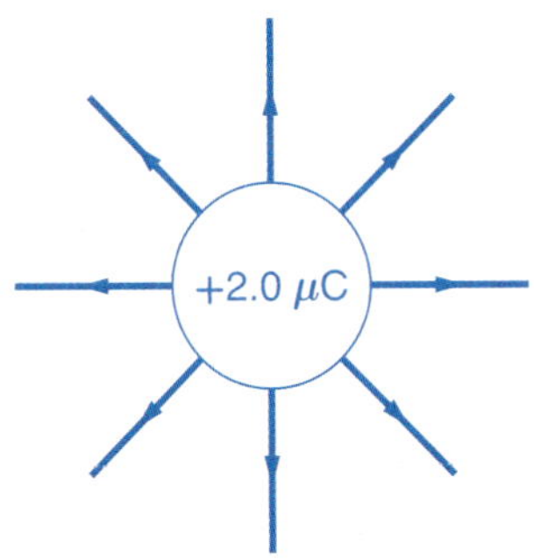

63. 3.0×10^{-4} N

64. 3.497×10^{19} N/C

65. a.

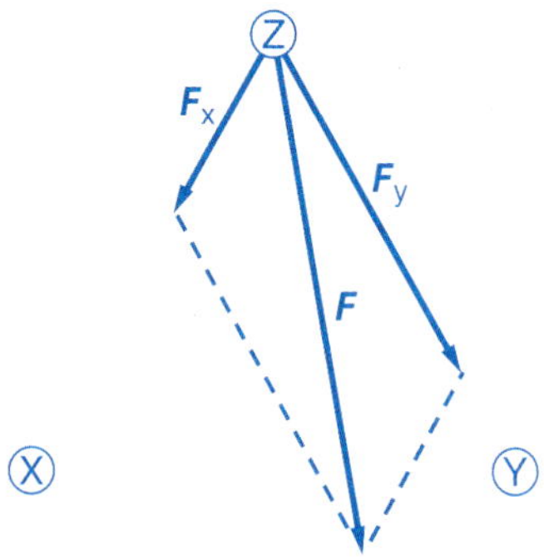

b.

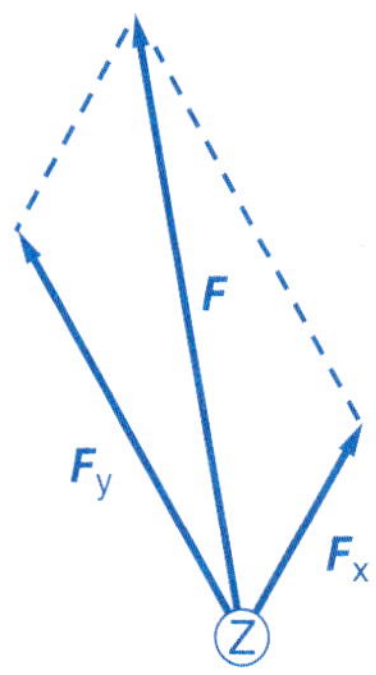

66. a. -1.60×10^{-14} N

b. -1.76×10^{16} m/s^2

67. a. 1.2×10^{13} N/C outward

b. -1.9×10^{-6} N toward the nucleus

SECTION 2

Mastering Concepts

68. joule, volt

69. A volt is the change in electric potential energy, ΔPE, resulting from moving a unit test charge q a distance d of 1 m in an electric field E of 1 N/C.

70. The charge is shared with the surface of Earth, which is an extremely large object.

71. The table is an insulator, or at least a very poor conductor.

72. The metal box shields the parts from external electric fields, which do not exist inside a hollow conductor.

Mastering Problems

73. 5.0×10^{1} V

74. 1.4 J

75. -7.2×10^{-17} J

76. 1.0×10^{2} C

77. 9.0×10^{1} V

78. 3500 N/C

79. 2.00 μF

80. a. 8.0×10^{-19} C

b. 5 electrons

81. 1.5×10^{2} V

82. 6.75×10^{-10} C

83. 4.4×10^{2} V

84. 0.45 J

85. a. 1.8×10^{-2} W

b. 4.5×10^{3} W

c. Power is inversely proportional to the time; the shorter the time is for a given amount of energy to be expended, the greater the power will be.

86. 5.6 μC

87. a. 3.1×10^{6} J

b. 3.1×10^{14} W

c. 3.1×10^{3} s

Applying Concepts

88. The electric potential energy of the particle will be converted into kinetic energy of the particle.

89. C

90. B > A = D > E > C

91. a. No; their masses could be different.

b. charge to mass ratio, q/m (or m/q)

92. José

93. Answers will vary, but a correct form of the answer is, "In a region of space with a uniform electric field, the potential changes by 9 V over a distance of 0.85 cm. What is the magnitude of the electric field in this region?"

94. Change the voltage across the capacitor.

Mixed Review

95. 6.4×10^{-6} J

96. 6.3 μC

97. 2×10^{-10} F

98. a. Answers will vary. Possible form of the correct answer, . ". . . in an electric field. If it experiences a force of 0.60 N, what is the magnitude of the electric field?"

b. Answers will vary. Possible form of the correct answer, ". . . and then moved to another location. If 0.35 J of work were done on the charge in moving it, what is the electric potential difference between the two locations?"

99. a. 5.6 μC

b. 4.8×10^{8} V/m

c. 7.7×10^{-15} N in the opposite direction to the electric field

100. 1.2×10^{-6} J

101. a. capacitance of the capacitor

b. 0.50 μF

c. work done to charge the capacitor

Thinking Critically

102. $E = 6.14\times10^{4}$ N/C at $\theta = -23.4°$

103. a. 1.2×10^{-10} N

b. 1.2×10^{3} m/s^2

c. 1.0×10^{-3} s

d. 0.60 mm

Writing in Physics

104. Student answers will vary depending on the scientist selected.

Cumulative Review

105. a. 2.3×10^{-4} s

b. 5.6×10^{2} rev/s

c. 1.1×10^{4} N

106. $h_{mountain} = 2000$ m

107. 104 cm

108. a. $F/9$

b. $3F$

c. $F/3$

d. $F/2$

e. F

Standardized Test Practice

student text p. 595

Multiple Choice

1. A

2. D

3. D

4. B

5. B

6. A

7. C

Free Response

8. $(18)(1.602\times10^{-19}\text{ C}) = 2.9\times10^{-18}\text{ C}$,

$$6.12\times10^{-14}\text{ N}\left(\frac{1.41\times10^{-2}\text{ m}}{2.88\times10^{-19}\text{ C}}\right) = 3.00\times10^{2}\text{ V}$$

Rubric

The following rubric is a sample scoring device for free response questions.

Points	Description
4	The student demonstrates a thorough understanding of the physics involved. The response may contain minor flaws that do not detract from the demonstration of a thorough understanding.
3	The student demonstrates an understanding of the physics involved. The response is essentially correct and demonstrates an essential but less-than-thorough understanding of the physics.
2	The student demonstrates only a partial understanding of the physics involved. Although the student may have used the correct approach to a solution or may have provided a correct solution, the work lacks an essential understanding of the underlying physical concepts.
1	The student demonstrates a very limited understanding of the physics involved. The response is incomplete and exhibits many flaws.
0	The student provides a completely incorrect solution or no response at all.

CHAPTER 22

Electric Current

About the Photo

Electrical Energy Ask students to identify the primary energy use in the photograph. Point out that the energy used to illuminate city streets and the interiors of buildings is delivered almost entirely as electrical energy. Before the electric light, artificial illumination depended directly on chemical energy. A great deal of electrical energy used in a city still comes from power plants that transform chemical energy from the burning of fuel such as coal, oil, and natural gas.

Chapter Overview

Current in electric circuits is discussed. Basic circuit components and their symbols are presented and used in schematic diagrams. Ohm's law is explained, as are power and the cost of using electrical energy.

Before students study the material in this chapter, they should study:

- Conservation of energy
- Electric charge
- Kinetic energy
- Potential difference
- Thermal energy
- Work, energy, and power

To solve the problems in this chapter, students will need a strong grasp of:

- Graphing data
- Scientific notation
- Significant figures
- Solving linear equations

Introduce the BIG IDEA

Show students two flashlights side by side: one powered by a battery, one powered by a hand crank. Ask students to identify similarities and differences between the two flashlights. Students may point out that the major difference is in the energy source. What happens to the bulb in each flashlight if the circuit is broken at any point, on either side of the battery or the crank? Remind students that electrical devices do not "use up" electric charge, but rather convert electrical energy into other

Go online!
connected.mcgraw-hill.com

Your online experience includes:

- **Video**
- **Concepts In Motion**
- **Virtual Investigations**
- **All Lab, Applying Practices activities, and PBL worksheets**
- **Online Practice**
- **Personal Tutor**
- **Multilingual eGlossary**
- **Presentation**

Use the LaunchLAB

In *Light the Bulb,* students can investigate electric circuits using a few basic components.

Use Physics T.V.

Video

CIRCUITS You might be surprised how many simple circuits are around you every day. Explore the physics of electricity and circuits.

1 Introduce

Bellringer

Power Connect a variable power supply to a 60-W light bulb. Use the meters on the power supply or external multimeters to monitor potential difference and current. Have students deliver increasing potential difference across the light bulb and calculate the power for several different potential differences. Ask them to draw a conclusion about the relationship between the brightness of the bulb and the power, *P*. The bulb will emit more light as the power increases. As the potential difference across the bulb increases, resistance remains constant and power increases. OL Visual-Spatial EL

Tie to Prior Knowledge

Energy Students will apply what they have learned regarding the concept of energy transformation. They will also apply the definition of power that they explored in their study of mechanics to electric devices.

2 Teach

Producing Electric Current

student text p. 598

Identify Misconceptions

Language It is not a good idea to use phrases such as "the voltage through this circuit." Students must realize that voltage is always measured as a potential difference between two points. Charges move through a circuit—not voltage and not current.

Electric Circuits

student text pp. 599–600

Use an Analogy

Current Ask students to describe how electric currents are similar to water currents. Current itself does not flow, but water and charges flow. Ask students to provide their own analogy to describe a circuit or current. For example, a model train moves around a circular track like charges move around a circuit. OL

Content Background

Energy Sources A hydropower plant converts the kinetic energy of falling water into electrical energy. The shaft of a turbine is used to turn a generator. The Hoover Dam has a total of 17 generators. The Hoover Dam facility generates about 4 billion kilowatt-hours of electricity, which is enough to serve 1.3 million people. Worldwide, hydropower plants provide about 24 percent of the world's electrical energy. More than 2000 hydropower plants operate in the United States, making hydropower the country's largest renewable energy source.

Critical Thinking

Battery Chargers Ask students to use what they have learned about potential differences and the flow of charges, to explain how a cell phone battery is recharged by plugging it into an electrical outlet. Ask them if this is different from plugging it into an automobile's cigarette lighter. Discussion on this point might include the flow of electrons from the car battery or from the household power system into the object being charged. This point can be revisited and expanded later when students are discussing batteries and chemical energy and again when AC/DC power conversion is presented. AL

Use Figure 2

Ask students to use the general theme shown in the figure to describe the operation of car headlights, starting with the gasoline in the tank. BL

Concept Development

MAIN IDEA The following analogy will help students visualize the flow of electric charge, known as current. A beach resort uses a water tank to supply its needs. Too many guests take showers in the late afternoon, and many of them complain about the weak water flow. Ask students how the resort could alleviate this problem. Raising the tank elevation would increase the water's potential energy, and larger pipes would decrease resistance. Alternatively, ask students to think of the flow of electric charges as if it were traffic on a highway at rush hour. Ask how you could increase the flow of cars. You could lower the resistance by widening the roadway and adding additional lanes, removing obstacles such as stoplights, or adding more exit ramps. You could increase the energy by raising the speed limit. OL

DIFFERENTIATED INSTRUCTION

Struggling Students Some students have ways of explaining difficult concepts that other students find appealing. If students seem to struggle with a concept, try forming small discussion groups. Seed the discussions with questions such as the following: Why is a complete circuit necessary for charges to flow? Why is an energy source necessary for charges to flow? Describe resistance and potential difference using everyday terms. What do potential difference and pressure have in common? BL COOP LEARN Interpersonal

Rates of Charge Flow and Energy Transfer

student text pp. 600–601

Activity

Energy and the Environment Ask students to investigate why wasting electrical energy has a negative environmental impact. Suggest that they list some examples of personal, national, and school activities that could be wasting electrical energy and then provide solutions to reduce the waste. OL Linguistic

ADDITIONAL IN-CLASS EXAMPLE

Use with Example Problem 1.

Problem A 120-V motor operates at 13 A. Determine the power and the energy used in one hour of operation.

Response $P = IV$, $P = 120\ \text{V} \times 13\ \text{A}$, $P = 1.6\ \text{kW}$; $E = Pt$, $E = 1.6\ \text{kJ/s} \times 60\ \text{min} \times 60\ \text{s/min}$
$E = 5.8\times10^6\ \text{J}$

DIFFERENTIATED INSTRUCTION

Visually Impaired The Bellringer with the battery and the lightbulb also may be performed with a battery and an electric bell (doorbell) or a buzzer. Additionally, you could demonstrate how to replace the bell on a door with a flashing light as is done for hearing-impaired individuals or in a soundproof location (such as a recording studio). Note: A variety of personal electronic devices exist to assist visually impaired individuals. Some students in your class may possess PDAs that are capable of scanning written text and translating it into Braille or ones that access an audible speech synthesizer.

Discussion

Question Show students D- and AAA-cell batteries. Point out that both are rated at 1.5 V and ask students to describe the significant difference between them.

Answer The D cell will last longer with a given load. Because the D cell has more mass (and consequently more chemicals (charges)), it can supply a given current for a longer period of time before the chemical energy is depleted. BL

Diagramming Circuits

student text pp. 602–603

Use the MiniLAB

In *Current Affairs,* students can draw, build, and investigate currents in circuits.

REAL-WORLD PHYSICS

Lightbulb Calculate the current and power at the moment that a room-temperature 100-W bulb is turned on. $I = \frac{V}{R} = \frac{120\ \text{V}}{10\ \Omega} = 10\ \text{A}$. $P = IV = (10\ \text{A})(120\ \text{V}) = 1\ \text{kW}$. This represents a rather large initial heating effect. If possible, obtain a clear 100-W bulb so students can see the size of the filament.

Resistance and Ohm's Law

student text pp. 604–607

Content Background

Resistors Resistors are made to create a precise quantity of resistance for insertion into a circuit. They are typically constructed of metal wire or carbon and engineered to maintain a stable resistance value over a wide range of environmental conditions. Unlike lamps, they do not produce light, but they transform electrical energy into thermal energy in a working circuit. Typically, though, the purpose of a resistor is to provide a precise quantity of electrical resistance, not to provide usable thermal energy. Because resistors transfer thermal energy to their surroundings as the electric charges going through them overcome the "friction" of their resistance, resistors also are rated in terms of how much thermal energy they can dissipate without overheating and sustaining damage.

Concept Development

Resistance v. Resistivity Although it is accurate to talk about the resistance of a copper wire, it does not make sense to talk about the resistance of the copper itself because resistance changes with length and cross-sectional area. Because resistance (R) is inversely proportional to a cross-sectional area (A) and directly proportional to length (L), resistivity (ρ) can be determined by multiplying resistance in a wire by the wire's area and dividing by the wire's length: $R = \rho\left(\frac{L}{A}\right)$.

Use an Analogy

Resistance and Walking To make the concept of resistance more concrete, try comparing electric resistance to the resistance a person might experience walking over different surfaces. Walking on asphalt is very easy (low resistance), walking across a muddy field is a bit more difficult, and walking through crowded hallways is extremely difficult (high resistance).

Reinforcement

Completing the Circuit Ask students to identify these items: (1) moving charges in wires, (2) a chemical charge pump, (3) converts electrical energy to thermal energy, (4) a law that states that the current in a given conductor is directly proportional to the potential difference across it, and (5) the rate of converting energy. (1) electrons, (2) battery or cell, (3) resistance or resistor, (4) Ohm's law, (5) power BL Linguistic

REAL-WORLD PHYSICS

Resistance and Resistivity Engineers use resistivity to predict resistance. Imagine current running through a copper wire that is 2.0 m in length and 2.0 mm in diameter. To predict the resistance of the wire, you can use the formula $R = \rho L/A$, where R is the resistance in ohms, ρ is the resistivity of copper in ohm meters, L is the length in meters, and A is the cross-sectional area in square meters. Substituting the appropriate values gives $R = \frac{\rho L}{A} = \frac{(1.68\times10^{-8}\ \Omega\cdot\text{m})(2.0\ \text{m})}{\pi(1.0\times10^{-3}\ \text{m})^2} = 1.1\times10^{-2}\ \Omega$

Using Models

A Car Battery Students can use the concept of resistance to model a partially drained battery. For example, a 12-V car battery might need to supply 200 A of current during engine cranking. If the resistance of the cranking motor is 0.060 Ω, the required current is supplied: $\frac{12\ \text{V}}{0.060\ \Omega} = 200$ A. A drained battery can be modeled as a 1-Ω resistor. Combined with the resistance of the cranking motor, the new resistance is 1.060 Ω. The current supplied would be about 11 A, not nearly enough to turn the engine. Point out that this model illustrates that measuring the potential difference across a battery with a meter that draws no current does not indicate whether the battery is able to do its job. For this reason, mechanics use a load test to evaluate a car's battery.

Content Background

Temperature and Resistance Almost all conductors have a positive temperature coefficient of resistivity, α. The temperature coefficient of resistivity can be determined by an equation similar to that for the coefficient of linear expansion. The following formula can be used to predict the change in resistance: $R_1 = \frac{R_2(1+\alpha T_1)}{(1+\alpha T_2)}$. R_1 is the resistance in ohms at temperature T_1 in °C, R_2 is the resistance in ohms at temperature T_2 in °C, and α is the temperature coefficient of resistivity in °C^{-1}.

Use the MiniLAB

In *Making Electrical Energy,* students can observe electrical energy from a series of cells.

ADDITIONAL IN-CLASS EXAMPLE

Use with Example Problem 2.

Problem A 9.0-V battery is connected to a 15-kΩ resistor. What is the current in this circuit?

Response $I = \frac{V}{R} = \frac{9.0\ V}{15\ k\Omega} = 0.60\ mA$

Quick DEMO

Ohm's Law

Estimated Time 10 minutes

Materials variable DC power supply; multimeters (2); 12-V lamp; lamp base or socket (as needed); 100-Ω, 2-W resistor; clip leads

Procedure Connect the circuit as shown in **Figure 5,** using the power supply rather than a battery. Start at 0 V and ask a student assistant to record the potential difference and current on the board. Increase the potential difference across the power supply in 2-V steps until 12 V is reached. Have the assistant record the potential difference and current at each step. Have a second student assistant draw a graph on the board. Repeat the entire process using the resistor in place of the lamp. Lead a class discussion using Ohm's law as the focus.

Parallel and Series Connections

student text p. 608

Quick DEMO

Alternating Current and Resonance

Estimated Time 15 minutes

Materials solar cell, amplifier, speaker, stroboscope

Procedure The following demonstration can show the beat production from constructive interference. Connect the solar cell to the amplifier and speaker. Expose the solar cell to fluorescent lights. Students should be able to hear the 60 Hz hum. Turn the lights on and off and ask students to listen to the difference. You can expand the demonstration by flashing a strobe light at 59 Hz or 61 Hz. Note the resulting beats. Refer to visual aids showing waves and constructive interference.

Use the PhysicsLAB

In *Voltage, Current, and Resistance,* students can investigate the relationship between potential difference and current and the relationship between resistance and current.

3 Assess

Assess the MAINIDEA

Electric Current Draw a simple circuit including a battery and a light bulb on the board or ask students to visualize one or draw it themselves. Imagine that the circuit is closed until the light bulb stops burning. How does the electric charge in the battery change with time? There is always the same amount of charge in the battery. What does the battery store? The battery stores electrochemical energy. It does not store electric charge.

Check for Understanding

Circuits Draw a complete circuit on the board in schematic form. Ask students to determine if the circuit is complete, to identify the symbols, to identify the source of electrical energy, to identify devices that transform electrical energy to other forms of energy, to indicate the direction of the current, to determine if Ohm's law applies, and to determine how power can be determined. **BL** **Visual-Spatial**

Extension

Batteries Ask students to explain rechargeable batteries from an energy point of view and to compare them to capacitors. A battery stores energy in chemical form, and a capacitor stores energy in an electric field. As a battery discharges, the current generated by the chemical reaction is through the electrolyte. In a car's battery, for example, chemical reactions involving lead peroxide and sulfuric acid produce lead sulfate and water as the battery discharges. There is no chemical change within a capacitor as it is discharging. Instead, the electric field resulting from the charge imbalance on the plates dissipates. **OL**

Reading Checks and Caption Checks

Reading Check, *student text p. 600*
Charges move through a circuit, but the total amount of charge in a circuit remains the same. The charges just move around, and they are not destroyed.

Reading Check, *student text p. 600*
$P = E/t$; $P = (q/t)\Delta V$; $P = I\Delta V$

Reading Check, *student text p. 602*
A schematic depicts the electrical properties of the elements of the circuit and the path or paths that current can take, but might omit details like the color of a lightbulb or its brightness. An artist's drawing would look more like what an observer would actually see, but does not convey as much information about a circuit's electrical properties.

Reading Check, *student text p. 602*
An ammeter measures current. A voltmeter measures potential difference.

Reading Check, *student text p. 604*
Resistance describes how much potential difference must exist across a charge pump to move electric charges through a circuit at a given rate.

Reading Check, *student text p. 605*
Devices that obey Ohm's law exhibit a constant resistance that does not depend on the potential difference across the device. Most metallic conductors obey Ohm's law.

Reading Check, *student text p. 605*
You can decrease the potential difference across the circuit, or you can increase the resistance. Either one will reduce the current.

Caption Check, *student text p. 606*
The potentiometer allows continuous change in the rotation of the motor rather than step-by-step changes.

Practice Problems
student text p. 601

1. 24 W

2. 0.60 A

3. 63 W

4. 2.5×10^4 J

5. 40 A

6. 0.30 A

7. The power increases by a factor of 6.

Practice Problems
student text p. 603

8. $I = 4.80$ A; see Online Solutions Manual for circuit diagram.

9. $R = 53\ \Omega$; see Online Solutions Manual for circuit diagram.

10. 60.0 V for Practice Problem 8, 4.5 V for Practice Problem 9; see Online Solutions Manual for circuit diagram.

11. Diagram should include correctly labeled symbols for a voltage source, a lamp, a switch, and a potentiometer. See Online Solutions Manual for circuit diagram.

12. Diagram will resemble that of Practice Problem 11 but will also include a voltmeter in parallel with the lamp. See Online Solutions Manual for circuit diagram.

Practice Problems
student text p. 607

13. 0.36 A

14. $1.5\times10^4\ \Omega$

15. 1.2×10^2 V

16. a. $2.4\times10^2\ \Omega$

b. 6.0×10^1 W

17. a. 0.60 A

b. $2.1\times10^2\ \Omega$

18. a. 6.3×10^1 V

b. $2.1\times10^2\ \Omega$

c. 19 W

Section 1 Review
student text p. 608

19. Charged particles move through an electric circuit. These moving charged particles are called an electric current. When charged particles move through a material such as a metal wire, the flow is impeded by the particles in the material. This flow impediment is called resistance.

20. Diagram should include a voltage source and a lamp in a closed circuit. See Online Solutions Manual for schematic.

21. No; resistance depends on the device. When V increases, so will I.

22. 4.0 W increase

23. Measure the current through the wire and the potential difference across it. Divide the potential difference by the current to obtain the wire resistance. See Online Solutions Manual for an example diagram.

24. 44×10^3 J

25. The potential energy of the charges decreases as they flow through the resistor. This potential energy is converted to thermal energy, and the thermal energy spreads out or dissipates in the surrounding environment.

1 Introduce

Bellringer

Rate of Change Submerge a 47-Ω, 10-W resistor in a small polystyrene cup half-filled with water. Use a thermometer to measure the water temperature. If time permits, run two trials, one with 10 V applied and one with 20 V applied. Note the rate of temperature increase. Ask students to announce the timed temperature readings, record the readings, and draw a graph on the board. BL Visual-Spatial EL

Tie to Prior Knowledge

Electrical Energy Throughout this section students will connect the concepts associated with electric current and power to everyday uses of electrical energy. They also will continue to explore the nature of the law of conservation of energy.

2 Teach

Electrical Energy, Resistance, and Power

student text pp. 609–611

ADDITIONAL IN-CLASS EXAMPLE

Use with Example Problem 3

Problem A water heater operates at 240 V, and the resistance of its heating element is 12 Ω. How much current does it demand, and how much thermal energy will it produce in 30 minutes?

Response $I = \frac{V}{R}$, $I = \frac{240\ \text{V}}{12\ \Omega}$, $I = 2.0\times10^{1}\ \text{A}$; $E = I^2Rt$, $E = (2.0\times10^{1}\ \text{A})^2 \times 12\ \Omega \times 30\ \text{min} \times \frac{60\ \text{s}}{1\ \text{min}}$, $E = 8.6\ \text{MJ}$

PHYSICS PROJECT ACTIVITY

Superconductors Have students prepare a report explaining the reasons that some materials begin to superconduct at very low temperatures. Students should discover that conductors hold their electrons loosely; moving electrons transform electrical energy into thermal energy. Electrons traveling in superconductors experience fewer energy transformations because they readily travel in pairs. While this advantageous pairing can take place at high temperatures, in superconductors low temperatures make it easier for electrons to pair and move quickly between atoms with essentially zero electrical-to-thermal-energy transformations. OL Linguistic

Use the PhysicsLAB

In *Energy Conservation,* students can measure and compare quantities of electrical energy and thermal energy.

Providing Electrical Energy

student text pp. 611–613

PHYSICS CHALLENGE ACTIVITY

Capacitance Use a 1000-μF, 25-V electrolytic capacitor; a 12-V DC power supply; and a 12-V lamp to demonstrate charge and energy storage. **CAUTION:** ***Check the wires in the circuit to make sure everything is connected correctly before turning on the power supply.*** Ask students to explain how there is sufficient current in this circuit to light the lamp when the capacitor has a DC resistance in the range of 106 Ω. **CAUTION:** ***Do not attempt to measure the resistance of the capacitor when it is charged.*** Short the capacitor leads for a minute or so before making such a measurement. AL Visual-Spatial

Quick DEMO

Energy Storage

Estimated Time 10 minutes

Materials 1-μF capacitor, 9-V battery, digital multimeter (DMM), 1-MΩ resistor

Procedure Use this demonstration to show energy storage in a capacitor. Set up the circuit, and time the discharge of the capacitor. Be sure to observe capacitor polarity. Be sure the digital multimeter is set to operate as a voltmeter. A typical DMM has a resistance of 10 MΩ, and the time required to discharge the capacitor is approximately 5 $RC = 5(10\ \text{M}\Omega)(1\ \mu\text{F}) = 50$ s. Repeat the demonstration with a 1-MΩ resistor in parallel with the meter leads, which will show an order-of-magnitude decrease in the discharge time.

Discussion

Question Why are high-voltage lines located on high towers?

Answer Locating high-voltage lines on towers is done for safety reasons. Potential differences of hundreds of thousands of volts are extremely dangerous. The insulation materials that would be required to locate the cables near the ground or underground would be impractical. The high towers also allow air to serve as a major insulator. OL

Use Figure 17

Ask students to suppose that you have a digital watt-hour meter. Also suppose that there is no current at this moment (everything in the building is turned off). Ask students whether the meter reading will go to zero. No; the last reading will hold because the meter is indicating the total energy used. OL

Content Development

MAIN IDEA The human body's resistance to current when the skin is dry is around $1.0\times10^5\ \Omega$. Perspiration reduces electrical resistance because it contains ions that readily conduct electric charge. This phenomenon is the basis of galvanic skin response tests or "lie detectors," which use a small current to detect changes in stress levels indicated by increased perspiration.

Concept Development

Stereos and Power Some people want sound systems that are very powerful in their vehicles. This is difficult to achieve with a 2-V system, given that speakers are usually 4.0 Ω. In such a system, the power delivered to one speaker is limited to 36 W (V^2/R). One solution is to use a bridge-type amplifier, which effectively doubles the potential difference across each speaker (for four times the power).

Reinforcement

Energy Use Display an immersion-type water heater and the heating element from an electric water heater. Ask students to explain the differences that exist between the two devices. Items that could be mentioned in the ensuing discussion include size, cost, resistance, voltage rating, and reliability.
OL Logical-Mathematical EL

3 Assess

Assess the MAIN IDEA

Resistance A calculator is rated at 0.10 W when connected to a 1.5-V battery. What is the resistance of the device? 22 Ω

Check for Understanding

Consumption and Cost To help students understand energy consumption and cost, ask them to compare the cost of operating various household appliances by explaining the relationships among power, current, and the cost of operating 1000-W, 250-W, and 50-W appliances at household voltages. Assuming that all other variables are equal and that only power changes, as the power increases, the current and operating cost increase. OL

Extension

Future Energy Production Assign students a research project relating to the possible future use of nuclear fusion for generating electricity. Students should compare fission, fusion, and combustion processes.
OL Linguistic

Reading Checks and Caption Checks

Caption Check, *student text p. 609*
Collisions among the flowing charges and the particles in the resistor convert electrical potential energy into thermal energy.

Reading Check, *student text p. 611*
A superconductor has zero resistance.

Reading Check, *student text p. 612*
To reduce thermal energy losses, the current in the transmission lines is reduced and the voltage is increased.

Reading Check, *student text p. 612*
One kilowatt-hour equals 3.6×10^6 J. It is a unit of energy that is convenient to use when describing the rate of power consumption and the duration over which power is used.

Practice Problems

student text p. 611

26. a. 8.0 A
b. 2.9×10^4 J
c. 2.9×10^4 J

27. a. 1.2 A
b. 1.6×10^4 J

28. a. 1.3×10^3 J
b. 4.7×10^3 J

29. a. 2.0×10^1 A
b. 1.3×10^5 J
c. 17°C

30. 1.1 h

Practice Problems

student text p. 613

31. a. 1.8 kW
b. 270 kWh
c. $32.40

32. a. 9.6×10^{-3} A
b. 1.1 W
c. $0.10

33. 9.5 h

34. 8.2 h

Physics Challenge

student text p. 613

1. 15 V
2. It remains 15 V because there is no path for the charge to be removed.
3. 15 V and 13 mA
4. The capacitor voltage remains at 15 V because there is no path to discharge the capacitor; the current remains at 13 mA because the battery voltage is constant at 15 V. However, if the battery and capacitor were real components instead of ideal circuit components, the capacitor voltage would eventually become zero due to leakage and the current would eventually become zero due to battery depletion.

Section 2 Review

student text p. 613

35. Mechanical energy from the engine converted to electrical energy in the generator; electrical energy stored as chemical energy in the battery; chemical energy converted to electrical energy in the battery; electrical energy converted to light and thermal energy in the headlamps.

36. Hot draws more power, $P = IV$, so the fixed voltage current is larger. Because $I = V/R$, the resistance is smaller.

37. Some potential benefits: lower cost of electricity, if less power were lost during transmission, less coal and other power-producing resources would be used, which would improve the quality of our environment.

38. For the same power, at twice the voltage, the current would be halved. The I^2R loss in the circuit wiring would be dramatically reduced because it is proportional to the square of the current.

39. $333.31

40. a. 29 Ω
b. 500 W

41. Power, not energy; most devices will have to run longer.

A CLOSER **LOOK**

Get Smart

Background

Energy conservation is a potentially confusing topic in the context of a physics class. After all, one of the major messages of physics is that energy is always conserved. It is really not a question of energy, but of entropy. In the transformation of energy from one form to another, some amount is converted to unrecoverable thermal energy. Consider the heating of wires due to electric current, the heat from a car engine, or the thermal energy removed from a refrigerator. It is the minimization of this unusable energy that is at the heart of energy conservation discussions.

Teaching Strategies

To reinforce the idea that energy is not created in an electrical power plant, trace the transformations that take place with any form of energy. Coal and oil can both be traced back to the solar energy first stored by living things long ago. Nuclear energy can be traced back to exploding stars that formed the heavy, unstable elements in nuclear reactors. Geothermal energy is partly due to the decay of radioactive elements and partly due to the potential energy present at the formation of Earth.

GOING**FURTHER** >>>

Expected Outcomes A smarter grid will provide energy at all times to the consumer without brownouts or electrical failure during periods of extremely high demand. The loss of electrical energy to consumers will be instantly reported so that energy delivery can be restored quickly. A smarter grid will allow for the efficient use of surplus energy provided to the grid by consumers. A smarter grid will provide a sufficient supply of affordable energy to its customers.

SECTION 1

Mastering Concepts

42. 1 A = 1 C/s

43. The positive voltmeter lead connects to the left-hand motor lead, and the negative voltmeter lead connects to the right-hand motor lead.

44. Break the circuit between the battery and the motor. Then connect the positive ammeter lead to the positive side of the break (the side connected to the positive battery terminal) and the negative ammeter lead to the negative side (the side nearest the motor).

45. from left to right through the motor

46. **a.** 4

b. 1

c. 2

d. 3

47. A larger-diameter wire has a smaller resistance because there are more electrons to carry the charge.

48. **a.** Diagram should include a voltage source and a resistor. See Online Solutions Manual.

b. The ammeter must be connected in series.

c. The voltmeter must be connected in parallel.

49. **a.** The potentiometer on an electric motor makes continuous changes in the speed of the motor instead of a step-by-step change in the motor's speed.

b. The potentiometer in a game joystick helps translate the motion of the gamer's hand to a position on the computer screen.

Mastering Problems

50. **a.** electrical energy to thermal energy and light

b. electrical energy to thermal energy and kinetic energy

c. electrical energy to light and sound

d. chemical energy to light and thermal energy

51. **a.** 18 W

b. 1.6×10^4 J

52. **a.** 1.5 A

b. 27 V

c. 41 W

d. 1.5×10^5 J

53. 9.6×10^2 W

54. 1.4×10^2 W

55. **a.** 3.0 A

b. 27 V

c. 81 W

d. 2.9×10^5 J

56. **a.** 0.50 A

b. 9.0 V

c. 4.5 W

d. 1.6×10^4 J

57. **a.** 6.0×10^1 W

b. 1.8×10^4 J

58. **a.** 2.5×10^3 J/s

b. 2.5×10^3 W

59. 19 A

60. **a.** 4.5 W

b. 3.0×10^3 J

61. 24 V

62. 6.0 V

63. 1.2×10^2

64. 5.0 A

65. **a.** $R = 143\ \Omega$, $R = 148\ \Omega$, $R = 150\ \Omega$, $R = 154\ \Omega$, $R = 159\ \Omega$, $R = 143\ \Omega$, $R = 143\ \Omega$, $R = 154\ \Omega$, $R = 157\ \Omega$, $R = 161\ \Omega$

b. Graph should indicate an approximately straight line of constant slope increasing from left to right and passing through the origin. See Online Solutions Manual for graph.

c. The resistance of the nichrome wire increases somewhat as the magnitude of the voltage increases, so the wire does not quite obey Ohm's law.

66. $V = 28$ V; See Online Solutions Manual for circuit diagram.

67. **a.** No; the voltage is increased by a factor of 9.0/6.0 = 1.5, but the current is increased by a factor of 75/66 = 1.1.

b. 0.40 W

68. 1.08×10^5 J; 9.5×10^4 J

69. **a.** $3.0\times10^2\ \Omega$

b. $6.0\times10^1\ \Omega$

c. 2.0 A

70. **a.** $32\ \Omega$

b. $1.2\times10^2\ \Omega$

71.

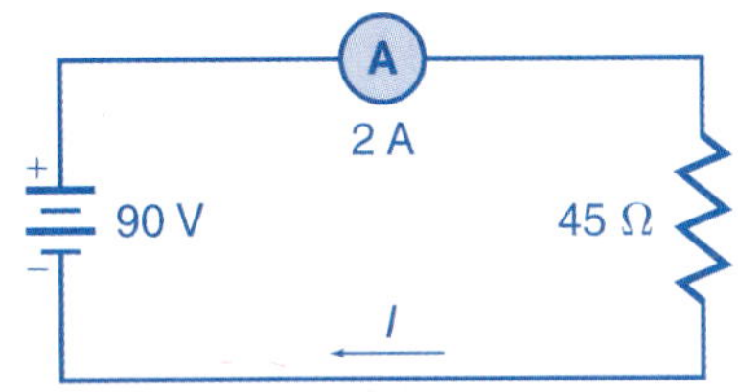

$I = 2$ A

SECTION 2

Mastering Concepts

72. The low resistance of the cold filament allows a high current initially and a greater change in temperature, subjecting the filament to greater stress.

73. The short circuit produces a high current, which causes more electrons to collide with the atoms of the wire. This raises the atoms' kinetic energy and the temperature of the wire.

74. the resistance of the wire and the current in the wire

75. $\mathrm{W} = \frac{\mathrm{C}}{\mathrm{s}} \cdot \frac{\mathrm{J}}{\mathrm{C}} = \frac{\mathrm{J}}{\mathrm{s}} = \frac{\mathrm{kg} \cdot \frac{\mathrm{m}^2}{\mathrm{s}^2}}{\mathrm{s}} = \frac{\mathrm{kg} \cdot \mathrm{m}^2}{\mathrm{s}^3}$

Mastering Problems

76. \$510/kWh

77. 0.15 A

78. 1.2×10^6 J

79. **a.** 1.1 A

b. 45 V

80. 13 A

81. \$220

82. **a.** \$18/kWh

b. \$0.02

Applying Concepts

83. A potential difference is felt over the entire circuit as soon as the battery is connected to the circuit. The potential difference causes the charges to begin to flow. Note: The charges flow slowly compared to the change in potential difference.

84. By touching the fence and the ground, the cow encounters a difference in potential and conducts current, thus receiving a shock.

85. No potential difference exists along the wires, so no current flows through the bird's body.

86. Increase the voltage or decrease the resistance.

87. 50-W bulb; $P = \frac{V^2}{R}$, so $R = \frac{V^2}{P}$; therefore, the lower P is caused by a higher R.

88. If the resistance is doubled, the current is halved.

89. No effect; $V = IR$, so $I = \frac{V}{R}$, and if the voltage and the resistance are both doubled, the current will not change.

90. yes, because the current is the same everywhere in this circuit

91. No; at 1.5 V, $R = 3.3 \times 10^4\ \Omega$. At 3.0 V, $R = 120\ \Omega$. A device that obeys Ohm's law has a resistance that is independent of the applied voltage.

92. The wire with the smaller resistance; $P = V^2/R$; smaller R produces larger power, P, dissipated in the wire, which produces thermal energy at a faster rate.

Mixed Review

93. 200 h

94. 2.2×10^4 J

95. **a.** 2.5 A

b. 2.3×10^4 J

96. **a.** 3.0 A

b. 12 A

c. the instant it is turned on

97. The range is 10 Ω to 600 Ω.

98. **a.** 5.0 A

b. 40%

99. **a.** 9×10^5 J

b. 8°C

c. \$7

Thinking Critically

100. Diagram should include a voltage source and three resistors or lamps in series, and the current will be clockwise. See Online Solutions Manual.

101.

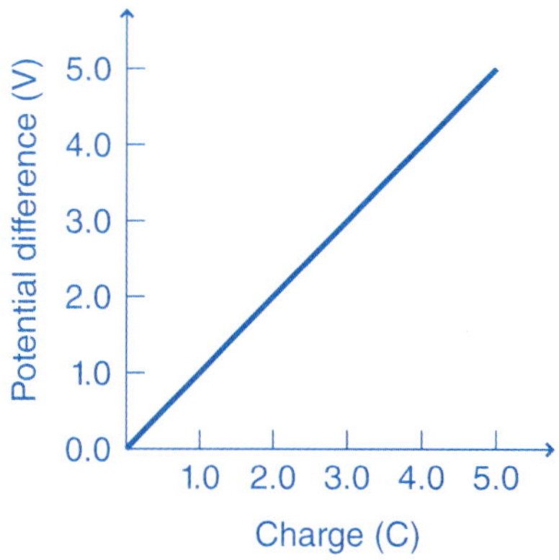

Voltage = 5.0 V. Energy = 13 J.

No; graphically, total charge times final potential difference is exactly twice the area under the curve. Physically, it means that each coulomb would require the same maximum amount of energy to place it on the capacitor. Actually, the amount of energy needed to add each charge increases as charge accumulates on the capacitor.

102. B > C > A > D > E

103. Answers will vary, but a correct form of the answer is, "A 60-W lightbulb is connected to a 110-V outlet. When the bulb is turned on, how much current flows through it?"

104. Answers will vary. A possible form of the correct answer would be, "... and the current through it is 250 mA. What is its resistance?"

105. The physical size of a resistor is determined by its power rating. Resistors rated at 100 W are much larger than those rated at 1 W.

106. The volt-ampere graph for a resistor obeying Ohm's law is a straight line and is seldom necessary.

107. Two possible parabolic graphs could plot watts dissipated versus the voltage across a 10-Ω resistor, and watts dissipated versus the current through the same resistor. See Online Solutions Manual for graph.

Writing in Physics

108. The student's answer should include the ideas that, for devices obeying Ohm's law, the voltage drop is proportional to current through the device and that the formula $R = V/I$, the definition of resistance, is a derivation of Ohm's law.

109. Answers will vary, but students should determine that transmission lines can become hot enough to expand and sag when they have high currents. Sagging lines can be dangerous if they touch objects beneath them, such as trees or other power lines.

Cumulative Review

110. 2.7×10^{4} J/K; melting ice: 2.4×10^{4} J/K

111. 1.9 kPa or about 2/100 of the total air pressure

112. 2.0 cm

113. 1.4×10^{-4} m

114. 0.41 N

Standardized Test Practice

student text p. 621

Multiple Choice

1. A
2. D
3. C
4. D
5. C
6. B
7. C
8. D

Free Response

9. $I = 14$ A; $E = 2.5\times10^5$ J

Rubric

The following rubric is a sample scoring device for free response questions.

Points	Description
4	The student demonstrates a thorough understanding of the physics involved. The response may contain minor flaws that do not detract from the demonstration of a thorough understanding.
3	The student demonstrates an understanding of the physics involved. The response is essentially correct and demonstrates an essential but less-than-thorough understanding of the physics.
2	The student demonstrates only a partial understanding of the physics involved. Although the student may have used the correct approach to a solution or may have provided a correct solution, the work lacks an essential understanding of the underlying physical concepts.
1	The student demonstrates a very limited understanding of the physics involved. The response is incomplete and exhibits many flaws.
0	The student provides a completely incorrect solution or no response at all.

CHAPTER 23

Series and Parallel Circuits

About the Photo

Building Lights Have students examine the photograph. Ask them what each light in the lamp shop needs to light up. electricity or electrical current Have students predict what would happen if each lamp were linked to others in a chain and one wire broke. All the lamps would go out. Ask students whether this would actually happen. no Point out that the wiring in the lamp shop must be arranged to prevent this.

Chapter Overview

This chapter discusses the principles of series and parallel circuits. The first section explains equivalent resistances in series and parallel resistors, circuit currents, potential difference, and series-parallel circuits. The second section describes how circuits are used, and it explains how circuit breakers, fuses, voltmeters, ammeters, and ground-fault interrupters operate.

Before students study the material in this chapter, they should study:

- Conductors and insulators
- Conservation of energy
- Electric charge
- Electric current
- Electric potential
- Ohm's law

To solve the problems in this chapter, students will need a strong grasp of:

- Scientific notation
- Significant figures
- Solving linear equations

Introduce the BIG IDEA

Display several small light bulbs wired in series. Remove one bulb. Ask students what happens to the lights. The entire group turns off. Explain that this occurs because the lights are wired in serial fashion. Ask students if the same thing happens when a light bulb in the class burns out. no Explain that classroom lights are wired in parallel and behave differently.

Go online!
connected.mcgraw-hill.com

Your online experience includes:

- **Video**
- **Concepts In Motion**
- **Virtual Investigations**
- **All Lab, Applying Practices activities, and PBL worksheets**
- **Online Practice**
- **Personal Tutor**
- **Multilingual eGlossary**
- **Presentation**

Use the LaunchLAB

In *Fuses and Electric Circuits,* students will demonstrate how a fuse operates to protect an electric circuit.

Use Physics T.V.

Video

HOLIDAY LIGHTS Why do some strings of lights stop working when a lightbulb burns out while others continue working? The answer is physics!

1 Introduce

Bellringer

Series Circuits Connect a variable power supply to a 12-W lightbulb. Use external multimeters to monitor potential difference and current. Adjust the potential difference across the supply to 10 V, and note the current. Turn off the supply and add a second 12-W bulb in series. Turn on the supply, and note the current and the brightness of the two bulbs. Measure the potential difference across each bulb. Adjust the potential difference across the power supply to 20 V, and note the current, brightness, and potential differences across the bulbs. Discuss the results. OL Visual-Spatial

Tie to Prior Knowledge

Potential Energy Conservation of energy applies to electric circuits. An energy source, such as a battery, raises the potential energy of charges flowing through it. That potential energy drops as the energy is transformed into thermal and light energy by the lamps, resistors, and other devices. The energy of the charges returns to its original value when the charges reenter the battery. The sum of the potential difference drops must equal the increase in potential difference across the battery.

2 Teach

A River Model and Series Circuits

student text pp. 624–630

PHYSICS PROJECT ACTIVITY

Series Circuits Have students think about and explain how a flashlight is set up. Have a large and small type available to take apart, investigate, and put back together. (Make sure the batteries are working well.) With each battery placed end-to-end, the electric current generated by one battery passes to the next battery or batteries in the series. Discuss how the brightness of a flashlight bulb would change if you could add more batteries to the series. OL Visual-Spatial

Identify Misconceptions

Equivalent Resistance Students sometimes think that the order of the resistors or loads in a series can affect the behavior of a series circuit. Explain that changing the order of the resistors has no effect on the current or the total power dissipation. In the case of unequal series resistors, the location of each potential difference will depend on the location of each resistor, but the total of the potential differences will always equal the potential difference across the power source. Students might better grasp this point by first constructing a series circuit, measuring the current, and calculating the power. They should then change the order of the resistors and measure the current and power again.

Discussion

Problem Ask students to determine the equivalent resistance of two series resistors where one is quite small and the other is quite large.

Response In a series circuit, the equivalent resistance is the sum of all the resistances in the circuit, no matter how large or small the individual resistances are. You add the two resistances to get the total.
OL Logical-Mathematical

Reinforcement

Series Circuits and Equivalent Resistance Have students form several small discussion groups. Have each group prepare lists of the characteristics of series circuits and describe equivalent resistance for a series circuit that contains three resistors. Ask students to include all the appropriate equations, create schematic diagrams, and see if they can think of applications for series circuits. Ask each group to exchange its list with another group for discussion. Later, merge the groups for a broader discussion. BL Interpersonal

Use Figure 4

Voltage Drop Ask students how the concept of voltage drop in a voltage divider could be applied as a volume control in a CD player or a radio. If the voltage source represents an audio signal, a voltage divider can deliver an adjustable fraction of the signal voltage to the next stage of the circuit. OL Logical-Mathematical

Use the PhysicsLAB

In *Resistors In Series Circuits,* students will measure the potential difference across each resistor in a series circuit.

Critical Thinking

MAINIDEA Holiday light sets are less expensive to manufacture using a series arrangement of bulbs. Yet such sets are not acceptable to many consumers because when one bulb burns out, the entire set shuts down. Even when consumers know how to troubleshoot this situation (for example, by moving a known good bulb from position to position until the set comes on), they don't want to take the time to do it. Have students investigate and explain what type of bulb holiday light designers have created to deal with this situation. By applying the physical principle that the potential difference of the entire line will drop across a burned-out bulb, designers have developed a special bulb that shorts out when 120 V appear across its terminals. This type of bulb will no longer light, but the rest will function at a slightly higher voltage. If too many bulbs reach the shorted condition, a series fuse will blow.
OL **Logical-Mathematical**

PHYSICS PROJECT ACTIVITY

Photoresistors In addition to the use of photoresistors in photographic light meters, these devices are commonly used as light sensors in security lights that automatically turn on when it gets dark. At night, or even during daytime stormy weather, the resistance in the equipment is very high—in the megaohm range. During the day when the sensor is illuminated, the resistance drops, typically to a few hundred ohms. Have students research the development and use of photoresistors in security lights or other equipment and graphically depict how they are set up and operated.
OL **Kinesthetic** **Visual-Spatial**

ADDITIONAL IN-CLASS EXAMPLE

Use with Example Problem 1.

Problem If 15 V are applied across three resistors in series (15.0 Ω, 22.0 Ω, and 47.0 Ω), determine the current in the circuit and the potential difference across the 47.0-Ω resistor. How will the current change by using a 60.0-Ω resistor instead of the 47.0-Ω resistor? Also demonstrate that the total power is equal to the sum of the individual power amounts used in the 15.0-Ω, 22.0-Ω, and 47.0-Ω resistors. If there is time, derive the equation $V_A = (V)(R_A)/(R_A + R_B + R_C)$.

Response

$$I = \frac{V}{R} = \frac{15\ \text{V}}{15.0\ \Omega + 22.0\ \Omega + 47.0\ \Omega}$$
$$= 0.18\ \text{A}$$

$$V = IR = (0.18\ \text{A})(47.0\ \Omega) = 8.5\ \text{V}$$

$$I = \frac{V}{R} = \frac{15\ \text{V}}{15.0\ \Omega + 22.0\ \Omega + 60.0\ \Omega}$$
$$= 0.15\ \text{A}$$

The current will decrease when the resistance increases.

$$P = IV = (0.18\ \text{A})(15.0\ \text{V}) = 2.7\ \text{W}$$
$$P_{(15)} = I^2R = (0.18\ \text{A})^2(15.0\ \Omega) = 0.49\ \text{W}$$
$$P_{(22)} = I^2R = (0.18\ \text{A})^2(22.0\ \Omega) = 0.71\ \text{W}$$
$$P_{(47)} = I^2R = (0.18\ \text{A})^2(47.0\ \Omega) = 1.5\ \text{W}$$
$$P = 0.49\ \text{W} + 0.71\ \text{W} + 1.5\ \text{W} = 2.7\ \text{W}$$

Reinforcement

Series Circuit Setup Ask students to compare a series circuit to a chain. They should mention that the links are connected end-to-end, one after another. Also, a chain fails if one link breaks (opens). **BL**

Content Background

Potential Differences in Wire Installation One of the main concerns when installing long lengths of wire is a low potential difference. There can be a significant amount of electrical energy transformed to thermal energy between the power supply and the equipment that needs the energy if the proper wire is not used. When designing the electrical systems for buildings, electrical engineers must specify the proper wire for each installation to make sure there is enough potential difference for the intended load. To do this, they must consider the length of the wire from the power supply to the load, as well as the return of the line from the load, the amount of electrical energy transformed for each type of wire, and the load current. National electric codes standardize this process for residential and commercial buildings by specifying the wire diameter that must be used for a circuit of a given current capacity. **OL** **Linguistic**

ADDITIONAL IN-CLASS EXAMPLE

Use with Example Problem 2.

Problem A voltage divider consisting of two 1.5-MΩ resistors is connected to a 12.0-V source. Determine the potential difference across one resistor before and after a voltmeter is connected, assuming the voltmeter's resistance is $1.0\times10^7\ \Omega$.

Response Before the voltmeter is connected, the potential difference across each 1.5-MΩ resistor will be half the supply potential difference, or 6.0 V. When the voltmeter is connected, it acts as a parallel resistance:

$$R = \frac{1}{\frac{1}{1.5\ \text{M}\Omega} + \frac{1}{1.0\times10^{7}\ \Omega}} = \frac{1}{\frac{1}{1.5\times10^{6}\ \Omega} + \frac{1}{1.0\times10^{7}\ \Omega}} = 1.3\ \text{M}\Omega$$

The voltage drop across the parallel combination is $V = (12.0\ \text{V})(1.3\ \text{M}\Omega)/(1.3\ \text{M}\Omega + 1.5\ \text{M}\Omega) = 5.6\ \text{V}$

Quick DEMO

Loaded and Unloaded Voltage Dividers

Materials variable DC power supply; multimeter; three 100-Ω, 1-W resistors; clip leads

Procedure Connect two of the resistors in series, and connect the power supply across them. Turn on the power supply to 6 V. Measure the potential difference across one of the resistors. Adjust the power supply for 12 V and again measure the potential difference. Begin to discuss parallel circuits by connecting the third resistor in parallel with the resistor where the potential difference measurements have been taken. Again measure the potential difference.

Discussion

Problem Draw a circuit diagram of a two-resistor voltage divider. Is it possible to make the divided voltage stable when varying loads will be connected?

Response Yes; it is possible, but not just by using resistors. When a voltage divider is loaded, the potential difference drops. It is possible to replace one of the voltage divider resistors with an integrated circuit voltage regulator, which can make the divided voltage more stable. A transistor can change its resistance and act as a regulator to hold the output more constant. **AL** **Visual-Spatial**

Parallel Circuits

student text pp. 630–633

PHYSICS CHALLENGE ACTIVITY

Parallel Circuits in Automobiles Interested students can investigate why the various devices in an automobile are wired in parallel and rated at 12 V. Have them evaluate the power source (a 12-V battery) and all the devices that require electrical power simultaneously, such as the ignition, headlights, taillights, and music player. They also can estimate what will happen to the current and combined resistance of the parallel circuit if another load is added. Researching car circuit diagnostics, overloads, and fuses also would be useful. Have them graphically depict a hypothetical car's circuitry. **AL** **Visual-Spatial**

REAL-WORLD PHYSICS

Measuring Resistance The resistance that an ohmmeter measures is the resistance between any two points. For example, in a circuit, it puts current through a wire and then measures the potential difference across the wire. Some practical applications for ohmmeters are to test the electrical resistance of motors, transformers, metal-to-wire connections, environmental sensors, high-voltage circuit breakers, and other disconnecting switches. Discuss other practical applications for this electrical testing equipment.

DIFFERENTIATED INSTRUCTION

Visually Impaired Prepare several lengths of light rope by tying several knots spaced about 5 cm apart. Tie several other ropes into parallel-circuit configurations, adding knots where necessary. Ask students to identify which ropes represent series circuits and which ones represent parallel circuits. Using a board that is perforated with rows of evenly spaced holes, pegs, and light rope or string, have students configure series and parallel circuits. Have students explain each of the configurations, including where the resistors and power source would be and where there would be current.
OL **Kinesthetic**

ADDITIONAL IN-CLASS EXAMPLE

Use with Example Problem 3.

Problem Four resistors, 50.0 Ω, 40.0 Ω, 30.0 Ω, and 20.0 Ω, are connected with a parallel circuit across a 120-V battery. Find the current through each branch of the circuit, the equivalent resistance of the circuit, and the current through the battery.

Response

$$I_A = \frac{V}{R_A} = \frac{120\text{ V}}{50.0\ \Omega} = 2.4\text{ A}$$

$$I_B = \frac{V}{R_B} = \frac{120\text{ V}}{40.0\ \Omega} = 3.0\text{ A}$$

$$I_C = \frac{V}{R_C} = \frac{120\text{ V}}{30.0\ \Omega} = 4.0\text{ A}$$

$$I_D = \frac{V}{R_D} = \frac{120\text{ V}}{20.0\ \Omega} = 6.0\text{ A}$$

$$R = \frac{1}{\frac{1}{R_A} + \frac{1}{R_B} + \frac{1}{R_C} + \frac{1}{R_D}}$$

$$= \frac{1}{\frac{1}{50.0\ \Omega} + \frac{1}{40.0\ \Omega} + \frac{1}{30.0\ \Omega} + \frac{1}{20.0\ \Omega}} = 7.8\ \Omega$$

$$I = \frac{V}{R} = \frac{120\text{ V}}{7.8\ \Omega} = 15\text{ A}$$

Content Background

Parallel-Circuit Fault-Current Protection A fault inside an electrical product can energize the conductive metal housing and present a shock hazard to a user. Insulation can fail and allow a wire to contact the housing. Three-prong power cords provide user protection because the third prong connects the housing to ground. When insulation fails, the third prong conducts the fault current directly to ground over a low-resistance path, which prevents any significant current from reaching the user's body. The characteristics of parallel circuits explain why this is true—the parallel branch with the lowest resistance (in this case, the third prong) will support most of the current.

Use the PhysicsLAB

In *Circuit Configurations,* students will investigate how circuit configuration affects the voltages and currents in lightbulbs in the circuit.

Use the PhysicsLAB

In *How do parallel resistors work?,* students will measure equivalent resistance in parallel circuits.

Use the MiniLAB

In *Parallel Circuit Measurements,* students will construct a simple parallel circuit and monitor the effect of adding resistors to the circuit.

Use the MiniLAB

In *Parallel Resistance,* students will show that potential differences across parallel components are the same.

3 Assess

Assess the MAINIDEA

Series and Parallel Circuits Show students a battery connected to three small lightbulbs in series. Ask students how they could rearrange the circuit so that the bulbs would be brighter. Arrange the three lightbulbs in parallel rather than series.

Check for Understanding

Comparing Series and Parallel Circuits Draw series and parallel circuits on the board. Ask students to identify and compare the two. Ask what electric quantity is constant in each case. Ask how equivalent resistance is determined in each case. Finally, ask how current, potential difference, resistance, and total and individual power dissipations are related. OL Visual-Spatial

Reteach

Series and Parallel Circuits Demo Review series and parallel circuits by displaying wires, switches, lamps, resistors, power supplies, and meters. Connect two resistors in series and measure the total resistance. Connect the same resistors in parallel and again measure the resistance.

Reading Checks and Caption Checks

Reading Check, *student text p. 625*

A series circuit is an electric circuit in which there is only one path for the current.

Caption Check, *student text p. 627*

The arrow is pointing in the direction that the conventional current flows.

Caption Check, *student text p. 630*

V symbolizes a voltmeter and A symbolizes an ammeter.

Reading Check, *student text p. 631*

The current through a resistor depends on the resistance of and the potential difference across the resistor.

Reading Check, *student text p. 633*

Electric potential is conserved in a circuit.

Reading Check, *student text p. 633*

The sum of increases of electric potential equals the sum of decreases in electric potential in an electric circuit.

Practice Problems

student text p. 626

1. 66 Ω, 2.9 A

2. 32 Ω, 2.3 A

3. 2.0×10^{2} Ω

4. a. It will increase.

b. It will decrease.

c. No; it does not depend on the resistance.

5. $V_1 = 28$ V; $V_2 = 35$ V; $V_3 = 12$ V

$V_1 + V_2 + V_3 = 75$ V, the voltage of the battery

student text p. 629

6. R_2 has failed. It has infinite resistance, and the battery voltage appears across it.

7. a. 0.031 A

b. 17 V

c. $P_{total} = 0.52$ W, $P_{R1} = 0.24$ W, $P_{R2} = 0.28$ W

d. Yes. The law of conservation of energy states that energy cannot be created or destroyed; therefore, the rate at which energy is converted, or power dissipated, will equal the sum of all parts.

8. If not for the shorting mechanism, the entire set would go out when one lamp burns out. After several lamps fail and then short, the reduced total resistance of the remaining working lamps results in an increased current sufficient to blow the fuse.

9. The resistor with the lower resistance will dissipate less power, and thus will be cooler.

10. 7.5 V

student text p. 630

11. a. 55 Ω; **b.** 2.2 A; **c.** 48 V, 72 V

12. a. 11.9 kΩ

b. 1.0×10^{-3} A

c. 11.9 V

13. 5.3 kΩ

student text p. 633

14. a. 5.00 Ω; **b.** 6.00 A; **c.** 2.00 A

15. a. It gets smaller.

b. It gets larger.

c. It remains the same. Currents are independent.

16. a. 20.0 Ω

b. 0.600 A

c. 0.100 A; 0.200 A; 0.300 A

17. 240 Ω in parallel with the 150-Ω resistance

Section 1 Review

student text p. 634

18. Students' answers should include the following ideas: (1) In a series circuit, the current in each device is the same, and the sum of the voltage drops is equal to the source voltage. (2) In a parallel circuit, the voltage drop across each device is the same, and the sum of the currents through each loop equals the source current.

19. 2.9 A

20. 810 mA. Current is the same everywhere in a series circuit.

21. a. 0 V; **b.** 0 V

22. When you walk around in a loop on the side of a hill and return to the starting point, the sum of the increases in height up the hill and the sum of the decreases in height down the hill are equal. When an electric charge travels around a loop in an electric circuit, the sum of the increases in electric potential equals the sum of decreases in electric potential.

23. The total number of charges is conserved. In an electric circuit, the total number of charges into a section of that circuit must equal the total number of charges out of that same section of circuit.

24. a. 0 A; the potentials of points A and B are the same.

b. Nothing

c. Nothing

d. Nothing

1 Introduce

Bellringer

Comparing Combined Series-Parallel Circuit Configurations Construct two series-parallel circuits. Each will use a 12-V power supply and three lightbulbs of the same type. One circuit will have two lightbulbs in series that are in parallel with the third. The other circuit will have two bulbs in parallel that are in series with the third. Energize both circuits at the same time to make the comparisons easy. OL Visual-Spatial

Tie to Prior Knowledge

Energy Conservation Review energy conservation and relate it to series-parallel circuits. Solve a series-parallel circuit for the total and individual power dissipations. Show that the total dissipation is equal to the sum of the individual dissipations.

2 Teach

Safety Devices

student text pp. 635–636

Identify Misconceptions

Circuit Breaker Operation There are several types of circuit breakers, some of which automatically reset when the breaker cools sufficiently. Discuss how a lack of understanding of this concept can lead people to draw false conclusions about the integrity of this type of circuit breaker. For example, a circuit might be overloaded and the circuit breaker will trip and turn everything off. Once the circuit breaker cools, electric charges will flow again. Once everything on the circuit turns on, the circuit breaker will trip again. This cycle might lead people to believe that the circuit breaker is not working properly, while it is actually working as it was designed to work.

Concept Development

Arc-Fault Circuit Interrupters An arc-fault circuit interrupter is designed to protect against fires caused by arcing faults in electrical wiring. Usually, circuit beakers respond to circuit overloads or short circuits. They are not designed to respond to arcing conditions. An arc-fault circuit interrupter monitors current and when an unwanted arcing condition occurs it trips. Arc-fault circuit interrupters use electronic detectors designed to respond to the electrical signals associated with arcing. They can shut down a circuit in many cases where an ordinary circuit breaker will not react. If a 120-V circuit is protected by a 15.0-A breaker, the maximum power is $P = IV = (15.0\ \text{A})(120\ \text{V}) = 1.80\ \text{kW}$. An arc at that power level is easily hot enough to ignite drapes, bedcovers, carpeting, and so on; a frayed and arcing power cord can start a fire.

Use Figure 12

Appliances Residential appliance circuits are often fused or breaker-protected at 15 A. Ask students to determine the maximum number of 400.0-W, 120-V appliances that can operate simultaneously on such a circuit. $I = \frac{P}{V} = \frac{400\ \text{W}}{120\ \text{V}} = 3.3\ \text{A}$ (each appliance)

Maximum number of appliances $= \frac{15.0\ \text{A}}{3.3\ \text{A}} = 4$ appliances

OL Logical-Mathematical

Reinforcement

Safety Devices Have students list the types of circuit protectors used in building wiring and how each one operates to protect household wiring. Also, ask which devices are more user-friendly.
OL Logical-Mathematical

Use the PhysicsLAB

In *Series and Parallel Circuits,* students will investigate the relationships among current, potential difference, and resistance in series and parallel circuits.

Combined Series-Parallel Circuits

student text pp. 637–638

Discussion

Problem Residential wiring uses parallel circuits. Why is it necessary to understand series-parallel behavior to fully understand residential wiring?

Response The resistance of the wires acts in series with the parallel loads. This action is what causes lights to dim when a heavy load is switched on. Also, the circuit breakers or fuses are in series and thus can deactivate the parallel loads. OL Logical-Mathematical

Critical Thinking

Operation of SPDT Switches Single-pole, double-throw switches can be used to control a load from two locations. For example, a light in a stairwell can be controlled by a switch located at the top and the bottom of the stairwell. Draw a diagram for a SPDT switch on the board with a load and a source, and explain its two-way operation. Ask students to think of other ways this type of circuit might be used, such as in a doorbell that has a push button at two different doors.
OL Logical-Mathematical

Use Models

MAIN IDEA Suppose a dozen different current values in R_3 had to be analyzed for the circuit included in the Problem-Solving Strategies. In such cases, the use of a model could save a lot of time. At the points where R_3 is connected, find the potential difference across the points if R_3 were disconnected.

$$V_{R2} = \frac{(60\ \text{V})(25\ \Omega)}{8.0\ \Omega + 25\ \Omega} = 45.5\ \text{V}$$

Next, find the equivalent parallel resistance of R_1 and R_2.

$$\frac{1}{R_P} = \frac{1}{R_1} + \frac{1}{R_2} = \frac{1}{8.0\ \Omega} + \frac{1}{25\ \Omega}$$

$$R_P = 6.1\ \Omega$$

In this case, the model is a 45.5-V battery in series with a 6.1-Ω resistor. If R_3 is now connected to the model, its current is found by $\frac{V}{R} = \frac{45.5\ \text{V}}{6.1\ \Omega + 15\ \Omega} = 2.2\ \text{A}$, which is the same value that can be obtained by calculating I_3 in the figure. Now write equations for the potential difference across and current in R_3. These equations are a mathematical model of the circuit at R_3 and allow you to quickly calculate V_3 and I_3 for a dozen, or more, different values of R_3.

$$V_3 = IR_3 = (2.2\ \text{A})R_3$$

$$I_3 = \frac{V_3}{R_3} = \frac{55\ \text{V}}{R_3}$$

PHYSICS CHALLENGE ACTIVITY

Balancing Circuits One model for the circuit shown in the Problem-Solving Strategies is appropriate when R_3 can take on various values. A different model is needed when R_2 can take on various values. Have students find the model that can be used to calculate the value of R_2. At the points where R_2 is connected, determine the voltage if R_2 were to become disconnected. $V_{R3} = (60\ \text{V})(15\ \Omega)/(8.0\ \Omega + 15\ \Omega) = 39\ \text{V}$. Next, find the equivalent parallel resistance of R_1 and R_3. $1/R_P = 1/R_1 + 1/R_3 = 1/8.0\ \Omega + 1/15\ \Omega$; $R_P = 5.2\ \Omega$. In this case, the model is a 39-V battery in series with a 5.2-Ω resistor. If R_2 is connected to the model, its current is found by $V/R = 39\ \text{V}/(5.2\ \Omega + 25\ \Omega) = 1.3\ \text{A}$.

AL **Logical-Mathematical**

ADDITIONAL IN-CLASS EXAMPLE

Use with Example Problem 4.

Problem An air conditioner with a resistance of 50.0 Ω (R_1) and a computer with a resistance of 20.0 Ω (R_2) are connected in parallel to a 120.0-V source through a 2.00-Ω resistor (R_3) in series, as shown below. Find the current through the computer when the air conditioner is and is not running.

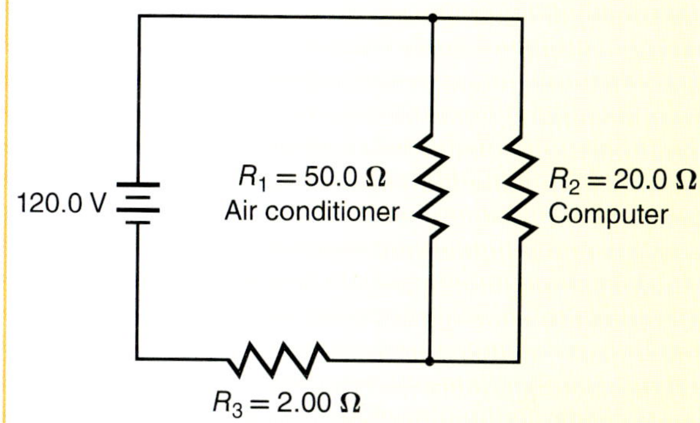

Response Current through computer when air conditioner is running:

$$R_P = \frac{1}{\frac{1}{R_1} + \frac{1}{R_2}} = \frac{1}{\frac{1}{50.0\ \Omega} + \frac{1}{20.0\ \Omega}} = 14.3\ \Omega$$

$$R = R_3 + R_P = 2.00\ \Omega + 14.3\ \Omega = 16.3\ \Omega$$

$$I = \frac{V_{source}}{R} = \frac{120.0\ \text{V}}{16.3\ \Omega} = 7.36\ \text{A}$$

$$V_3 = (2.00\ \Omega)(7.36\ \text{A}) = 14.7\ \text{V}$$

$$V_1 = 120.0\ \text{V} - 14.7\ \text{V} = 105.3\ \text{V}$$

$$V_2 = V_1 = 105.3\ \text{V}$$

$$I_2 = \frac{V_2}{R_2} = \frac{105.3\ \text{V}}{20.0\ \Omega} = 5.26\ \text{A}$$

Current through computer when air conditioner is not running:

$$R_2 = 20.0\ \Omega$$

$$R = R_3 + R_2 = 2.00\ \Omega + 20.0\ \Omega = 22.0\ \Omega$$

$$I = \frac{V_{source}}{R} = \frac{120.0\ \text{V}}{22.0\ \Omega} = 5.45\ \text{A}$$

$I = I_2$ because the current only has one path.

REAL-WORLD PHYSICS

Load Leveling Power companies must make investments in electric energy generation to meet peak demands. Since the demand often is not at peak, their equipment is not always used near full capacity. This makes it more difficult for power companies to realize a profit and pay for equipment. As such, power companies charge many of their customers a premium rate during periods of peak use, a practice that encourages customers to level their use and lower their energy bills. Computerized load levelers can be used to control heating and cooling loads in large buildings and avoid high costs. For example, water can be heated after midnight when costs are at their lowest point.

Ammeters and Voltmeters

student text p. 639

Quick DEMO

Series-Parallel Circuits

Estimated Time 10 minutes

Materials power supply, resistors or lamps, multimeter

MAIN IDEA Build a series-parallel circuit. Use the mulitmeter to measure all potential differences and currents. Demonstrate that potential differences in series add and that currents in parallel add.

3 Assess

Assess the MAIN IDEA

Circuit Effects Using a power supply and four light bulbs, build a series-parallel circuit similar to that shown in **Figure 13.** Ask students to describe what would happen if various wires were cut. Possible answers: If the wire directly adjacent to the series bulb is cut, all lights will turn off. If any wire directly adjacent to one of the parallel bulbs is cut, that individual bulb will turn off, but the rest of the bulbs will stay on.

Check for Understanding

Equivalent Resistance Ask students to calculate the equivalent resistance for the voltmeter circuit in **Figure 16.** (Ignore significant digits.) 19.99 Ω Ask how it would change if the voltmeter's resistance were 100 Ω. 19.09 Ω Ask them to explain the advantage of a large resistance for the voltmeter. The larger the voltmeter's resistance is, the less measuring a potential difference within a circuit affects that circuit. OL

Extension

Circuit Design Ask groups of students to design a circuit using switches that can turn a light off or on from three or more locations. Give each group a three-way switch and a double-pole, double-throw switch and an ohmmeter to get them started. OL Visual-Spatial

Reading Checks and Caption Checks

Caption Check, *student text p. 636*
15 A

Caption Check, *student text p. 637*
The electrons moving through the wire collide with each other and with other atoms. These collisions cause resistance to the movement of the electrons.

Practice Problems
student text p. 638

25. 0.54 A

26. The 11 lights in series will burn brighter. The parallel lights each will conduct half of the current of the series lights, and they will burn at one-fourth the intensity of the series lights since $P = I^2R$.

27. 2.5 A

Physics Challenge
student text p. 636

1. Yes; yes. You also can balance this circuit by adjusting the resistance values so $R_2/R_3 = R_4/R_5$ remains equal.
2. $R_3/R_2 = R_5/R_4$
3. any resistor but R_1
4. R_1; a galvanometer can be damaged by too much current. If R_1 is adjustable, it is set for a high value before the circuit is energized. This limits the current through the galvanometer. As the balancing resistor is adjusted and as the meter reading approaches zero, the sensitivity is then increased by decreasing R_1.

Section 2 Review
student text pp. 639

28. A combination series-parallel circuit contains segments that are in series and segments that are in parallel.

29. Bulbs 2 and 3 are equal in brightness but dimmer than bulb 1.

30. 0.6 A

31. Both dim equally. The current in each is reduced by the same amount.

32. 8.0 V

33. No; identical bulbs in series would have identical potential differences across them because their currents are the same.

34. fuses, circuit breakers, and ground-fault interrupters

35. Yes; you could arrange the circuit so that all the bulbs are in series with each other. Alternatively, you could arrange the circuit so that all the bulbs are in parallel with each other.

ON THE **JOB**

Live Wire

Background

Electricity in the home is both useful and safe when wiring is designed with physics principles in mind. Electrical energy in current through a wire always transforms some energy to thermal energy because of resistance. The more current drawn, the more thermal energy transformed. Proper design of wires, circuit breakers, and resistors ensures that the amount of current drawn will not exceed safe operating ranges.

Teaching Strategies

A powerful example of series and parallel circuit is the melting wire demonstration. In this demonstration, a single, thin wire is connected in series to a set of resistors, all connected to one another in parallel. As each resistor (perhaps a lightbulb) is switched on, the current through the single wire grows larger and larger until the wire glows and finally melts.

GOING**FURTHER** >>>

Expected Outcomes Answers will vary. While educational programs are available, many electricians learn their trade through apprenticeship programs while working with licensed electricians.

SECTION 1

Mastering Concepts

36. When one bulb burns out, the circuit is open and all the bulbs go out.

37. Each new resistor provides an additional path for the current.

38. The equivalent resistance will be less than that of any of the resistors.

39. Appliances in parallel can be run independently of one another.

40. In a series circuit, the current is opposed by each resistance in turn. The total resistance is the sum of the resistors. In a parallel circuit, each resistance provides an additional path for current. The result is a decrease in total resistance.

41. The amount of current entering a junction is equal to the amount of current leaving.

Mastering Problems

42. a. 0.20 A, because current is constant in a series circuit

b. 0.20 A, because current is constant in a series circuit

43. 13 kΩ

44. 40 kΩ

45. 12.4 V

46. 4.45 A

47. Potential increase = 1.5 V + 1.5 V − 1.5 V = 1.5 V.
Potential decrease across bulb = I(15 Ω).
So, $I = \frac{1.5\ \text{V}}{15\ \Omega} = 0.10$ A.

48. a. The energy increase across the battery is $q\Delta V$. That equals the energy decrease across the resistor, qIR. Therefore, there is no net change in energy despite how many times it goes around the circuit. Energy is conserved.

b. Its increase, $q\Delta V$, is larger than its decrease, qIR, so its energy increases each time it goes around the circuit. Energy would be created, not conserved.

49. a. 37 Ω

b. 7.4 V

c. 0.88 W

d. 1.5 W

50. a. 11 V

b. 7.5 V

c. 19 V

51. a. 26 Ω

b. 1.7 A

c. 37 V, 7.6 V

d. 63 W, 13 W

52. 8.40 V

53. 2.45 A

54. a. 0.20 A

b. 7.0 V

c. 3.0 V

d. 120 J

e. $R = R_1 + R_2 + 35\ \Omega + 15\ \Omega = 50\ \Omega$

55. a. 50 Ω; since $P = I^2R$ and I is constant in a series circuit, the largest value of resistance will produce the most power.

b. 15 Ω; since $P = I^2R$ and I is constant in a series circuit, the smallest value of resistance will produce the least power.

c. 2.0 A

d. 4×10^2 W

56. a. 62 Ω

b. 6.0 V

57. 1.6×10^2 Ω

58. 13.45 W

59. a. 10.0 Ω; since $P = \Delta V^2/R$ and ΔV is constant in a parallel circuit, the smallest resistor will dissipate the most power.

b. 50.0 Ω; since $P = \Delta V^2/R$ and ΔV is constant in a parallel circuit, the largest resistor will dissipate the least power.

c. 19 A

d. 5.5 A

e. 2.2 A

f. 11 A

60. a. 2.0×10^1 V

b. 3.4 A

c. 1.0 A

d. 2.0 A

61. down

62. a. 2.3×10^2 Ω

b. 13 Ω

c. 3.6 W

63. a. 2.2×10^2 Ω

b. 65 W

c. It increased.

64. a. 8.89 Ω

b. 4.50 A

c. 2.50 A

65. 15 V

66. 66 Ω

67. a. 52 Ω

b. 110 V

c. 9.8 Ω

d. 96 V

SECTION 2

Mastering Concepts

68. The purpose of a fuse is to prevent conductors from being overloaded with current, causing fires due to overheating. A fuse is simply a short length of wire that will melt from the heating effect if the current exceeds a certain maximum.

69. A short circuit is a circuit that has extremely low resistance. A short circuit is dangerous because any potential difference will produce a large current. The heating effect of the current can cause a fire.

70. An ammeter must have low resistance because it is placed in series in the circuit. If its resistance were high, it would significantly change the total resistance of the circuit and thus serve to reduce the current in the circuit, thereby changing the current it is meant to measure.

71. A voltmeter is placed in parallel with the portion of the circuit whose difference in potential is to be measured. A voltmeter must have very high resistance for the same reason that an ammeter has low resistance. If the voltmeter had low resistance, it would lower the resistance of the portion of the circuit it is across and increase the current in the circuit. This would produce a higher potential difference across the part of the circuit where the voltmeter is located, changing the voltmeter's measurement.

72. An ammeter is connected in series; a voltmeter is connected in parallel.

Mastering Problems

73. 45.0 Ω

74. 360 mW

75. 11 mA

76. 15 mA

77. a. 2.0 A

b. 3.0 A

c. 15 A

d. yes

78. a. $I_{30.0\,\Omega} = I_{20.0\,\Omega} = I_{10.0\,\Omega} = I_{40.0\,\Omega} < I_{25.0\,\Omega}$

b. $V_{10.0\,\Omega} < V_{20.0\,\Omega} < V_{30.0\,\Omega} < V_{40.0\,\Omega} < V_{25.0\,\Omega}$

79.

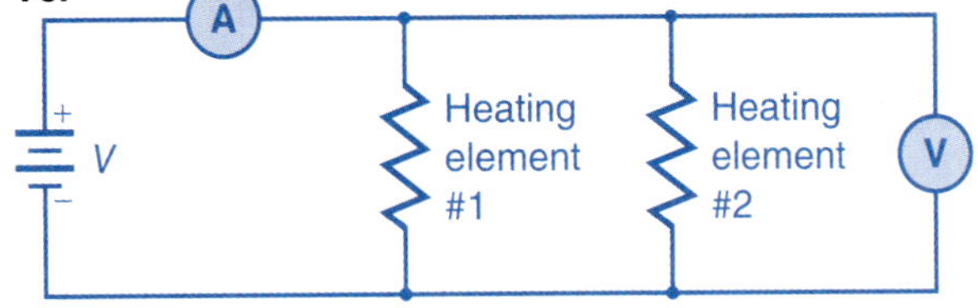

80. 1.1×10^3 s

81. a. 0.24 kΩ

b. 0.50 A

c. 6.0×10^1 W

Applying Concepts

82. If one of the lamp filaments burns out, the current will cease and all the lamps will go out.

83. As R_1 increases, ΔV_2 will decrease.

84. Circuit A: There will be no current in the resistor. Circuit B: The current in the resistor will remain the same.

85. If one of the filaments burns out, the resistance and the potential difference across the other lamps will not change; therefore their currents will remain the same.

86. No, the 30-Ω resistors can be used in parallel. Three 30-Ω resistors in parallel will give a 10-Ω resistance. Two 30-Ω resistors in parallel will give a 15-Ω resistance.

87. Connect four of the bulbs in series. The potential difference across each will be $(6.0\ \text{V})/4 = 1.5$ V.

88. a. The lamp with the lower resistance: $P = I\Delta V$ and $I = \Delta V/R$, so $P = \Delta V^2/R$. Because the potential difference is the same across both lamps, the smaller R means larger P, and thus will be brighter.

b. The lamp with the higher resistance; $P = I\Delta V$ and $\Delta V = IR$, so $P = I^2R$. Because the current is the same in both lamps, the larger R means larger P, and thus will be brighter.

89. The 30-A fuse allows more current through the circuit, generating more heat in the wires, which can be dangerous.

90. a. series

b. series

c. parallel

d. series

e. parallel

f. series

Mixed Review

91. a. 6.0 V

b. 4.7 V

c. 6.0 V

92. 7.0×10^{1} V

93. 11 W

94. 21 V

Thinking Critically

95. a. See Online Solutions Manual for circuit diagram.

b. See Online Solutions Manual for circuit diagram.

c. See Online Solutions Manual for circuit diagram.

d. See Online Solutions Manual for circuit diagram.

96. a. 0.137 A

b. 0.395 W

c. 0.014 W

97. a. 6.0 kΩ

b. 2.0 kΩ

98. Answers will vary, but a correct form of the answer is, "A 6.0-V battery is connect to two parallel branches. One branch contains a 500 Ω resistor, while the other has a 100 Ω resistor in series with a 200 Ω resistor. Find the current in each branch of the circuit."

99. Answers will vary. A possible form of the correct answer would be, "If you connected it to a 12-V battery, what current would the battery produce?"

Writing in Physics

100. Key ideas are:

(1) Kirchhoff's Voltage Law (KVL) is conservation of energy applied to electric circuits.

(2) Kirchhoff's Current Law (KCL) is conservation of charge applied to electric circuits.

(3) KVL states that the algebraic sum of voltage drops around a closed loop is zero. In a series circuit there is one closed loop, and the sum of voltage drops in the resistances equals the source voltage. In a parallel circuit, there is a closed loop for each branch, and KVL implies that the voltage drop in each branch is the same.

(4) KCL states that the algebraic sum of currents at a node is zero. In a series circuit, at every point the current in equals current out; therefore, the current is the same everywhere. In a parallel circuit, there is a common node at each end of the branches. KCL implies that the sum of the branch currents equals the source current.

Cumulative Review

101. a. 1.00 m

b. 340 Hz

c. 170 m/s

d. 680 Hz

e. 230 Hz

102. −7.41 cm

103. 1.71

104. 0.40 m

105. a. $E/9$

b. $3E$

c. $E/3$

d. E

e. $E/3$

Standardized Test Practice

student text p. 647

Multiple Choice

1. C
2. D
3. A
4. D
5. C
6. D
7. D
8. C

Free Response

9. Chris must disconnect 5 lamps.
10. $I = 0.20$ A; $P = 1.6$ W

Rubric

The following rubric is a sample scoring device for free response questions.

Points	Description
4	The student demonstrates a thorough understanding of the physics involved. The response may contain minor flaws that do not detract from the demonstration of a thorough understanding.
3	The student demonstrates an understanding of the physics involved. The response is essentially correct and demonstrates an essential but less-than-thorough understanding of the physics.
2	The student demonstrates only a partial understanding of the physics involved. Although the student may have used the correct approach to a solution or may have provided a correct solution, the work lacks an essential understanding of the underlying physical concepts.
1	The student demonstrates a very limited understanding of the physics involved. The response is incomplete and exhibits many flaws.
0	The student provides a completely incorrect solution or no response at all.

CHAPTER 24
Magnetic Fields

Chapter Overview

This chapter defines and explains magnetic poles, magnetic fields, and magnetic forces. The first section presents the right-hand rules and establishes the relationship between electric current and magnetic fields. The second section covers forces exerted on current-carrying wires in magnetic fields and the force exerted on a charged particle as it travels through a magnetic field. Applications such as loudspeakers, computer storage media, and motors are also explained.

Before students study the material in this chapter, they should study:

- Electric fields
- Vectors v. scalars

To solve the problems in this chapter, students will need a strong grasp of:

- Scientific notation
- Significant figures
- Solving linear equations

Introduce the BIG IDEA

Using experiments very much like the ones discussed in this chapter, Oersted, Faraday, and others showed that magnetism and electricity were linked—a link that involves, of all things, light! Today magnetic fields and the forces they exert on moving charges or other magnetic materials are used all around us, from junkyards to hospitals. Ask your students to think of devices that use magnetic fields and forces. Many devices use magnetism, including computer hard drives, MRI equipment, electromagnets, electric motors and generators, and speakers.

(l)Kevin Schafer/The Image Bank/Getty Images, (r)Index Stock Imagery, Inc.

About the Photo

Aurora Borealis The aurora borealis, shown in the photo, is a dramatic effect arising from the interaction of Earth's magnetic field with the solar wind. Charged particles, mostly electrons, travel in spiral paths along the lines of Earth's magnetic field and get funneled toward the poles. Discuss with students why understanding the physics of such phenomena might be important. Besides being of great general interest, this phenomenon can perturb radio and satellite communications. It also occurs at altitudes at which satellites orbit, so it affects manned and unmanned spaceflight. Magnetism in general is used in many technologies, such as MRI, electric motors, etc.

Go online!
connected.mcgraw-hill.com

Your online experience includes:

- **Video**
- **Concepts In Motion**
- **Virtual Investigations**
- **All Lab, Applying Practices activities, and PBL worksheets**
- **Online Practice**
- **Personal Tutor**
- **Multilingual eGlossary**
- **Presentation**

Use the LaunchLAB

In *Direction of Force on a Magnet,* students can observe the effects of magnets on a compass.

Use Physics T.V.

Video

MAGNETIC FIELDS How are magnetic fields used in your life? You might be surprised! Explore the physics of everyday magnets and magnetic fields.

1 Introduce

Bellringer

Creating a Magnet You can use a pistol-type soldering gun to illustrate demagnetization. Magnetize a screwdriver by stroking it with a permanent magnet. Show students that it can pick up small metal objects such as pins. Now demagnetize the screwdriver blade as follows: Squeeze and hold the soldering-gun trigger and insert the blade between the wires that hold the tip. Withdraw the blade before releasing the trigger. The screwdriver will no longer attract metal objects because its magnetic domains are now randomly arranged.

Tie to Prior Knowledge

Fields and Electric Current The chapter on gravitation introduced the concept of fields, and students studied this concept again in the chapter on electric fields. This chapter introduces magnetic fields to help explain attraction and repulsion between magnets.

2 Teach

Properties of Magnets

student text pp. 650–653

Identify Misconceptions

Magnetized Steel Students might believe that all steel alloys can be magnetized. You can have students test this belief by placing a long stainless steel bolt or screw in contact with the north pole of a permanent magnet. You can test to see whether the other end is acting as a magnetic pole by trying to pick up iron filings. Some stainless alloys are magnetic, albeit more weakly magnetic than other steels. Test different items and let students observe the differences.

Quick DEMO

Magnetic Repulsion

Estimated Time 5 minutes

Materials pencil, two disk magnets

Procedure Hold a pencil vertically. Slide the disk magnets onto the pencil with like poles facing each other. The top magnet will float above the bottom magnet. Ask students what keeps the upper magnet floating. Magnetic repulsion between the two like poles pushes the two magnets apart. The upper magnet floats in a position where the magnetic force from the bottom magnet balances the gravitational force from Earth. Ask students what determines the size of the gap. Gravity and the strength of the magnets **OL** **Kinesthetic**

Concept Development

Mutual Forces Remind students that interactions always come in pairs. For example, if one pole exerts a force on a second pole, the second pole exerts a force of equal strength that pushes the first pole in the opposite direction. This is another example of Newton's third law.

Using an Analogy

Reluctance in Magnetic Circuits A magnetic circuit is the closed path described by magnetic flux. Reluctance (magnetic) is analogous to resistance (electric) in that reluctance is a measure of the opposition to magnetic flux offered by a magnetic circuit. A magnetic circuit with reluctance is analogous to an electric circuit with resistance: flux, reluctance, and magnetomotive force are analogous to electric current, resistance, and electromotive force. Use resistance in an electric circuit to help students understand reluctance in a magnetic circuit. As resistance goes up, current goes down and as reluctance goes up, magnetic field strength goes down.

DIFFERENTIATED INSTRUCTION

Visually Impaired Have students handle two common ceramic magnets. Ask them to move the magnets together, face-to-face, and notice if there is a force of attraction or repulsion. If there is repulsion, have them describe any changes they observe as they try to bring the magnets closer together. Next, have them turn one of the magnets so the opposite face is presented and repeat the experiment. Students should observe a force in the direction opposite to what they first experienced. Ask them if the face of a ceramic magnet is a pole. If it attracts or repels other magnets, then it is a magnetic pole. **OL** **Kinesthetic** **EL**

Use the MiniLAB

In *Magnetic Domains,* students can observe the order of magnetized objects and the disorder of unmagnetized objects.

Magnetic Fields Around Magnets

student text pp. 653–654

Reinforcement

Induced Polarity Have students reflect on what they know about induced charges. Refer them to the chapter about static electricity. Ask students to predict the induced polarity of a metal object as it is brought near a pole of a permanent magnet. The end of the metal object that is closest to the pole of the permanent magnet becomes an opposite pole because opposite poles attract. **OL**

Use the MiniLAB

In *3-D Magnetic Fields,* the 3-D shape of magnetic fields is explored.

Quick DEMO

Induced Polarity

Estimated Time 2 minutes

Materials compass, permanent magnet with the poles identified, two nails

Procedure Place the head of one of the nails into contact with the north pole of the permanent magnet. Place the opposite end of the other nail in contact with the north pole of the permanent magnet. Test the polarity of both nails using the compass. Discuss induced polarity. Ask students if the nail will stay magnetized when removed from the permanent magnet. The nail will only be magnetized while in contact with the permanent magnet. **OL** **Kinesthetic**

REAL-WORLD PHYSICS

How a Stud Finder Works Before beginning renovations or when hanging heavy paintings, it is important to locate the studs in a wall without damaging the wall. The capacitive stud finder is able to distinguish between the densities of building materials such as wallboard and wooden framing components (i.e., studs). The two materials have different dielectric constants, so a stud that lies behind a wall can be detected as a difference in capacitance. The old-fashioned magnetic stud finder uses a small magnet that can rotate. The magnet turns upright when it is over a nail in a stud.

Use Figure 8

Ask students how the concept of repulsion could be applied in a transportation system to improve energy efficiency. With maglev trains, strong magnets (electromagnets) are used to produce magnetic repulsion between the train and the track. By eliminating physical contact between the train and the track, friction is eliminated. With friction eliminated, less energy is required to propel the train. **OL**

Identify Misconceptions

Metal Surgical Implants Are Not Magnetic A metal plate is sometimes used to fill a defect in the skull, which may arise from trauma, surgery, or another cause. Thanks to Hollywood and some fiction writers, students might think that it is possible to use a magnetic field to attract these metal plates. However, like many other metal surgical implants, these plates are made of titanium, which is nonmagnetic.

Electromagnetism

student text pp. 655–657

Discussion

Question Oersted observed the rotation of a compass needle in response to current in a nearby wire. Ask students how Oersted's observations might have changed if he had used a variable resistor in series with the wire and the power supply. He probably would have noticed a relationship between the circuit resistance and the deflection of the needle. He might have concluded that there was an indirect relationship between resistance and deflection and a direct relationship between current and deflection. **AL**

Critical Thinking

Constantly Changing Fields Ask students to predict what would happen to an iron bar placed into a magnetic field with constantly changing polarity. Then ask them to elaborate, assuming the domains resisted reorientation. The domains will constantly realign with the alternating field. Since the domains resist this change, heat results just as friction results in heat in mechanical systems. The heat from domain realignment is called *hysteresis loss.* Designers of motors and transformers use silicon steel alloys to minimize this loss. Since the domains in silicon steel are easily reoriented, it is not useful as a permanent magnet. **OL** **Logical-Mathematical**

REAL-WORLD PHYSICS

MAINIDEA Electromagnetic cranes can lift and release on command. Waste management workers use such cranes to move wrecked cars and trucks in junkyards, among other things. Challenge students to build a working model of an electromagnetic crane that will lift and release a toy car.

PHYSICS CHALLENGE ACTIVITY

Determining Magnetization Steel Without a Magnet Ask students how they would identify which of several steel bars were magnetized and which were demagnetized, using only the steel bars. Only the magnetized steel bars would show a force of repulsion. The process could begin by randomly selecting two bars, bringing the bars together end-to-end, and then flipping one bar end-for-end. Eventually, a force of repulsion will be discovered. This would identify the selected bars as permanent magnets. Then, each remaining bar could be tested at each end against one end of a magnetized bar. The demagnetized bars would show attraction at both ends. Also, if two demagnetized bars are selected, no force would be detected as the ends are brought together. **AL** **Kinesthetic**

Use the PhysicsLAB

In *Current and Field Strength,* students can observe the magnetic field around a current-carrying wire and they can explore the properties of the magnetic field.

Use the PhysicsLAB

In *Make an Electromagnet,* students can learn more about the properties of electromagnets.

3 Assess

Assess the MAINIDEA

Ask the students what would happen if the current in the wires of an electromagnet were reversed. The poles of the electromagnet would reverse, but ferromagnetic materials would still be attracted to the electromagnet. Ask the students what would happen if the wiring of a solenoid-type electromagnet were wrapped in both directions (i.e., counter-clockwise and clockwise). The wires in each direction would produce opposing magnetic fields, which would act against each other, resulting in a weaker overall magnetic field.

Check for Understanding

Earth's Magnetic Poles The colored (often blue) end of a compass needle is called the north-seeking pole or simply the north pole. Ask students to draw a conclusion about the identification and location of Earth's magnetic poles. The blue end of the needle points to Earth's magnetic north pole. In principle, you could walk along, following the needle, until it pointed down into the ground. This point would be Earth's magnetic north pole. Then ask students how they would use a compass to verify the right-hand rule for a magnetic field around a solenoid. Hold the solenoid in your hand as shown in **Figure 12** and allow current to flow through the solenoid as shown in the figure. If you hold the compass near the end of the solenoid that your thumbs points to, the north-seeking pole of the compass should point toward your thumb. **OL**

Reading Checks and Caption Checks

Caption Check, *student text p. 651*

The head of the nail is its north pole because it is attracted to the magnet's south pole.

Reading Check, *student text p. 653*

Both electric fields and magnetic fields are regions in space in which a magnet or electric charge exerts a force on another object.

Reading Check, *student text p. 655*

The field would go in the opposite direction, but it would remain at a right angle to the current.

Caption Check, *student text p. 656*

Inside the solenoid, the contributions to the magnetic field from all sides of the solenoid add up, creating a stronger overall magnetic field. Outside the solenoid, the contributions to the overall magnetic field from the near and far sides of the solenoid are in opposite directions, making a much weaker overall magnetic field.

Practice Problems

student text p. 652

1. a. repulsive

b. attractive

2. south, north, south, north

3. Earth is like a giant magnet. Earth's geographic North Pole is actually its magnetic south pole. The north end of a compass needle, therefore, points to Earth's magnetic south pole.

4. When students bring compasses near magnets, the magnetization of the compass flips.

Practice Problems

student text p. 657

5. a. twice as strong

b. three times as strong

6. a. from south to north

b. west

7. the pointed end

8. Use the iron rod. Iron would be attracted to a permanent magnet and take on properties of a magnet, whereas aluminum or glass would not. This effect would support the magnetic field in the wire coil and thus make the strongest electromagnet.

9. Yes. Connect the potentiometer in series with the power supply and the coil. Adjusting the potentiometer for more resistance will decrease the current and the field strength.

Section 1 Review

student text p. 657

10. You could connect either end of a wire to a source of current. The strongest electromagnets are solenoids, where wire in a circuit is wrapped around a ferromagnetic rod, such as iron, which increases field strength.

11. Field lines represent the strength and the direction of a magnetic field.

12. Student answers may vary. Answers could include magnets on a refrigerator and Earth's magnetic field. The effects of these forces can be demonstrated by bringing another magnet, or a ferromagnetic material, nearby.

13. at the poles

14. If the currents were in the same direction, the magnetic field would be approximately twice as large; if the currents were in opposite directions, the field would be approximately zero.

15. If you grasp the wire with your right hand with your thumb pointing in the direction of the conventional current, your fingers curl in the direction of the field.

16. None; the filings would show the same field pattern. However, a compass would show that the magnetic polarity had reversed.

17. The domains return to a random arrangement because they no longer align with the domains of the field of the permanent magnet.

18. a. The metal rods could be magnets with their axes parallel. If the top magnet is positioned so that its north and south poles are above the north and south poles of the bottom magnet, it will be repelled and float above. If the top magnet is turned end-for-end, it will be attracted to the bottom magnet.

b. ferromagnetic

1 Introduce

Bellringer

Electrons and Magnetic Fields Older computer monitors used cathode-ray tubes (CRT) that sent charged particles to the screen. Liquid-crystal display (LCD) monitors work with changing crystal structures, not moving charges. Use a small ceramic magnet in conjunction with a CRT monitor to demonstrate the interaction between the electrons and the magnetic field. Repeat the demonstration with an LCD monitor to show that there is no interaction. Follow up by asking students why CRT displays often have demagnetizing coils wound around their envelopes. It shields the display from changing magnetic fields. Other nearby appliances could be generating these fields. OL Visual-Spatial

Tie to Prior Knowledge

Magnetic Fields Review how a static charge produces an electrostatic field that, in turn, produces a force on a nearby second static charge. Discuss how permanent magnets and current-carrying wires produce magnetic fields. A moving charge produces a magnetic field that produces a force on another moving charge. Help students draw an analogy between the static charges and the electrostatic field versus moving charges and the magnetic field. OL

2 Teach

Forces on Current-Carrying Wires

student text pp. 658–662

Concept Development

Right-Hand v. Left-Hand Rules The right-hand rules are appropriate for analyzing magnetism from conventional currents (i.e., current is in the direction in which positive charge flows). Those who think about electron current often use the left-hand rules. Demonstrate the left-hand rule for students by showing them a "hitchhiker's" fist (left handed, of course). You know that the magnetic field produced by an electric current is always oriented perpendicular to the direction of the current. The left-hand rules says that, if the thumb points in the direction of the electron current, the curled fingers point in the direction of the magnetic flux lines produced by the electron current.

Reinforcement

Changing Directions Using **Figure 14,** have students consider what happens to the force when the current is in the opposite direction, when the magnetic field is in the opposite direction, or when both are in opposite directions. For the first two cases, the direction of force reverses. For the last case, the direction of force doesn't change.

Quick **DEMO**

Brute Force on Wire

Estimated Time 10 minutes

Materials permanent magnet, lightweight wire, 6-V battery

Procedure Lay the magnet on a table and position the middle portion of the wire approximately 1 cm from one of the poles of the magnet. Connect the ends of the wire to the battery terminals. The wire should move. If not, the magnetic field may be parallel to the wire, so change the orientation of the magnet with respect to the wire. Now, without changing the orientation of the wire in the vicinity of the magnet, reverse the wire ends on the battery terminals to show that the force between the magnet and wire reverses.

Critical Thinking

Do a Thought Experiment with a Speaker Ask students what would happen if you took a sturdy loudspeaker and plugged it directly into the wall socket. **WARNING: Do this only as a thought experiment!** If the speaker is not sturdy enough, it may explode. If the speaker's coil is of sufficiently low resistance to allow an excessive current, the current may be high enough to overheat and damage the speaker or trip the circuit breaker. OL Auditory-Musical

Quick DEMO

MAIN IDEA

Estimated Time 10 minutes

Materials 1.5-V alkaline D-cell battery, wire, loudspeaker

Procedure Demonstrate how the speaker converts electrical energy to sound energy. Stick a piece of ferromagnetic material on the back side of the speaker first, to show that the speaker has a magnet. With the wires connected to the speaker contacts, brush one wire along one of the battery contacts to produce a crackling sound. Then have students observe the direction of cone movement when there is a steady current. Reverse the battery polarity and have students notice that the cone moves in the opposite direction. Discuss with the students how the current in the magnetic field produces a force on the speaker.

ADDITIONAL IN-CLASS EXAMPLE

Use with Example Problem 1.

Problem What is the force on a 12-cm straight wire in a 1.9-T magnetic field when the current in the wire is 25 A?

Response $F = ILB = (25\ \text{A})(0.12\ \text{m})(1.9\ \text{T}) = 5.7\ \text{N}$

PHYSICS PROJECT ACTIVITY

Uses of Superconducting Magnets Have each student research a particular application of superconducting magnets and write a brief (two-page) report on that application. Accept both current and future applications as topics. Some possibilities include magnetic levitation, fusion reactors, nuclear magnetic resonance spectroscopy, MRI, particle accelerators, and high-efficiency propulsion systems. OL Linguistic Intrapersonal

Identify Misconceptions

Forces at the Ends of the Loop Students might be concerned about forces that they believe would be exerted on either end of the loop in **Figure 17.** Assuming the ends of the loop are inside the magnetic field, the currents at the ends are either in the same direction as the magnetic field or in the opposite direction. In both cases, there is no force because a magnetic field does not produce a force on a current that is parallel to the magnetic field. It only produces a force when the current is nonparallel to the magnetic field lines.

Discussion

Question Why are digital meters now more popular than analog meters?

Answer There are several reasons: (1) digital meters are easier to interpret, (2) analog meters are delicate mechanical devices with many moving parts and are thus more fragile, and (3) integrated circuit technology often makes all-electronic solutions to problems less expensive. OL

DIFFERENTIATED INSTRUCTION

Struggling Students Form small groups. Have each group identify several applications of magnetism. Have each group give its list to a different group. The groups should select an item from the list that they were given and identify the units and the equations that would be relevant for that item. Have each group share the item, units, and equations with the class. BL Interpersonal EL

Concept Development

Magnetic Circuit Breakers Magnetic circuit breakers trip when the current in a coil is high enough to move an iron armature, which then opens the contact points. One problem with high-energy circuit breakers is that an arc can form when the contacts open. A magnetic field can be used to "blow out" the arc. This is sometimes called a magnetic quench.

Using Models

IGRF Magnetic models of our planet are often used for investigative research, navigation, and surveys. One such model is known as the international geomagnetic reference field (IGRF). IGRF users must be aware of its limitations. Earth's magnetic field is extremely complicated in both space and time. The IGRF model does not account for local magnetizations. Many geological formations and rocks are partially magnetized.

Content Background

Magnetic Bearings Magnetic bearings eliminate friction and wear and allow high speeds of rotation. They work by suspending a rotating steel shaft in a magnetic field. The shaft is attracted by the field but is not allowed to touch the poles of the attracting electromagnet. This is accomplished by sensing shaft position and using that information to control the amount of current in the electromagnet. As the shaft moves closer to the electromagnet, the current is decreased to weaken the field. As the shaft moves away from the electromagnet, the current is increased to strengthen the field. This is an example of a negative feedback system where the error between a desired parameter value and the actual parameter value adjusts a control parameter to decrease the error.

Careers

Electrical Engineers Electrical engineers use the scientific principles of magnetism and electromagnetism to design motors, transformers, generators, data storage units, relays, circuit breakers, and a wide variety of other devices. They often work in conjunction with physicists and mechanical engineers. They sometimes use computers to model magnetic circuits and devices.

Forces on Single Charged Particles

student text pp. 662–665

ADDITIONAL IN-CLASS EXAMPLE

Use with Example Problem 2.

Problem An engineer is designing a deflection system for an electron-beam device and needs a force of 2.8×10^{-14} N on each electron in the beam, which travels at 1.7×10^{6} m/s. Determine the required field strength.

Response $B = \frac{F}{qv}$

$$= \frac{2.8 \times 10^{-14}\ \text{N}}{(1.602 \times 10^{-19}\ \text{C})(1.7 \times 10^{6}\ \text{m/s})}$$

$$= 0.10\ \text{T}$$

Reinforcement

Magnetic Flicker Light Obtain a magnetic flicker light, and ask how it works. A magnetic flicker light has a moveable filament and a permanent magnet. When there is current through the filament, a field is produced that interacts with the field of the permanent magnet, resulting in a force on the filament. OL Visual-Spatial EL

REAL-WORLD PHYSICS

Northern Lights The aurora borealis, also known as the northern lights, are produced when charged particles hit air particles in Earth's atmosphere. The charged particles come from the Sun. Earth's magnetic field funnels these charged particles toward Earth's magnetic poles. At the poles, these charged particles collide with air particles, exciting the atoms of those air particles. The air particles emit light when their atoms return to a nonexcited state. This light is visible and is called the northern lights.

3 Assess

Assess the MAIN IDEA

Have the students consider two wires. Assume the current in each wire flows in opposing directions. Have the students determine the direction of the magnetic field in the vicinity of the right-hand wire that is created by the current in the left-hand wire, and vice versa. If there is a current up the page in the left-hand wire, it will produce a magnetic field into the page at the right-hand wire. The right-hand wire will have a current down the page and produce a magnetic field into the page at the left-hand wire. Ask the students to determine the direction of the force exerted between the two wires. If the current in the wires is in the opposite direction, the force between the wires will be repulsive.

Check for Understanding

The Three Right-Hand Rules Verify that students understand the three right-hand rules. Draw various wire, coil, and field examples on the board. Ask them to predict the fields, poles, and force directions depending on the conditions that you provide. OL

Extension

Centripetal Acceleration When the initial velocity of a charged particle is perpendicular to a uniform magnetic field and no other forces are involved, the charged particle will move in uniform circular motion. Thus, the force of the magnetic field on the charged particle is a centripetal force. Ask students to determine the centripetal acceleration. The centripetal acceleration is determined by $F = ma$. Thus, $qvB = ma$, or $a = qvB/m$. OL

Reading Checks and Caption Checks

Caption Check, *student text p. 658*

The force would reverse direction, so it would push down instead of up.

Reading Check, *student text p. 661*

For an ammeter, connect the resistance in parallel to the galvanometer. For a voltmeter, connect the resistor in series with the galvanometer.

Practice Problems

student text p. 660

19. You would use the right-hand rule for magnetic force on a wire. When you point the fingers of your right hand in the direction of the magnetic field and your thumb in the direction of the wire's conventional (positive) current, the palm of your hand will face in the direction of the force acting on the wire. To use this method, you would need to know the direction of the current and the direction of the field.

20. 1.6 N

21. 0.13 T

22. 0.15 T

23. 7.8 A

24. One pole should be held as close to the coil as possible so that the field lines are perpendicular to both the wires and the direction of motion of the plate.

Practice Problems

student text p. 664

25. down

26. 3.2×10^{-12} N, up

27. 8.6×10^{-16} N

28. 1.7×10^{-13} N

29. 4.2×10^{6} m/s

30. 0.05 T

Section 2 Review

student text p. 665

31. An armature in a magnetic field rotates 360° as a split-ring commutator changes the direction of current, producing mechanical energy.

32. up away from the surface of Earth

33. **a.** The magnetic field must be up, at a right angle to the protons' velocity. The electric fields should be in the direction of the velocity—clockwise.

b. Neither field needs to be changed.

34. Both the galvanometer and the electric motor use a loop of wire positioned between the poles of a permanent magnet. When a current passes through the loop, the magnetic field of the permanent magnet exerts a force on the loop. The loop in a galvanometer cannot rotate more than 180°. The loop in an electric motor rotates through many 360° turns. The motor's split-ring commutator allows the current in the loop to reverse as the loop becomes vertical in the magnetic field, enabling the loop to spin in the magnetic field. The galvanometer measures unknown currents; the electric motor has many uses.

35. Not necessarily; if the coil is already in rotation, then rotational inertia will carry it past the point of zero torque. It is the coil's acceleration that is zero, not the velocity.

36. 28 kΩ

37. Because the force is attractive, the currents are in the same direction. That is, an up current in the first wire creates a magnetic field that intersects the second wire. If the current in the second wire is in the same direction, the force on it will pull the wires together.

A CLOSER **LOOK**

Leaf me alone

Biomagnetic Monitoring

Background

Particulate pollution, especially particulate matter with a diameter less than 2.5 μm, poses a significant health risk. These particles can carry toxic compounds deep into the lungs, where they can cause respiratory infections. The particles can find their way into the bloodstream and eventually lead to heart and liver problems, certain types of cancer, and possibly Alzheimer's and Parkinson's diseases. Worldwide, small-particle particulate pollution is estimated to be responsible for 800,000 deaths per year.

Teaching Strategies

- Before reading the feature, ask students to review the material in the chapter on temporary magnets and magnetic domains. The biomagnetic monitoring process described in the feature uses a strong magnetic field to turn iron-containing pollutant particles into tiny temporary magnets.
- After reading the feature, tell students that biomagnetic monitoring measurements taken from both sides of a road on a hill show significantly more particulate pollution on the side of the road closer to the traffic going uphill. Ask students to explain these results. The vehicles' engines require more energy to propel the vehicle up the hill. In doing so, the engines consume more fuel and emit more particles.

GOING**FURTHER** >>>

Expected Outcomes Student responses to the Going Further activity should include a map of the chosen area clearly identifying different tree species. In order to have samples for comparison, students should choose a species found close to roads as well as far removed from roads.

SECTION 1

Mastering Concepts

38. Like poles repel each other; opposite poles attract.

39. A temporary magnet is like a magnet only while under the influence of another magnet. Its domains will revert to a random arrangement when the magnet is removed. The domains of a permanent magnet are permanently aligned.

40. iron, cobalt, and nickel

41.

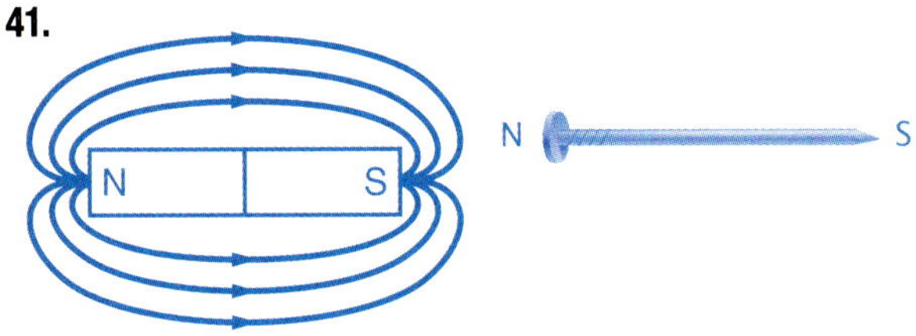

The bar magnet's field is stronger closer to the bar magnet, so the attractive force on the closer pole is stronger than the repulsive force on the farther pole, making a net attractive force.

42.

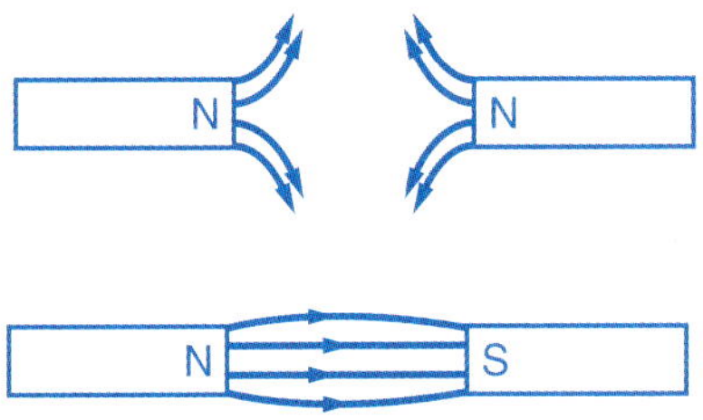

43. No; new poles would form on each of the broken ends.

44. Grasp the wire with your right hand, keeping your thumb pointing in the direction of the conventional current through the wire. Your fingers will encircle the wire and point in the direction of the magnetic field.

45. The magnetic field is concentrated inside the loop because the direction of the fields from the individual loops is always the same, and the fields add together.

46. Grasp the electromagnet with your right hand, keeping your fingers encircling the electromagnet in the direction of the conventional current flow through the loops. The thumb of your right hand will point toward the north pole of the electromagnet.

47. The poles of the atoms do not necessarily align in the same direction. When not in the presence of a strong magnet, the iron is not magnetized; the poles of its atoms point in random directions. If the iron were placed near a strong magnet, however, the poles of the atoms would align.

48. Its domains are jostled out of alignment.

SECTION 1

Mastering Problems

49. a. 4 and 2

b. 2

c. 4

50. Move to the left or begin to turn.

51. Move to the right.

52. At the right end.

53.

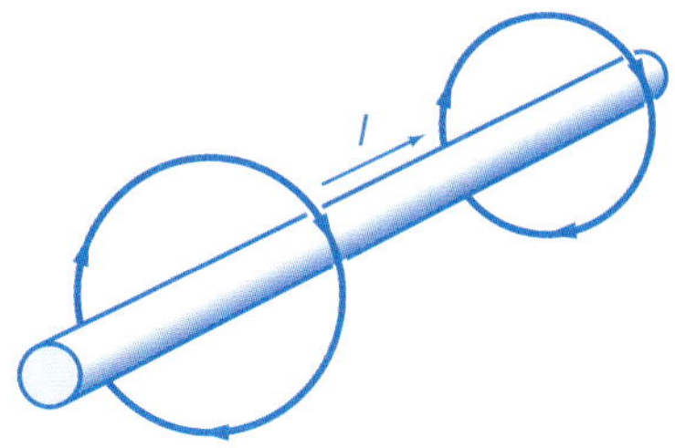

54.

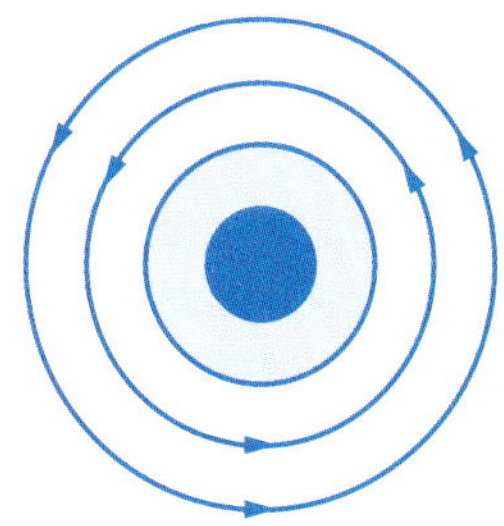

55. a. down (into the page)

b. up (out of the page)

56. a.

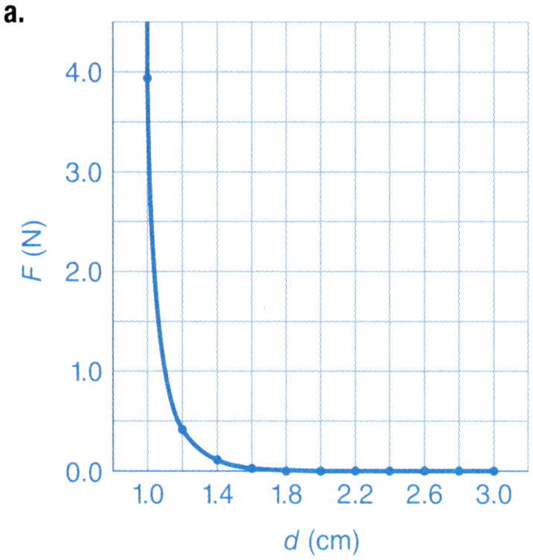

b. No.

Standardized Test Practice

student text p. 673

Multiple Choice

1. D

2. B

3. A

4. D

5. C

6. A

7. B

8. A

9. D

Free Response

10. $B = \frac{F}{qv} = \frac{\text{N}}{\text{C·m/s}} = \frac{\text{kg·m·s}}{\text{C·m·s}^2} = \frac{\text{kg}}{\text{C·s}}$

$B = \frac{F}{IL} = \frac{\text{N}}{\text{A·m}} = (\text{kg·m/s}^2)(\text{s/C})(1/\text{m}) = \frac{\text{kg}}{\text{C·s}}$

11. 35°

Rubric

The following rubric is a sample scoring device for free response questions.

Points	Description
4	The student demonstrates a thorough understanding of the physics involved. The response may contain minor flaws that do not detract from the demonstration of a thorough understanding.
3	The student demonstrates an understanding of the physics involved. The response is essentially correct and demonstrates an essential but less-than-thorough understanding of the physics.
2	The student demonstrates only a partial understanding of the physics involved. Although the student may have used the correct approach to a solution or may have provided a correct solution, the work lacks an essential understanding of the underlying physical concepts.
1	The student demonstrates a very limited understanding of the physics involved. The response is incomplete and exhibits many flaws.
0	The student provides a completely incorrect solution or no response at all.

CHAPTER 25

Electromagnetic Induction

About the Photo

Wind Farm Have students examine the photograph. Ask students what wind farms do. They produce electric power from generators turned by the wind. Have students describe how the picture shows this might happen. Possible answer: Wind turns the blades on the rotor at the top of a tower. The blades turn an axle, which is connected to a generator. The generator turns the rotational energy of the blades into electric current.

Chapter Overview

The chapter opens with a discussion of the process by which an electric current is induced from a changing magnetic field. The concept of induced electromotive force in generators is explored. Lenz's law is introduced and several applications presented. The chapter ends with a discussion of self-inductance and mutual inductance and their role in transformers.

Before students study the material in this chapter, they should study:

- Electric current and fields
- Electrical energy and power
- Magnetic fields
- Potential difference
- Work, energy, and power

To solve the problems in this chapter, students will need a strong grasp of:

- Scientific notation
- Significant figures
- Sine, cosine, and tangent
- Solving linear equations

Introduce the BIG IDEA

Induction Ask students how electric current and magnets are related. They exert a force on each other. Point out that objects such as electric guitars use magnets and coils to generate current. Ask students what that current might do in a guitar. It might change frequencies to make different sounds.

Go online!
connected.mcgraw-hill.com

Your online experience includes:

- **Video**
- **Concepts In Motion**
- **Virtual Investigations**
- **All Lab, Applying Practices activities, and PBL worksheets**
- **Online Practice**
- **Personal Tutor**
- **Multilingual eGlossary**
- **Presentation**

Use the LaunchLAB

In *Changing Magnetic Fields,* students explore how a changing magnetic field affects a coil of wire that passes through the field.

Use Physics T.V.

Video

GENERATING ELECTRICITY Have you ever used a flashlight or radio that had a hand crank? How does spinning a crank provide enough energy to power your device? Explore some applications of electromagnetic induction.

1 Introduce

Bellringer

Electric Generator This activity uses a bicycle generator-light combination. Rotate the generator by hand to simulate the mechanical energy provided by the rider and have students observe the light. Then have students write a brief explanation of the energy conversion that is taking place within the generator. the conversion of mechanical energy to electrical energy Have students brainstorm other forms of energy that could be input energy for the generator.
BL Visual-Spatial EL

Tie to Prior Knowledge

Magnetic Fields and Electric Charges Students will use the concepts of potential difference, current, and magnetic fields to analyze the force on charges in a conductor moving through a magnetic field.

2 Teach

Changing Magnetic Fields

student text pp. 676–679

Content Background

Michael Faraday Michael Faraday discovered many fundamental laws of physics and chemistry, despite the fact that he had virtually no formal education. The son of an English blacksmith, he was apprenticed at age 14 to a bookseller and bookbinder. Among his many discoveries, Faraday found that a changing magnetic field can induce a current. He performed a series of experiments that showed clearly that the induced *EMF* is equal to the rate of change of magnetic flux. Also, generalizing from the patterns formed by iron filings around magnets, he developed the concepts of magnetic and electric field lines. Faraday also devised the first electric generator and discovered the correct laws of electrochemistry.

Use Figure 2

Right-hand Rule Have students compare **Figure 2** with the right-hand rules presented before. Encourage students to use their hands to act out the rule. Ask students the following questions: What does *B* represent in the figure? the magnetic field strength What is its direction? into the page, away from the reader What is the direction of *F*? up, toward the top of the page On what does *F* act in **Figure 2?** charge What is in motion in the figure? the wire OL

Discussion

Problem Will current be produced if you drop a magnet down a long tube of conducting material, such as copper? Explain.

Response Current will be produced in the conducting path of the wall of the tube when a *variable* magnetic flux passes though the cross-sectional area of the tube. Because the magnet is falling, the field experienced by any point is constantly changing, inducing a current.

Use an Analogy

EMF, Batteries, and Potential Difference Point out that the induced *EMF* across a wire segment produces a potential difference between two points—the two ends of the wire. The potential difference is analogous to that produced by a cell or a battery. Have students describe where a battery produces a potential difference. between the two terminals Explain that neither the battery nor the wire must be part of a circuit to produce the potential difference. OL

Concept Development

Loudspeaker and Microphone Compare the loudspeaker and the microphone: the loudspeaker converts electrical energy into sound energy, while the microphone converts sound energy into electrical energy. Some home and school intercom systems use a loudspeaker for both functions. This is a good example of the symmetry of many electromagnetic effects. Wire loops moving in a magnetic field induce motion of charges (current) through the loops, and current-carrying wire loops in magnetic fields produce motion of the loops.

DIFFERENTIATED INSTRUCTION

Struggling Students Use the following table to help students understand how to derive the appropriate unit for *EMF*.

Variable	SI Unit	Related Units
B	T	N/A•m
I	A	C/s
L	m	
v	m/s	
V	V	J/C
W	J	N•m

$$(\mathrm{T})(\mathrm{m})\left(\frac{\mathrm{m}}{\mathrm{s}}\right)$$
$$=\left(\frac{\mathrm{N}}{\mathrm{A{\cdot}m}}\right)\left(\frac{\mathrm{m}}{1}\right)\left(\frac{\mathrm{m}}{\mathrm{s}}\right)$$
$$=\left(\frac{\mathrm{N}}{\left(\frac{\mathrm{C}}{\mathrm{s}}\right)\left(\frac{\mathrm{m}}{1}\right)}\right)\left(\frac{\mathrm{m}}{1}\right)\left(\frac{\mathrm{m}}{\mathrm{s}}\right)$$
$$=\left(\left(\frac{\mathrm{N}}{\mathrm{C{\cdot}m}}\right)\left(\frac{\mathrm{s}}{1}\right)\right)\left(\frac{\mathrm{m}}{1}\right)\left(\frac{\mathrm{m}}{\mathrm{s}}\right)$$
$$=\frac{(\mathrm{N{\cdot}s{\cdot}m^2})}{(\mathrm{C{\cdot}m{\cdot}s})}$$
$$=\frac{(\mathrm{N{\cdot}m})}{\mathrm{C}}$$
$$=\frac{\mathrm{J}}{\mathrm{C}}$$
$$=\mathrm{V}$$

BL Logical-Mathematical

ADDITIONAL IN-CLASS EXAMPLE

Use with Example Problem 1.

Problem A straight wire, 0.30 m long, moves at a constant speed of 10.0 m/s perpendicular to a 0.20-T magnetic field.

a. What is the induced *EMF* in the wire?

b. What is the current in the wire if it is part of a circuit with a resistance of 25 Ω?

Response

a. $EMF = BLv(\sin\theta)$

$= (0.20\ \mathrm{T})(0.30\ \mathrm{m})(10.0\ \mathrm{m/s})(1)$

$= 0.60\ \mathrm{V}$

b. $I = \frac{V}{R} = \frac{EMF}{R}$

$= \frac{0.60\ \mathrm{V}}{25\ \Omega} = 0.024\ \mathrm{A}$

Identify Misconceptions

Stationary Wire Coil Ask students if an *EMF* can be induced in a stationary wire. Most will respond no because $v = 0$ in the equation $EMF = BLv$. Point out that the equation is a special case for a wire moving in a constant magnetic field. A stationary wire experiences an induced *EMF* if the magnetic field is changing around it. For example, a wire between the poles of an electromagnet that is being switched on will experience an induced *EMF*. In this case the induced *EMF* is related to $\frac{\Delta B}{\Delta t}$, not *B*.

PHYSICS CHALLENGE ACTIVITY

Drawing Circuit Diagrams Have students create circuit diagrams for part b of Example Problem 1 and Practice Problems 1 and 3. Have students explain how the wire moving at a constant speed perpendicular to a constant magnetic field acts as a source of potential difference in a circuit to produce the *EMF*. OL Visual-Spatial EL

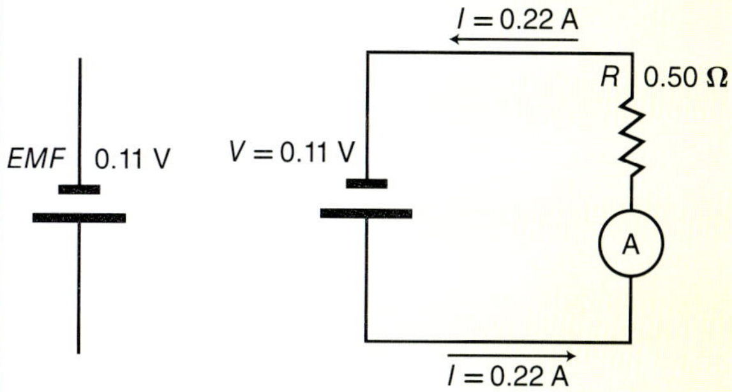

Electric Generators

student text pp. 680–683

Reinforcement

AC Generator Symbol Have students recall the circuit symbol for a DC generator, and compare it to that for an AC generator. BL Visual-Spatial EL

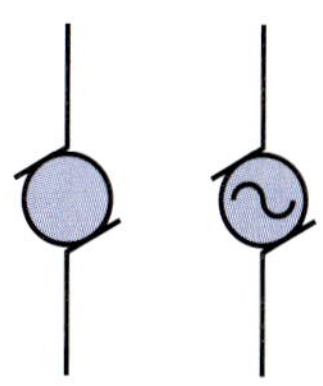

Reinforcement

Generators and Motors Reinforce the concepts of generators and motors. Have students compare a generator and a motor. Point out that a generator converts mechanical energy into electrical energy. Conversely, a motor converts electrical energy into mechanical energy. OL

Content Background

Faraday's Law Faraday discovered the way a changing magnetic field can produce an electric field. The induced *EMF* around the edge of an imaginary surface through which there are changing magnetic fields is equal to the negative of the time-rate-of-change of the magnetic flux through the surface. The magnetic flux, in the case of a uniform field, is expressed as $\Phi = BA \cos \theta$, where θ is the angle between the field and the perpendicular to the surface. You can use Faraday's law to calculate the potential difference, and thus the current, that results when a circular or rectangular coil is in a changing magnetic field or rotates in a constant magnetic field, as is found in an electric generator.

Quick DEMO

MAIN IDEA

Estimated Time 5 minutes

Materials 2 identical galvanometers, 2 wires

Procedure Connect the two galvanometers in series. Pick up one galvanometer and wiggle it, with students observing its needle. Repeat with the second galvanometer. Then, have students describe what they observed about the needles. The needles of both galvanometers moved. Explain that by wiggling the first galvanometer, you made the coil rotate within a magnetic field. An *EMF* was induced in the coil. Thus, this galvanometer acted as a generator, converting mechanical energy to electrical energy. The second galvanometer had an induced current in the coil, and the coil with its attached needle was rotated by the magnetic field surrounding the coil. Thus, the second galvanometer acted like a motor, converting electrical energy to mechanical energy.

Discussion

Problem What is the significance of the mark inside the AC generator symbol that distinguishes it from the DC generator symbol?

Response The mark is intended to indicate the shape of the current output, as the Current-Time plot in **Figure 5.** BL

Reinforcement

Closed Circuit Have students describe what indicates that the bulb in **Figure 8** is part of a closed circuit. It is lit. Have students sketch the closed circuit.
OL Spatial-Visual EL

Critical Thinking

Generator Anatomy Have students identify which parts of the generator shown in **Figure 8** are rotating and which are stationary. The armature (wire loop) and slip rings are moving; the brushes and magnets are stationary. Have students describe how the number of moving parts could be reduced. by rotating the magnets OL

Use the MiniLAB

In *Motor and Generator,* students make a generator and investigate how it works.

Concept Development

Rotating Field Magnets Because the coils and armature can be very heavy in some generators, it becomes more efficient to have the magnets revolve around the armature.

DIFFERENTIATED INSTRUCTION

Struggling Students To help students relate the value of the effective alternating current to the maximum current, write the following relationships on the board:

$$P_{AC} = \frac{1}{2}P_{AC\ max};\ I_{eff}^2R = \frac{1}{2}I_{max}^2R$$

$$I_{eff} = \sqrt{\frac{1}{2}I_{max}^2}$$

$$= \frac{\sqrt{1}}{\sqrt{2}}I_{max}$$

$$= \frac{\sqrt{2}}{2}I_{max}$$

$$= 0.707I_{max}$$

Using the above relationship and $I = \frac{V}{R}$, have students relate V_{eff} to V_{max}.

BL **Logical-Mathematical**

DIFFERENTIATED INSTRUCTION

Visually Impaired To help visually impaired students better understand the function of a generator, pair a visually impaired student with a sighted student to construct a simple circuit with a battery, a switch, and a buzzer. Have students close the switch and listen to the buzzer. Next, have students replace the dry cell with a hand-crank generator and have each repeat the demonstration, first without cranking the generator, and then with cranking it. Ask students to compare the simple buzzer circuit with the hand-cranked generator. Possible answer: The hand-cranked generator in the circuit produced effects that were similar to those of the battery. Ask students to describe the energy conversions that they observed. The generator circuit converted mechanical energy from the cranking to electrical energy. Then the circuit converted the electrical energy to sound energy when the buzzer made noise.

BL **Kinesthetic** **EL**

Quick DEMO

Current Effects

Estimated Time 10 minutes

Materials length of wire approximately 1 m, a galvanometer, 1 strong horseshoe magnet, a coil of wire or a solenoid

Procedure Connect the ends of the length of wire to the posts of the galvanometer. Loop the wire to make 3 or 4 overlapping loops. Using the horseshoe magnet, allow students to observe as you move it closer to and farther from the wire loops. Have students describe what happens on the galvanometer. The relative motion between the magnet and the wire produces a current shown when the needle on the galvanometer moves. The direction of the current relates to the direction of the motion of the magnet. Ask students what they could do that might increase the current. Possible answers: Increase the number of loops in the wire; move the magnet closer. Show students what happens when the 1-m wire is replaced with a solenoid or ready-made coil. Have students describe the effect on the current. The current was greater than when the wire had just a few loops in it.

OL **Visual-Spatial** **EL**

3 Assess

Assess the MAINIDEA

Induction Have students describe three ways an *EMF* can be induced in a loop of flexible wire in a magnetic field. Possible answers: change the strength of the magnetic field; change the shape of the loop; move the loop in or out of the field; rotate the loop in the magnetic field.

Check for Understanding

***EMF* in an AC Generator** Ask students to explain why the *EMF* in an alternating current generator has two zero values in one cycle. The value of the induced *EMF* changes sign twice in each cycle. Because the *EMF* changes continuously, its value is zero when it changes signs. Ask what the coil's position is when the induced *EMF* is zero. Its motion is parallel to the magnetic field.
OL Visual-Spatial EL

Reteach

$v(\sin\theta)$ Review with students the equation $EMF = BLv(\sin\theta)$ and have them identify the quantities B, L, and $v(\sin\theta)$. B is magnetic field strength, L is the length of the conductor, and $v(\sin\theta)$ is a component of the coil's velocity perpendicular to the magnetic field. Have students identify the one quantity that varies in most generators, explain why it is not constant, and state when its value is zero. The quantity is $v(\sin\theta)$; it is not constant because the coil is rotating in the magnetic field; it is zero when $\theta = 0, \pi, 2\pi, 3\pi, \ldots$
OL Logical-Mathematical

Answers to Reading Checks and Caption Checks

Caption Check, *student text p. 680*
Counterclockwise. Using a right-hand rule, you would point your fingers in the direction of the magnetic field and your thumb in the direction the wire is moving. Your palm points in the direction of the current.

Reading Check, *student text p. 681*
Potential difference is greatest when the wire loop is horizontal, or perpendicular to the magnetic field, so $\sin 90° = 1$.

Caption Check, *student text p. 682*
When the armature is vertical and thus moving parallel to the magnetic field, the current is zero.

Practice Problems
student text p. 679

1. a. 4 V
  b. 0.7 A

2. 0.16 V

3. a. 0.10 T
  b. 1.2 A

4. Using a right-hand rule, the north pole is at the bottom.

Practice Problems
student text p. 683

5. a. 120 V
  b. 0.49 A
  c. $2.4\times10^{2}\ \Omega$

6. 165 V; 7.8 A

7. a. 3.01×10^{2} V
  b. 0.60 A

8. 1.5×10^{2} W

Section 1 Review
student text p. 683

9. An *EMF* is induced in the armature of a generator as it is turned—by a mechanical force—in a magnetic field. When the generator is in a circuit, the *EMF* induces a current. As the armature rotates through 180°, the induced *EMF*—and current—reverse direction.

10. Yes; only relative motion between the coil and magnetic field is important. Note, this generator would not have much power as the relative velocities of the magnets and coil will be very small.

11. The rider provides the mechanical energy that turns the generator's armature.

12. A current is induced in the coil.

13. You could increase the number of magnetic pole pairs or make the armature spin faster.

14. The magnitude of the induced *EMF* is directly related to the strength of the magnetic field. A greater potential difference is induced in the conductor(s) if the field strength is increased. Because $EMF = BLv(\sin\theta)$, you can also increase output potential difference by increasing the length of the wire or the velocity of the wire.

15. Power is the rate at which energy is transferred. Power is the product of I and V. When I is positive, so is V, and therefore, P is positive. When I is negative, so is V; thus, P is positive. Energy is always transferred through the lamp.

SECTION 2 Applications of Induced Currents

student text pp. 684–691

1 Introduce

Bellringer

Small-Electronics Transformers Set up several small electronic devices (such as a CD player and a calculator) operating with transformers visibly plugged into a power strip. Have students infer the function of the devices plugged in. They supply electrical power to the small electronics. Have students describe what evidence indicates that the devices are not generators. Visibly and audibly, they seem to have no moving parts, and they must be connected to another energy source in order to function. BL Visual-Spatial EL

Tie to Prior Knowledge

Inductance Students employ previously learned right-hand rules and the concept of induced *EMF* to explore mutual inductance and self-inductance, as well as the operation of transformers.

2 Teach

Lenz's Law

student text pp. 684–687

Use Figure 10

Current Direction Explain to students that the induced current will be in the opposite direction if the bar magnet moves away from the coil. Have students describe what will happen to the induced current if the bar magnet stops. The induced current will fall to zero. OL

PHYSICS CHALLENGE ACTIVITY

Motors A motor depends on the electromagnetic force on a current-carrying wire in a magnetic field. While in principle, Ampère's law can be used to calculate that magnetic field, in practice the law is seldom used. Have students describe how Faraday's law is useful in describing why, when a motor is rotating slowly, it draws more current than it does when it rotates rapidly. When a motor rotates rapidly, an *EMF* is produced in the coils as a result of the changing magnetic flux through them that acts in a direction opposite to the potential difference put across the motor by the external power source. Therefore, the current through the motor is reduced. When a motor stalls, it draws very large currents, which can blow a fuse or overheat the motor. AL

Use the PhysicsLAB

In *Swinging Coils,* students observe the effect of Lenz's law on a generator.

Use the MiniLAB

In *Slow Motor,* students observe the effect of turning on a motor.

Extension

Motor Power Have students calculate the power for the two sets of data in the *Slow Motor* MiniLAB. Have students describe how the power is dissipated in a motor that is under load. The power is dissipated as thermal energy. OL Logical-Mathematical

Critical Thinking

Eddy Currents in Electrical Devices Have students explain why it is beneficial to reduce eddy currents in devices such as motors and transformers. Reducing eddy currents reduces heating effects. OL

Reinforcement

Cutting *B*-Lines Point out that the statement "a wire cuts the lines of a magnetic field" implies that there is relative motion between the *B*-field and the wire.

Use Models

***I-t* Graph of Self-Inductance** Drawing the following graph on the board will help students understand the effect of self-inductance.

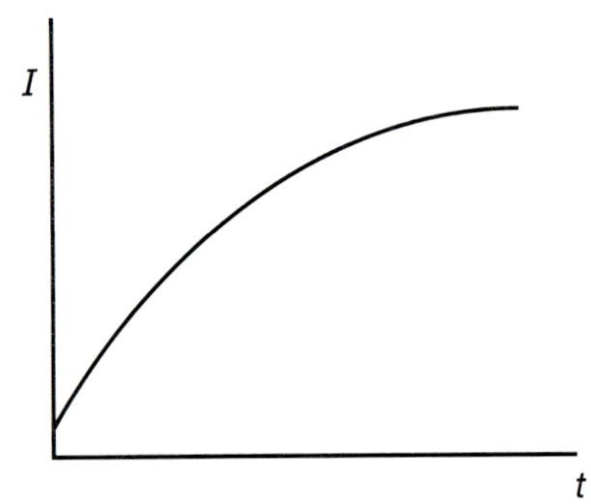

Explain that the graph models the effect of a self-induced *EMF* on the increase of current in a circuit with constant resistance immediately after the switch is open. Have students recognize that the induced *EMF* decreases the rate at which the current increases. OL Visual-Spatial EL

PHYSICS PROJECT ACTIVITY

Teacher to Teacher

Induced Magnetic Fields First show students that a horseshoe magnet does not attract a copper plate. Then tie one end of a string to the magnet and spin it above the copper plate. The induced current in the copper plate generates a magnetic field, resulting in the copper plate being attracted to the magnet.
OL Visual-Spatial EL

Content Development

DC Generator Help students relate generators and motors through the following activity. Have students suggest ways to modify the AC generator shown in **Figure 8** to produce a DC output. Students can produce a diagram of their modified generator and include a written report outlining the improvements and why each is beneficial. One possible modification is to replace the slip rings with a split ring similar to the commutator in a DC motor. In this arrangement, one of the coil wires touches one end of the split ring, and the other wire end touches the other half of the split ring. Ask students to explain what happens with the split ring. The split ring prevents the electric current from changing direction each time the rotating wire flips over.
OL Visual-Spatial EL

Quick DEMO

Slow Magnet

Estimated Time 10 minutes

Materials 1 m of 1/2-inch-diameter copper tubing, pair of neodymium magnets, stopwatches, 9-mm steel marble

Procedure Drop the marble through the vertical copper tubing, and have students time the fall. Repeat the drop and timing with the pair of magnets. Catch the magnets, or let them drop onto a resilient surface such as foam rubber, so they do not strike the tabletop or floor and fracture. Students should note a significant increase in the time it takes for the magnets to fall. Be sure students recognize that the magnets induce a current in the copper that, in turn, induces a magnetic field that retards the motion of the falling magnets.

REAL-WORLD PHYSICS

James West An African American experimental physicist, James West, is the co-inventor of the foil electret microphone. This device uses a paper-thin piece of plastic that is coated with metal on one side to convert sound into electric signals. It is used in hearing aids and small microphone. As a result of this development, these devices can be made much smaller and still be effective.

Reinforcement

MAIN IDEA Have students describe how the operation of generators and motors is affected by induced-current phenomena. Possible answers: When a generator is off, you can easily turn the armature; when you turn a generator on, turning the armature causes induced current. This produces a magnetic field that makes the armature harder to turn. When you turn a motor on, there is a current through it. As the coil in the motor rotates, it induces an EMF that reduces the current through the motor. When a motor is under load, it slows down, decreasing the induced EMF which, in turn, increases the net current through the motor so it can do more work.

Transformers

student text pp. 688–691

Identify Misconceptions

Transformer Efficiency Students assume that real transformers, like ideal ones, are 100 percent efficient. You can ask students what evidence there is to the contrary. Transformers usually feel warm. Therefore, the transformer must be converting some electrical energy to thermal energy, lessening the transformer's efficiency.

Content Background

Energy-Efficient Transformers Although many distribution transformers are more than 98 percent efficient, they do suffer from core (no-load) and winding (load) losses. Core losses occur whether or not there is a load on a transformer; they are always present because transformers are constantly energized to be able to meet demand. Winding losses occur only when a transformer is actually being used, and they result from resistance in the windings. These losses vary in proportion to the square of the load. Since millions of distribution transformers are in use, even small improvements in efficiency can save significant amounts of energy. OL

Use the PhysicsLAB

In *Induction and Transformers,* students investigate how a transformer works.

ADDITIONAL IN-CLASS EXAMPLE

Use with Example Problem 2.

Problem A transformer has a primary coil winding consisting of 400 turns and a secondary coil winding of 200 turns. The secondary voltage is 6.0 V, and the secondary current is 0.30 A.

a. What is the voltage in the primary circuit?

b. What is the current in the primary circuit?

Response

a. $\frac{V_s}{V_p} = \frac{N_s}{N_p}$

$V_p = \frac{V_s N_p}{N_s} V$

$V_p = \frac{(6.0\text{ V})(400)}{200} = 12\text{ V}$

b. $P_p = P_s$

$V_p I_p = V_s I_s$

$I_p = \frac{I_s V_s}{V_p} = \frac{(6.0\text{ V})(0.30\text{ A})}{12\text{ V}}$

$= 0.15\text{ A}$

3 Assess

Assess the MAINIDEA

Induced Current Have students describe how electric generators, motors, and transformers use induced current. In a generator, an armature rotates in a magnetic field, inducing an electric current. In an electric motor, current in the armature interacts with a magnetic field, causing the armature to rotate. A transformer uses mutual inductance to increase or decrease voltage.

Check for Understanding

DC and Transformers Ask students if a transformer could be operated with steady direct current. No; direct current that is not changing would not generate a changing magnetic field, which is what produces the electromagnetic induction required for the operation of a transformer. OL

Extension

Transformer Frequency Ask students what property I_p and I_s have in common in any transformer. frequency OL

Reading Checks and Caption Checks

Caption Check, *student text p. 685*
The force would be the same. The magnet and coil would still repel each other. The only difference is that the part of the coil closest to the magnet would become a south pole instead of a north pole.

Reading Check, *student text p. 688*
In a step-up transformer, the voltage on the secondary coil is larger than that on the primary voltage. In a step-down transformer, it is the reverse.

Practice Problems

student text p. 690

16. 1.2×10^{2} V; 0.60 A

17. 1.80×10^{4} V; 1.6×10^{-1} A

Section 2 Review

student text p. 691

18. Away from the magnet; the changing magnetic field induces a current in the coil, producing a magnetic field. This field opposes the field of the magnet, and thus, the force between coil and magnet is repulsive.

19. The inductance of the motor creates an *EMF* that causes the spark. The bulb has very low self-inductance, so there is no *EMF*.

20. Transformers rely on changing currents to induce changing magnetic fields. DC always produces the same magnetic field and thus can't induce a current in another wire.

21. More current can go through the coil with fewer turns, so thick wires with capacity for large currents are needed. Also, resistance must be kept low to prevent voltage drops and I^2R power loss and heating.

22. According to the transformer equations, the ratio of primary to secondary current is equal to the ratio of turns and doesn't change. Thus, if the secondary current increases, so does the primary.

23. No; induced *EMF* depends on a changing magnetic field through the core. Permanent magnets are "permanent" because they are made of materials that resist such changes.

HOW IT **WORKS**

Magnetic Money

Magnetic Card Readers

Purpose

This article demonstrates the operation of magnetic-stripe cards and their readers.

Background

A magnetic stripe card is another example of the use of electromagnetic induction in our everyday lives. Consider that without this principle it would be impossible to translate electrical energy into motion, and vice versa. Electric motors and generators are the most obvious applications of this idea, but there are many other applications, including these card readers.

Teaching Strategies

Have students describe where they see card readers used. Ask students if they have ever had a card not work in a card reader. Ask what, if anything, they did to try to make the card work in the reader—such as wrapping a piece of paper or plastic around it. Have student speculate why wrapping a card in paper or plastic sometimes helped it read.

GOING**FURTHER** >>>

Expected Outcomes Answers will vary. As the magnetic stripes age, the adjoining magnetic fields may begin to smear into one another. By slightly increasing the distance between card and reader (the paper or plastic wrapping), the places where magnetic particles have smeared into nonmagnetic regions are less likely to cause a reading, since only the strongest fields along the stripe are now registered.

Bar codes (found on many library cards and shopper saver cards) rely on visible light reflecting off a series of bright and dark lines. RFID readers use radio waves to activate a sensor in the card. The sensor then sends back information about itself to the reader. Note that this allows the cards to work without a power source such as a battery, and be sure students understand why. Smart cards are increasingly replacing credit cards with magnetic stripes. On smart cards, a microprocessor replaces the magnetic stripe.

CHAPTER 25 ANSWERS

SECTION 1

Mastering Concepts

24. You would set up a magnetic field (north and south poles facing). You could generate a potential difference (*EMF*) by moving the wire within the field or moving the field and keeping the wire still. In both cases, the wire and field are at right angles. Because the wire is connected in a circuit, the *EMF* generates current.

25. DC generators connect using commutators so current is in one direction; AC generators use slip-ring devices in which one ring is connected to one end of the wire on the armature, and the other ring is connected to the other end of the wire. As the armature rotates in the magnetic field through 180°, the induced *EMF* reverses direction.

26. Iron is used in an armature to increase the strength of the magnetic field.

27. The minimum amount of potential difference (0 V) is generated when the conductor is moving parallel to the magnetic lines of force.

28. The magnetic field is out of the page, and the charges move to the right as the wire is pulled to the right. According to a right-hand rule, this means that any current in the wire would be up, toward the top of the page.

29. Increasing the wire length results in a net increase in induced potential difference because $EMF = BLv(\sin\theta)$.

30. They are similar in that each showed a relationship between electricity and magnetism. They are different in that Oersted found that electric currents generated magnetic fields, while Faraday found that magnetic fields can generate *EMFs* and currents.

31. Electromotive force; it is not a force but an electric potential difference (energy per unit of charge). It was named before related concepts were understood.

32. Generators and motors are constructed in similar ways, but in a generator, mechanical energy turns an armature in a magnetic field. The induced potential difference produces current, thus producing electric energy. In a motor, potential difference is placed across an armature in a magnetic field. The potential difference produces current in the coil and the armature turns, transforming electrical energy into mechanical energy.

33. An AC generator consists of a permanent magnet, an armature (made of wire loops), a set of brushes, and a slip ring device to connect to the circuit.

34. As the armature turns in an alternating-current generator, the generated power varies between some maximum value and zero. The average power is equal to one-half the maximum power. The effective current is the constant value of current that would cause the average power to be dissipated in the load, *R*.

35. Stored water has gravitational potential energy. As it falls, it gains kinetic energy (KE). The KE is transferred to the turbines and then to the generator's armature. The generator transforms KE into electrical energy. In addition, there are frictional losses in the turbine and generator resulting in thermal energy.

Mastering Problems

36. 0.5 T

37. 0.89 V

38. a. 3.6 V

b. 0.33 A

39. 20 m/s

40. 4.00×10^{2} V

41. a. 110 V

b. 21.2 A

c. 2.3 kW

42. a. 340 V

b. 22 A

43. 23 m; this is the shortest length of wire assuming that the wire and the direction of motion are each perpendicular to the field ($\sin\theta = 90°$).

44. 2.3 μA

45. a. 0.13 V

b. 0.13 A

c. Point A is negative relative to point B.

46. a. 0.039 T

b. 0.23 V

47. a. 440 MW input

b. 4.4×10^{8} J each second

c. 2.0×10^{6} kg

48. 0.8 V

SECTION 2

Mastering Concepts

49. When a change in a magnetic field induces a current, the current is in a direction that opposes the change in the original magnetic field. It is consistent with the law of conservation of energy because it prevents a changing magnetic field to grow without limits, and thus it prevents current from increasing to infinity. Otherwise, energy would be created.

50. This is Lenz's law. Once the motor starts turning, there is more and more current through its wires, strengthening the wires' magnetic field. According to Lenz's law, the wires generate an *EMF* to oppose the strengthening current.

51. When a saw and lightbulbs are in parallel circuit, a large current in the wires of the saw increases the potential difference across the wires but decreases the potential difference across the lightbulbs. Thus, they dim.

52. 3.3

53. Self-inductance only plays a role when current (and its induced magnetic field) are changing. An alternating current is always changing in magnitude and direction. A direct current eventually becomes steady, and thus, after a short time, there is no changing magnetic field.

54. As Faraday discovered, only a changing magnetic field induces *EMF* and current. And only a changing current generates a changing magnetic field.

55. The ratio of *EMF*s in a transformer is set by the ratio of the turns of wire in the primary coil to the number of turns of wire in the secondary coil.

56. The changing magnetic field in the track induces eddy currents that create a magnetic field that exerts a repulsive force on the coils in the train. When the train stops, the magnetic field no longer changes so there is no longer a repulsive force.

Mastering Problems

57. a. 71.25 turns, which rounds to 71

b. 18.75 mA

58. 72 V

59. a. 2 to 1

b. 5 A

60. a. 1.8 kV

b. 3.0×10^{1} A

c. 3.6 kW; 3.6 kW

Applying Concepts

61. No current is induced because the direction of the velocity is parallel to the magnetic field.

62. The current is larger several hundredths of a second after the connection is made. A decreasing self-induced *EMF* will cause the current to grow gradually to its maximum value.

63. a. A right-hand rule will show the current moving left.

b. The force will act in an upward direction.

64. $B = \frac{F}{IL} \quad \left(\frac{\text{N}}{\text{A}\cdot\text{m}}\right)$

L is measured in m, and v is measured in m/s.

Therefore,

$$\left(\frac{\text{N}}{\text{A}\cdot\text{m}}\right)(\text{m})(\text{m/s}) = (\text{N}\cdot\text{m})(\text{A}\cdot\text{s}) = \text{J/C} = \text{V}$$

65. E < B < A = D < C

66. current only

67. The bulb will light because there is a current in the secondary circuit. This will happen whenever the primary current changes, so the bulb will glow either when the switch is closed or when it is opened. However, it will glow only for a moment. The secondary coil will only generate current when current in the primary coil is changing.

68. If the armature (conductors) are not rotating, no lines of force are being cut, and no potential difference is induced. Therefore, the induced *EMF* is zero. Since there is no current in the armature, no magnetic field is formed around the stationary conductor. It should be noted that this explanation only holds true at the instant of startup, at time just greater than 0. The instant the armature begins to rotate, it will be cutting the lines of force and will have an induced potential difference. This potential difference, the induced *EMF*, will have a polarity such that it produces a magnetic field opposing the field that created it. This reduces the current in the motor. Therefore, the motion of the motor increases its apparent resistance.

69. second law of thermodynamics

70. The current is from west to east.

71. a. The changing magnetic field in the pipe induces eddy currents. No matter whether the field is increasing or decreasing, there are always eddy currents produced.

b. The eddy currents cause the pipe to get warmer. As the magnet falls, its gravitational potential energy decreases but its kinetic energy does not. The thermal energy of the pipe increases and as it gets hotter than the surrounding air, heat is transferred to the air.

c. A non-conducting (or poorly conducting) pipe would have no or very small eddy currents, and so the magnet wouldn't slow down.

72. Answers will vary, but a correct form of the answer is, "A metal bar of length 0.80 m is moving to the left through a 0.15-T magnetic field directed straight up. If the *EMF* induced between the ends is 6.5 V, what is the speed at which the wire is moving?"

Mixed Review

73. Answers will vary. A possible form of the correct answer is, ". . . moves perpendicularly to a 0.06-T magnetic field at a speed of 5.0 m/s. What is the *EMF* induced between its ends?"

74. a. 3.6×10^{3} V

b. 9.0×10^{1} A

c. 1.1×10^{4} W; 1.1×10^{4} W

75. 20 m/s

76. 1×10^{1} m/s

77. 170 V

78. 3.5 A

79. 407 V

80. 15.03 A

81. primary : secondary = 545 : 1

82. 96 A

83. The motor becomes a generator. It produces electrical energy that charges the battery. This energy comes from the kinetic energy of the car.

84. Large current; the efficiency doesn't change, so if the output power is increased, so is the difference between input and output power, and thus the rate of heating.

85. 4.7 A

86. a. 1.0×10^{2} kW

b. 200 A

87. a. 6.4 V

b. 1.0 A

88. 41 mA

89. 69.4 W

90. a. 500 turns

b. 0.40 A

Thinking Critically

91. It would violate the law of conservation of energy. It would allow a changing magnetic field to grow without limits. The current would increase without any work done. A generator would create energy, not just change it from one form to another.

92. a. Efficiency: $e = \frac{P_s}{P_p}\times 100$

b. 6.05 A

c. Thermal energy will be produced at the rate of 57 J/s. This is the difference between the input power (757 W) and the output power (700 W).

93. a. 67 kW is supplied to the ovens in eight homes.

b. The difference between the secondary and primary power is the power that heats up the transformer, 4 kW.

Writing in Physics

94. A series DC motor uses both an armature and series coil. When operated on alternating current, the polarity on both fields changes simultaneously. Therefore, the polarity of the magnetic field remains unchanged, and hence the direction of rotation is constant.

Cumulative Review

95. 4.60×10^{14} Hz

96. 4.73×10^{7} m/s

97. 1.1×10^{-3} C

98. 7.4 V

99. 71 Ω

100. $F = 2.7\times10^{-13}$ N; $a = 3.0\times10^{17}$ m/s^2

Standardized Test Practice

student text p. 699

Multiple Choice

1. B
2. D
3. D
4. C
5. A
6. A

Free Response

7. $I = P/V$, so the current $I = 182$ A. The power lost in the line because of its resistance $P = I^2R = (182\text{ A})^2(0.2\ \Omega) = 6.6$ kW, so the station must deliver 86.6 kW. If the voltage is raised to 12 kV the current is now 6.7 A, and the power loss in the line is 8.9 W. Therefore, the station must deliver 88.0 kW.

Rubric

The following rubric is a sample scoring device for free response questions.

Points	Description
4	The student demonstrates a thorough understanding of the physics involved. The response may contain minor flaws that do not detract from the demonstration of a thorough understanding.
3	The student demonstrates an understanding of the physics involved. The response is essentially correct and demonstrates an essential but less-than-thorough understanding of the physics.
2	The student demonstrates only a partial understanding of the physics involved. Although the student may have used the correct approach to a solution or may have provided a correct solution, the work lacks an essential understanding of the underlying physical concepts.
1	The student demonstrates a very limited understanding of the physics involved. The response is incomplete and exhibits many flaws.
0	The student provides a completely incorrect solution or no response at all.

CHAPTER 26

Electromagnetism

About the Photo

Why are these parabolic dish antennas all oriented in the same direction? They are oriented towards a geostationary satellite. What are they trying to capture? They are capturing signals sent on electromagnetic waves. This photo shows to what extent the understanding of electromagnetism has influenced our lives and culture. Ask students to list other technologies that use electromagnetic waves. Wi-Fi, cell phones, radar, lasers

Chapter Overview

The concepts of electric and magnetic fields are explored further in this chapter. Electromagnetic waves, such as radio and gamma waves, are explained in terms of oscillating electric and magnetic waves. Although most of the radiation in the universe is beyond the limits of direct human perception, instruments have been developed to produce and measure all types of electromagnetic waves.

Before students study the material in this chapter, they should study:

- Electric charge
- Electric current
- Electric fields
- Fundamentals of waves
- Magnetic fields
- Potential difference

To solve the problems in this chapter, students will need a strong grasp of:

- Scientific notation
- Solving linear equations

Introduce the BIGIDEA

In previous chapters, students learned that electricity and magnetism are related. In this chapter, this relationship is explored further by looking at electromagnetic radiation. Time-varying electric fields create time-varying magnetic fields, which in turn recreate the time-varying electric fields. This interconnected phenomenon propagates through space as an electromagnetic wave.

Go online!
connected.mcgraw-hill.com

Your online experience includes:

- **Video**
- **Concepts In Motion**
- **Virtual Investigations**
- **All Lab, Applying Practices activities, and PBL worksheets**
- **Online Practice**
- **Personal Tutor**
- **Multilingual eGlossary**
- **Presentation**

Use the LaunchLAB

In *Broadcasting Radio Waves,* students can relate radio broadcast signals to electromagnetic wavelengths.

Use Physics T.V.

Video

MISSION: WI-FI POSSIBLE! What happens when you need to surf the Web, but you're out of range of your wireless Internet signal? If you understand the physics of Wi-Fi, you'll have no problem!

SECTION 1 Electric and Magnetic Forces on Particles

student text pp. 702–709

1 Introduce

Bellringer Activity

Electron Path Bring in an old, nonfunctioning oscilloscope. Remove the cover and discharge all of the capacitors. (Briefly connect a wire between the two terminals of the capacitor to create a short circuit and cause the voltage across it to become nearly zero.) Have students draw a diagram showing the setup that directs the electron beam. Drawings should include the vertical and horizontal deflection plates. Ask students how the electron path could be altered. The electron path may be controlled by changing the applied voltage on the deflection plates. **OL** **Visual-Spatial**

Tie to Prior Knowledge

What Students Need to Know Students should understand the concepts of mass, charge, forces from electric fields, and waves before beginning this section. Students will use equations for circular motion.

2 Teach

Thomson's Experiments

student text pp. 702–705

DIFFERENTIATED INSTRUCTION

Struggling Students Help students understand the significance of Thomson's experiments by having them carefully examine **Figure 1.** Ask them whether the cathode-ray tube allows scientists to see the electrons. No; only the light emitted as a result of electrons striking the fluorescent coating is visible. Electrons are much smaller than the wavelength of visible light, and any interaction with light changes the state of the electron. Their existence is known from indirect evidence that shows changes in their state. Ask students to draw the path of the electrons under different conditions, such as without the positive charge-deflection plate or the negative charge-deflection plate. **BL** **Visual-Spatial**

Identify Misconceptions

Charged Particles Ask students whether a stationary charged particle located between two bar magnets experiences a net force. No; provided the charged particle remains stationary with respect to the magnets, no net force is exerted on the particle by the magnetic field.

Use Figure 1

Ask students what is the purpose of the slits in the anode plates. The slits are used to block all of the electrons except for those traveling in line with the slits. The electrons passing through the slits form a relatively collimated beam.

Content Background

Cathode-Ray Tube (CRT) A cathode-ray tube is a sealed glass tube in which a metal electrode is placed and most of the air is removed. Inside this tube, electrons are emitted by the cathode and then accelerated and focused by a strong electric field. The electrons then pass through a region with charged horizontal and vertical metal plates. Changing the voltage applied to the plates bends the electron beams and causes the electrons to strike the inside end of the tube in different places. In an older CRT-type TV, the picture is a result of turning the electron beam on and off quickly as the beam is moved vertically and horizontally across the entire front of the picture tube.

Use the MiniLAB

In *Moving Charged Particles,* students can use a cathode-ray tube and a magnet to explore how charged particles are affected as they move through electric and magnetic fields.

Use the PhysicsLAB

In *Mass of an Electron,* students apply physics to determine the mass of the electron.

ADDITIONAL IN-CLASS EXAMPLE

Use with Example Problem 1.

Problem An electron with a mass of 9.11×10^{-31} kg moves vertically and horizontally through a television tube at 3.0×10^{5} m/s to a magnetic field of 6.3×10^{-2} T. Assume that the electric field is off. Determine the radius of the electron's circular path.

Response $Bqv = mv^2/r$

$$r = \frac{mv}{Bq} = \frac{(9.11\times10^{-31}\text{ kg})(3.0\times10^{5}\text{ m/s})}{(6.3\times10^{-2}\text{ T})(1.602\times10^{-19}\text{ C})}$$

$$= 2.7\times10^{-5}\text{ m}$$

Quick **DEMO**

Electron Distortion

Estimated Time 10 minutes

Materials old oscilloscope (functioning), strong magnet

Procedure

1. Adjust controls to obtain a steady trace on the screen.
2. Bring a strong magnet near the oscilloscope to demonstrate how the electron trajectories curve.

The Mass Spectrometer

student text pp. 706–709

PHYSICS CHALLENGE ACTIVITY

Mass Spectrometer Stick Diagrams Have students research the typical output from a mass spectrometer, often simplified into a "stick diagram," which charts the relative abundance of current against the mass-to-charge ratio. This type of diagram shows the relative current produced by ions of varying mass-to-charge ratio. The vertical scale is related to the current measured by the recording device and so to the number of ions arriving at the detector. The greater the current, the more abundant the ion. Ask students to bring in an example of a stick diagram for an assigned element. They should explain to the class the meaning of the chart and demonstrate how the data are used to determine the average atomic mass of the element. AL Linguistic

Content Background

Mass Spectrometers Because mass spectrometers have so many uses, they come in many different varieties. One kind of mass spectrometer, called a time-of-flight spectrometer, measures the time it takes particles to reach the detector. The lower-mass ions have a greater acceleration and so reach the detector sooner.

Concept Development

MAIN IDEA Help students understand the importance of a particle's charge in a mass spectrometer by asking them to compare the paths of a proton and an electron traveling through a magnetic field. Have students use their knowledge of how an electron is affected by a magnetic field. Then point out that the charge of a proton is opposite to that of an electron. The magnitude of the electric and magnetic forces would be the same in either case, although the electrons and protons would be deflected in opposite directions. Because the mass of a proton is much greater than that of an electron, the proton's trajectory would bend much less. OL

Critical Thinking

Acceleration of Hydrogen Atoms As an extension, ask students whether Thomson could have chosen to accelerate neutral hydrogen atoms in a vacuum tube instead of electrons or protons. Explain that, because neutral atoms are not affected by electric or magnetic fields, this would not have been possible. OL

REAL-WORLD PHYSICS

Separation of Uranium Isotopes In 1941, Ernest Lawrence suggested that researchers could separate isotopes of uranium by means of a huge mass spectrometer. They completed such a device, called a calutron, by December 1941 and they used it to separate U-235 from U-238. Scientists used hundreds of these devices from 1943 to 1945 to produce the highly enriched U-238 essential to the Manhattan Project, which resulted in the development of the atomic bomb. Today, these same devices are used to produce isotopes for medical purposes.

Reinforcement

Acceleration of Objects Students can relate the accelerations of electrons to the acceleration of objects encountered in their everyday lives. Have them write up a short comparison that shows estimates of the accelerations of different objects, such as a sports car, a tennis ball (from a serve), and an electron in an electric field of 100 V/m. Ask students to compare the accelerations of these objects.

Typical values:

Sports car 8 m/s^2

Tennis ball 500 m/s^2

Electron 1×10^{13} m/s^2

OL Linguistic

ADDITIONAL IN-CLASS EXAMPLE

Use with Example Problem 2.

Problem A certain mass spectrometer produces a beam of doubly ionized (2+) neon atoms with a mass of 3.3×10^{-26} kg that are accelerated by a potential difference of 30.0 V. If the ions pass through a magnetic field of 0.025 T, determine the radius of their path.

Response First, derive the equation:

$$\frac{q}{m_{\text{neon}}} = \frac{2V}{B^2r^2} \Rightarrow r = \frac{1}{B}\sqrt{\frac{2Vm_{\text{neon}}}{q}}$$

Next, solve the equation:

$$r = \frac{1}{0.025\ \text{T}}\sqrt{\frac{2(30.0\ \text{V})(3.3\times10^{-26}\ \text{kg})}{1.602\times10^{-19}\ \text{C}}} = 0.14\ \text{m}$$

DIFFERENTIATED INSTRUCTION

Struggling Students Students can understand how a mass spectrometer works by having them act out the effect of the magnetic field on two isotopes.

- Obtain two balls of the same approximate diameter but different weights.
- Have a student hold one ball in each hand and assume that both balls are composed of the same material (identical chemical properties—in particular, charge).
- Explain that, by making use of magnetic force on a moving charged particle, a mass spectrometer can measure the masses and relative concentrations of atoms and molecules and thus separate chemically identical isotopes that have different mass.
- Point out that the magnetic force will pull (deflect) both particles in the same direction because they have the same charge.
- Have students move both balls in an arc. Then point out that the lighter ball will deflect slightly more because the magnetic force gives it a greater centripetal acceleration.
- Help students curve the lighter ball through a tighter arc than the heavier ball.
- Explain that the difference in where the balls (particles) hit the detector reveals their different masses. **BL** **Kinesthetic**

Careers

Broadcast Technician Serving as the technical expert for a large TV station requires knowledge of electronics and computer programming for monitoring and adjusting equipment, as well as troubleshooting and repairing it in high-pressure situations. Necessary qualifications include an associate's degree (or appropriate vocational- or technical-school background), knowledge of current FCC rules, and appropriate licensure or certification.

3 Assess

Assess the MAINIDEA

To separate U-235 from the U-238, workers at the Manhattan Project used a calutron with a magnetic field of 0.350 T and an acceleration voltage of 35.0 kV. Assuming each uranium isotope was singly ionized, calculate the distance between the isotope beams at the output of the calutron. Note that one atomic mass unit is 1.66×10^{-27} kg. Solving the formula $q/m = 2V/(B^2r)$ for r gives $r = 2Vm/(B^2q)$. Inserting the values gives

$$r_{235} = \frac{2(3.50\times10^4\ \text{V})(235)(1.66\times10^{-27}\ \text{kg})}{(0.350\ \text{T})^2(1.602\times10^{-19}\ \text{C})} = 1.39\ \text{m}$$

$$r_{238} = \frac{2(3.50\times10^4\ \text{V})(238)(1.66\times10^{-27}\ \text{kg})}{(0.350\ \text{T})^2(1.602\times10^{-19}\ \text{C})} = 1.41\ \text{m}$$

The difference is

$r_{238} - r_{235} = 1.41\ \text{m} - 1.39\ \text{m} = 2\ \text{cm}$

Check for Understanding

Radius of Curvature On the board, write the equation for the radius of curvature for a charge moving across a magnetic field, $r = mv/(qB)$. Divide the class into four groups. Assign each group one of the variables on the right side of the equation. Ask each group to identify each variable in the equation and give the units in which the variable will be expressed. Then ask students how halving m, doubling v, tripling B, or halving q would alter the radius of curvature of a charged particle in a magnetic field. Halving m would halve r, $(r = \frac{mv}{2qB})$. Doubling v would double r, $(r = \frac{2mv}{qB})$. Tripling B would reduce r by one third, $(r = \frac{mv}{3qB})$. Halving q would double r, $(r = \frac{2mv}{qB})$. **OL** **Interpersonal**

Extension

Mass Spectrometry Ask students to research real-world applications of mass spectrometry. Have them assemble their findings into a classroom media presentation. **OL** **Linguistic**

Reading Checks and Caption Checks

Reading Check, *student text p. 703*
You would reverse the direction of the magnetic field, but still maintain it at right angles to the electric field.

Caption Check, *student text p. 706*
The north pole is on the left side in the left-hand figure, and on the top in the right-hand figure.

Reading Check, *student text p. 707*
4.806×10^{-29} C

Reading Check, *student text p. 709*
To produce a double ionized ion, you would remove two electrons from an atom.

Caption Check, *student text p. 709*
52.06 amu

Practice Problems

student text p. 705

1. 9.8 mm
2. a. 8.0×10^{6} m/s
   b. 1.2×10^{2} V
   c. 1.5 cm
3. 1.5×10^{5} m/s
4. an inward spiral

Practice Problems

student text p. 708

5. 2.7×10^{-26} kg
6. 6.8×10^{-26} kg
7. 0.46 m
8. It must be only a single elementary charge.

Section 1 Review

student text p. 709

9. Assuming all of the ions have the same charge, the only variable that is not constant in the equation is the ion mass m. As m increases, the radius of the ion's path also increases. This results in separate paths for each unique mass.
10. Electrons are emitted by the cathode, accelerated toward the (positive) anodes by a potential difference, and pass through slits to form a beam at the end of the tube.
11. Because $r = \frac{1}{B}\sqrt{\frac{2V_{accel}m}{q}}$ then B must increase if m increases. If m is raised by a factor of 4, B would have to increase by a factor of 2 because, to keep r constant, B must increase as $\sqrt{m}$.
12. 7.3 cm
13. 2.7×10^{-26} kg
14. 3 mm

student text pp. 710–719

1 Introduce

Bellringer Activity

Perpendicular Waves Visually depict the two perpendicular forces of an electromagnetic wave. Have one student wave a 4- to 6-m gymnastics ribbon up and down in a sine wave, parallel with the walls. Be sure that everyone stands clear and there are no other obstructions to the flow of the ribbon. Have another student stand next to the first student and, stooping down, wave a second ribbon sideways, parallel to the floor. Again, make sure there are no obstructions and that the two ribbons don't interfere with each other. Explain that electric and magnetic waves vibrate perpendicular to each other and perpendicular to the direction in which they are moving through space. Remind students that, unlike the ribbon demonstration, electric and magnetic waves have the same frequency and a fixed phase relationship to each other.
BL Visual-Spatial EL

Tie to Prior Knowledge

Wave Properties Students will need to recall the definition of a wave and the definitions of frequency and wavelength. It will also be useful for students to refresh their knowledge about other wave properties, including speed, amplitude, phase, and period.

2 Teach

What are electromagnetic waves?

student text pp. 710–713

Use Figure 5 and Figure 6

You can help students understand the connection between electric and magnetic fields by having them look at the combined fields in **Figure 5.** Ask them to describe the orientation of the electric field with respect to the magnetic field. perpendicular You can then use this understanding to help students with **Figure 6.**
OL Visual-Spatial

Reinforcement

Wavelength, Frequency, and Energy Explain to students that it is accurate to describe electromagnetic waves by their wavelengths, frequencies, or energies. Scientific convention, however, is to categorize radio waves in terms of their frequencies (such as millions of hertz), optical and infrared light in terms of their wavelengths (in nanometers, which is billionths of a meter; 10^{-9} m), and X-rays and gamma rays in terms of their energies (in electron volts; eV).

Transmitting Electromagnetic Waves

student text pp. 714, 715

Concept Development

Wave Travel Ask students to explain why electromagnetic waves travel through empty space, but sound waves cannot. Sound waves are pressure waves that are carried by vibrations of particles. Space is a vacuum and has few atoms or molecules of gas that can vibrate. Electromagnetic waves, on the other hand, can propagate through empty space via their mutual interaction. That is, they do not require a medium in which to travel.

Using Models

Radio-Wave Propagation Students can create a model that reflects how radio waves propagate from a point source. Ask students to glue a glass marble to the center of a piece of Bristol board or heavy paper, then draw concentric circular rings of ever greater radii with the marble at the center. After applying glue along these lines, students should lightly sprinkle sand onto the glue. The pattern is very similar to how ripples form on a still pond when a stone is dropped into it.
OL Kinesthetic EL

Use the MiniLAB

Students gain experience with electromagnetic waves in *Wave Signals.*

Content Background

Maxwell James Clerk Maxwell developed a model of how field lines could transmit electric and magnetic forces and was able to translate them into equations describing E and B fields. These equations encompassed Coulomb's, Ampère's, and Faraday's laws, and a law relating magnetic fields due to the fact that there are no magnetic monopoles. When he combined the equations representing the laws, they predicted that an oscillating electric charge would produce electric and magnetic fields that moved in space at a velocity given by $v = \sqrt{1/(\varepsilon_0\mu_0)}$. Substituting numeric values gave a velocity of 3×10^8 m/s—the speed of light. Maxwell's work led to the predictions of the existence of electromagnetic waves and that light was just such a wave.

Quick DEMO

Radio Waves

Estimated Time 15 minutes

Materials portable AM radio, chicken wire

Procedure

1. Turn on and tune a portable AM radio to a local music station.
2. Make a cover out of chicken wire that will fit over the radio.
3. Place the cover over the radio. The music should turn to static.
4. Move the cover so that the antenna pokes out. The music will resume. Ask students whether different types of electromagnetic waves have different abilities to penetrate materials. yes Ask whether a radio wave can be stopped easily. Students should conclude that radio waves are relatively easy to stop. Also point out that different wavelengths have varying abilities to penetrate materials. Try this demonstration using a cell phone to see whether these wavelengths get blocked.

DIFFERENTIATED INSTRUCTION

Physically Impaired You can convey the nature and range of the electromagnetic spectrum without the use of ropes and springs by having students visualize waves with references to the everyday world. First have them imagine a wave that is as long as the school, moving so fast that it can circle Earth seven times per second. After soliciting ideas from students, reveal that this image describes a radio wave. Have them repeat the exercise with a wave that travels at the same speed, but has a wavelength approximately equal to the width of their finger. This is a microwave signal. Students can go several more rounds with this activity, brainstorming various other visual references to electromagnetic waves of different wavelengths.
BL Visual-Spatial

Use Figure 7

Ask students to look closely at **Figure 7** and state which types of waves have the highest frequency. gamma rays Which have the longest wavelength? radio waves What is the relationship between frequency and wavelength? Frequency and wavelength are inversely related: $c = f\lambda$. Ask students to verify that $c = f\lambda$ by choosing values for f and λ from the table and multiplying them. Because of a lack of detail in the table, calculations will only confirm that $f\lambda \approx 10^8$ m/s (the actual value of c is 3.0×10^8 m/s). OL

Reinforcement

Biology Connection Divide the class into small groups. Ask students to explain the effects of X-rays on living tissue. Ask students how X-rays can be beneficial in the medical field. X-rays can ionize atoms in the body, which can lead to damage in a variety of ways. One way is by creating free radicals, which are molecules with unpaired electrons. These molecules can damage the body; for example, by reacting with a DNA molecule to cause a mutation in which the cell loses control of cell division. This may lead to a cancer. The beneficial use of X-rays in medicine include X-ray imaging of bones, X-ray tomography, which gives three-dimensional images of the body, and X-ray crystallography, in which X-rays are used to in medical research to study the shape of proteins and other molecules.
OL COOP LEARN Interpersonal

Critical Thinking

Laser Light Ask students what the similarities are in the use of lasers in DVD players and laser printers. Both laser printers and DVD players use the electromagnetic waves of a laser to convey information. In a DVD, the laser converts binary information encoded as pits on the disk surface to electrical signals based as variations in the intensity of reflected light. In a laser printer, the laser exposes a photosensitive drum, which changes the electrical charge of a spot on the drum. This, in turn, determines whether a spot on the drum will attract toner ink to produce a spot on the paper. OL

Producing Electromagnetic Waves

student text pp. 715–717

Careers

Hertz Heinrich Hertz used two metal spheres, separated by a small gap and connected to a high-voltage transformer, to produce electromagnetic waves and thereby test Maxwell's theory. When ΔV between the spheres was large enough, a spark jumped, and ions oscillated between them. A few meters away, Hertz used a spark-gap receiver consisting of two metal spheres, separated by a tiny gap, on each end of a circular wire to see if the oscillating ions produced detectable electromagnetic waves.

He proposed that the waves would induce currents in the wire creating a ΔV between the spheres, resulting in a tiny spark. Sparks occurred even when the transmitter was many meters away. The waves' speed was 3×10^8 m/s—the speed predicted by Maxwell. Less than 20 years later, Marconi transmitted electromagnetic waves across the Atlantic, leading to the fields of radio and television transmission.

Concept Development

Energy, *B* Fields, and *E* Fields Throughout the section, you can emphasize the relationship between coils and capacitors. The first involves energy storage in a magnetic field, and the latter involves energy storage in an electric field.

Applying Physics

As previously noted, the electromagnetic spectrum is employed by a wide variety of transmitting and receiving devices. To prevent interference, the FCC (Federal Communications Commission) assigns a specific band of frequencies to each transmitter.

Discussion

Question Ask students to think about how a transformer works. Ask them to identify the electric part and the magnetic part.

Answer Transformers convert AC power of given voltage levels to different voltage levels at the same frequency. Inside a transformer, multiple coils (called windings) are wrapped around a magnetic core. The input lines connect to a primary coil. The secondary coils are connected to output lines, such as the power lines leading from a substation to a neighborhood. The AC current in the primary coil induces a magnetic flux around the magnetic core, alternating direction during each electric cycle. The flux in the core then induces an AC current in the secondary coils. The change in voltage is dependent upon the number of windings of each of the coils. OL

Identify Misconceptions

Microwave Ovens Some students may think that the microwaves in a microwave oven excite a natural resonance in water. The frequency of a microwave oven is well below any natural resonance in an isolated water molecule, and in water those resonances are so smeared out that they are barely noticeable. It is somewhat like playing a violin under water—the strings will not emit well-defined tones in water because the water impedes their vibrations. A microwave oven exposes the water molecules to the intense electromagnetic fields in strong, nonresonant microwaves.

Content Background

Semiconductor Microwave Sources Semiconductor sources have, for the most part, replaced other low- to medium-power microwave sources. One such source is the Gunn Oscillator. An n-doped semiconductor can be made to oscillate when a strong electric field is placed across it and when it is biased so that it has a negative resistance (increasing voltage yields a reduction in current). Other semiconductor microwave sources include the microwave transistor and the impact avalanche and transit time (IMPATT) diode. High-power oscillators, such as those in a microwave oven, use vacuum tubes.

Reality Check

MAIN IDEA Have the students list as many examples as possible of electromagnetic waves interacting with matter. For each example, have them list the source and the receiver of the wave. Examples include human sight, where the source may be the Sun and the receiver is the rods and cones of the human eye, radio waves where the source is the transmitting antenna and the receiver is the receiving antenna in your car or radio. Wi-Fi is another example, where the transmitter-receiver pair is an antenna in a computer and one in the router. Finally, the process of photosynthesis starts with visible electromagnetic waves transmitted from the Sun to Earth where they are absorbed by organic molecules in plants to create photochemical reactions.

Quick **DEMO**

Induction of Current Between Coils

Estimated Time 5 minutes

Materials large magnet, demo galvanometer, coil of wire

Procedure

1. Connect the coil to the galvanometer.
2. Have students observe the readings on the galvanometer as the coil is moved with respect to the magnet.
3. First move the coil near one end of the magnet, then pass the coil above or below the magnet, and finally move the coil while surrounding the magnet.

Generating a reading on the galvanometer indicates that current is induced when the coil is passed between the poles of the magnet. Students should note how the readings change as you move the coil around.

Reinforcement

Destructive Interference In one of his experiments, Heinrich Hertz beamed radio waves onto a metal sheet and compared the strength of the signal at different distances from the metal sheet. He found places with weak signals at distances that were evenly spaced from the plate. Ask students to offer an explanation for these results. The weak spots were places of destructive interference (nodal points). OL

Activity

Piezoelectric Effect Demonstrate the piezoelectric effect using a piezoelectric wafer that you have connected to a small neon lamp. Turn off the room lights and squeeze the wafer to light the lamp. Explain that the work done by squeezing the wafer was transformed into electric energy, causing the light to shine. Some miniature flashlights operate on the same principle.
OL Visual-Spatial

PHYSICS PROJECT ACTIVITY

Seeing the Invisible Most astronomy research is done with wavelengths that are invisible to the human eye. Have students research images of galaxies, star clusters, and other phenomena that were captured using radio, infrared, or X-ray receivers. Ask students to bring the images to class to share with other students. They should identify the image, the type of astronomical body, how the image was obtained, the distance from Earth, and any other interesting or pertinent information.
OL Linguistic

Receiving Electromagnetic Waves

student text pp. 718, 719

Discussion

Question Ask students why an antenna that is tuned to the desired frequency has a larger effective area and therefore better performance than one that is not tuned to that frequency.

Answer There is always noise at the front end of the receiver from interfering sources as well as from background radiation present in our universe. The energy that the antenna picks up must be great enough to overcome the energy associated with all that noise. The desired incoming waves always carry a certain amount of energy and frequency.

An antenna tuned to the specific desired energy or frequency has a larger effective area than one that is not tuned to that frequency because it resonates with the desired frequency. The closer the antenna is tuned to the desired frequency, the greater the coupling of the energy and the resonance. Resonance peaks when the antenna is precisely tuned. (Note that antenna performance is also a function of the orientation of the antenna in relation to the direction of the incoming wave.) OL

Using an Analogy

Catching Waves The large parabolic ears of many felines catch faint sound waves and are analogous to large dish antennas that collect and focus weak radio signals. Many animals, including dogs, owls, and hares, have ears that rotate like some parabolic dishes to determine just where a sound is coming from.

Use the PhysicsLAB

In *Blocking Waves,* students can investigate what types of materials shield electromagnetic waves.

DIFFERENTIATED INSTRUCTION

Struggling Students Connect a solar cell to an amplifier and speaker. As students observe, expose the solar cell to fluorescent lights. A 60-Hz hum should be audible from the speaker. Switch the lights off and on, or cover and uncover the solar cell, noting any difference. Try flashing a strobe light at the solar panel at 59 or 61 Hz, having students listen to the beats. Refer to visual aids showing waves and constructive interference. (Note that the oscillation is at 120 Hz, not 60 Hz, so try flashing the strobe light somewhere near 120 Hz.) BL Auditory-Musical EL

3 Assess

Assess the MAINIDEA

Imagine holding a fist-full of charge and moving it up and down as fast as possible. This accelerated bundle of charge will produce an electromagnetic wave. In what direction will the electric field oscillate? The electric field will oscillate in the vertical direction. In which direction will the magnetic field oscillate? in the horizontal direction To capture these waves with an antenna, in which direction should it be oriented, and why? It should be oriented vertically so that the electric field of the electromagnetic wave can more efficiently accelerate the electrons in the antenna.

Check for Understanding

Show students several photographs of parabolic dish antennas, or ask students if they use parabolic antennas for their television reception. Ask them whether the positioning of the receiving antenna should be different for different wavelengths. No; all waves reflect in the same manner and are focused to the same point. OL Visual-Spatial

Reteach

Hold up an AM radio and a cellular phone. Ask students to compare the speed and wavelengths used by these devices. Both long and short wavelengths travel at the same speed. BL

Reading Checks and Caption Checks

Reading Check, *student text p. 711*
$\lambda_{new} = c/(1.25f) = 0.8(c/f) = 0.8\lambda$

Caption Check, *student text p. 712*
gamma wave

Reading Check, *student text p. 713*
Long radio waves can reflect off ions in the atmosphere and can thus travel long distances. Shorter radio waves do not reflect off atmospheric ions; they travel in straight lines and so must be relayed from station to station across Earth's curved surface.

Caption Check, *student text p. 716*
zero

Reading Check, *student text p. 717*
They can't be used because infrared waves are produced within molecules. The resonant cavity would have to be smaller than it possibly can be—molecular size.

Reading Check, *student text p. 719*
Antennas are most efficient when they are half the size of the wavelength they are designed to detect. Because AM waves are longer than FM waves, AM antennas are longer than FM antennas.

Practice Problems
student text p. 711

15. 5.26×10^{-7} m

16. 3.7×10^{-7} m

17. 1.4×10^{10} Hz

18. up and down

Physics Challenge
student text p. 713

1. red

2. All electromagnetic waves travel at the same velocity in a vacuum.

3. Red light diffracts the most, violet the least.

4. violet: 6.59×10^{14} Hz to 7.69×10^{14} Hz
blue: 6.10×10^{14} Hz to 6.59×10^{14} Hz
green: 5.20×10^{14} Hz to 6.10×10^{14} Hz
yellow: 5.03×10^{14} Hz to 5.20×10^{14} Hz
orange: 4.82×10^{14} Hz to 5.03×10^{14} Hz
red: 4.29×10^{14} Hz to 4.82×10^{14} Hz

Practice Problems
student text p. 715

19. 2.99712×10^{8} m/s

20. 2.25×10^{8} m/s

21. 1.52

22. 2.51034 s

Section 2 Review
student text p. 719

23. The changing electric field induces a changing magnetic field, and the changing magnetic field induces a changing electric field. The waves propagate as these two fields regenerate each other.

24. Electromagnetic waves can be described by frequency and wavelength. They behave similarly to other waves in that they reflect, refract, diffract, and can be Doppler-shifted. Differences between the electromagnetic waves and other waves are that electromagnetic waves can travel through a vacuum and can be polarized.

25. 2.0×10^{13} Hz

26. They must also be horizontal.

27. Parabolic dish antennas are able to receive signals only within a very narrow range of angles.

28. Longer; lower-frequency waves would have longer wavelengths, so they use a longer antenna.

29. 2.30

30. If the entire ozone layer is thinning, the amount of UV radiation from the Sun that is blocked by the ozone layer decreases, allowing more UV rays to reach Earth's surface. The wavelengths of UV waves are short enough and their energies are high enough to damage skin molecules. Thus, the resulting increase in UV rays might increase the prevalence of skin cancer in humans.

FRONTIERS IN **PHYSICS**

In the Zone

Cell Phones

Purpose

for students to gain an understanding of cell phones and cell phone networks

Background

A cell phone is essentially a two-way radio. Unlike walkie-talkies, cell phones can send and receive messages at the same time because they use separate frequencies for each of these operations. Cell phones operate in the microwave region of the electromagnetic spectrum. Microwaves are less energetic than infrared waves but more energetic than radio waves.

Teaching Strategies

Set up a model of a cell phone network in your classroom. Use a foam ball or other soft ball. Students standing in place are towers. A small number of students (maybe only one to start) move through the class while "talking" on a cell phone. Towers must hand off the call by passing the ball from one student to the next. Adding more students on cell phones adds to the complexity. A real cell phone network can handle millions of calls at once.

GOING**FURTHER** >>>

Expected Outcomes The FCC banned the use of cell phones on airplanes in 1991 out of concern that the transmissions from cell phones could disrupt sensitive equipment on the aircraft. At around the same time, hospitals began banning the use of cell phones in certain areas. There is no FCC ban on the use of cell phones in hospitals. Rather, these are hospital-specific regulations that may be due to privacy issues as well as equipment concerns.

Both the FCC and the FAA uphold the ban on cell phones on airplanes, but many groups are actively researching ways to allow limited cell phone use during some portions of airplane flights.

NOTE: The U.S. Congress upheld the airline ban on cell phones in 2008.

ANSWERS

SECTION 1

Mastering Concepts

31. 9.11×10^{-31} kg and -1.602×10^{-19} C

32. Isotopes are atoms of the same element that have different masses.

33. It must be accelerating.

34. You would produce and separate gases from the sample and then insert each gas into the ion source, where an electron beam would strip off electrons, yielding positive ions. The ions would accelerate in an electric field and then curve in a magnetic field. From the curvature *r* of each ion, you could determine the charge-to-mass ratio and then, by dividing the charge on the ion by the charge-to-mass ratio, you could determine the mass of each ion. You would then compare the values to masses of other known ions to determine the composition of the material.

Mastering Problems

35. 0.16 T

36. 6.9×10^{5} m/s

37. 3.9×10^{-26} kg

38. a. 7.8 kV

b. 2.4×10^{-15} J

c. 8.7×10^{5} m/s

39. Answers will vary, but a correct form of the answer is "Singly ionized carbon-12 is accelerated by 515 V in a mass spectrometer. If the magnetic field is 75 mT, what is the radius of curvature *r* of the ions? $r = 0.15$ m"

40. $D < E < C < A < B$

41. 5.7×10^{-26} kg

42. 4.5×10^{-3} T

SECTION 2

Mastering Concepts

43. An oscillating potential difference across the antenna (sent from the transmitter) accelerates electrons in the antenna's metal. The acceleration generates an oscillating electric field that propagates away from the antenna and that also produces an oscillating magnetic field. The result is a radio wave. When an antenna captures a radio wave, electrons accelerate in the antenna and the antenna converts the wave's oscillating electric fields back to potential differences, which oscillate at the frequency of the wave.

44. An oscillator supplies the changing electric field, which in turn generates a changing magnetic field. A battery or DC generator would not generate a changing electric field except briefly when turned on or off.

45.

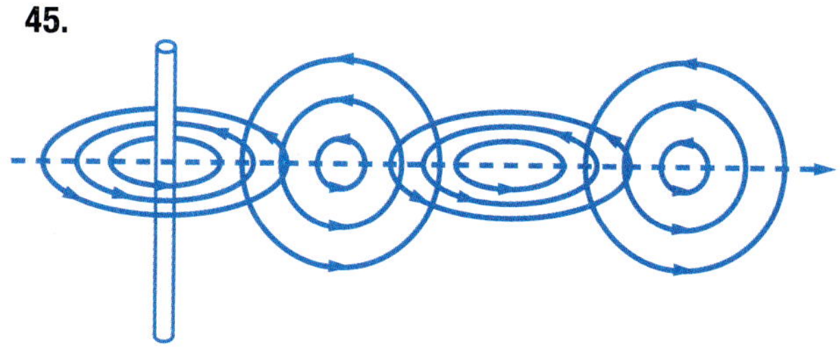

46. Quartz crystals bend or deform when voltage is placed across them. The crystals then vibrate at a set frequency.

Mastering Problems

47. 1.0 cm

48. 4.6×10^{14} Hz

49. 31 MHz

50. 1.48 m

51. 1.98×10^{8} m/s

52. 0.0938 m

Applying Concepts

53. Answers will vary, but a correct form of the answer is, "A charged particle moves at 2.8×10^5 m/s into a region with a magnetic field of 0.065 T. If it follows a circular path of radius 0.045 m, what is its charge-to-mass ratio?"

54. top plate

55. $\frac{E}{B} = \frac{\frac{\text{N}}{\text{C}}}{\frac{\text{N}}{\text{A•m}}} = \frac{\text{A•m}}{\text{C}}$

Because 1 A is 1 C/s, this becomes

$\frac{E}{B} = \frac{\text{C•m}}{\text{s•C}} = \frac{\text{m}}{\text{s}}$

56. The magnetic field would be directed out from and perpendicular to the plane of the paper.

57. You can change either both fields or neither field, but you cannot change only one field.

58. a. radio waves

b. X-rays

c. All travel at the same velocity.

59. The eyes would be much larger because the wavelength of microwave radiation is much larger than that of visible light.

60. 2.45 GHz, 2.05 GHz, 1.90 GHz, 1.575 GHz, 0.90 GHz

61. Answers will vary. A possible form of the correct answer would be, " . . . and enters a region where the electric field is 500 N/C down. What must the magnetic field in this region be so that the proton travels through undeflected?"

Mixed Review

62. 1.59 m

63. 9.0×10^8 Hz

64. 1.25×10^7 C/kg

Thinking Critically

65. 26.4 m/s

66. To see, you must detect the light, which means the light will be absorbed or scattered. Essentially, an invisible person would be completely transparent so light would just pass through the eye without ever being absorbed or scattered.

67. The charge-to-mass ratio for isotopes in a mass spectrometer is $\frac{q}{m} = \frac{2V_{\text{accel}}}{B^2r^2}$, so the radius of the isotope's path is $r = \frac{1}{B}\sqrt{\frac{2V_{\text{accel}}m}{q}}$.

The difference in the radii for the two isotopes is

$$0.10\times10^{-3}\ m = r_{176} - r_{175}$$
$$= \frac{1}{B}\sqrt{\frac{2V_{\text{accel}}}{q}}\left(\sqrt{m_{176}} - \sqrt{m_{175}}\right)$$
$$= \frac{1}{B}\sqrt{\frac{2V_{\text{accel}}}{q}}\left(\sqrt{176m_{\text{P}}} - \sqrt{175m_{\text{P}}}\right)$$
$$= \frac{1}{B}\sqrt{\frac{2V_{\text{accel}}m_{\text{p}}}{q}}\left(\sqrt{176} - \sqrt{175}\right)$$

The magnetic field, then, is

$$B = \frac{\sqrt{176} - \sqrt{175}}{0.10\times10^{-3}m}\sqrt{\frac{2V_{\text{accel}}m_{\text{p}}}{q}}$$
$$= \frac{\sqrt{176} - \sqrt{175}}{0.10\times10^{-3}m}\sqrt{\frac{(2)(500.0\text{ V})(1.67\times10^{-27}\text{ kg})}{1.602\times10^{-19}\text{ C}}}$$
$$= 1.2\text{ T}$$

The radius for the isotope with a mass equal to 176 proton masses is

$$r_{76} = \frac{1}{B}\sqrt{\frac{2V_{\text{accel}}(176m_{\text{p}})}{q}}$$
$$= \frac{1}{1.2\text{ T}}\sqrt{\frac{(2)(500.0\text{ V})(176)(1.67\times10^{-27}\text{ kg})}{1.602\times10^{-19}\text{ C}}}$$
$$= 3.6\times10^{-2}\text{ m}$$

When designing the spectrometer, you can choose any value of V_{accel} and B, provided V_{accel} is at least 500 V. However, since q/m is constant, V_{accel} will be proportional to $B^2\,r^2$.

Writing in Physics

68. There are several types of mass spectrometers in use at airports. Time-of-flight instruments accelerate ions to a specific kinetic energy using an electric field. The time it takes for an ion to reach a detector is measured. The heavier an ion, the longer it takes to travel. In quadrupole mass-filter instruments, oscillating potential differences are applied to four metal rods connected to AC potential source differences. Only ions with certain charge-to-mass ratios pass through without deflection. Ion trap instruments have a system of three electrodes—a metal ring with two metal end caps that are near but not touching the ring. The caps are electrically connected. An AC potential difference is applied between the ring and caps. Only ions having a specific charge-to-mass ratio are trapped.

Cumulative Review

69. 7.1×10^{-5} m

70. The spheres share charges equally when they are touched together, so they each carry $-1q$ of charge. The force varies with the product of the charges, so the new force is to the old force as $1q \times 1q$ is to $4q \times 2q$, or 1:8. Therefore, the new force is $F/8$. The direction of the new force will be repulsive rather than attractive.

71. 3.8×10^{3} V/m or N/C

72. $1.68

73. 0.016 T

74. north

Standardized Test Practice

student text p. 725

Multiple Choice

1. D
2. A
3. D
4. C
5. A
6. C

Free Response

7. $r = \frac{vm}{Bq}$

$$= \frac{(2.8\times10^{5}\ \text{m/s})(3.34\times10^{-27}\ \text{kg})}{(0.150\ \text{T})(1.602\times10^{-19}\ \text{C})}$$

$$= 4.00\ \text{cm}$$

Rubric

The following rubric is a sample scoring device for free response questions.

Points	Description
4	The student demonstrates a thorough understanding of the physics involved. The response may contain minor flaws that do not detract from the demonstration of a thorough understanding.
3	The student demonstrates an understanding of the physics involved. The response is essentially correct and demonstrates an essential but less-than-thorough understanding of the physics.
2	The student demonstrates only a partial understanding of the physics involved. Although the student may have used the correct approach to a solution or may have provided a correct solution, the work lacks an essential understanding of the underlying physical concepts.
1	The student demonstrates a very limited understanding of the physics involved. The response is incomplete and exhibits many flaws.
0	The student provides a completely incorrect solution or no response at all.

CHAPTER 27
Quantum Theory

About the Photo

Are quantum phenomena useful for everyday life? yes How many people in the picture are using a quantum phenomenon? You can tell by counting the flashes. Digital cameras use the photoelectric effect, by which photons knock electrons out of a light-sensitive material to be collected in little bins (one per pixel). The charge accumulated in each bin is then read out to produce the picture. In what other ways do we use quantum phenomena in our lives? Quantum phenomena are exploited in computer circuits, lasers, and MRI.

Chapter Overview

The first section introduces the concept of the quantum by discussing the emission spectrum of an incandescent body. The photoelectric effect and Compton scattering are discussed and used to support the photon model of light. The chapter closes by presenting evidence of the wavelike properties of small-scale matter.

Before students study the material in this chapter, they should study:

- Conservation of energy
- Conservation of momentum
- Diffraction
- Electromagnetic waves
- Potential difference

To solve the problems in this chapter, students will need a strong grasp of:

- Graphing data
- Scientific notation
- Slope
- Solving linear equations

Introduce the BIG IDEA

In the early 20th century, scientists expected that increasing the intensity of light shining on a material would result in electrons being knocked "farther" out of the material. But this didn't happen—more electrons came out, but they were not more energetic. This phenomenon could only be explained if light were thought of as particles instead of as waves. This was the first encounter with the particle-wave duality of the quantum world.

Go online!
connected.mcgraw-hill.com

Your online experience includes:

- **Video**
- **Concepts In Motion**
- **Virtual Investigations**
- **All Lab, Applying Practices activities, and PBL worksheets**
- **Online Practice**
- **Personal Tutor**
- **Multilingual eGlossary**
- **Presentation**

Use the LaunchLAB

Students will observe the change in the color of light from an incandescent bulb as they increase the brightness in *Lightbulb Spectrum*.

Use Physics T.V.

Video

NIGHT-VISION GOGGLES You've probably seen night-vision gear in spy movies, on television, or in video games, but do you know how they work? Are all night-vision goggles the same? Explore the physics of spy tech.

1 Introduce

Bellringer

WARNING: Students should be careful not to touch hot lightbulbs.

Light and Color Connect an ordinary clear lightbulb of any size to a variable source of AC power; for example, an autotransformer. Increase the voltage. Have students note the change in color of the filament—at first a dull red, then changing gradually to white. Have them conjecture about changes in the color of light emitted from the filament and changes in its temperature. Possible answer: The color of the light changes from red to white as the filament becomes hotter.
OL Visual-Spatial EL

Tie to Prior Knowledge

Photoelectric Effect In previous chapters, students learned about the wave description of light and the relationship between work, charge, and potential difference. Students will use this knowledge in explaining the photoelectric effect.

2 Teach

A New Model Based on Packets of Energy

student text pp. 728–730

Use the MiniLAB

This section focuses primarily on radiation emitted by all bodies and whose total power output is proportional to a body's temperature. But not all radiation emitted by objects is produced by virtue of their temperature. In *Glows in the Dark,* students will explore a substance that emits radiation (in particular visible light) when light is shined on it.

Use Figure 2

Have students answer the following questions: How does the frequency of the peak-intensity radiation of an incandescent body vary as a function of temperature? The frequency increases with temperature. How does the peak intensity of the radiation vary with temperature? The peak intensity increases with temperature.

Critical Thinking

Electromagnetic Radiation Ask students the following question: If an incandescent object glows red, most of its energy is likely being emitted as what type of electromagnetic radiation? infrared radiation OL

Content Background

Temperature and Color of Fire The color of a fire tells us about its temperature. As a fire heats up, its emission shifts toward the green and blue part of the spectrum. The red areas of a candle flame are about 800°C, whereas the blue areas are about 1400°C. The bright yellow-white part of the flame above the wick is around 1200°C. A fire emits light over a broad range of frequencies, and its peak-intensity radiation does not match the curves shown in **Figure 2** because the substance burning is normally not an ideal black body. Often, a nonideal radiator is described by its color temperature. This is the color of the peak-intensity radiation that would be emitted by an ideal black body at the given temperature. For many light sources, the actual temperature is less than the color temperature. For example, the color temperature of a candle is about 1650°C, but the actual peak temperature is about 1400°C.

Applying Physics

In 1964–65, radio astronomers Arno Penzias and Robert Wilson were measuring radio waves from a certain part of the Milky Way galaxy when they unexpectedly detected radiation of wavelength 7.35 cm that was independent of direction and time of year. They did not offer an explanation when they reported their results but suggested that certain astrophysicists might have a "possible explanation." The modern explanation, developed from the late 1940s and supported by observation several times since the mid-1960s, is that Penzias and Wilson observed just one small part of the electromagnetic spectrum of radiation emitted by matter about 13 billion years ago, when the universe was incomparably hotter and denser than it is today. The emission spectrum for this cosmic radiation (the same kind of spectrum as that shown in **Figure 2**) corresponds to a temperature of 2.7 K. (The reason this temperature is so low despite its having been produced under conditions of extraordinarily high temperature is that the radiation "cooled" as the universe expanded.) Have students classify an electromagnetic wave with a wavelength of 7.35 cm using the electromagnetic spectrum chart. microwave OL

Critical Thinking

Color and Temperature Ask students if a white-hot ember is cooler than a red-hot ember. No; the white ember is hotter than the red one. The hotter the ember, the more the peak of the intensity-frequency curve shifts toward higher frequencies, as shown in **Figure 2.** When the peak is in the infrared, there is more energy in the low-frequency end of the visible spectrum than the high-frequency end, so the object will appear red (assuming it is hot enough to be visible to the naked eye). When the temperature is high enough that the peak is in (or nearly in) the visible part of the spectrum, then the energy of the radiation is more evenly distributed across the visible spectrum and the object will appear white. **OL**

Activity

f_{max}-T Graph Have students approximate the maximum-intensity frequency f_{max} for the radiation emitted by a lightbulb filament at a temperature of 2500 K. Suggest that students use linear regression or extrapolation of an f_{max} -T plot constructed from the data in **Figure 2.** 1.6×10^{14} Hz. **AL** **Logical-Mathematical** **EL**

REAL-WORLD PHYSICS

Analyzing Structure by Infrared Spectroscopy Analyzing the infrared absorption spectrum of materials is useful for understanding the structure of their molecules. A beam of infrared radiation from an incandescent source is projected onto a solution of the material to be analyzed. A prism separates the wavelengths of the radiation after it passes through the solution, and a detector moves across the spectrum to record the wavelengths at which the solution absorbs infrared energy. The absorbed energy, which increases the vibrational energy of the molecules, depends on the mass of the atoms in the molecules and the strength of the bonds that connect the atoms together. Typically, wavelengths from 5 to 10 μm are used.

Quick **DEMO**

Quantized Potential Energy

Estimated Time 10 minutes

Materials 1-m-long U-channel, 14 similar-sized books, ball

Procedure Prop one end of the U-channel on four books to make a ramp. Next to the ramp, stack one, two, three, and four books side by side to make a set of descending steps. Have the students observe the ball as you release it from the top of the ramp. Lead them to understand that the gravitational potential energy of the ball decreases uniformly along the ramp. Now tell them that you are going to model the ramp as a series of steps. Ask them to observe the ball and think about its potential energy as it rolls down the new "quantized" ramp, then place the ball on the top step and gently nudge it so it descends the steps.

Discuss with students that the gravitational potential energy of the ball remains constant as it rolls along a given step because the steps are level. However, because each step is at a different height, the ball's potential energy decreases when it drops to a new step. Point out the analogy with electrons and ions in a hot object, which are not allowed to occupy a continuous range of vibrational energy states but only discreet ("quantized") vibrational energy states.

Using Models

Electron Energy Point out that the ball-on-a-step model of vibrational energies of an electron is limited because the ball goes through an essentially continuous range of potential energy as it drops from one step to another. When an electron makes a transition from one vibrational energy level to another, its change in energy is emitted as a chunk (i.e., quantum) of energy as opposed to a continuous stream of energy.

Use the MiniLAB

In *Modeling the Quantum,* students are given four opaque canisters, the knowledge that the canisters contain quarters, and hints regarding the numbers of quarters. By weighing the canisters the students should be able to find the mass of one "coin-quantum."

The Photoelectric Effect

student text pp.731–734

Use the PhysicsLAB

Students will release steel balls from different heights on the long end of a hockey-stick-shaped ramp to represent the different energies of photons impinging on a surface. Waiting at the bottom are steel-ball "valence electrons." From what height must the steel balls be released to send those valence electrons on their way? Students will find out in *Modeling the Photoelectric Effect.*

Content Background

Discoverer of the Photoelectric Effect In 1887 Heinrich Hertz discovered (what would later be called) the photoelectric effect during his classic experiments in which he generated radio waves and showed that they moved with a finite speed. He also demonstrated that they could be reflected, refracted, and polarized. Hertz generated radio waves by producing sparks that oscillated back and forth between two metal balls at the same frequency as that of the waves generated. He detected the radio waves with a device that responded to the waves by producing sparks that oscillated between a second set of metal balls.

Hertz observed, first by accident, and then by careful investigation, that the sparks in the receiving device were stronger when they were exposed to the sparks in the generating device. Hertz determined that, as the radio-wave generator was producing sparks, it was emitting ultraviolet light, which impinged on the metal balls of the detector, increasing the strength of the detector sparks, which are light produced by electrical currents in the air. Hertz considered this "new and very puzzling phenomenon" to be important but was unable to explain it. He investigated the phenomenon for several months and wrote an article on it. Other physicists continued investigating this effect, the explanation of which was published in 1905 by Albert Einstein.

Using an Analogy

Water Waves Students may better understand the photoelectric effect by contrasting it with the effects of water waves. Suppose small ripples in a lake strike a beach ball on the surface of the water. You would expect the effect on the beach ball to be small. Then imagine large waves from a motorboat wake striking the ball. You would expect the effect on the beach ball to be greater, and it is. This is what students may intuitively expect with light, but light behaves differently in the photoelectric effect. Increasing the intensity of light (analogous to going from ripples to motorboat waves) can increase the number of ejected electrons, but all electrons starting at the same energy level will have the same kinetic energy after ejection, regardless of the intensity of the incident light (analogous to the beach ball being moved the same vertical distance by the ripples as by the motorboat waves).

DIFFERENTIATED INSTRUCTION

Struggling Students Using the data from the MiniLAB *Modeling the Quantum,* have students make bar graphs of the mass data in ascending order for the four canisters, labeling the canisters *A, B, C,* and *D,* respectively, on the graph. Have students recall the assumptions of the activity: each canister contains two or more coins, and at least one canister has an even number and one has an odd number of coins. Have students look for differences in the heights of the histograms and lead them to realize that each change represents a change in mass due to a difference of one, two, or some other integer number of quarters. BL Visual-Spatial EL

Quick DEMO

Photoelectric Effect

Estimated Time 15 minutes

Materials electroscope, materials to charge electroscope both positively and negatively, 2 cm × 10 cm strip of zinc cleaned of its oxide coat with steel wool, UV light source

Procedure Attach the zinc strip securely to the terminal of the electroscope. Positively charge the electroscope and the attached zinc. Shine the UV light on the zinc. **WARNING: UV light is harmful to the eyes. Do not allow anyone to look directly into the source.** Discharge the scope and then charge it negatively. Shine the UV light on the zinc strip. The positively charged scope did not discharge. The negatively charged scope discharged, causing the leaves to collapse. The incident UV light caused electrons to be ejected from the surface of the negatively charged zinc, but the positively charged ions in the positively charged zinc could not be ejected.

PHYSICS CHALLENGE ACTIVITY

Work Functions and Ionization Energies Have students use a source such as the *Handbook of Chemistry and Physics* to look up the work functions and first-ionization energies of several elements, such as sodium, barium, copper, gold, and zinc. Have students compare the values and explain the differences. Ionization energies are measured essentially for individual atoms. However, work functions are measured for solid-state materials. Because atoms in the solid state interact with neighboring atoms, their electronic energy levels change, so the work function and ionization energy are different. **AL** **Logical-Mathematical**

Concept Development

Surface Condition Note that the cleanliness of a metal's surface has a significant impact on its experimentally measured work function.

Concept Development

Photon Energy Consider the experimental demonstration of the photoelectric effect illustrated in **Figure 4.** Explain that many of the photons from the light source do not have sufficient energy to cause electron emission. These photons excite vibrational modes in the solid, thereby increasing its thermal energy.

Use the PhysicsLAB

The connection between photon energy and the voltage (or energy) applied to an LED is explored in *Relating Color and LED Voltage Drop.*

Identify Misconceptions

Albert Einstein's Nobel Prize Ask students to cite the work for which Einstein received the 1921 Nobel Prize in Physics. Most students will respond that it was his theory of relativity. Explain that this theory was considered too revolutionary at the time. He received the prize for his explanation of the photoelectric effect, which—although also controversial—had been experimentally confirmed by Robert Millikan. **OL**

PHYSICS PROJECT ACTIVITY

Photovoltaic Devices Students will investigate devices that employ photovoltaic cells. Suggest that students first brainstorm to generate a list of situations where these devices are probably used along with their likely functions. After determining if a given device uses a photovoltaic cell, have students research the cell's function. Students can present their findings as written reports, posters, or models. In all cases, the function of the photovoltaic cell in the device should be clearly identified. **OL** **Linguistic**

Discussion

Problem Why would one expect metals to demonstrate the photoelectric effect more than crystalline solids?

Response Electrons in metals are less strongly bound than those in crystalline solids. **OL**

Planck and Einstein

student text, pp. 734–738

Content Background

Importance of the Photoelectric Effect The discovery of the photoelectric effect was important for several reasons. First, it provided more information about electromagnetic radiation, helping to solidify the concept that electromagnetic radiation can consist of individual particles (or quanta) called photons that deliver their energy only in certain discrete quantities. Second, it was another line of evidence that electrons are contained in metals. Einstein showed that, for a given metal, there is a minimum frequency of light for which the photon energy is equal to the work function and, no matter how intense, light at a lesser frequency will not cause photoemission. Third, it helped to validate quantum theory. Robert Millikan helped confirm quantum theory by measuring Planck's constant to within 0.5%.

Concept Development

Intensity and Photocurrent Increasing the intensity of a light source (assuming the frequency is above the threshold) does not increase the energy of the ejected electrons. The increased intensity increases the photocurrent, which is the rate at which photoelectrons are ejected. For example, dim blue light will cause the emission of photoelectrons that are more energetic than those emitted by bright red light. However, the bright red light will cause the emission of photoelectrons at a faster rate.

Reinforcement

Conservation of Energy Emphasize the conservation of energy in the photoelectric effect. Begin by stressing that input radiation is the cause, while the emission of electrons is the effect. Then write a statement such as the following on the board: $E_{input} = E_{output}$. Have students identify the energy input as hf, the energy of the photon. Ask students what effect the energy has on an electron in a metal and to use conservation of energy to formulate E_{output}. Some of the photon energy "rips" the electron loose from the metal and the remaining photon energy is converted to the kinetic energy of the free electron. The energy needed to "rip" the electron from the metal is the work function and is represented by hf_0. Thus, hf = hf_0 + KE. **OL** **Linguistic**

ADDITIONAL IN-CLASS EXAMPLE

Use with Example Problem 1.

Problem A potassium metal cathode has a stopping potential of 2.24 V. What energy in joules is given to the electrons by the incident light?

Response
$$KE = -qV_0$$
$$= -(-1.60\times10^{-19}\text{ C})\times(2.24\text{ V})$$
$$= 3.58\times10^{-19}\text{ J}$$

ADDITIONAL IN-CLASS EXAMPLE

Use with Example Problem 2.

Problem The work function of a cathode is 2.24 eV. What energy (in eV) will photoelectrons have if the cathode is illuminated with light that has a wavelength of 425 nm?

Response
$$E = \frac{1240\text{ eV}\cdot\text{nm}}{\lambda}$$
$$= \frac{1240\text{ eV}\cdot\text{nm}}{425\text{ nm}}$$
$$= 2.92\text{ eV}$$
$$KE = E - W$$
$$= 2.92\text{ eV} - 2.24\text{ eV}$$
$$= 0.68\text{ eV}$$

REAL-WORLD PHYSICS

Automatic Door Openers Many public buildings have doors that open automatically as people approach. These doors are often operated by circuits that contain photovoltaic cells. In operation, a light beam shines across the entrance onto a photovoltaic cell, producing a current in a circuit. The circuit contains a relay, which is an electromagnet that controls a switch in a second circuit that opens the doors. When the electromagnet is magnetized, it attracts the switch, which keeps the second circuit open and the doors closed. When the beam is broken, the current stops and the electromagnet ceases to attract the switch. A spring then pulls the switch, which closes the secondary circuit, and the doors open.

The Compton Effect

student text pp. 739, 740

Reinforcement

MAIN IDEA An electron loses momentum by slowing down. However, a photon cannot slow down but must travel at the speed of light. A decrease in momentum is accompanied by an increase in wavelength. Students may wonder how a massless particle such as the photon can have momentum. The answer to this must await the theory of special relativity.

Use an Analogy

Compton Effect Have students consider the following mechanical analogy of the Compton effect: The interaction of an X-ray photon and an atom resembles a pea bouncing off a bowling ball—the bowling ball is barely affected by the collision because of its much greater mass. However, the interaction between an X-ray photon and an electron is more like a collision between two billiard balls. Their momenta are of the same order of magnitude.

Concept Development

Photon Evidence The photoelectric effect and the Compton effect are the two major phenomena that reveal the particle properties of electromagnetic radiation. The quantization that Planck introduced as a mathematical solution to a perplexing problem was used by Einstein to model light as a particle.

Content Background

Compton's World War II Contributions Compton was one of the leading scientists on the Manhattan Project, which developed the atomic bomb. He directed research on methods of producing plutonium.

DIFFERENTIATED INSTRUCTION

Struggling Students To help students understand that photons scattered from essentially free electrons have less energy than unscattered photons, consider the two I-λ graphs in **Figure 10.** Furthermore, for the lower graph, consider only the right-hand peak, which corresponds to photons scattered from electrons (the left-hand peak corresponds to photons scattered from nuclei—thus the absence of a noticeable shift in wavelength). The wavelength of the scattered photons is greater than that of the unscattered photons, so the energy of the scattered photons is less than that of the unscattered photons because energy and wavelength are inversely proportional. **BL** **Visual-Spatial**

Discussion

Problem In what way might Robert Louis Stevenson's fictional character Dr. Jekyll/Mr. Hyde be used as a metaphor for the behavior of light?

Response Just as Dr. Jekyll and Mr. Hyde are never seen at the same time, the wave and particle aspects of light are never observed at the same time. So, in that sense, Jekyll and Hyde might be thought of as a metaphor for the behavior of light (not to be confused with a serious description or explanation).

3 Assess

Assess the MAIN IDEA

Young's double-slit experiment reveals the wave-like properties of light. Ask the students how they might modify this experiment to detect the particle-like properties of light. Reduce the intensity of the light source until it emits photons one at a time. The detector will show spots for each photon, revealing the particle-like nature of photons. The initial pattern of spots on the detector will look quite random, but after many photons hit the detector, the interference pattern (i.e., the wavelike property of light) becomes clear.

Check for Understanding

Photoelectric Effect Ask students these questions: If a single-frequency light source shining on a photosensitive surface produces a current of emitted electrons, what happens to the current if the intensity of the light source increases? The current also increases. If you decrease the frequency of the light shining on the surface, what happens to the current? At a particular frequency, called the threshold frequency, the photocurrent abruptly falls to zero. **OL**

Reteach

Conservation of Energy Remind students that energy is conserved in the photoelectric effect. Write a statement to the effect: *Initial energy equals final energy.* Point out that the initial energy is the photon energy hf. Now have students identify the final energy. The final energy is the energy needed to eject the electron (the work function hf_0) plus the kinetic energy of the ejected electron (KE).

$$E_{initial} = E_{final}$$

$$hf = hf_0 + KE$$

If the photon energy hf is less than the work function hf_0, no electrons are ejected, yet energy is still conserved. If the photon is absorbed, its energy increases the thermal energy of the cathode. If the photon is reflected, its energy is unchanged. **OL** **Linguistic**

Reading Checks and Caption Checks

Reading Check, *student text p. 729*

The power P is proportional to T^4, so doubling T would lead to a $2^4 = 16$-fold increase in P.

Caption Check, *student text p. 730*

Electromagnetic wave theory is a good model for low-frequency black-body emission.

Caption Check, *student text p. 731*

Ultraviolet radiation has a shorter wavelength than visible radiation, so the frequency of ultraviolet radiation is greater than the frequency of visible radiation. Because frequency is proportional to photon energy, ultraviolet radiation is more energetic than visible radiation.

Reading Check, *student text p. 734*

The kinetic energy is the difference between the photon energy hf and the threshold energy hf_0. If the photon energy is just slightly larger than the threshold energy, this difference (i.e., the kinetic energy) will be smaller than the threshold energy.

Caption Check, *student text p. 739*

Wavelength and energy are inversely proportional; $1/\lambda = E/(hc)$. If the wavelength increases, the corresponding energy must decrease.

Practice Problems

student text p. 734

1. 2.41 eV
2. 611 nm
3. $E_C < E_D < E_B < E_A$
4. 1.8 eV to 3.1 eV

Practice Problems

student text p. 736

5. 3.7×10^{-19} J
6. 9.0×10^{5} m/s
7. 1.1×10^{2} eV
8. 5.7 eV
9. 8.2×10^{-19} J
10. 4.7 V
11. 5.1×10^{-19} J

Practice Problems

student text p. 738

12. 9.7×10^{14} Hz, 4.0 eV
13. 0.968 eV
14. 2.9 eV
15. 4.54 eV; silver

Section 1 Review

student text p. 740

16. Each incident photon interacts with a single electron. If the incident photon does not have sufficient energy, it cannot eject an electron. Energy is directly related to frequency, low-frequency light does not have sufficient energy to eject an electron, whereas high-frequency light does.
17. Both frequency of peak intensity and total energy radiated increase. The peak frequency increases as T, whereas the total energy increases as T^4.
18. The Compton effect is the scattering of a photon by matter, resulting in a photon of lower energy and momentum. The photoelectric effect is the emission of electrons from a metal when radiation of sufficient energy is incident on it.
19. It is a result of the photoelectric effect, which is the capture of a photon by an electron in matter and the transfer of the photon's energy to the electron.
20. 0.89 eV
21. 1.9 eV
22. The scattered X-ray has a longer wavelength.
23. 6×10^{4} eV
24. The answer to both questions is no.

1 Introduce

Bellringer

Properties of Waves Have students label three columns: *particle properties, wave properties,* and *particle-like properties of waves* and list examples under each category.

After discussing the lists, point out that the third column suggests another category based on the first two: *wave-like properties of particles.* Tell students that this section of the chapter deals with these properties. BL Linguistic

Tie to Prior Knowledge

Diffraction—Evidence of Wave Behavior Students have learned about the relationship between diffraction and wavelength and the relationship between wavelength and the momentum of photons. They will use this knowledge to investigate the consequences of the diffraction of particles.

2 Teach

de Broglie Waves

student text pp. 741–742

Critical Thinking

Interference Ask students why it would be impossible to observe single-slit interference patterns of baseballs. The slit width would have to be in the range of 10^{-34}–10^{-32} m. For comparison, the diameter of an electron is approximately 10^{-18} m. AL

Concept Development

MAIN IDEA Point out that both particles and electromagnetic waves have energy and momentum. In particles, the two are related to mass; in electromagnetic waves, they are related to frequency (or wavelength).

ADDITIONAL IN-CLASS EXAMPLE

Use with Example Problem 3.

Problem What is the de Broglie wavelength, in nanometers, of an electron accelerated by a potential difference of 0.90 V?

Response

$\frac{1}{2}mv^2 = -qV, p = mv, \lambda = \frac{h}{p}$

$\lambda = 1.3 \text{ nm}$

Identify Misconceptions

Wave and Particle Ask students if it would be correct to say that an electron is a particle under some circumstances and a wave under others. It is not correct to say that an electron "really is" a particle or "really is" a wave. The more accurate statement is that an electron displays particle-like properties under some circumstances (e.g., a well-defined position or trajectory), and it displays wave-like properties under other circumstances (e.g., interference, including diffraction around objects). Both aspects are needed to accurately describe electron behavior. OL

Location and Momentum

Concept Development

Uncertainties in Measurements Heisenberg's uncertainty principle states that the product of the uncertainties in measuring certain pairs of physical quantities, such as momentum and position, is approximately equal to Planck's constant.

3 Assess

Assess the MAIN IDEA

Ask students to think of an experiment that might reveal the wave-like nature of particles (e.g., electrons). If they need help, you may indicate that they have already discussed such an experiment in this chapter for looking at the wave-like nature of light. Young's double-slit experiment may be used to see the wave-like nature of electrons. Instead of shining a light on the double slits, shine a beam of electrons.

Check for Understanding

Matter Waves Ask students why we do not see examples of the wave-like behavior of everyday objects. We do not detect the wave-like properties of everyday objects because the de Broglie wavelength of common objects is very, very small. OL

Reading Checks and Caption Checks

Caption Check, *student text p. 741*
The wavelength would decrease if the baseball's velocity increases.

Reading Check, *student text p. 743*
A collision between a high-energy photon and a particle would result in a greater change in the particle's momentum.

Caption Check, *student text p. 743*
The scattered photon has a longer wavelength, which means that the scattered photon has less energy than the incident photon.

Practice Problems

student text p. 742

25. 7.8×10^{-11} m, 9.4×10^{6} m/s

26. a. 1.1×10^{-35} m

b. The wavelength is too small to show observable effects.

27. 96.5 V

28. 4.2×10^{-2} eV

Reading Checks and Caption Checks

Section 2 Review

student text p. 743

29. The de Broglie wavelengths of most objects are much too small to be detected.

30. 0.84 nm

31. 0.110 nm

32. The Heisenberg uncertainty principle states that you cannot simultaneously know the precise position and momentum of a particle. Thus, if you know the precise position of a photon or an atom as it passes through the slit, you cannot know its precise momentum. In practice, any experiment you do to measure the position of the photon or the atom as it passes through one slit or the other changes the momentum of the atom or the photon such that the interference pattern is destroyed.

FRONTIERS IN **PHYSICS**

Pushing the Limits

Quantum Touchscreens

Purpose

A strange result of the wave properties of matter is quantum tunneling. Electrons and other small objects can actually pass through what to classical physics are insurmountable barriers. This effect has real applications.

Background

The pressure-sensitive touchscreens described in the article are an interesting mixture of quantum and classical physics. Quantum tunneling is crucial to their operation but so is the classical effect of increasing surface area. The spikes are much like the spikes on some seeds that led directly to the invention of hook-and-loop fasteners. Many other physical effects, such as dust-generated grain elevator explosions, depend on increased surface area.

Teaching Strategies:

- Before reading the feature, review the terms conductor, insulator, and current.
- The feature describes an electrician twisting two wires together to form a connection. Ask students to explain why electricians sometimes rub the exposed ends of the wires with sandpaper before making the connection. *Sanding the wires removes some of the oxidation, making a better connection between the wires.*
- Ask students to identify touchscreen technology they use or have seen used in everyday situations. Possible examples include mobile phones, desktop and tablet computers, credit card machines, and cash registers.

GOING**FURTHER** >>>

Expected Outcomes Touchscreen technologies may use resistive, capacitive, or surface-wave sensing. Resistive sensing uses the pressure of the user's touch to bring two surfaces together, creating an electric current at the spot of the touch. Capacitive sensing transfers a built-up electric charge to the user's finger. The change in electric charge on the screen tells where the touch occurred. Surface wave sensing detects disturbances in acoustic waves sent across the surface of the screen. Note that none of these systems is pressure sensitive but rather is either on (touched) or off (not touched).

CHAPTER 27 ANSWERS

SECTION 1

Mastering Concepts

33. The light becomes redder.

34. Quantized energy means that energy can exist only in multiples of some minimum value (the "quantum").

35. The vibrational energy of the incandescent atoms is quantized.

36. a photon

37. Each photon ejects a photoelectron. Light with greater intensity contains more photons per second; thus, it causes the ejection of more photoelectrons per second.

38. Photons below the threshold frequency do not have sufficient energy to eject an electron. If the intensity of the light increases, the number of photons increases but their energy does not; the photons are still unable to eject an electron.

39. Red photons do not have enough energy to cause the chemical reaction that exposes film.

40. Elastic collisions transfer both momentum and energy. Only if photons have momentum can the equations be satisfied.

Mastering Problems

41. 8.21×10^{14} Hz

42. 3.0 V

43. 1.7×10^{-27} kg•m/s

44. 1.07×10^{-19} J

45. 2.9×10^{-19} J

46. 3.7×10^{-19} J

47. a. 5.0 eV

b. 8.0×10^{-19} J

48. 1.8 eV

49. a. 1×10^{10} J

b. 2×10^{2} m^2

SECTION 2

Mastering Concepts

50. No; using the equation yields a photon momentum of zero because photons are massless. This result is incorrect because massless photons have nonzero momenta.

51. a. Balance the force of gravity against the force of an electric field on the charge.

b. Balance the force of an electric field against that of a magnetic field to find *m/q,* and then use the measured value of *q*.

c. Scatter electrons off a crystal and measure the angles of diffraction.

52. a. Measure the *KE* of the electrons ejected from a metal for at least two different wavelengths, or measure the *KE* of the electrons ejected from a known metal at only one wavelength.

b. Measure the change in wavelength of X-rays scattered by matter.

c. Measure the angle of diffraction when light passes through two slits or a diffraction grating, measure the width of a single-slit diffraction pattern, or measure the angle through which the light is bent when it passes through a prism.

Mastering Problems

53. 0.24 nm

54. 2.4×10^{6} m/s

55. a. 4.2×10^{7} m/s

b. 0.017 nm

56. a. 2.19×10^{6} m/s

b. 0.332 nm

c. 3.26 nm; the circumference is approximately equal to 10 complete wavelengths.

57. $\lambda_a > \lambda_b > \lambda_c > \lambda_d$

58. a. 47 V

b. 0.025 V

Applying Concepts

59. a. the rod glowing bright orange

b. the rod glowing bright orange

60. Not necessarily; the number of ejected electrons is proportional to the number of incident photons or the brightness of the light, not the frequency of the light.

61. a. Blue light has a lower frequency and energy than UV light. Thus, tungsten has the higher threshold frequency.

b. tungsten

62. The diameter of the baseball is about 0.10 m, whereas the de Broglie wavelength is 10^{-34} m; the baseball is about 10^{33} times larger than the wavelength.

Mixed Review

63. 3.8 eV

64. 5.3×10^{-19} J

65. 5.3×10^{-19} J

66. 8.0×10^{-12} m

67. a. 2.6×10^{2} nm

b. 3.6 eV

68. 501 nm

69. a. 1.82×10^{3} m/s

b. 9.42×10^{-6} eV

70. a. 2.5×10^{-7} m

b. 8.0×10^{-19} J

c. 2.4 eV

Thinking Critically

71. a. 5.8×10^{-16} W

b. 1600 photons/s

72. A possible form of the correct answer would be, "and ejects electrons with a kinetic energy of 1.56 eV. What is the metal's work function?"

73. The work function is $W = hf_0 = 3.31\times10^{-19}$ J. The threshold wavelength is $\lambda = 601$ nm. $h/e = 4.18\times10^{-15}$ J/Hz•C, which is close to the accepted value of $h/e = 4.14\times10^{-15}$ J/Hz•C.

74. Answers will vary, but a correct form of the answer is, "A photon's momentum is 1.19×10^{-27} kg•m/s. What is the photon's wavelength?"

Writing in Physics

75. This is an active field of research, but at the time of this text's publication the most massive particles shown to exhibit interference effects are molecules consisting of up to 430 atoms. An international team of physicists published results in 2011 describing an experiment in which a beam of these molecules, which can be up to 6 nm in diameter, create an interference pattern when passed through a virtual diffraction grating made of laser light.

Cumulative Review

76. 3×10^{3} N/m

77. The pitch of a wind instrument depends on the speed of sound in the air within it. The colder the air, the smaller the speed of sound and the flatter the pitch.

78. 5×10^{-7} C

79. 1.0×10^{1} A

80. 2×10^{1} A

Standardized Test Practice

student text p. 749

Multiple Choice

1. A
2. D
3. D
4. C
5. B
6. D
7. B

Free Response

8. $\lambda = h/(mv)$

   $m = h/(\lambda v)$
   $= (6.63\times10^{-34}\ J{\cdot}s)/((2.3\times10^{-34}\ m)\ (45\ m/s))$
   $= 0.064\ kg$

9. $\lambda = h/(mv)$ so $v = h/(m\lambda)$

   $E = mv^2/2 = h^2/(2m\lambda^2)$
   $= (6.63\times10^{-34}\ J{\cdot}s)^2/(2(9.11\times10^{-31}\ kg)(20\times10^{-9}\ m)^2)$
   $= 6.02\times10^{-22}\ J = 3.76\times10^{-3}\ eV$

Rubric

The following rubric is a sample scoring device for free response questions.

Points	Description
4	The student demonstrates a thorough understanding of the physics involved. The response may contain minor flaws that do not detract from the demonstration of a thorough understanding.
3	The student demonstrates an understanding of the physics involved. The response is essentially correct and demonstrates an essential but less than thorough understanding of the physics.
2	The student demonstrates only a partial understanding of the physics involved. Although the student may have used the correct approach to a solution or may have provided a correct solution, the work lacks an essential understanding of the underlying physical concepts.
1	The student demonstrates a very limited understanding of the physics involved. The response is incomplete and exhibits many flaws.
0	The student provides a completely incorrect solution or no response at all.

CHAPTER 28

The Atom

About the Photo

Laser at Work Have students look at the photograph. Have students describe how the laser in the picture differs from laser pointers they may see in school. Possible answers: it looks like it's producing sparks; it spreads out more than a laser pointer; it is a different color. Tell students that lasers like these can be used to cut metal. Have students predict a reasonable purpose for this kind of laser. to do work in a manufacturing plant Have students speculate what it is making. a metal part for a machine

Chapter Overview

The chapter begins with a discussion of two early models of the atom: Thomson's atomic model and Rutherford's nuclear model. The Bohr model of the atom also is presented, related to the spectra of the elements, and used in various calculations. The second section focuses on the modern quantum model of the atom and on lasers.

Before students study the material in this chapter, they should study:

- Electromagnetic waves
- Fundamentals of waves
- Wave-particle duality

To solve the problems in this chapter, students will need a strong grasp of:

- Significant figures
- Solving linear equations
- Scientific notation

Introduce the BIG IDEA

Students will have some knowledge of the idea that an atom has a center surrounded by other parts. Have students describe how what they know of the structure of an atom compares to that of the solar system of which Earth is a part. Possible answer: An atom has a central part like the Sun in the center of the solar system. Other parts of the atom move around the center like planets around the Sun.

Go online!
connected.mcgraw-hill.com

Your online experience includes:

- **Video**
- **Concepts In Motion**
- **Virtual Investigations**
- **All Lab, Applying Practices activities, and PBL worksheets**
- **Online Practice**
- **Personal Tutor**
- **Multilingual eGlossary**
- **Presentation**

Use the LaunchLAB

In *Identifying Samples,* students will explore the use of inference to identify objects they cannot see.

Use Physics T.V.

Video

LASERS How are lasers used? A better question might be how *aren't* lasers used. You might be surprised to discover how many applications of lasers are around you every day.

1 Introduce

Bellringer

Rutherford's Planetary Model of an Atom To help students visualize the Rutherford model of the atom, have them draw the solar system with the Sun and the planets. Have them label the Sun as a positively charged nucleus and the planets as negatively charged electrons. Clarify that although planets take elliptical paths around the Sun mostly in one plane, the Rutherford model represents electrons as not being limited to traveling in one plane. Depict these additional pathways on the board. **OL** **Visual-Spatial** **EL**

Tie to Prior Knowledge

Bohr Atom The concepts of quantized energy, photon emission and absorption, and spectroscopy presented in previous chapters are used to develop the Bohr model of the atom, to explain the characteristics of emitted and absorbed photons, and to make calculations of electron orbital radii, electron energy levels, and emitted photon energy and wavelength.

2 Teach

The Nuclear Model and Revising Rutherford's Nuclear Model

student text pp. 752–757

Concept Development

Explaining Rutherford's Results After students have read the Rutherford experiment, reinforce the concepts by having them explain the results using Newton's second and third laws of motion. Ask students how Newton's third law relates to the Rutherford experiment. According to the third law, the force an alpha particle exerts on a gold nucleus is equal to the force the gold nucleus exerts on the alpha particle. Next, ask students to explain how Newton's second law relates to the Rutherford experiment. According to the second law, the acceleration of the alpha particle is much greater because it has much less mass than a gold nucleus. As a result, the alpha particles can deflect through a large angle, while the gold foil remains for the most part unchanged by the collisions. **BL** **Linguistic**

Use an Analogy

Alpha-Particle Motion The motion of the alpha particle in the gold-foil experiment is analogous to the motion of a marble on a surface that encounters a steeply rising hill. A marble that is rolling past the hill at a great distance undergoes little or no deflection. A marble that rolls past the hill at a closer distance undergoes a deflection that is more pronounced. A marble heading directly toward the center of the hill will roll back on its path (if it lacks sufficient kinetic energy to overcome the rise). The rise of the hill (h) can be modeled by the equation $h \propto 1/r$, where r is the distance from the center of the hill. As r goes to zero, the height of this hill becomes infinite. A marble cannot roll over the infinitely high hill.

Content Background

The Scattering Experiment The origin of the alpha-particle scattering experiments was a suggestion by Rutherford to an associate who had asked for a research topic for his assistant. Rutherford suggested to his colleague, Hans Geiger, that he should see if any alpha particles are scattered through a large angle. Geiger then relayed the suggestion to his assistant, Ernest Marsden. After Geiger and Marsden had conducted several experiments, they found that they had been able to detect some of the alpha particles coming back.

Concept Development

Diffraction Have students recall that white light passing through the thin slits of a diffraction grating interferes to form a spectrum. Monochromatic light passed through a diffraction grating produces a pattern.

Quick DEMO

Continuous Spectrum

Estimated Time 10 minutes

Materials slide projector, slide-sized piece of cardboard, craft knife, screen or white wall, diffraction grating

Procedure Prior to the demonstration, cut a 2-mm slit in the cardboard. Place the cardboard into the slide projector. Darken the room. Turn on the projector and project the light onto a screen or wall. Adjust the cardboard slide if necessary so that the slit makes a white vertical line on the screen or wall. Have students observe the screen as you place a diffraction grating over the front of the projector. Adjust the grating to get the spectrum on the wall (and not the ceiling or floor).

Explain to students that the white light from the projector comes through the slit in the cardboard. When the diffraction grating is placed in front of the projector, the light from the projector is broken into its component colors, producing a continuous spectrum on both sides of the white line on the screen or wall. Point out that the spectrum from an incandescent object, such as the filament of the projector bulb, contains light with a range of wavelengths that lie in the visible spectrum and in the infrared region. Ask students what they could infer if a diffraction grating were placed in front of a light source and two bands of distinctly different colors were observed. The light produced by the light source has only two visible wavelengths.

DIFFERENTIATED INSTRUCTION

Visually Impaired Model continuous and discrete spectra with a sound spectrum from a piano and a trombone or violin. On a piano play the 13-note interval in the octave C–C, one note at a time (this includes black keys as well as white keys). Explain that just as light waves of different wavelengths make up a spectrum, sound waves of different wavelengths make up the notes in an octave. When you play individual notes on the piano, a discrete spectrum of sound spans from the high-frequency sound to the low-frequency sound. On instruments such as the trombone or violin, you can model a continuous spectrum by changing notes in arbitrarily small increments. To model a discrete spectrum, depress a variety of piano keys in the range. OL Auditory-Musical EL

Reinforcement

Emission and Absorption Spectra Stress the concept that each element emits a unique spectrum; that is, it emits light of certain frequencies. Emission spectra—the bright lines—are generated when an element is heated until it emits photons. Absorption spectra—the spectra with the missing frequencies and black lines—are generated when light passes through relatively cool gases, and the gas absorbs photons at particular frequencies. The black lines represent absorbed frequencies.

Discussion

Problem Why does a fluorescent lightbulb emit both a continuous spectrum and a series of emission lines?

Response Students may suggest that the bulb contains a glowing solid and a glowing gas. The emission lines are produced by a heated gas (mercury), but the continuous spectrum is produced by fluorescence. In this process, a phosphor powder on the inside of the bulb absorbs ultraviolet photons emitted by the mercury and then emits a continuous spectrum of visible light.

Content Background

Classifier of Stellar Spectra Willamina Paton Stevens Fleming (1857–1911) was an American physicist who was born in Scotland. After moving to Boston, she started working for the director of Harvard University Observatory as a maid. He was impressed with her intellect and offered her part-time work doing clerical and computational tasks. In 1886, five years after permanently joining the observatory staff, she became administrator of a new program to classify stars by their spectra. She analyzed many of the photographs and supervised a staff of female research assistants. In 1908 the Royal Astronomical Society elected her a member, one of the few women selected.

Identify Misconceptions

Accelerating Electron Some students might have heard or read the following statement when studying the atom: "An electron in an orbit is constantly accelerated toward the nucleus." Some students may infer from this statement that the electron is moving radially inward. Point out that any object moving in a constant circular orbit has a constant radial acceleration, that is, a centripetal acceleration caused by a net force. This net force does not cause the object to move inward; it causes the direction of the object's velocity to change in such a manner that the object moves in a circle at constant speed. **OL**

Use the PhysicsLAB

In *Emission Spectra,* students observe the emission spectra of various elements in the gaseous state.

Concept Development

Spectroscopy Explain to students that spectroscopy can be used to detect the motion of planets, stars, and galaxies. As most stars contain hydrogen, a frequent component of stellar spectra should be the four lines shown in **Figure 6.** However, early observers of distant galaxies did not observe these lines but observed two lines at 615 nm and 651 nm. It was soon determined that these two lines represent the two shortest wavelengths of light in the visible spectrum of hydrogen (410 nm and 434 nm) that had shifted toward the red region of the spectrum. Astronomers suggested that the change was a Doppler shift due to the motion of the galaxy away from Earth. Have students recall the Doppler effect for sound. Write the following equation on the board and explain that it is the special relativity equation for the Doppler shift, where λ_0 is the emitted wavelength as seen in a reference frame at rest with respect to the source and λ is the wavelength measured in a frame moving with a velocity v away from the source along the line of sight.

$$\lambda = \lambda_0 \frac{\sqrt{1 + v/c}}{\sqrt{1 - v/c}}$$

Show that the redshift accounts for the observed 615-nm and 651-nm lines in the stellar spectrum of hydrogen by showing that

$$\frac{\Delta\lambda}{\lambda_{410}} = \frac{\Delta\lambda}{\lambda_{434}}$$

$$\frac{615\text{ nm} - 410\text{ nm}}{410\text{ nm}} = \frac{651\text{ nm} - 434\text{ nm}}{434\text{ nm}} = 0.5$$

AL **Visual-Spatial** **Logical-Mathematical**

PHYSICS CHALLENGE ACTIVITY

Stellar Spectrum of Hydrogen Using the relationship that $\Delta\lambda/\lambda$ = constant for the Doppler shift of light, have students calculate the shifted wavelengths of the two remaining wavelengths in the visible spectrum of hydrogen shown in **Figure 6.** Ask students whether these wavelengths are in the ultraviolet, visible, or infrared region of the stellar spectrum. 486-nm line: 729 nm; 656-nm line: 984 nm. They are in the infrared region.

AL **Logical-Mathematical**

Quantized Energy

student text, pp. 757–758

Use Figure 8

MAIN IDEA Bohr considered the electron energy levels in an atom to be quantized. To reinforce this concept, draw students' attention to the stair steps in **Figure 8.** Have students recall when they might have walked off a stalled ascending escalator. Have them think how the vertical spacing between the treads—which can represent the spacing of the gravitational potential energy levels in a Bohr atom—became less and less as they walked off the top of the escalator. Ask students what situation in an atom is modeled as they walk onto and begin to descend the stalled escalator. energy-emitting transitions for excited electrons

OL **Visual-Spatial**

Discussion

Problem What happens when an element's electrons drop from a higher quantized energy level to a lower quantized level?

Response As the electrons in an atom fall back to a lower quantized level, photons are released. As a result of these transitions, the element's atomic emission spectrum is observed. **BL** **Visual-Spatial**

REAL-WORLD PHYSICS

Chemical Composition of Stars and Planets Spectroscopy has frequently been used to determine the composition of planets and stars. In spectroscopy, light collected by a telescope is directed to a spectroscope instead of the observer's eye or film. The element helium was identified as making up part of the solar spectrum before helium was discovered on Earth. Analysis of the solar spectrum reflected by Jupiter indicated that Jupiter's atmosphere contained methane. This prediction was confirmed years later when space probes approached the giant planet.

Predictions of the Bohr Model

student text pp. 759–764

Quick DEMO

Electron Motion

Estimated Time 5 minutes

Materials electric table fan

Procedure Turn on the fan at high speed and have the students observe the rotating blades. **CAUTION:** ***Do not allow anyone to touch the moving blades.*** The fan blades are blurred. Point out that by simple observation it is impossible to determine the configuration of each blade while the fan is running. Point out that the blurred image is the result of the eye's inability to detect rapid motion. The inability to locate an electron exactly in an atom is a consequence of the inability to observe the electron without disturbing it. Recall the Heisenberg uncertainty principle—the more precisely the position of a particle is determined, the less precisely its momentum is known. To study the atom, what is most important is to know how much energy an electron has, not so much where the electron actually is.

Use the MiniLAB

In *Bright Line Spectra,* student will examine how the emission spectra of elements differ.

Reinforcement

Photon Energy Remind students that light waves near the red end of the visible spectrum have longer wavelengths than light waves near the violet end. Ask students which light color will have photons of greater energy and why.

$E_{photon} = hf$ Because the frequency of light waves of violet light is greater than the frequency of light waves of red light, the energy of a photon of violet light is greater than the energy of a photon of red light.

OL **Visual-Spatial**

Reinforcement

Bohr's Planetary Model Point out that Bohr's planetary model of a hydrogen atom has an electron orbiting a proton, much as a planet orbits a star, or closer to home, as a satellite orbits Earth. Have students recall that a satellite in circular orbit around Earth has a centripetal acceleration. The net force causing the acceleration is the gravitational force (F_g) acting on it, described by Newton's law of universal gravitation. $F_g = F_{net}$; $\frac{Gm_S m_E}{r^2} = \frac{m_S v^2}{r}$ Explain that in Bohr's model, F_{net} is the force between the electron and the proton as described by Coulomb's law: $F = K((q_A q_B)/r^2)$.
BL **Logical-Mathematical**

PHYSICS CHALLENGE ACTIVITY

Total Energy of the Atom You can show students that the quantized energy equation for an atom follows from the statement that
$E_{total} = \frac{-Kq^2}{2r}$ and $r = \frac{h^2n^2}{4\pi^2Kmq^2}$.

$$E = \frac{-Kq^2}{2r} = \frac{-Kq^2}{2}r^{-1} = \left(\frac{-Kq^2}{2}\right)\left(\frac{h^2n^2}{4\pi^2Kmq^2}\right)^{-1}$$

$$E = \left(\frac{-Kq^2}{2}\right)\left(\frac{4\pi^2Kmq^2}{h^2n^2}\right)$$

$$E = \frac{-2\pi^2Kmq^4}{h^2} \times \frac{1}{n^2}$$

Have students show that E_n has the unit joules.

$$\frac{K^2mq^4}{h^2}; \frac{(N\cdot m^2/C^2)(kg)(C)^4}{(J\cdot s)^2} = \frac{(N^2)(m^4)(kg)(C^4)}{(C^4)(J^2)(s^2)}$$

$$= \frac{(kg)(m^2)}{(s^2)} = (kg)(m/s^2)(m) = N\cdot m = J$$

OL **Logical-Mathematical**

Use Figure 11

Series Transitions Ask students what common characteristic all the transitions in a particular series (Balmer, Lyman, or Paschen) possess. the final energy level Ask which series contains the highest energy photon, then the lowest. highest: Lyman; lowest: Paschen **OL**

Critical Thinking

Energy Level Transitions If each of the series shown in **Figure 11** has a transition from an excited electron in the $n = 6$ energy level, have students explain why each series doesn't contain the same wavelength of light. The energy of the photon emitted does not depend on the initial energy level of the electron, but on the transition between the electron's initial and final energy levels. Because the final energy level in each series is different, ΔE of the electron emitted from the $n = 6$ energy level is different. Because the wavelength of the emitted photon is related to ΔE, the wavelength of the light from each series will be different. **AL** **EL**

DIFFERENTIATED INSTRUCTION

Struggling Students Reinforce understanding of atomic structure, by carefully showing students the step-by-step substitutions used to derive the equation for the orbital radii of the electron in hydrogen.

$$\frac{Kq^2}{r^2} = \frac{mv^2}{r}; \frac{Kq^2}{r} = mv^2$$

substitute $v = \frac{nh}{2\pi mr}$

$$Kq^2 = rm\left(\frac{nh}{2\pi mr}\right)^2 = \frac{rmn^2h^2}{4\pi^2m^2r^2} = \frac{n^2h^2}{4\pi^2mr}$$

Solving for r in the equation:

$$Kq^2 = \frac{n^2h^2}{4\pi^2mr}$$

yields

$$r_n = \frac{n^2h^2}{4\pi^2Kmq^2}$$

BL **Logical-Mathematical**

Use the PhysicsLAB

In *Electron Transitions,* students will examine how the emission spectra of elements differ.

ADDITIONAL IN-CLASS EXAMPLE

Use with Example Problem 1.

Problem Calculate the energy difference between the E_3 and E_1 in the hydrogen atom.

Response $E_1 = -13.6 \text{ eV} \times \frac{1}{1^2}$

$E_3 = -13.6 \text{ eV} \times \frac{1}{3^2}$

$\Delta E = E_3 - E_1 = -1.51 \text{ eV} - (-13.6 \text{ eV})$
$= 12.1 \text{ eV}$

Content Background

Elliptical Orbits Bohr's quantized atomic theory does not explain the groups of narrow, closely spaced lines that make up the structure of hydrogen because he assumed circular orbits. A.J.W. Sommerfield provided a mathematical description of elliptical orbits using quantum theory and then proceeded to demonstrate that only certain types of ellipses were possible if the relativistic change in mass of the orbiting electron is considered. This combination of circular and elliptical electron orbits successfully explains some of the structure of the hydrogen spectrum.

ADDITIONAL IN-CLASS EXAMPLE

Use with Example Problem 2.

Problem What wavelength photon is emitted when an electron in a hydrogen atom drops from the third energy level ($n = 3$) to the first energy level ($n = 1$)?

Response

$\Delta E = E_1 - E_3 = -13.6 \text{ eV} - (-1.51 \text{ eV})$
$= -12.1 \text{ eV}$

$\lambda = hc/\Delta E = \frac{1240 \text{ eV·nm}}{12.1 \text{ eV}} = 102 \text{ nm}$

3 Assess

Assess the MAINIDEA

Bohr's Atom Have students explain how the Bohr model of the atom takes into account both the emission and the absorption spectra of an element. Each line represents and corresponds to a jump from one energy level to another.

Check for Understanding

Bohr Model Have students explain how the Bohr model of the hydrogen atom accounts for the absorption of photons of only certain wavelengths. Because an electron in a hydrogen atom is in a quantized energy level, a transition from a lower to a higher energy level takes place only by the absorption of a photon that has an energy equal to or greater than the two energy levels. Because the wavelength of light is related to this energy difference by the equation $\lambda = hc/\Delta E$, only photons of light with certain minimum wavelengths can be absorbed. OL Linguistic

Extension

Transition Between the Ground and Excited States To aid people exiting a building during power blackouts, staircases and stairwells are sometimes outlined with phosphorescent paints that continue to glow after illumination by natural or artificial visible light ceases. Have students describe an atomic model that could account for phosphorescence. Electrons in atoms of the material absorb photons of light with wavelengths in the visible spectrum and move from the ground state to an excited state. The electrons maintain their excited states for a few seconds or hours and emit photons of light with wavelengths in the visible spectrum as they return to the ground state. AL

Reading Checks and Caption Checks

Reading Check, *student text p. 753*

He was expecting the α-particles to be slightly deflected according to Thomson's model of the atom, in which the positive charge was evenly distributed throughout the atom. Instead, the particles were mostly undeflected, but a few were deflected at large angles.

Reading Check, *student text p. 756*

They compare the absorption spectra of the stars with known emission spectra of various elements.

Caption Check, *student text p. 756*

As the photon energy increases, the difference between adjacent wavelengths decreases.

Caption Check, *student text p. 758*

In the diagram, the transition from E_3 to E_1 produces the photon with the greatest energy.

Reading Check, *student text p. 760*

The electron orbital radius increases as the square of the principal quantum number. The atom's energy depends on the inverse square of the principle quantum number.

Reading Check, *student text p. 761*

From E_6 to E_2 results in a violet line. From E_5 to E_2 results in a blue line. From E_4 to E_2 results in a green line. From E_3 to E_2 results in a red line.

Practice Problems

student text p. 762

1. $E_2 = -3.40$ eV
   $E_3 = -1.51$ eV
   $E_4 = -0.850$ eV
2. 1.89 eV
3. 2.55 eV
4. $r_2 = 2.1\times10^{-10}$ m or 0.21 nm
   $r_3 = 4.8\times10^{-10}$ m or 0.48 nm
   $r_4 = 8.5\times10^{-10}$ m or 0.85 nm

student text p. 763

5. $\lambda_{3\text{ to }2} = 656$ nm (red line)
   $\lambda_{4\text{ to }2} = 486$ nm (green line)
6. **a.** 2.15 eV
   b. 577 nm
7. -50.3 eV

Physics Challenge

student text p. 764

1. $\Delta E = E_f - E_i = (-13.6\text{ eV})\left(\frac{1}{3^2} - \frac{1}{5^2}\right)$
   $= -0.967$ eV
2. $\lambda = \frac{hc}{-\Delta E} = \frac{1240\text{ eV}\cdot\text{nm}}{0.967\text{ eV}} = 1280$ nm
3. percent discrepancy
   $= \left(\frac{\text{Accepted value} - \text{Predicted value}}{\text{Accepted value}}\right) \times 100$
   $= \left(\frac{632.8\text{ nm} - 1280\text{ nm}}{632.8\text{ nm}}\right) \times 100 = 102\%.$

   The calculated wavelength is roughly twice as large as the actual wavelength.

Section 1 Review

student text p. 764

8. The initial sum of the energy of the electron in the atom plus the energy of the incident photon equals the final energy of the electron in the atom.
9. In Rutherford's nuclear model, all of an atom's positive charge and most of its mass are in a tiny, centrally located nucleus around which negatively charged electrons orbit.
10. To obtain an absorption spectrum, white light is passed through a sample of gas and then a spectroscope. Because the gas absorbs specific wavelengths, the normally continuous spectrum of white light contains dark lines.
11. Incandescent solids produce spectra consisting of a continuous band of colors, whereas gases produce a set of discrete lines. All spectra are created by energy level transitions in atoms.
12. The radius depends on n^2, so the second level would have a radius four times the first, or 0.106 nm.
13. 4.63×10^7 nm $= 4.63$ cm; the wavelength indicates the radiation is a microwave.
14. The scale is 5 cm $= 1.5\times10^{-15}$ m, or 1 cm $= 3.0\times10^{-16}$ m. The Bohr radius is 5.3×10^{-11} m. In our model this would be
    $\frac{5.3\times10^{-11}}{3.0\times10^{-16}} \times 1\text{ cm} =$
    1.8×10^5 cm, or 1.8 km. This is beyond the classroom and probably off school property.

SECTION 2 The Quantum Model of the Atom

student text pp. 765–769

1 Introduce

Bellringer

Electron Cloud Model of an Atom Ask students to sketch or write on a piece of paper their concepts of the electron cloud model of an atom that they had been exposed to in chemistry or an earlier physical science class. Allow the students to discuss their concepts. Use the discussion as a springboard for describing the properties of an electron in an atom in terms of a probability wave. **OL** **Visual-Spatial** **EL**

Tie to Prior Knowledge

Developing the Quantum Model of the Atom The de Broglie relationship and the Heisenberg uncertainty principle discussed in the previous chapters are used to introduce students to the quantum mechanical model of the atom.

2 Teach

From Orbits to an Electron Cloud

student text pp. 765–766

Critical Thinking

Unstable Orbits Ask students what wave phenomenon accounts for the unstable condition shown in **Figure 12.** destructive interference **OL**

Use Figure 13

MAIN IDEA Remind students that the maximum density of the probability cloud for an energy level ($n = 1, 2, 3 \ldots$) occurs at a distance from the nucleus corresponding to the electron orbital radius for that energy level. Ask students the following questions about **Figure 13:** What is the most probable distance from the nucleus for an electron in the $n = 1$ energy level? The $n = 2$ energy level? $r_1 = 0.053$ nm; $r_2 = n^2 r_1 = (4)(0.053 \text{ nm}) = 0.21$ nm **OL**

Lasers

student text pp. 766–768

Content Background

Masers Amplification of stimulated emission was first achieved with microwaves, an effect called MASER. MASER is an acronym for **M**icrowave **A**mplification of **S**timulated **E**mission of **R**adiation. This effect was first demonstrated in 1953 by the American physicist Charles Hard Townes. The same principle was later achieved with light (LASER) in 1959 by Theodore Harold Maimen. Townes, along with Russian physicists Nicolay B. Basov and Alexsandr M. Prokhorov, was awarded the 1964 Nobel Prize in Physics for "fundamental work in quantum electronics, which led to the construction of oscillators and amplifiers based on the maser-laser principle."

Identify Misconceptions

Constructive Interference Have students sketch a wave model of the light coming from a HeNe laser. Most will draw a series of parallel, coherent waves, similar to the coherent light in **Figure 14.** Point out that such a situation would not exist because of a property of waves—interference. Explain that the waves would not exist independently but would undergo constructive interference. When the crest or trough of one wave passes through the crest or trough of another, a single light wave that has an amplitude equal to the sum of the amplitude of each wave is produced.

Reinforcement

Coherent Photons Emphasize that the photon that stimulates an excited atom has the same energy and wavelength as the emitted photon because an excited electron in the same energy level in an identical atom emitted it. **OL**

Concept Development

Reflection and Transmission Explain that both the reflection of light inside the laser device and the transmission of the laser beam outside the device can be explained by both wave and particle pictures. However, it might be easier for students to visualize the reflection of light back and forth in the lasing gas as an illustration of light's particle-like characteristics and the transmission of light through the partially reflective mirror as an illustration of light's wavelike characteristics. **OL** **Visual-Spatial**

PHYSICS PROJECT ACTIVITY

Quantum and Classical Mechanics Interested students can research quantum mechanics, a scientific study developed to explain submicroscopic behavior because the mechanics based on Newton's laws, classical mechanics, failed to do so. Ask students to discuss the validity of the traditional laws of classical mechanics as approximations of the laws of quantum mechanics and to investigate the wave-particle duality and energy considerations that are so significant at the submicroscopic level and insignificant at the macroscopic level. AL Linguistic

Careers

Laser Technician A laser technician may produce, test, operate, and/or repair lasers. A technician might be employed by a hospital, a fiber-optics company, a research lab, a manufacturing plant, the military, the space program, or at construction sites. Some laser technicians are responsible for identification of flaws in machine parts, diagnosis of medical problems, or creation of holograms for applications such as on debit cards. A two- to four-year degree in laser technology is required to become certified as a laser technician.

Use the MiniLAB

In *Diffraction of Laser Light,* students investigate the wavelengths of laser light.

Laser Applications

student text pp. 768–769

Discussion

Problem What are examples of laser technology applications that depend on laser light being (a) monochromatic, (b) coherent, (c) highly directional, and (d) intense?

Response Answers may include (a) spectroscopy, (b) holography, fiber optics, and communications, (c) surveying, and (d) laser surgery and drilling.
OL Logical-Mathematical

PHYSICS CHALLENGE ACTIVITY

Laser Technology Have students research two categories of laser technology. The first is the use of lasers to enhance an older technology, such as the use of laser light in surveying, drilling, and surgery. The second area is the use of lasers to generate new technology, such as holography and optical communications. Have each student select a topic in each category and present the research in the form of a poster, TV news script, or written report.
OL Linguistic

3 Assess

Assess the MAINIDEA

Quantum Model Have students describe what the quantum model of the atom predicts about an electron. An electron can most likely be found in a specific region of the electron cloud around an atom.

Check for Understanding

Density of Electron Cloud Ask students what the density of the distribution of dots in an electron cloud model represents. The density represents the probability of finding an electron at that location.
OL Visual-Spatial

Reteach

Electron Clouds and Probability Have students recall seeing the water spray from a rotating lawn sprinkler in which the sprinkler arms are shaped like the letter z and the z is parallel to the ground. Sketch the shape of the spray as a horizontal figure eight on the board. Shade the figure—increasing then decreasing, then increasing and decreasing again—symmetrically from left to right. Point out that the density of shading represents the density of the droplets in the spray as well as the students' chances of getting drenched if they passed through. Similarly, the density of the distribution of dots in the electron cloud represents the chances of detecting an electron. OL Visual-Spatial

Reading Checks and Caption Checks

Reading Check, *student text p. 767*
Although each electron exists at a specific place, its position and momentum cannot be simultaneously measured. Instead, we describe the probability of their existence at various places. The electron cloud is the region where there is a high probability of finding electrons.

Reading Check, *student text p. 766*
Coherent light waves are in phase. Incoherent light waves are not in phase.

Caption Check, *student text p. 767*
The wavelength of the incident and emitted photos are the same.

Reading Check, *student text p. 767*
They have the same frequency (wavelength), and they are coherent.

Reading Check, *student text p. 768*
The laser beams are narrow and highly directional. They may also be monochromatic.

Section 2 Review

student text p. 769

15. The Bohr model could not predict the behavior of any other atom besides hydrogen. The model also could not explain why the laws of electromagnetism do not apply within the atom.

16. The uncertainty principle does not allow a particle to have both a precisely known position and a precisely known momentum. The Bohr orbits would require both these quantities.

17. Yes; green photons have enough energy to excite atoms into energy levels from which the atom can emit red light. Red photons do not have enough energy to put the atoms in energy levels high enough to enable them to emit green photons.

18. When a photon strikes an atom in the excited state, it stimulates the excited atom to emit a photon of the same energy and in step with the incident photon. The incident photon remains unchanged and these two photons in turn strike other excited atoms, producing more and more in-step, coherent light.

19. concentrated, high power; directional; single wavelength; coherent light

20. The smaller the electron cloud, the more precisely we know the position of the electrons. If a particle's position is well known, its momentum must be uncertain. The uncertainty of the momentum can be large only if momentum itself is large. Therefore, the kinetic energy of the electron also must be large, and it takes lots of energy to do this.

FRONTIERS IN **PHYSICS**

Future Looks Bright

Glowing Nanoparticles in the Brain

Purpose

This feature discusses an application of two very different physical properties of atoms to brain surgery.

Background

Fluorescent materials absorb light of higher energy (often invisible ultraviolet light) and emit light of lower energy (often visible light). Fluorescent materials are found in a variety of everyday materials, including fluorescent lightbulbs, cleaning and hygiene products, and paper. Ferromagnetic materials respond to a magnetic field by temporarily becoming magnet themselves. This happens because electrons align their spins with the external magnetic field. Iron, nickel, and cobalt are common ferromagnetic materials.

Teaching Strategies

One way to get a sense of the relative size of the nanometer-scale world is to compare with other measurements, such as time. For instance, a nanometer is to a meter as a second is to about 31.7 years. Encourage students to find other novel measurement comparisons.

GOING**FURTHER** >>>

Expected Outcomes Possible answers include barium and gadolinium. Barium, a metal that reflects X-rays within the body, is toxic in most forms but is safe as barium sulfate, an extremely insoluble solid. Toxic gadolinium is made safe in a process called chelation, in which the toxic metal is surrounded by organic compounds.

CHAPTER 28 ANSWERS

SECTION 1

Mastering Concepts

21. He directed a beam of α particles at a thin metal sheet and measured the number of particles deflected at various angles. The small but significant number deflected at wide angles indicates a concentrated nucleus.

22. Bohr said the energy of an emitted photon or an absorbed photon is equal to the change in energy of the atom, which can have only specific values.

23. As the electrons undergo centripetal acceleration, they would lose energy and spiral into the nucleus. In addition, all atoms should radiate at all wavelengths, not discrete wavelengths.

24. Bohr's model correctly predicted values for hydrogen's spectrum, but was unable to predict other elements' spectra. Bohr's assumptions include quantized energy levels, atom emits or absorbs radiation only when it changes states, and angular momentum is quantized.

25. Energy is supplied to the gas, which causes the electrons to excite and move to higher energy levels. The electrons then give off the difference in energy between energy levels as they drop back down to a less excited state. The energy differences between levels correspond to spectral lines.

26. Photon wavelengths are determined by the difference in energies of allowed levels as electrons jump inward to stationary states.

27. Each element has a different configuration of electrons and energy levels.

Mastering Problems

28. 709 nm

29. 3.91 eV; this energy corresponds to level E_6.

30. 1.91 eV

31. 0.49 eV

32. a. 2.72 eV

b. 3.06 eV

33. 1.24 eV; 2.99×10^{14} Hz

34. $E_2 = -3.40$ eV; $E_3 = -1.51$ eV; $E_4 = -0.850$ eV; $E_5 = -0.544$ eV; $E_6 = -0.378$ eV

35. B < A < D < C < E

36. A correct form of the answer is, "What is the energy of a photon emitted when a hydrogen atom transitions from the $n = 3$ state to the ground state?"

37. a. 4.77×10^{-10} m

b. 1.01×10^{-9} N

c. 1.11×10^{21} m/s^2

d. 7.28×10^{5} m/s, or 0.24% of c

SECTION 2

Mastering Concepts

38. Light is concentrated into a narrow beam rather than being spread over a wide area.

39. Lasers are directional and single, pure colors.

Mastering Problems

40. 1.5 eV

41. a. 1.97

b. 2.1×10^{16} photons/s

42. a. 1.96 eV, 2.28 eV, and 1.08 eV

b. red, green, and infrared, respectively

Applying Concepts

43. Line spectrum; the light comes from gas mixture made of specific elements.

44. No; as white light passed through Earth's atmosphere, certain energies would be absorbed by the gases composing the atmosphere. Its spectrum, therefore, would have black lines on it.

45. Yes; money comes in only certain discrete values. No; water seems to come in any possible quantity.

46. Six lines are possible. $E_4 \rightarrow E_1$ produces the highest energy photon.

47. The maximum energy is 13.6 eV. This is also the ionization energy for hydrogen. The electron would have enough energy to leave the nucleus.

48. The Bohr model has fixed orbital radii. The quantum model gives a probability of finding an electron at a location. The Bohr model allows for calculation of only hydrogen atoms. The quantum model can be used for all elements.

Mixed Review

49. 0.4 eV kinetic energy

50. a. 3.74 eV Yes, the atom is ionized.

b. 5.4×10^{-20} J

51. 12.1 eV

52. 243 nm

Thinking Critically

53. 2.84 eV (E_6 to E_3); 2.27 eV (E_6 to E_4), 2.14 eV (E_6 to E_5)

54. No. The three highest energy lines leave the atom in states at least 4.64 eV above the ground state. A photon with this energy has a wavelength of 267 nm in the ultraviolet. The change from E_4 to E_2 involves an energy change of only 0.79 eV, resulting in light with a wavelength of 1570 nm in the infrared.

55. The radii would be twice as large because *m* appears in the denominator of the equation. The energies would be half as large because *m* appears in the numerator. Therefore, the wavelengths would be twice as large. Thus, the light emitted from E_2 to E_1 would be twice that for hydrogen: (2)(122 nm) = 244 nm.

56. A possible form of the correct answer would be, ". . . when it absorbs a photon and goes to the third excited state. What was the energy of the photon?"

Writing in Physics

57. Students should describe the "raisin pudding" model, a classical planetary model, the Bohr model, and the quantum model. The first explains how atoms can have electrons and mass but fails to describe the results of Rutherford's experiments. The planetary model explains electrons and Rutherford's results but is unstable and would collapse in about 1 ns. Bohr's model explains known spectra and fits Rutherford's nuclear model but has unexplained assumptions and fails the uncertainty principle. It is also unable to describe atoms with more than one electron. The quantum model can explain all known facts but requires computers to solve the equations.

58. It uses a pulsed Nd laser at 1064 nm. The IR is put into a "frequency doubling" crystal. Light with half that wavelength, or 532 nm, results

Cumulative Review

59. 0.20 mV

60. 9.0×10^{4} N/C

61. Yes; put four 1 Ω resistors in series.
$R_T = R_1 + R_2 + R_3 + R_4$

62. 5.0×10^{1} μT

63. **a.** No; $v = E/B$, so the ratio is the same for a given *v*.

b. For a larger mass, *B* must be larger to keep *r* constant.

64. 1.2×10^{-18} J

Standardized Test Practice

student text p. 775

Multiple Choice

1. B

2. C

3. D

4. B

5. D

6. C

Free Response

7. $\Delta E = -13.6 \text{ eV}\left(\frac{1}{5^2} - \frac{1}{2^2}\right) = 2.86 \text{ eV}$

$$\Delta E = \frac{1240 \text{ eV}\cdot\text{nm}}{\lambda}$$

$$\lambda = \frac{1240 \text{ eV}\cdot\text{nm}}{\Delta E} = \frac{1240 \text{ eV}\cdot\text{nm}}{2.86 \text{ eV}} = 434 \text{ nm}$$

Rubric

The following rubric is a sample scoring device for free response questions.

Points	Description
4	The student demonstrates a thorough understanding of the physics involved. The response may contain minor flaws that do not detract from the demonstration of a thorough understanding.
3	The student demonstrates an understanding of the physics involved. The response is essentially correct and demonstrates an essential but less-than-thorough understanding of the physics.
2	The student demonstrates only a partial understanding of the physics involved. Although the student may have used the correct approach to a solution or may have provided a correct solution, the work lacks an essential understanding of the underlying physical concepts.
1	The student demonstrates a very limited understanding of the physics involved. The response is incomplete and exhibits many flaws.
0	The student provides a completely incorrect solution or no response at all.

CHAPTER 29

Solid-State Electronics

About the Photo

Electronics Ask students to identify the technologies in the cafe. cell phones, lap tops Bring in an old computer and cell phone for students to take apart. Assist students in identifying the major electronic components. Inform students that the development of semi-conductor components from the diode and transistor to the integrated circuit has significantly impacted the size and speed of electronics.

Chapter Overview

The band theory of solids is presented and used to explain the behavior of insulators, conductors, and semiconductors. The difference between extrinsic and intrinsic semiconductors is explained. Applications of semiconductor technology, such as thermistors and light meters, are presented. The second section of this chapter details how diodes and transistors work and how integrated circuits are made.

Before students study the material in this chapter, they should study:

- Circuit diagrams
- Conductors and insulators
- Electric charge
- Electric current
- Mass and weight
- Ohm's law
- Potential difference
- Structure of the atom

To solve the problems in this chapter, students will need a strong grasp of:

- Scientific notation
- Solving linear equations

Introduce the BIG IDEA

Have students search the Internet for images of early and modern computers and cell phones. What is the trend among all of these items? getting smaller How? Semiconductors now may precisely control conductivity, which requires fewer electrons and a much lower power input.

Go online!
connected.mcgraw-hill.com

Your online experience includes:

- **Video**
- **Concepts In Motion**
- **Virtual Investigations**
- **All Lab, Applying Practices activities, and PBL worksheets**
- **Online Practice**
- **Personal Tutor**
- **Multilingual eGlossary**
- **Presentation**

Use the LaunchLAB

In *Conduction in a Diode,* students explore how light-emitting diodes (LEDs) work.

Use Physics T.V.

Video

ROBOT PHYSICS Integrated circuits and microchips are everywhere! Look around you and discover how solid-state electronics allow us to do what used to be impossible.

1 Introduce

Bellringer

Resistance and Temperature You can use a thermistor and an ohmmeter to demonstrate how resistance changes with temperature in some materials. Measure the resistance of the thermistor at room temperature. Then immerse the thermistor in hot water for 30 s, and measure the resistance of the thermistor again. Most thermistors have a negative temperature coefficient, which means their resistance drops with increasing temperature. Semiconductors exhibit this characteristic. Conductors exhibit a positive temperature coefficient. For example, a miniature lamp might measure 15 Ω at room temperature but conduct 1.00 A at 120 V. Ohm's law tells us that the lamp's resistance increases to 120 Ω at operating temperature. **BL** **Visual-Spatial** **EL**

Tie to Prior Knowledge

Voltage-Current Graphs Students have learned that the voltage-current graph of a material or a device that obeys Ohm's law is a straight line. Voltage-current graphs of devices made from semiconductor materials have distinctly non-ohmic behaviors.

2 Teach

Band Theory of Solids

student text pp. 778–779

Concept Development

MAIN IDEA As an analogy for the three types of solids—conductors, semiconductors, and insulators—set up three broad jump stations with masking tape on the floor. Mark the lines "valence band" and "conducting band." Choose a group of four students to be "electrons." Model each type with the "electrons" broad jumping or not between the gaps. Insulators have a large gap and full valence level, conductors have no gap and partially filled valence, and semiconductors have an in-between gap and filled valence level.

Identify Misconceptions

Conductivity and Electrons Students might think that materials with more electrons are better conductors. You can point out that sulfur contains 1.07 times as many electrons as copper per unit of mass, yet copper is a much better conductor.

Use Figure 1

Band Gap for Copper Ask students how this figure would look for a good conductor, such as copper. The broken vertical line would intersect where the bands overlap, and there would be no energy gap between the conduction band and the valence band. **OL** **Visual-Spatial**

Reinforcement

Conductivity Electrical conductivity and thermal conductivity in metals are both enhanced by free electrons, although the electron interactions influencing each type of conductivity are quite different. In general, metals are both the best electrical and the best thermal conductors. Silver has the highest electrical and thermal conductivity of all metals. Copper, the second best electrical conductor, also is a very good heat conductor. For metals at a given temperature, thermal and electrical conductivities form a fixed ratio. But because the mechanisms influencing each type of conduction differ, increasing the temperature of a metal increases the thermal conductivity but decreases the electrical conductivity.

Concept Development

Valence Electrons Ask students to elaborate on the relationship between valence electrons and conduction electrons. Those who have taken chemistry should remember that valence electrons occupy the highest energy levels of atoms and are the ones involved in all chemical reactions. Conduction electrons are valence electrons in solids that have gained sufficient energy to move freely between atoms. **OL** **Interpersonal**

Careers

Solid-State Physics The study of solid-state physics is a gateway to an amazing array of careers. On one end of the career spectrum are scientists who do basic research on condensed matter and thin films. These scientists study the growth of crystals, development of polycrystalline alloys, and deposition of metals. Important advances in electronic device technology often follow this basic research. On the other end of the spectrum are the engineers and technicians involved in the massive worldwide industries that design and manufacture these solid-state devices.

Identify Misconceptions

Conductivity and Valence Electrons Students might think that materials with more valence electrons will make better conductors. Point out that silver and copper are the best electrical conductors, and yet the atoms of those metals have only one valence electron each.

Conductors and Insulators

student text pp. 780–781

ADDITIONAL IN-CLASS EXAMPLE

Use with Example Problem 1.

Problem How many free electrons are there in 2.55 cm^3 of iron? Iron has two valence electrons.

Response

$$\text{free e}^- = \left(\frac{2.55\text{ cm}^3}{1}\right)\left(\frac{7.86\text{ g}}{1\text{ cm}^3}\right)\left(\frac{1\text{ mol}}{55.8\text{ g}}\right)\left(\frac{6.022\times10^{23}\text{ atoms}}{1\text{ mol}}\right)\left(\frac{2\text{ free e}^-}{1\text{ atom}}\right)$$

$$= 4.33\times10^{23}\text{ free e}^-$$

Content Background

Conductivity and Resistivity The quantity used to express the ratio of current density per unit of applied electric field is called conductivity. Its symbol is σ. Conductivity is a property of a material, and generally it will vary with temperature. Resistivity (ρ) is the reciprocal of conductivity. The range in resistivity between conductors and insulators represents more than 20 orders of magnitude. The resistance of a material in a particular geometric configuration is directly proportional to resistivity. A copper rod 1 cm in diameter and a billion times as long as the distance from Earth to the Sun would have less resistance than a piece of quartz of the same diameter that is only 0.1 mm thick.

Semiconductors

student text pp. 782–783

Quick DEMO

Photo Cell

Estimated Time 2 minutes

Materials cadmium sulfide photo cell, ohmmeter

Procedure A cadmium sulfide cell is a type of photoconductive detector in which incoming photon energy that exceeds the energy gap between the valence and conduction bands produces additional carriers and a drop in resistance. Some carriers are available at room temperature, even with zero light input, giving a dark resistance that is high but not infinite.

Cover the CdS cell window and measure its dark resistance. Typically, the dark resistance will be close to 500 kΩ. Expose a portion of the cell window to light and again measure its resistance. This resistance will be less than the dark resistance. Expose the entire cell window to light and again measure its resistance. This resistance will be close to 10 kΩ in bright light.

Discussion

Problem Some resistors are made of carbon. Carbon also is used to make electrodes and sliding contacts such as motor brushes. If carbon conducts well enough for these applications, why is diamond an insulator?

Response Diamond in its pure form is crystalline carbon. In the noncrystalline, amorphous form of the element, not all of the valence electrons are captured in covalent bonds. This allows amorphous carbon to conduct. The valence electrons in diamond, however, are tightly bound and not easily moved to the conduction band. Therefore, diamonds behave as insulators.

OL **Logical-Mathematical**

REAL-WORLD PHYSICS

Optical Lithography Optical lithography is the process that is now used to make most integrated circuits. This process has evolved from being unable to resolve features smaller than 10 μm wide to now being able to resolve features that are less than 100 nm wide. The characteristics of light dictate that shorter wavelengths allow finer resolutions. Hence, ultraviolet lithography provides greater circuit density than visible-light lithography, and this allows integrated circuits to be made smaller and less expensive. Current research is examining other approaches, such as electron-beam lithography and the use of excimer lasers and X-ray sources. Someday, circuit features smaller than 10 nm wide may be possible in high-volume production devices.

ADDITIONAL IN-CLASS EXAMPLE

Use with Example Problem 2.

Problem How many free electrons are there in a layer of pure silicon that is 1.03 mm × 15.9 mm × 3.33 mm at 0 K? How many are there at room temperature?

Response None at absolute zero because all the electrons are in the ground state. At room temperature,

$$(0.103\text{ cm})(1.59\text{ cm})(0.333\text{ cm})\left(\frac{1.45\times10^{22}\text{ free e}^-}{\text{cm}^3}\right)$$

$$= 7.91\times10^{20}\text{ free e}^-$$

Careers

The Semiconductor Industry Mathematicians, physicists, chemists, and engineers work as a team in the semiconductor industry. They collaborate on research, new applications, and improved manufacturing techniques. There is intense world competition in the semiconductor industry because it contributes a large share to the gross domestic product of many countries and because it is an enabling technology—that is, it drives development and efficiency in other sectors.

Doped Semiconductors

student text pp. 783–786

Reinforcement

Use of Dopants Electrical conductivity is a critical property of materials used to fabricate semiconductor devices. Intrinsic semiconductors are not very useful in many applications because their conductivities are not high enough. Through the addition of donor (*n*-type) or acceptor (*p*-type) atoms, called dopants, the conductivity of a semiconductor can be greatly increased. Ask students to explain why adding only a few dopant atoms per million semiconductor atoms greatly increases semiconductor conductivity. In the intrinsic semiconductor, only a few electrons are in the conduction band at any time, so a relatively small percent of dopant added to the material greatly increases the supply of conduction electrons. **OL**

Critical Thinking

Choosing Dopants Ask students why materials such as thallium (Group 13) and bismuth (Group 15) are not used to make *p*-type and *n*-type silicon. Thallium and bismuth atoms are much larger than silicon atoms, and they would therefore cause disruptions (flaws) in the crystalline structure if they were used as dopants. **OL**

Concept Development

Using the Periodic Table To help students understand why certain elements are used for donors and acceptors, you can ask them to investigate a periodic table and locate potential acceptor materials in Group 13 and potential donor materials in Group 15. Common acceptor materials for silicon or germanium are boron, gallium, and indium. Common donor materials are phosphorus and arsenic. Also make sure they note that silicon and germanium are Group 14 elements. **OL** **Visual-Spatial**

REAL-WORLD PHYSICS

Diamonds and Electronics Diamond semiconductors are being intensely studied. Diamond in its pure form is a nonconductive material. Boron-doped diamond thin films, however, exhibit characteristics similar to *p*-type silicon semiconductors. It is more difficult to construct *n*-type diamond semiconductors. Nonetheless, several prototype diamond transistors and light-emitting diodes have already been made. Silicon devices can't work at temperatures higher than 150°C, and they are limited by power dissipation. Diamond devices could work at much higher temperatures and at much greater power levels. Also, it might be possible to make diamond devices that are smaller than silicon devices.

ADDITIONAL IN-CLASS EXAMPLE

Use with Example Problem 3.

Problem A typical silicon substrate is doped with 1 0^{15} arsenic atoms per cubic centimeter when making *n*-type silicon. What is the ratio of doped carriers to thermal carriers at room temperature?

Response The ratio of doped carriers to thermal carriers

$$\frac{\left(\frac{1.0\times10^{15}\text{ doped e}^-}{\text{cm}^3}\right)}{\left(\frac{1.45\times10^{10}\text{ thermal e}^-}{\text{cm}^3}\right)}=6.9\times10^4\,\frac{\text{doped e}^-}{\text{thermal e}^-}$$

Problem What is the ratio of silicon atoms to arsenic atoms?

Response The ratio of silicon atoms to arsenic atoms is

$$\frac{\left(\frac{4.99\times10^{22}\text{ Si atoms}}{\text{cm}^3}\right)}{\left(\frac{1.0\times10^{15}\text{ As atoms}}{\text{cm}^3}\right)}=5.0\times10^7\,\frac{\text{Si atoms}}{\text{As atoms}}$$

Use the MiniLAB

In *Optoisolator,* students use an LED, a photo cell, and an ohmmeter to demonstrate that it is possible to transfer electric signals with almost infinite electrical isolation from one circuit to another.

PHYSICS PROJECT ACTIVITY

Ultrapure Silicon Have students research how ultrapure silicon is prepared for the manufacture of semiconductor devices. Chemical-vapor deposition is used to create a pure silicon layer on top of a silicon wafer that has been sawed from an ingot. That layer is then used to build the devices. Consequently, increasing the purity of silicon ingots has become less important in microelectronics manufacturing. Ask students to write an outline to prepare for a class presentation about semiconductor manufacturing. They should use the outline to organize any visuals they will use, such as a poster display or computer slide show. **AL** **Linguistic**

3 Assess

Assess the MAINIDEA

Reflection Bring in an inexpensive solar panel, place it under a desk lamp, and operate a tiny fan or other device. Have students write a short essay relating what they have learned about band theory and how it relates to the solar cell. Have a few students read their responses. If time permits have students research the challenges in increasing efficiencies of solar cells.

Check for Understanding

Potential Difference Ask students what would happen if an *n*-type chunk of silicon were joined to a *p*-type chunk and a voltmeter were connected across the joint. Because there would be no net charge in either chunk, the voltmeter would read zero. **OL**

Extension

Heat Sinks Remove the side panel from a desktop computer. Point out the heat sink on the processor chip. Ask students what purpose it serves. A heat sink provides a path of low thermal resistance. Electronic heat sinks increase the rate of thermal energy transfer from a device, such as a transistor or an integrated circuit, to the surrounding air. Because thermal conductivity increases with temperature, silicon computer chips make errors or stop working when they get too hot. **OL**

Reading Checks and Caption Checks

Caption Check, *student text p. 778*
about 3 eV

Reading Check, *student text p. 779*
Electrons in the valence band have lower energy and can form bonds. Electrons in the conduction band have higher energy and can move between atoms.

Reading Check, *student text p. 780*
The forbidden gap is too large.

Reading Check, *student text p. 782*
An increase in temperature decreases the conductivity of metals but increases the conductivity of semiconductors.

Caption Check, *student text p. 784*
an *n*-type semiconductor

Reading Check, *student text p. 786*
A thermistor's conductivity increases with increasing temperature. The thermistor is very sensitive to temperature changes.

Practice Problems
student text p. 781

1. 1.31×10^{23} free e^-/cm^3
2. 5.86×10^{22} free e^-/cm^3
3. 5.90×10^{22} free e^-/cm^3
4. 1.81×10^{23} free e^-/cm^3
5. 1.90×10^{26} free e^-/cm^3 free in the tip

Practice Problems
student text p. 783

6. 5.10×10^{-10} free e^-/atom
7. 2.63×10^{-13} free e^-/atom
8. 7.87×10^{-23} free e^-/atom
9. 1.09×10^{-11} free e^-/atom
10. 1.85×10^{-32} free e^-/atom; -173°C
11. 3.78×10^{-18} free e^-/atom; -73°C
12. 4.84×10^{7} free e^-/cm^3

Practice Problems
student text p. 785

13. 3×10^{-9} As atoms/Si atom
14. 2.59×10^{-6} As atoms/Ge atom
15. 38.4
16. 5.35×10^{-10}
17. 1.10×10^{4}
18. Germanium devices do not work well at such temperatures because the ratio of doped carriers to thermal carriers is small enough that temperature has too much influence on conductivity. Silicon is much better.

Section 1 Review
student text p. 786

19. insulator
20. insulator
21. insulator
22. Leave it.
23. An *n*-type semiconductor is doped with a material with five valence electrons. One of the electrons is a donor electron that can easily move into the conduction band. A *p*-type semiconductor is doped with a material with three valence electrons. This results in one missing binding electron, creating a hole into which electrons from the conduction band can easily drop.
24. An intrinsic one because all its conduction is from thermally freed electrons, whereas the doped semiconductor depends on the charges from the dopants, which depend little on temperature.
25. Even though rate of change for thermal carrier production is greater for silicon, at any given temperature silicon shows far fewer thermally liberated carriers.

1 Introduce

Bellringer

Transistors Connect a 1.5-V cell in series with a loudspeaker and a 390-Ω resistor. Make and break the circuit. The sound will be very soft, owing to the resistor and the resulting small current. Connect the positive terminal of the cell to one speaker lead and the other speaker lead to the collector of an NPN transistor. With the flat side of the transistor case facing you and the leads pointing down, the collector lead is at the right and the emitter lead at the left. Connect the emitter of the transistor to the negative terminal of the cell. No sound will be heard because the transistor is off. Connect the 390-Ω resistor from the base lead to the collector lead. The sound is relatively loud (when making and breaking the base connection) due to the current amplification of the transistor from the base circuit to the collector circuit. BL Auditory-Musical EL

Tie to Prior Knowledge

Resistors You can relate electronic components to what students have already learned about circuits by reviewing how rises and drops in electric potential add up in a series circuit. The collector resistor in a transistor amplifier drops part of the supply potential, and the other part is dropped across the collector-emitter circuit of the transistor.

2 Teach

The Birth of Solid-State Electronics and Diodes

student text pp. 787–790

Concept Development

MAINIDEA Draw the symbols for the diode and the transistor on the board and give brief summary of how they work. Let students go online to piece together a time line of images of how these devices have changed since the first one was built in 1947.

ADDITIONAL IN-CLASS EXAMPLE

Use with Example Problem 4.

Problem An LED, a 330-Ω resistor, and a 9.0-V battery are connected in series. What is the current, assuming that the drop across the LED is 2.2 V?

Response First, find the drop across the resistor: $V_R = 9.0\text{ V} - 2.2\text{ V} = 6.8\text{ V}$. Use Ohm's law to find the current:

$$I = \frac{V}{R} = \frac{6.8\text{ V}}{330\ \Omega} = 21\text{ mA}$$

Concept Development

- **Diode Symbols** The diode symbol uses a straight line to represent the cathode, or more negative side, in forward bias. Diode packages have a line or band at the cathode end. The anode, or more positive side in forward bias, is represented by an arrow.
- **Reverse Bias** Explain that reverse bias means the cathode is more positive than the anode of the diode. Under reverse bias, there is a tiny reverse saturation current through the diode that is independent of reverse voltage and is usually many orders of magnitude smaller than typical forward currents.
- **Avalanche Breakdown** Excessive reverse bias voltage applied to a diode will cause a sudden wild increase in reverse current. This condition is sometimes called avalanche breakdown. A device called a zener diode is designed to operate properly with a high reverse bias voltage.
- **Series Resistance** Diodes, including LEDs, must never be connected to a source of electric potential without a series resistance that limits current. Otherwise, runaway temperature increase will destroy the device.

PHYSICS CHALLENGE ACTIVITY

LED Flashlight LEDs are diodes that emit visible light. Have students propose a circuit design for a battery-powered LED flashlight. A typical white-light LED manufactured for this purpose will supply good lighting and long life at a current level of 25 mA and will have a forward-bias drop in electric potential of 3.5 V. Students should specify battery configuration, series resistance, and component power ratings for such a flashlight. An example series circuit containing 4, 1.5-V cells would provide a potential difference of 6.0 V. The potential difference across the series resistance would be 6.0 V − 3.5 V = 2.5 V. A 100.0-Ω series resistance would allow a 25-mA current. This resistor must be rated for at least $P = (0.025\ \text{A})^2(100.0\ \Omega) = 0.063\ \text{W}$.
AL **Visual-Spatial**

Use the PhysicsLAB

In *Diode Current and Potential Difference,* students compare the current-voltage characteristics of a diode, an LED, and a resistor.

Quick DEMO

LEDs

Purpose to demonstrate that a diode conducts only in one direction

Materials two 1.5-V cells, red LED, 100-Ω resistor

Procedure Put the cells in series to form a 3.0-V supply. Connect the red LED, forward biased, in series with the 100-Ω resistor and the supply. The LED will light. Reverse the supply polarity. There will be no light. Green and blue LEDs have a larger forward voltage drop (about 3.4 V versus 1.6 V for red ones). If a green or blue LED is available, use it to demonstrate that a diode will not conduct in either direction when the supply voltage is too low.

Applying Physics

Collimated Laser Beam Unlike a gas or ruby laser, a diode laser produces a beam that diverges significantly. This is usually corrected with a lens to produce a collimated beam. A typical beam width is 1 μm. Ask students to explain why a narrow beam width is important in a device such as a CD player or an optical fiber. In a CD player, the laser light reflects off pits on the surface of the compact disc. The pits typically have a spacing of 0.85 μm, and each pit has a width of about 0.5 μm. A 1-μm laser beam is narrow enough that it will reflect light from a single pit at a time. A typical optical fiber core has a diameter of 5–100 microns. Because the diode laser beam is narrower than this, the beam can be precisely directed into a single optical fiber.

Transistors

student text pp. 791–792

Use Figure 13 and Figure 14

When transistors are connected as amplifiers, the collector-base diode is always reverse biased and the base-emitter diode is usually forward biased. To convert **Figure 14** for use with a *pnp* transistor will require that both sources have reversed polarity.

Use the MiniLAB

In *Red Light,* students explore the electric properties of a diode.

Discussion

Problem Under what circumstances are vacuum tubes better than transistors?

Response There are several answers: (1) Vacuum tubes can operate at higher potential differences, making them capable of producing many kilowatts of power, and they are still used in high-powered radio, television, and microwave applications. (2) They can better tolerate surges in electric potential. (3) They can survive a large electromagnetic pulse. **OL**

DIFFERENTIATED INSTRUCTION

Struggling Students Amplification can be a difficult concept. Try making an analogy between amplification and some small action that controls something big. Power steering is a possibility. Because of the hydraulic amplifier in the car, the force that is required to turn the steering wheel is much less than the force that would be required to turn the front wheels of the car directly.
BL Logical-Mathematical

Critical Thinking

LEDs and White Light LEDs are monochromatic sources. Ask students how LEDs can be made to produce white light. One way is to combine the output of red, green, and blue LEDs. Another approach is based on conversion technology, in which blue InGaN chips are combined with a suitable fluorescent converter. OL

Identify Misconceptions

Collector Current Students sometimes don't understand why almost all of the current coming from the emitter lead ends up reaching the collector. Point out that the base is very thin and very lightly doped. Thus, the probability of a charge carrier finding its opposite is small.

DIFFERENTIATED INSTRUCTION

Visually Impaired Provide the student with a digital multimeter that has a diode test function with an audible beep. (Many digital multimeters offer this feature.) Have the student test various diodes, wires, resistors, and transistors. You can bend the resistor's leads into a U shape so that they can be differentiated from the diodes. Ask the student to draw conclusions about how semiconductor devices would affect the current and resistance of a circuit. BL Auditory-Musical

Reinforcement

Ohmic Behavior Ask students whether the collector circuit of a transistor obeys Ohm's law. It does not. The collector current is mostly determined by the base current and not by the collector voltage. OL

Integrated Circuits

student text pp. 792–793

Use Models

Circuit Simulation Computer simulation of electronic circuits is usually based on SPICE: *s*imulation *p*rogram with *i*ntegrated *c*ircuit *e*mphasis. SPICE models are available for diodes, transistors, and integrated circuits. SPICE simulation once required a mainframe computer, but now that it can be run on desktop computers, it has become a standard tool for anyone interested in modeling electronic circuits.

Concept Development

Moore's Law Gordon Moore made his famous observation in 1965, four years after the first planar integrated circuit was developed. Moore observed an exponential growth in the number of transistors per integrated circuit, and he predicted that this trend would continue. Moore's law, the doubling of the number of transistors per integrated circuit every couple of years, still seems to be in effect.

Use the PhysicsLAB

In *Computer Logic,* students investigate logic circuits and how computers make decisions.

3 Assess

Assess the MAINIDEA

Birth of the Transistor Bell Laboratories was the birthplace of the transistor. In an advertisement they claimed that, "Transistors promise smaller and cheaper electronic equipment and the spread of electronics where other equipment has not been able to do the job as economically." Discuss this claim and its validity today.

Check for Understanding

Transistor Construction Ask students how the construction of a transistor is similar to that of a diode. You can think of a transistor as two back-to-back diodes. What is the difference between *npn*-transistors and *pnp*-transistors? An *npn*-transistor has two *n* regions separated by a thin *p* region, while the *pnp*-transistor has two *p* regions separated by a thin *n* region. OL

Extension

MOS Transistors Interested students may want to research MOS (metallic oxide semiconductor) transistors, which are controlled by gate voltage rather than by base current. Ask students why they are efficient. They can operate at very low current, so they use very little power.
AL Linguistic

Reading Checks and Caption Checks

Caption Check, *student text p. 791*
The arrow shows the direction of conventional current.

Caption Check, *student text p. 791*
The circuit should show the *npn*-transistor replaced by a *pnp*-transistor, which has the emitter arrow reversed. The direction of conventional current should also be reversed.

Reading Check, *student text p. 792*
The conductivity of aluminum is greater.

Practice Problems

student text p. 789

26. 1.9 V

27. 6.0 V

28. 2.6 V

29. The anode of one connects to the cathode of the other and then the unconnected anode must be connected to the positive side of the circuit.

30. It would be impossible to obtain 2.5 mA of current with any reasonable power supply voltage because one of the diodes would be reverse biased.

Physics Challenge

student text p. 790

1. $I = \frac{\Delta V}{R} = \frac{1.8\ \text{V}}{180\ \Omega} = 1.0\times10^{1}\ \text{mA}$

2. $I = \frac{\Delta V}{R} = \frac{(1.8\ \text{V} - 0.70\ \text{V})}{180\ \Omega} = 6.1\ \text{mA}$

3. The straight line represents all the possible diode voltage drops for a diode, a 180-Ω resistor, and a 1.8-V battery (from a diode voltage drop of 0.0 V at 10.0 mA on one end to a 1.8-V drop at 0.0 mA on the other end). The solution is the intersection of both lines and occurs at $I = 6.3$ mA.

4. The first approximation error is $\frac{(10.0\ \text{mA} - 6.1\ \text{mA})}{6.1\ \text{mA}}$, or 64% (when the actual diode drop is 0.70 V).
This error decreases for greater battery voltages. For the second approximation, the error source is any deviation from 0.70 volts as the actual diode drop. It cannot be exactly determined, but it's much less than 64%. This error also decreases with higher battery voltages. For the third approximation, the error is due to graphic resolution and is not affected by the battery voltage.

Section 2 Review

student text p. 793

31. A diode will allow current in only one direction, and a transistor will amplify the current.

32. It conducts much better when forward-biased, so its resistance is much smaller when forward-biased than when reverse-biased.

33. The LED must be forward-biased, so the positive terminal must be connected to the *p*-end.

34. 120

35. 96 to 1

36. 2.0 V

37. **a.** Because the voltage across the diode is always 0.70 V, the voltage across the resistor increases by 1 V.

b. The current increases by $I = \frac{1\ \text{V}}{R}$.

38. No; the *p*-region of an *npn*-transistor must be thin enough to allow electrons to pass through the base to the collector.

A CLOSER **LOOK**

Robots in Space

Robotics

Purpose

This article discusses the history and future of uncrewed exploration of space.

Background

By one definition, a robot is a mechanical extension of a computer. Space probes fit this definition inasmuch as space probes are loaded with changeable software. Sometimes that software is changed mid-flight as problems or new opportunities arise. But as space probes take on more challenging missions, they must be able to change their own missions. They must become more intelligent.

Teaching Strategies

- The time delay between event and signal is a great challenge for scientists and engineers working with space probes. Practice this time delay by sending instructions to a "robot" (perhaps a team of students, each with a specific job) that must be followed exactly. Remember that there is a delay in each direction, from Mission Control to the robot and from the robot back to Earth.
- Artificial Intelligence (AI) is the ability of a machine to reason and learn functions normally associated with human intelligence. AI is desirable for creating autonomous robots in many fields. Have students investigate non-space exploration uses of AI and share their findings with the class.

GOING**FURTHER** >>>

Expected Outcomes Answers will vary. The distance to even the nearest stars makes the time lag a serious concern. A space probe in another star system must make all its choices on its own. These might involve locating sources of fuel, making choices about what to explore, and deciding when to take chances and when to avoid risks.

CHAPTER 29 ANSWERS

SECTION 1

Mastering Concepts

39. The energy levels of a single atom have discrete and unique values. The energy levels in a crystal have a small range around the values found in a single atom.

40. A higher temperature gives electrons additional energy, permitting more electrons to reach the conductive band.

41. positively charged holes

Mastering Problems

42. 2.54×10^{22} free e^-/cm^3

43. 3.22×10^{13} atoms/free e^-

44. $D < E < A = B = C$

SECTION 2

Mastering Concepts

45. Yes; one way you forward-bias the diode; the other way you reverse-bias it.

46. The arrowhead shows the direction of the conventional current for a forward-biased diode.

47. A forward-biased diode has a *p*-type and an *n*-type semiconductor layer, connected to wires on either end through metal caps. The *p*-type layer is connected to the positive terminal of the battery. New holes are created in the *p*-type layer, and those holes move toward the boundary between the two semiconductors. New electrons are added to the *n*-type layer, and those electrons move toward the boundary between the two semiconductors. As the holes and electrons combine, the circuit is completed and there is current. The current direction is from the *p*-type semiconductor to the *n*-type.

Mastering Problems

48. Answers will vary. A possible form of the correct answer would be, ". . . If a 0.70-V silicon diode is connected in series, what is the current through the resistor when the diode is forward-biased?"

49. 4.8 V

50. a. 2.0×10^{1} mA

b. 2.0×10^{1} mA

51. 160 Ω

52. a. 0; by inspection, the base circuit is off.

b. 0; by definition: when the base current is zero, so is the collector current

c. 15 V; with no current, the drop across the collector resistor is zero and the full 15-V drop across the transistor

53. a. 2.3×10^{-5} A

b. 5.1×10^{-3} A

c. 7.3 V

Applying Concepts

54. material C

55. material A

56. material B

57. one with an 8-eV gap

58. one with an 8-eV gap

59. circuit A: no; circuit B: no; circuit C: yes

60. L_1 is on; L_2 is off.

61. You make a resistor because there is no junction.

62. To have the three valence electrons needed, the dopant must be in group IIIA. B, Al, Ga

63. more like Si

64. The diode will have a higher resistance when it is reverse-biased.

65. higher potential, more positive

66. The graph should show a series of humps, all below the *x*-axis (negative).

Mixed Review

67. 1100 nm (in the near infrared)

68. 8.2 μA

69. 2.3 eV

70. 380 μA

71. a. 0.70 V; by inspection and the approximation that a silicon diode will drop 0.70 V when it is forward biased

b. 0 A; by inspection; 0.70 V is not enough to turn on two diodes in series.

c. 42 mA

Thinking Critically

72. 180 Ω, 160 Ω

73. 140 Ω

74. a. circuit A

b. circuit B

c. circuit C

d. Circuit A has the advantage of simplicity. It has the disadvantage of dropping 0.70 V, which can be important in low voltage circuits. Circuit B has the advantage of not wasting 0.70 V. It has the disadvantage of having to replace fuses. Circuit C has the advantage of always working, regardless of polarity. It has the disadvantage of wasting 1.4 V.

75. Answers will vary slightly, but a correct form of the answer is, "How many free electrons are there in 1 cm^3 of cadmium?"

Writing in Physics

76. Student answers will vary. Students could discuss his contributions to quantum mechanics, the Pauli matrices, his suggestion of a neutral, massless particle in beta decay (the neutrino), and his proof of the spin-statistics theorem.

77. Student answers will vary but should indicate that the Fermi energy is the energy of the highest occupied state at absolute zero. It applies when there are many electrons in a system, and thus, the electrons are forced into higher energy levels due to the Pauli exclusion principle.

Cumulative Review

78. 0.10 m

79. 5.0 V

Standardized Test Practice

student text p. 799

Multiple Choice

1. C
2. A
3. B
4. B
5. B
6. B
7. B
8. B

Free Response

9. $V_b = IR + V_d = (0.014\ \text{mA})(485\ \Omega) + 0.70\ \text{V} = 7.5\ \text{V}$

Rubric

The following rubric is a sample scoring device for free response questions.

Points	Description
4	The student demonstrates a thorough understanding of the physics involved. The response may contain minor flaws that do not detract from the demonstration of a thorough understanding.
3	The student demonstrates an understanding of the physics involved. The response is essentially correct and demonstrates an essential but less-than-thorough understanding of the physics.
2	The student demonstrates only a partial understanding of the physics involved. Although the student may have used the correct approach to a solution or may have provided a correct solution, the work lacks an essential understanding of the underlying physical concepts.
1	The student demonstrates a very limited understanding of the physics involved. The response is incomplete and exhibits many flaws.
0	The student provides a completely incorrect solution or no response at all.

CHAPTER 30

Nuclear and Particle Physics

About the Photo

Nuclear Fission Nuclear reactors such as this one harness the kinetic energy released in nuclear fission (as fast-moving neutrons) to heat water and turn it to steam. The result of the nuclear fission reaction is still radioactive but cannot sustain a chain reaction on its own.

Chapter Overview

Near the turn of the nineteenth century, certain substances were discovered to be unstable and to emit particles. Research led to an understanding of the nucleus, which is composed of protons and neutrons held together by the strong nuclear force. Three types of nuclear reactions are explored. Work on detectors and on high-energy accelerators has led to the Standard Model based on three types of elementary particles.

Before students study the material in this chapter, they should study:

- Conservation of energy
- Conservation of momentum
- Electric charge
- Electric fields
- Magnetic fields
- Electric Potential
- Structure of the atom
- Thermal energy

To solve the problems in this chapter, students will need a strong grasp of:

- Graphing data
- Scientific notation
- Significant figures

Introduce the BIGIDEA

Nuclear Energy Inform students that nuclear physics deals with a different class of energy from those they have studied so far—the energy of atomic nuclei.

Go online!
connected.mcgraw-hill.com

Your online experience includes:

- **Video**
- **Concepts In Motion**
- **Virtual Investigations**
- **All Lab, Applying Practices activities, and PBL worksheets**
- **Online Practice**
- **Personal Tutor**
- **Multilingual eGlossary**
- **Presentation**

Use the LaunchLAB

In *A Nuclear Model,* students will represent protons by cylindrical magnets on the sides of which are double-sided tape.

Use Physics T.V.

Video

SMOKE DETECTORS You probably have a smoke detector in your home. Many smoke detectors use radiation in order to function. Where does this radiation come from? And how does the radiation help the smoke detector work?

1 Introduce

Bellringer

Radioactive Substances Use a Geiger counter with a clearly audible clicking sound to demonstrate that radioactivity is present in some ordinary objects and substances. If possible, obtain a smoke detector containing americium, an old clock with a radium dial, a thorium-containing mantle for a camp lantern (which some manufacturers have phased out), an old piece of orange Fiesta pottery, and a rock containing uranium ore. Each of these items will cause the rate of clicking to increase as the Geiger tube is brought nearby.
BL Auditory-Musical EL

Tie to Prior Knowledge

Modern Physics Concepts Students will need to review the basic concepts of quantized charge, quantum theory, and the atomic theory of matter.

2 Teach

Nucleons, Isotopes, and The Strong Nuclear Force

student text pp. 802–804

Discussion

Question Why do X-ray emissions of varying wavelengths appear when electrons bombard target atoms of varying atomic numbers, as in Moseley's experiments?

Answer A beam of electrons that strikes a target can transfer energy to the most tightly bound electrons in the target's atoms, removing them from the atom—that is, ionizing the atom. The larger the nuclear charge (Z), the greater the binding energy of these electrons. Another electron in the atom quickly replaces the electron that was knocked out. In doing so it releases a photon of energy equal to the energy the electron lost. The photon is in the X-ray part of the spectrum. Thus the X-ray's wavelength is related to the nuclear charge of the atom. OL

Critical Thinking

X-ray Emissions from Materials Various techniques for probing substances involve scattering or emission of X-rays. One such process is known as X-ray fluorescence. This technique uses X-rays to excite electrons in lower energy levels. When the electrons return to their original state, they emit X-rays with wavelengths that depend on the energy levels of the atom. These X-rays can be used to identify elements (especially metallic ones) present in bulk samples, regardless of the chemical structure of the material. Ask students how this is possible. The X-ray fluorescence process excites lower-energy-level electrons whose energies are not affected by the chemical nature or structure of the material. Chemical bonds affect only the higher-energy (valence) electrons. OL
Logical-Mathematical

Reinforcement

Mass Number and Atomic Number Two numbers are required to specify an isotope, the atomic number and the mass number. The atomic number is equal to the number of protons. All atoms of an element have the same number of protons. Therefore, the chemical symbol and mass number are sufficient to fully specify the identity of an isotope. To make sure students understand the relationships, give them a few isotopes, specifying only the chemical symbol and mass number. For example, "What is the atomic number of the element and how many neutrons are in the nucleus of ^{137}Cs? ^{15}N? ^{143}Ce?" 55 and 82; 7 and 8; 58 and 85. Students may refer to a periodic table to determine the atomic number of the indicated elements. Indicating both mass number and atomic number for all isotopes is essential for keeping track of what is going on when writing nuclear reactions. OL

Identify Misconceptions

Isotopes and Natural Radioactivity Many people hold the false notion that radioactivity is always "unnatural." In fact, some isotopes found in nature are not stable. Several very common elements in nature do contain radioactive (unstable) isotopes. For example, the isotopes $^{39}_{19}K$, $^{40}_{19}K$, and $^{41}_{19}K$ occur in natural potassium with abundances of 93.08 percent, 0.01 percent, and 6.91 percent, respectively. The second of these is slightly radioactive, decaying to both stable $^{40}_{18}Ar$ and to stable $^{40}_{20}Ca$.

Content Background

Neutrons Rutherford and most other physicists of the time first thought that the nucleus consisted of protons and electrons. This hypothesis was supported by the emission of electrons from the nucleus in beta decay. Rutherford suggested in 1920 that the proton and electron were so tightly bound that they could be considered a single particle. This model was discarded for several reasons. Among them are that wave mechanics and the uncertainty principle show that an electron confined to a region the size of a proton has much greater kinetic energy than that exhibited by beta particles. Also, the angular and magnetic moments of nuclei cannot be explained with a proton-electron model.

Binding Energy, Mass Defect, and Nuclear Mass

student text pp. 805–807

Concept Development

$E = mc^2$ The theory of relativity shows that the energy (E) of an object *at rest* is related to the object's mass, m, by the equation $E = mc^2$, where c is the speed of light. This equation is actually a special case of the equation $E = mc^2 + E_{\text{kinetic}}$, where the right-most term is the object's kinetic energy (KE). When an object is at rest, then $E_{\text{kinetic}} = 0$ J, and the equation reduces to $E = mc^2$. It is common practice to call the term mc^2 the rest energy (E_0) such that the total energy (E) of the object is given by $E = E_0 + E_{\text{kinetic}}$, where $E_0 = mc^2$. One consequence of $E_0 = mc^2$ can perhaps be better appreciated if we rewrite the equation as $m = E_0/c^2$, which you can interpret as meaning that an object's mass is proportional to the object's rest energy. Example 1: A frying pan has a greater mass after it has been heated than before. Example 2: If you throw two pieces of putty toward each other so that they collide and stick together, the combined-putty's mass will be greater than the sum of the masses of the two original pieces. Why? The mass of each piece before the collision is proportional to its rest energy. In throwing the pieces toward each other, you are doing work on them, increasing their kinetic energies. During their collision, some of the kinetic energy is converted to thermal energy. The mass of the final object is proportional to its rest energy, which is greater than the sum of the rest energies of the original objects by an amount equal to the thermal energy added from the collision.

Activity

Investigate Binding Energy Have students set up a spreadsheet table with the following columns: Element Name, Chemical Symbol, Mass Number, Atomic Number, Number of Neutrons, Sum of Particle Masses, Atomic Mass, Mass Defect, and Binding Energy per Nucleon. They may obtain some data for these columns from the periodic table, but they will need to use a table of isotopes, or do additional research, to obtain some of the data. Students should create formulas to calculate Sum of Particle Masses, Mass Defect, and Binding Energy per Nucleon. Have them fill out about 10 rows of isotopes and see if they can reproduce points on the graph in **Figure 6.** AL Logical-Mathematical

PHYSICS PROJECT ACTIVITY

Four Fundamental Forces of Nature Have students prepare informational posters on each of the four fundamental forces. These are (in order of relative strength): the strong nuclear force, the electromagnetic force, the weak nuclear force, and the gravitational force. Individual students or small groups could take responsibility for research on each one. At least one student or small group should do research on the development of a unified field theory. They should include the unification of the electromagnetic and weak interactions into the electroweak interaction, Einstein's work with respect to gravity, and modern string theories. OL Linguistic

Content Development

MAIN IDEA Point out to students that both fusion of light elements and fission of heavy elements will release nuclear energy. Binding energy is often misinterpreted as the energy that must be added to form a more stable nucleus. The opposite is true. It is energy released as mass defect per nucleon increases and nuclei transform into more stable configurations. The most stable configurations are near $^{56}_{26}Fe$. OL Visual-Spatial EL

DIFFERENTIATED INSTRUCTION

Struggling Students Difficulty may arise for some students concerning mass and energy units. Use of certain non-SI units, such as the atomic mass unit (u) and the electron volt (eV) assumes mass-energy equivalence and can be very confusing. Emphasize that both mass and energy units can be used interchangeably. For example, the equivalence $1.0000 \text{ u} = 1.6605\times10^{-27} \text{ kg} = 931.49 \text{ MeV/c}^2 = 1.4924\times10^{-10}$ J is developed in the text. Ask students what the advantage is of expressing equivalent mass in u and energy in MeV. They are of a scale that is much more convenient in calculations. BL Logical-Mathematical

Discussion

Question Near the beginning of the periodic table, stable nuclei have approximately equal (though often not exactly the same) numbers of protons and neutrons. As atomic number increases, there tend to be more neutrons than protons in stable nuclei. Why?

Answer Protons repel each other, but neutrons do not. The energy required to add a proton increases substantially as *Z* increases. The energy required to add a neutron is not dependent on overcoming repulsive force, so more neutrons are added to a nucleus that is stable. In addition, adding more neutrons increases the space between protons and, thereby, decreases their repulsive forces. OL

ADDITIONAL IN-CLASS EXAMPLE

Use with Example Problem 1.

Problem Compare the mass defect and binding energy of deuterium, ${}^{2}_{1}H$, with that of helium, ${}^{4}_{2}He$. The mass of ${}^{2}_{1}H$ is 2.014102 u and the mass of ${}^{4}_{2}He$ is 4.002603 u.

Response The mass defect of ${}^{2}_{1}H$ is 0.0023880 u; the binding energy is −2.2244 MeV. The mass defect of ${}^{4}_{2}He$ is 0.030377 u; the binding energy is −28.296 MeV. There is a large change in binding energy from ${}^{2}_{1}H$ to ${}^{4}_{2}He$.

Concept Development

Comparing Binding Energy and Chemical Energy Challenge students to explain why chemical reactions do not ordinarily lead to nuclear reactions. For example, when hydrogen burns, the heat of combustion is −286 kJ/mol, which means that each mole of hydrogen burned releases 286 kJ. Ask students to calculate the energy in eV evolved per atom. 286 kJ/mol of atoms = 2.97 eV/atom. Have students compare this energy to the binding energy of hydrogen. The mass defect of hydrogen is 0.002388 u, so the binding energy is −2.224 MeV. The energy released by the combustion of hydrogen is much less than the binding energy of the hydrogen nucleus. AL Logical-Mathematical

3 Assess

Assess the MAIN IDEA

Isobars and Binding Energy Isotopes of different elements that have the same mass number are known as *isobars.* What is the difference in binding energy between the isobars ${}^{23}_{11}Na$ and ${}^{23}_{12}Mg$? How would you explain the difference in binding energy? The difference is 0.2101 MeV/nucleon (total = 4.838 MeV). Because ${}^{23}_{12}Mg$ has one more proton than ${}^{23}_{11}Na$, the overall repulsive force in the nucleus is greater, so less energy is required to break the Mg nucleus into its components.

Check for Understanding

Binding Energy The individual masses of two neutrons, two protons, and two electrons added together is 4.032980 u. But the mass of a helium atom consisting of these particles is measured to be 4.002603 u. Ask students where the extra mass went. The lost mass represents the work that would have to be done to overcome the attractive force while removing the nucleons from the nucleus. This is the nuclear binding energy of the atom. OL Logical-Mathematical

Extension

Table of Isotopes A table of isotopes provides information about all known isotopes of each element. It is an extended version of the periodic table. Ask students to use a table of isotopes to find the mass and percent abundance of the three naturally occurring isotopes of neon. ${}^{20}_{10}Ne$, 19.992440 amu, 90.48 percent; ${}^{21}_{10}Ne$, 20.9938467 amu, 0.27 percent; ${}^{22}_{10}Ne$, 21.991385 amu, 9.25 percent AL Logical-Mathematical

Reading Checks and Caption Checks

Reading Check, *student text p. 803*

Both have 6 protons. Carbon-12 has 6 neutrons, but carbon-14 has 8 neutrons.

Caption Check, *student text p. 803*

Hydrogen-1 has a mass number of 1 and a charge of +1. Hydrogen-2 has a mass number of 2 and a charge of +1. Hydrogen-3 has a mass number of 3 and a charge of +1. Helium-2 has a mass number of 3 and a charge of +2. Helium-4 has a mass number of 4 and a charge of +2.

Reading Check, *student text p. 805*

The equation $E = 931.49$ MeV/u gives the energy equivalent of one atomic mass unit. It simplifies calculations involving nucleons.

Caption Check, *student text p. 807*

The magnitude of the binding energy per nucleon of $^{131}_{52}I$ is larger.

Practice Problems

student text p. 804

1. 142, 143, and 146 neutrons
2. 120 neutrons
3. 7 neutrons
4. $^{1}_{1}H$, $^{2}_{1}H$, $^{3}_{1}H$

Practice Problems

student text p. 806

5. a. −0.098940 u
   b. −92.161 MeV
6. a. −0.002388 u
   b. −2.2244 MeV
7. a. 15.010109 u
   b. −106.18 MeV
8. a. −0.137005 u; −0.148410 u
   b. −127.62 MeV; −138.24 MeV
   c. The nucleus of $^{18}_{8}O$ has more nucleons than $^{16}_{8}O$ and therefore experiences a larger strong force. Thus, $^{18}_{8}O$ is more tightly bound and has a more negative binding energy. Both $^{18}_{8}O$ and $^{16}_{8}O$ have eight protons and experience the same repulsive electromagnetic force.

Section 1 Review

student text p. 807

9. a. −0.113196 u
   b. −105.44 MeV
10. The first pair has the same number of protons but a different number of nucleons. The second pair has the same number of nucleons but a different number of protons.
11. the tritium nucleus, because tritium emits a particle with mass and kinetic energy in its decay.
12. the tritium nucleus
13. The electric force is long range, so that all protons repel each other, even in large nuclei. The strong force is short range, so that only neighboring protons attract. The repulsive force grows, with increasing nuclear size, faster than the strong force.
14. 26, iron, because its binding energy is largest.

SECTION 2 Nuclear Decay and Reactions

student text pp. 808–810

1 Introduce

Bellringer

Radioactive Salt Obtain a Geiger counter and a container of potassium chloride. It can be purchased at the grocery store and is a very safe material often used as a salt substitute. You will need to be able to place the Geiger tube very close to about 100 g of the material. Turn on the counter and first hold the tube in the air in order to get the class used to the click rate. Now put the tube near the potassium chloride. Ask students the following questions: What has happened to the counting rate? What does this mean? The counting rate will increase. This proves that there are at least some unstable atoms in an ordinary sample of potassium chloride. BL Auditory-Musical EL

Tie to Prior Knowledge

Isotopes and Binding Energy This section builds on the concept of isotopes as it describes their transformations in nuclear reactions. Changes in nuclear binding energy explain the release of energy during fission and fusion reactions.

2 Teach

Radioactive Decay

student text pp. 808–811

Reinforcement

Alpha, Beta, and Gamma These are the names of the first three letters of the Greek alphabet and also the names of the three main types of nuclear radiation. Emission of an alpha particle radiation removes two protons and two neutrons from a nucleus, so the mass number decreases by four. An electron emitted from a nucleus is called a beta particle and removes one unit of negative charge, but the mass number does not change. This means a neutron changes into a proton and simultaneously emits an electron (beta particle). Gamma radiation is pure energy that is emitted as nuclear energy is transformed to radiant energy.

Use the PhysicsLAB

In *Common Sources of Radiation,* students will measure the nuclear activity of common substances.

Use the PhysicsLAB

In *Exploring Radiation,* students will learn how radiation levels vary with distance from a radioactive source.

Concept Development

MAIN IDEA In chemical reactions, the atoms within molecules are rearranged. Molecules change, but atoms do not. In nuclear reactions, protons and neutrons are rearranged within nuclei. In addition, neutrons and protons can change into one another. Atoms can thereby change into different elements.

DIFFERENTIATED INSTRUCTION

Visually Impaired and Hearing Impaired Radiation counting devices are good instruments for helping both visually impaired students and hearing impaired students study physics. Old-style Geiger counters usually provide audible clicks plus a small mechanical meter that indicates count rate. If you purchase a newer instrument, be sure to get one with both a speaker to produce audible clicks and a digital scale for visual output. A connection for an external speaker may be helpful as well, since such a speaker could provide a tactile response for a hearing impaired student.
BL Auditory-Musical EL

Quick DEMO

Stopping Radiation

Estimated Time 10 minutes

Materials sealed artificial radioactive sources (alpha and beta), Geiger counter, sheets of paper, aluminum foil, and lead

Procedure Place the Geiger tube close to the alpha source. Then, in sequence, move the paper, aluminum, and lead between the source and tube. Repeat for the beta source. Counts from the alpha source will slow markedly with all three barriers. Aluminum and lead will both slow the counts from the beta source.

Use the MiniLAB

Students will make use of the half-life concept when they analyze the results of a coin experiment, in *Modeling Radioactive Decay.* Starting with 50 coins, students will observe the "decay" of the coins and graph the number of coins as a function of coin half-lives.

ADDITIONAL IN-CLASS EXAMPLES

Use with Example Problem 2.

Problem Write the nuclear equation for the decay of radioactive $^{87}_{37}Rb$ to $^{87}_{38}Sr$ by the emission of a beta particle and an antineutrino.

Response $^{87}_{37}Rb \rightarrow ^{87}_{38}Sr + ^{0}_{-1}e + ^{0}_{0}\bar{\nu}$

Problem Write the nuclear equation for the decay of radioactive $^{222}_{86}Rn$ to $^{218}_{84}Po$ by the emission of an alpha particle.

Response $^{222}_{286}Rn \rightarrow ^{218}_{284}Po + ^{4}_{2}\alpha$

DIFFERENTIATED INSTRUCTION

Struggling Students Keeping track of details about the different types of radiation can be difficult for some students. To help, have them make their own tables or charts in which they summarize the names, symbols, characteristics, and accompanying changes in mass number and atomic number for the three common types of radioactive decay. Then ask them to use their reference charts to write several examples of equations for nuclear decay reactions.
BL **Linguistic** **Intrapersonal**

ADDITIONAL IN-CLASS EXAMPLE

Use with Example Problem 3.

Problem One type of neutron source used for research is known as a neutron howitzer. When beryllium target nuclei absorb energetic alpha particles from a plutonium source inside the neutron howitzer, they are transmuted into an excited form of $^{13}_{6}C$. When some of the excited $^{13}_{6}C$ atoms relax to a stable state, they transmute again through emission of a neutron. What new isotope is created?

Response $^{13}_{6}C \rightarrow ^{12}_{6}C + ^{1}_{0}n$ The new isotope is $^{12}_{6}C$.

Half-Life

student text pp. 812–813

Use Figure 9

Ask students to use **Figure 9** to compute the ratios of remaining undecayed nuclei for various time periods. First, have them use number of half-lives as the unit of time; then ask them to substitute a value for half-life from **Table 2** and redo the exercise. For example, by what ratio does the fraction of remaining undecayed nuclei decrease from 1.5 to 2.5 half-lives? $(3/16)/(3/8) = 1/2$ Now substitute the half-life of $^{14}_{6}C$ for number of half-lives. 1.5 half-lives is 8595 years; 2.5 half-lives is 14,325 years. The fraction by which remaining nuclei will decrease from 8595 years to 14,325 years is 1/2. **OL**
Visual-Spatial

Content Background

Radioactive Dating Shortly after the beginning of the twentieth century, geologists started to make careful measurements of the radioactive materials contained in rocks. Rutherford presented a lecture in 1904 suggesting that heat released inside Earth due to radioactive decay made age estimates accepted at that time far too small. Rutherford also proposed that the ages of uranium minerals could be determined by measuring decay products inside rocks. Today, geologists use a variety of decay systems beginning with natural radioactive isotopes, including $^{87}_{37}Rb$, $^{40}_{19}K$, and several uranium and thorium isotopes, to obtain more accurate ages for rocks.

Identify Misconceptions

Natural Radioactivity and Half-Life Students are often surprised to learn that some common elements have radioactive isotopes that exist in nature. They have long half-lives and low levels of radioactivity. The naturally occurring isotopes $^{87}_{37}Rb$, $^{40}_{19}K$, and $^{238}_{92}U$ have half-lives of 4.75×10^{10} years, 1.23×10^{9} years, and 4.51×10^{9} years, respectively. To exist in nature now, these isotopes must have very long half-lives because they have been around since the formation of the materials that make up Earth.

Applying Physics

Cobalt in Medicine The cobalt isotope $^{60}_{27}Co$ is used in medicine because it emits two strong gamma rays in the higher portion of the energy spectrum (1.33 MeV and 1.17 MeV). It is obtained by bombarding $^{59}_{27}Co$ with neutrons in a nuclear reactor. It is very dangerous to handle. A pellet 1 mm in diameter requires hundreds of kilograms of shielding to handle safely.

Using an Analogy

Half the Distance In football, penalties against a team that backs the ball up within a certain range of its own goal line are enforced as "half the distance to the goal." Ask students to imagine a series of such penalties, each one closing the distance to the goal by half. A population of radioactive nuclei decreases by half with each half-life in a similar manner. But the analogy is not completely perfect because nuclei come in discrete rather than continuous numbers. Ask students whether a series of penalties in football can drive a team to its own goal line (zero point). No; because the measured distance is continuous, there always will be a point halfway between the goal and any location in front of it. Ask whether a discrete population of unstable nuclei will ever decay to zero. Maybe; if a population of unstable nuclei is small enough, and the decay probability is high enough, decay of the last unstable nucleus in the population is possible. However, the number of nuclei is usually so large that this does not occur.
OL **Visual-Spatial**

Careers

Medical Physics and Health Physics Even though the names of these two fields sound similar, they represent different career paths. Medical physicists use the concepts and methods of physics to help diagnose and treat human disease. They use radioactive isotopes and medical imaging equipment to provide doctors with the information they need for diagnoses and treatments. On the other hand, health physicists specialize in the effects of and protection from radiation exposure of all kinds. They may work in nuclear power facilities, nuclear materials industries, or medical settings.

Fission and Fusion

student text pp. 814–817

Using Models

Chain Reaction Demo Obtain a dozen or more spring-type mousetraps, the same number of table-tennis balls plus one extra, and a box large enough to lay out the traps. If available, use an aquarium so that students are better able to see the action. Carefully set the traps and "bait" each one on the swing arm with a table-tennis ball. Dropping one table-tennis ball (representing a neutron) in the box will start off a model of a nuclear chain reaction. Using more traps will make the model more impressive.

Critical Thinking

Uranium Enrichment Nuclear reactors will run on a controlled chain reaction in uranium fuel that has its $^{235}_{92}U$ content boosted only slightly, perhaps 3 percent–5 percent from the 0.7 percent content in natural uranium. Nuclear weapons explode in an uncontrolled chain reaction. It is very difficult to manufacture weapons-grade uranium because it must be enriched to 90 percent $^{235}_{92}U$. Ask students why highly enriched uranium is needed to make a uranium-based nuclear bomb. Because the reaction is uncontrolled, it must happen very quickly. Too much $^{238}_{92}U$ present in the fuel would absorb too many slow neutrons without fission and prevent a runaway chain reaction.
OL **Logical-Mathematical**

Concept Development

Criticality Have students study the conditions required to sustain controlled and uncontrolled chain reactions. For example, a nuclear weapon based on highly enriched uranium requires that subcritical amounts of $^{235}_{92}U$ be brought together in order to start an uncontrolled chain reaction. Ask students to explain why. If a piece of $^{235}_{92}U$ is not large enough (if it is subcritical), too many neutrons leave the material before they can initiate fission in other atoms. When enough $^{235}_{92}U$ is brought together, the initial slow neutrons will produce more fission reactions and more neutrons before leaving the material and thus get the chain reaction started.
AL **Logical-Mathematical**

REAL-WORLD PHYSICS

Formation of the Elements The formation of elements heavier than hydrogen requires a nuclear-fusion process. Ask students where all the matter in the universe comes from. On a routine basis, all stars produce nuclear reactions in an extremely hot, explosive furnace where hydrogen fuses into helium. Nuclei as heavy as silicon are produced in stellar interiors. Near the end of the life of a massive star, the collapsing star will explode. Such an explosion is called a nova or supernova. Heavy elements up to the mass of uranium are formed in such explosions. **OL**

3 Assess

Assess the MAIN IDEA

Origin of Radium $^{226}_{88}Ra$ has a half-life of about 1.6×10^3 years. Our solar system is about 5 billion years old, but this radium isotope is still found in nature. How is this possible? Other elements with unstable isotopes and with a longer half-life might decay into $^{226}_{88}Ra$. In fact, this radioactive isotope is part of the decay series of $^{238}_{92}U$, which has a half-life of about four-and-a-half billion years.

Check for Understanding

Nuclear Versus Chemical Energy The ratio of the energy released by the nuclear fusion of one helium nucleus to that released by one molecule of the chemical explosive trinitrotoluene (TNT) is about 1.25×10^6. Ask students how many grams of TNT (molar mass 227 g/mol) would be required to equal the explosive energy of the fusion of 1.00 g of helium. (The molar mass of helium is 4.00 g/mol.) 1.00 g of He represents 0.250 mol. The same explosive energy would come from $(0.250\ \text{mol})(1.25\times10^6) = 3.13\times10^5$ mol of TNT. This is $(3.13\times10^5\ \text{mol})(227\ \text{g/mol}) = 7.09\times10^7$ g, or about 71,000 kg of TNT. To understand how much energy that is: One kg of TNT = 4.7 megajoules (million joules) of energy and one joule of energy can lift a small apple one meter. 4.7 million joules then could lift 4.7 million apples one meter. Multiply that by 71,000 and you could lift 3.3 billion apples one meter. **OL** **Logical-Mathematical**

Extension

Radon Naturally occurring isotopes of the heavy elements uranium and thorium are the parents of several radioactive decay series. The uranium isotope $^{238}_{92}U$ (half-life 4.51×10^9 years) decays through several steps to $^{222}_{86}Rn$ (half-life 3.82 days), which is gaseous at room temperature. Homes sometimes have problems with this radioactive gas entering and building up in the interior spaces where residents breathe it. Ask students how it is that a home can have a persistent problem with radon, even though its half-life is not even 4 days. Because radon is a "daughter" of uranium, which has a very long half-life, radon is always being produced in the subsurface near an affected home. Have students research the concept of radioactive equilibrium or secular equilibrium for more information.
AL **Visual-Spatial** **EL**

Reading Checks and Caption Checks

Reading Check, *student text p. 810*
Burning methane is a chemical reaction that rearranges atoms into different compounds, but the atoms involved in the reaction are unchanged. Alpha decay is a nuclear reaction because it involves changes in the nucleus of an atom.

Reading Check, *student text p. 814*
The energy released is (0.186 u)(931.49 MeV/u) =173 MeV.

Reading Check, *student text p. 815*
Each fission releases neutrons that can collide with other radioactive isotopes, causing a chain reaction.

Caption Check, *student text p. 815*
Rods that absorb free neutrons can be added to a reactor to stop or slow a chain reaction.

Reading Check, *student text p. 816*
The water slows the neutrons so that they can more easily react with nuclei and maintain the chain reaction. It also absorbs heat that can be used to heat other water to produce steam that turns a turbine. The rods can absorb neutrons to slow or stop a chain reaction. Together the water and rods control the chain reaction.

Caption Check, *student text p. 817*
Each of the two reactions starts with two hydrogen-1 isotopes, each of which has a +1 charge. The total initial charge, then, is +4. The ending charge is +2 from the helium-4 isotope and +1 from each of the two hydrogen-1 isotopes. The total ending charge, then, is +4.

Practice Problems
student text p. 811

15. ${}^{234}_{92}\text{U} \rightarrow {}^{230}_{90}\text{Th} + {}^{4}_{2}\text{He}$

16. ${}^{230}_{90}\text{Th} \rightarrow {}^{226}_{88}\text{Ra} + {}^{4}_{2}\text{He}$

17. ${}^{241}_{95}\text{Am} \rightarrow {}^{237}_{93}\text{Np} + {}^{4}_{2}\text{He}$

18. ${}^{226}_{88}\text{Ra} \rightarrow {}^{222}_{86}\text{Rn} + {}^{4}_{2}\text{He}$

19. ${}^{214}_{82}\text{Pb} \rightarrow {}^{214}_{83}\text{Bi} + {}^{0}_{-1}\text{e} + {}^{0}_{0}\bar{\nu}$

20. ${}^{14}_{6}\text{C} \rightarrow {}^{14}_{7}\text{N} + {}^{0}_{-1}\text{e} + {}^{0}_{0}\bar{\nu}$

Practice Problems
student text p. 812

21. a. ${}^{33}_{16}\text{S}$

b. ${}^{55}_{25}\text{Mn}$

22. ${}^{263}_{106}\text{Sg} \rightarrow {}^{259}_{104}\text{Rf} + {}^{4}_{2}\text{He}$

23. ${}^{15}_{7}\text{N} + {}^{1}_{1}\text{H} \rightarrow {}^{12}_{6}\text{C} + {}^{4}_{2}\text{He}$

24. a. ${}^{210}_{82}\text{Pb} \rightarrow {}^{210}_{83}\text{Bi} + {}^{0}_{-1}\text{e} + {}^{0}_{0}\bar{\nu}$

b. ${}^{210}_{83}\text{Bi} \rightarrow {}^{210}_{84}\text{Po} + {}^{0}_{-1}\text{e} + {}^{0}_{0}\bar{\nu}$

c. ${}^{234}_{90}\text{Th} \rightarrow {}^{234}_{91}\text{Pa} + {}^{0}_{-1}\text{e} + {}^{0}_{0}\bar{\nu}$

d. ${}^{239}_{93}\text{Np} \rightarrow {}^{239}_{94}\text{Pu} + {}^{0}_{-1}\text{e} + {}^{0}_{0}\bar{\nu}$

Practice Problems
student text p. 813

25. 0.25 g

26. 0.25 g

27. 5×10^{5} Bq

28. Six years is about half of tritium's half-life of 12.3 years. Thus, the brightness is $(1/2)^{1/2}$, or about 7/10 of the original brightness.

Section 2 Review
student text p. 817

29. In the nucleus, a neutron changes into a proton and emits an electron and an antineutrino.

30. ${}^{210}_{84}\text{Po} \rightarrow {}^{206}_{82}\text{Pb} + {}^{4}_{2}\text{He}$

31. From the graph, 3/8 of the original sample is left at about 1.4 half-lives. From the table the half-life is 8.07 days, so it would take about 11 days.

32. Lead is used as a radiation shield because it absorbs radiation, including neutrons. A moderator should only slow the neutrons so they can be absorbed by the fissile material.

33. The thermal energy of the helium nuclei is much too low to overcome the repulsive forces that would prevent fusion. Nuclei in the molecule must be moving very fast to undergo fusion.

34. 0.931 MeV

35. The plate being struck by the α particles has the more positive potential because the positively charged α particles move positive charge from the emitter plate to the struck plate.

student text pp. 818–821

1 Introduce

Bellringer

Deflecting Particles Use an old cathode ray tube (CRT) television or monitor and a magnet to demonstrate the ways in which particle beams are controlled in accelerators. Use an old black-and-white television, if available. A black-and-white television will revert back to normal, while a color television or monitor will not. If a black-and-white television is not available, use a color television, but be prepared to degauss (demagnetize) it. Hand-held degaussing units are available in electronics supply stores. **OL** **Visual-Spatial** **EL**

Tie to Prior Knowledge

Atomic and Nuclear Physics Have students review concepts of nuclear decay and parts of the atom, as well as electric fields and magnetic fields, velocity, acceleration, and conservation of energy and momentum.

2 Teach

Producing and Detecting Particles

student text pp. 818–821

Concept Development

Particle Energy Point out that the electron volt (eV) is defined to be equal to 1.6×10^{-19} J. The maximum energy a linear accelerator can impart is often given in GeV, where 1 GeV $= 10^9$ eV.

Circular Accelerators Linear accelerators and circular accelerators utilize similar accelerating methods; however, in circular machines, magnetic fields are used to bend the track of the particles into a circle. Circular accelerators are more compact.

Using an Analogy

Swings Particles in a synchrotron receive boosts from an alternating electric field at exactly the right times so that their energy progressively increases. This aspect of a synchrotron is analogous to a child being pushed on a swing at the resonant frequency. At this frequency, the pushes enhance the maximum acceleration of the gravitational field with each cycle. The result is a progressive increase in energy of the child-swing system.

Critical Thinking

Particle Sources The usual source of protons used in accelerators is hydrogen gas, which is ionized. Ask students what might constitute a convenient source of alpha particles. atoms of helium gas, which become alpha particles when doubly ionized **OL** **Logical-Mathematical**

REAL-WORLD PHYSICS

Proton Therapy The first circular accelerator, the cyclotron, was invented in 1929 by Ernest Lawrence, who later won a Nobel Prize for his invention. The cyclotron is still used today in proton therapy for the treatment of tumors. Like X-ray and cobalt-60 therapy, proton therapy uses an external beam to deliver ionizing radiation to diseased tissue inside the body. The proton beam can be targeted more precisely than with the other techniques because at a given speed the penetration of the protons varies very little. In addition, the proton beam is highly focused and can be shaped in three dimensions to match the tumor, which reduces damage to surrounding healthy tissue.

Reinforcement

Detection A basic strategy for particle detection is to arrange some way to collect charge produced when particles interact with matter. Ask students what the physical processes are that produce the audible click in the Geiger counter's speaker when a charged particle or gamma ray ionizes an atom in the attached gas-filled tube. The electric field accelerates the electron produced by the first molecule ionized. That electron then strikes and ionizes the next, eventually producing an avalanche. The charge produces a current pulse that is amplified and shaped, and eventually moves the voice coil in the speaker. **OL**

PHYSICS PROJECT ACTIVITY

Solid-State Detectors A solid-state detector uses a silicon or germanium crystal manufactured to function like a large diode. When a reverse voltage is applied to such a detector, gamma rays that enter it will generate electron-hole pairs that can be transmitted as current pulses to a pulse height analyzer. The analyzer counts pulses, sorts them by magnitude, and displays a spectrum of gamma-ray energy. Have students report on such a system. Students should explain why it is important to know the precise energies of gamma rays present in a sample. This allows identification of the isotopes that are the sources of the gamma rays.
OL **Logical-Mathematical**

Quick **DEMO**

Cloud Chamber

Estimated Time 10 minutes

Materials simple cloud chamber, alcohol, 2 pounds dry ice, alpha source on a needle

CAUTION: Avoid extended contact with a radioactive source. Handle with extreme caution.

Procedure Prepare a cloud chamber by soaking the cloth ring in alcohol. Place the radioactive needle into the side of the cloud chamber, and then place the chamber on a block of dry ice. After the chamber cools down, you should be able to observe the tracks of the radiation. Alpha particles will shoot outward from the needle, leaving momentary contrails behind in the alcohol vapor.

Some Aspects of Particle Theory

student text pp. 822–827

Identify Misconceptions

Beyond the Atom For most students, previous science classes will have emphasized the basic atomic particles: the proton, the neutron, and the electron. Since the 1940s, many subnuclear particles have been discovered.

Content Development

MAIN IDEA Students may have difficulty reconciling their concept of force with the concepts of interactions and mediating field particles. To help develop an intuition for interactions and force carriers, ask students how an electric charge "knows" that there is another charge nearby? When a photon is emitted by an electron, it changes the momentum of the electron. The photon could be reabsorbed by the electron as if nothing had happened, but if the photon is absorbed by a different nearby electron, the momentum of that electron is also changed, with the result that the two electrons move away from one another.

Content Background

Leptons and Hadrons Leptons are particles and antiparticles that do not respond to the strong nuclear force but do respond to the gravitational, electromagnetic, and weak forces. The electron, neutrino, and muon are leptons. Hadrons respond to the strong nuclear force and are subdivided into baryons and mesons. Examples of baryons are the proton and neutron. Examples of mesons are the pion and the kaon. Hadrons are of measurable size and are composed of combinations of quarks.

Discussion

Question Why don't protons and neutrons decompose into quarks under ordinary conditions?

Answer Quarks are held together within neutrons and protons by forces that increase with the separation of quarks like the force exerted by a spring. If enough energy is added, the "spring" effectively breaks, and a quark-antiquark pair is produced at the location of the break. OL

Concept Development

Mass of the Photon It does not appear to be possible to do an experiment to verify that photons, or other bosons, have zero rest mass. However, it is possible through experiment to place an upper limit on the photon mass. Experiments have shown this upper limit to be on the order of 10^{-54} kg.

Content Background

Matter and Antimatter Astrophysicists have discovered that the universe presently contains much more matter than antimatter, dispelling the notion that if there were an inherent symmetry between matter and antimatter at the Big Bang, mutual annihilation would proceed until only energy remained. This asymmetry, which may have begun as soon as 10^{-35} seconds after the beginning of the universe as we know it, was postulated by Russian physicist Andrei Sakharov to be the result of the violation of charge conjugation (C), the reversing of the electric charge and all the internal quantum numbers, and parity (P) invariance, the reversal of space coordinates, excluding time.

Use Figure 24

The figure suggests conditions under which particles may be created. Ask students whether it is possible for a single particle to materialize out of energy. No; a single particle cannot materialize out of radiant energy. However, if the available energy is greater than or equal to $2mc^2$ (in this formula, m is the mass of the particle), a particle and its corresponding antiparticle can be created. Because the particle and antiparticle have equal but opposite charges, all conservation laws are satisfied. OL

Reinforcement

Energy Conversion Ask students to compute the energy required for the production of a neutron-antineutron pair.

$$E = 2(1.67\times10^{-27}\ \text{kg})(3.00\times10^{8}\ \text{m/s})^2 = 3.01\times10^{-10}\ \text{J}$$

$$E = (3.01\times10^{-10}\ \text{J})(1\ \text{eV}/1.60\times10^{-19}\ \text{J}) = 1880\ \text{MeV}$$

OL **Logical-Mathematical**

Careers

Particle Physicists The goal of particle physics is to investigate and understand the constituents of matter and their interactions. There are three kinds of particle physicists—theorists, experimenters, and computational physicists. Colleagues in all three areas cooperate to develop and test theoretical ideas. For example, particle physicists have produced evidence in theories and experiments for the existence of quarks. Other areas of research that are of interest to particle physicists include theories about the formation process of the universe and the nature of the fundamental forces.

PHYSICS CHALLENGE ACTIVITY

Scintillation and Beta Detection Have students report on the methods used to detect beta particles. A method known as *liquid scintillation counting* allows for measuring beta emitters that are in liquid solution. Beta decay electrons emitted by an isotope in the solution excite a solvent molecule. Energy is transferred to the solute, or "fluor." A tiny "flash" of photons is detected and converted into an electrical signal by a photomultiplier. Ask students how a liquid scintillation system prevents noise from overwhelming the real counts. The photomultiplier greatly amplifies the tiny flash into a readable signal. Coincidence counting with two photomultipliers is used to reduce noise. AL Linguistic

Concept Development

Modes of Beta Decay There are three different modes of beta decay: electron (ß−) emission, electron capture, and positron (ß+) emission. Beta decay is commonly discussed as electron emission along with emission of an antineutrino after a neutron changes to a proton. But nuclei also can capture an orbital electron, changing a proton to a neutron. Positron emission from the nucleus also occurs, accompanied by the transformation of a proton into a neutron along with emission of a neutrino.

Use the MiniLAB

In *What's in the Background?,* students will use a Geiger tube to try to detect radiation.

Critical Thinking

Stopping Neutrinos Ask students if it is possible to construct a shield against all neutrinos. This would be very difficult because neutrinos can travel through matter with only a very small probability of interacting with a particle of matter. OL

REAL-WORLD PHYSICS

Neutrino Detection Physicists have built huge water-filled tanks in mine shafts and outfitted them with sensitive detectors in hopes of seeing evidence of neutrinos. One purpose of building such an instrument is to investigate a theoretical implication of unified field theories—that the proton should decay with a half-life on the order of 10^{31} years. No definitive results have come from these experiments. Neutrino detectors have observed and verified neutrino emissions from a distant supernova explosion.

3 Assess

Assess the MAINIDEA

Electrons and Leptons A proton is made up of three quarks, each with a fractional charge, held together by the strong force. What evidence can you present that electrons are not also made of quarks, but instead are truly elementary particles, like neutrinos? Particles composed of quarks interact by means of the strong nuclear force. If electrons were affected by the strong nuclear force, it would be difficult to explain how atoms could form as a small nucleus surrounded by a distant electron cloud.

Check for Understanding

Building Blocks Ask students if all known matter is composed of quarks and leptons. The answer is yes, except that bosons should probably also be included. Physicists refer to leptons and composites that contain quarks as the building blocks of matter. But the force carriers, photons, gluons, and the W and Z bosons can be observed as particles, so they too could be considered to be matter. OL Linguistic

Reteach

Particles and Forces The characteristics and relative magnitudes of the fundamental forces and interactions between subatomic particles are the basis of the universe as presented by physics. Have students make a list of these forces and the particle that carries each one, in order from strongest to weakest. strong, gluon; electromagnetic, photon; weak, W^-, W^+, and Z^0 bosons; gravitation, graviton AL Logical-Mathematical

Reading Checks and Caption Checks

Reading Check, *student text p. 819*

Electric fields are used to accelerate the particles, but the neutrons are neutral and cannot respond to the applied field.

Caption Check, *student text p. 820*

The paths of the charged particles are curved by a magnetic field. Because the particles have different charges and masses, their paths are bent in different directions.

Reading Check, *student text p. 821*

The word *neutrino* means "little neutral one" in Italian, and a neutrino is an almost-massless particle that has no charge.

Reading Check, *student text p. 823*

$\bar{u}d$

Reading Check, *student text p. 824*

about 1.88×10^{9} eV

Caption Check, *student text p. 825*

The W^- boson must have a –1 charge.

Practice Problems

student text p. 824

36. a. electron

b. antineutron

c. proton

d. antimuon

37. a. 1.50×10^{-10} J

b. 9.36×10^{8} eV

c. 1.87×10^{9} eV

38. 0.438 MeV

39. a. 939.56 MeV

b. 1879.1 MeV

40. 105.2 MeV

Physics Challenge

student text p. 826

1. ${}^{238}_{92}\text{U} \rightarrow {}^{234}_{90}\text{Th} + {}^{4}_{2}\text{He}$

${}^{234}_{90}\text{Th} \rightarrow {}^{234}_{91}\text{Pa} + {}^{0}_{-1}\text{e} + {}^{0}_{0}\bar{\nu}$

${}^{234}_{91}\text{Pa} \rightarrow {}^{234}_{92}\text{U} + {}^{0}_{-1}\text{e} + {}^{0}_{0}\bar{\nu}$

2. $A = 234$

Section 3 Review

student text p. 827

41. Answers include: The Standard Model has several parameters obtained only from experiment. The Higgs particle that would set the energy scale has not been found. It is not really a theory that explains, and so is not complete. Supersymmetry and String Theory are two possible replacements.

42. The proton and nucleus both have positive charge, so they repel each other. The proton must have enough kinetic energy to overcome the potential energy caused by the repulsion. The uncharged neutron does not experience this repulsive force.

43. down, into Earth

44. The electron/positron pair at the bottom has more kinetic energy, and therefore, a greater velocity.

45. Because a proton has one more u quark than a neutron, the equation would be ${}^{1}_{1}\text{p} \rightarrow {}^{1}_{0}\text{n} + {}^{0}_{+1}\text{e} + {}^{0}_{0}\nu$.

FRONTIERS IN **PHYSICS**

Wishing On A Star

Nuclear Fusion

Background

Solar energy, nuclear fusion, and nuclear fission are all related. With solar energy, we harvest the radiant energy from the Sun. In nuclear fusion, we recreate the Sun's energy source here on Earth. Nuclear fission extracts the energy present in atomic nuclei created, not in the Sun, but in ancient stars that exploded before Earth formed.

Teaching Strategies

What is the difference between a nuclear reactor and a nuclear bomb? A simple demonstration can show this. Give each member of the class two corks. These are neutrons. First have the students spread throughout the room. One student tosses a cork randomly. If it strikes another student, that student releases two corks, and so on. If the students are far apart, a chain reaction cannot occur—the fuel is subcritical. If the students are just the right distance apart, a slow, sustained chain reaction may occur—critical mass has been achieved, such as that in a nuclear reactor. If the students are very close together, a runaway chain reaction is sure to be the result—this is the super-critical reaction that occurs in a nuclear bomb. NOTE: encourage students to wear eye protection for this activity.

GOING**FURTHER** >>>

Expected Outcomes An important theme among most current energy experts is that there is no magic bullet to our energy future. A multifaceted approach that capitalizes on the advantages of each energy source is the one most favored today.

CHAPTER 30 ANSWERS

SECTION 1

Mastering Concepts

46. the repulsive electric force; the strong nuclear force

47. The mass defect is the difference between the sum of the masses of the individual particles of the nucleus and the mass of the nucleus. It is related to binding energy through $E = mc^2$.

48. Large nuclei are generally more unstable. The greater number of protons makes the repulsive electric force overcome the attractive strong force.

49. They both have the same number of protons.

Mastering Problems

50. 47 electrons, 47 protons, 62 neutrons

51. ${}^{64}_{30}\mathrm{Zn}$

52. a. −0.29177 u

b. −271.78 MeV

c. −8.4931 MeV/nucleon

53. a. −6.1556 MeV/nucleon

b. It requires more energy to remove a nucleon from ${}^{14}_{7}\mathrm{N}$.

54. 58 N

55. 4.00 u

SECTION 2

Mastering Concepts

56. Transmutation is the change of one element into another by means of a nuclear reaction. For example, U-238 decays into Th-234 and an α particle.

57. An α particle is a helium nucleus, a β particle is an electron, and γ radiation is a high-energy photon.

58. the atomic number, to conserve charge; the mass number, to conserve the number of nucleons

59. For a chain reaction to take place, an average of one or more neutrons must be produced by the fission of each nucleus and be absorbed by neighboring nuclei, leading to further fissions.

60. The moderator slows the fast neutrons, increasing their probability of being absorbed.

61. When a large atom undergoes fission, the mass of the products is less than that of the original nucleus. An amount of energy equivalent to the difference in mass is released. When small nuclei fuse into a larger nucleus, the mass of the more tightly bound, large nucleus is less, and the extra mass appears as energy.

Mastering Problems

62. ${}^{222}_{86}\mathrm{Rn} \rightarrow {}^{218}_{84}\mathrm{Po} + {}^{4}_{2}\mathrm{He}$

63. ${}^{89}_{36}\mathrm{Kr} \rightarrow {}^{89}_{37}\mathrm{Rb} + {}^{0}_{-1}\mathrm{e} + {}^{0}_{0}\overline{\nu}$

64. a. ${}^{225}_{89}\mathrm{Ac} \rightarrow {}^{4}_{2}\mathrm{He} + {}^{221}_{87}\mathrm{Fr}$

b. ${}^{227}_{88}\mathrm{Ra} \rightarrow {}^{0}_{-1}\mathrm{e} + {}^{227}_{89}\mathrm{Ac} + {}^{0}_{0}\overline{\nu}$

c. ${}^{65}_{29}\mathrm{Cu} + {}^{1}_{0}\mathrm{n} \rightarrow {}^{66}_{29}\mathrm{Cu} \rightarrow {}^{1}_{1}\mathrm{p} + {}^{65}_{28}\mathrm{Ni}$

d. ${}^{235}_{92}\mathrm{U} + {}^{1}_{0}\mathrm{n} \rightarrow {}^{96}_{40}\mathrm{Zr} + 3({}^{1}_{0}\mathrm{n}) + {}^{137}_{52}\mathrm{Te}$

e. ${}^{27}_{13}\mathrm{Al} + {}^{1}_{0}\mathrm{n} \rightarrow {}^{24}_{11}\mathrm{Na} + {}^{4}_{2}\mathrm{He}$

65. a. 25%

b. 13%

c. 6.3%

66. For the activity to fall to 1/8 its present amount, you must wait three half-lives, or 9 days.

67. a. carbon

b. ${}^{11}_{5}\mathrm{B} + {}^{1}_{1}\mathrm{p} \rightarrow {}^{11}_{6}\mathrm{C} + {}^{1}_{0}\mathrm{n}$

c. ${}^{11}_{6}\mathrm{C} \rightarrow {}^{11}_{5}\mathrm{B} + {}^{1}_{+0}\mathrm{e} + {}^{0}_{0}\overline{\nu}$

68. 1.2 kg

69. a neutron

70. about 500 years

71. 1/16

SECTION 3

Mastering Concepts

72. It accelerates particles using electrical force, and the neutron has no charge.

73. a. electromagnetic, weak, gravitational

b. strong, electromagnetic, gravitational

c. weak

74. The atomic number decreases by 1, and there is no change in the mass number. $Z \rightarrow Z - 1$; $A \rightarrow A$

75. It would annihilate with an equivalent amount of matter, producing an extremely large amount of energy.

Mastering Problems

76. +2 elementary charges

77. a. +1 elementary charge

b. −1 elementary charge

78. a. 2.1×10^{-5} s

b. 1.6×10^{5} revolutions

c. 3.4 s

d. 1.0×10^{9} m, or about 1 million km

79. a. 0

b. −1

80. A track would curve more if a particle had less mass, lower speed, or greater charge.

Applying Concepts

81. It is not true. Iron is the most tightly bound element, and thus the most stable nucleus. It cannot decay by fission.

82. The initial binding energy is less than the final binding energy and, therefore, the reaction is energetically possible.

83. A natural radioactive material is one that is found in nature and occurs normally in ores. Artificial radioactive isotopes are produced by bombarding naturally occurring materials with nuclear particles.

84. The water that passes through the core is at high pressure, so it doesn't boil. A second loop carries water at lower pressure, producing steam.

85. a. the fission of a uranium nucleus

b. the fusion of a kilogram of hydrogen

c. Although the fission of one uranium nucleus produces more energy than the fusion of four hydrogen nuclei to produce helium, there are more than 200 times as many hydrogen nuclei in a kilogram as there are uranium nuclei.

Mixed Review

86. a. ${}^{14}_{7}\mathrm{N} + {}^{4}_{2}\mathrm{He} \rightarrow {}^{18}_{9}\mathrm{F}$

b. ${}^{27}_{13}\mathrm{Al} + {}^{4}_{2}\mathrm{He} \rightarrow {}^{31}_{15}\mathrm{P}$

87. ${}^{1}_{0}\mathrm{n} + {}^{1}_{1}\mathrm{p} \rightarrow {}^{2}_{1}\mathrm{H}$; 1.7130 MeV

88. 0.2 MeV

89. Answers will vary. A possible form of the correct answer would be, "… If a particular isotope of radon contains 138 neutrons, what is the mass defect of this nucleus?"

Thinking Critically

90. Because the initial momentum is zero, this must be the final momentum. Thus, the two gamma rays must have equal and opposite momentums. The magnitude of the momentums is 2.72×10^{-22} kg•m/s. They move in opposite directions.

91. To conserve momentum, the three gamma rays must be in the same plane with angles of 120° between them.

92. 10^{38} reactions/s

93. about 4 min

94. B > D > A > C > E

95. Answers will vary, but a correct form of the answer is, "When a thorium nucleus containing 120 neutrons transmutes by alpha decay, what is the resulting nucleus?"

Writing in Physics

96. About 25 percent of the universe is dark matter. It is needed to explain galactic rotation and the expansion of the universe. According to one theory, dark matter is not made of ordinary matter that is covered by the Standard Model. It may interact with ordinary matter only through gravity and weak nuclear forces.

97. Theorists had predicted the existence of different flavors of quarks. They realized quarks occur in pairs. The bottom quark was found in 1977. The top quark was not found until 1995.

Cumulative Review

98. 5.0×10^{-3} T = 5.0 mT

99. a. 1.82×10^{3} m/s

b. 9.43×10^{-6} eV

100. 0.4 eV

101. a. 5.97 V

b. 43.6 Ω

Standardized Test Practice

student text p. 833

Multiple Choice

1. A
2. B
3. A
4. B
5. A
6. D
7. C
8. B
9. D

Free Response

10. $(20{,}000\ \text{T})\left(\frac{4\times10^{9}\ \text{J}}{\text{T}}\right)\left(\frac{1\ \text{nucleus}}{3.2\times10^{-11}\ \text{J}}\right) = 2\times10^{24}$ nuclei

Rubric

The following rubric is a sample scoring device for free response questions.

Points	Description
4	The student demonstrates a thorough understanding of the physics involved. The response may contain minor flaws that do not detract from the demonstration of a thorough understanding.
3	The student demonstrates an understanding of the physics involved. The response is essentially correct and demonstrates an essential but less than thorough understanding of the physics.
2	The student demonstrates only a partial understanding of the physics involved. Although the student may have used the correct approach to a solution or may have provided a correct solution, the work lacks an essential understanding of the underlying physical concepts.
1	The student demonstrates a very limited understanding of the physics involved. The response is incomplete and exhibits many flaws.
0	The student provides a completely incorrect solution or no response at all.

Math Handbook Answers

Practice Problems

1. a. 4 significant figures

b. 3 significant figures

c. 2 significant figures

d. 5 significant figures

e. 2 significant figures

f. 3 significant figures

2. a. 1400 m

b. 2.5 km

c. 0.003 m

d. 12.0 kg

3. a.

$$\begin{array}{r} 5.012 \text{ km} \\ 3.4 \quad \text{km} \\ +2.33 \quad \text{km} \\ \hline 10.7 \quad \text{km} \end{array}$$

b. 45 g − 8.3 g = 37 g

c. $3.40 \text{ cm} \times 7.125 \text{ cm} = 24.2 \text{ cm}^2$

d. (54 m)/(6.5 s) = 8.3 m/s

4. a. $\frac{1}{x} + \frac{y}{3} = \left(\frac{1}{x}\right)\left(\frac{3}{3}\right) + \left(\frac{y}{3}\right)\left(\frac{x}{x}\right)$

$= \frac{3}{3x} + \frac{xy}{3x} = \frac{3 + xy}{3x}$

b. $\frac{a}{2b} - \frac{3}{b} = \left(\frac{a}{2b}\right) - \left(\frac{3}{b}\right)\left(\frac{2}{2}\right)$

$= \frac{a}{2b} - \frac{6}{2b} = \frac{a - 6}{2b}$

c. $\left(\frac{3}{x}\right)\left(\frac{1}{y}\right) = \frac{3}{xy}$

d. $\left(\frac{2a}{5}\right) \div \left(\frac{1}{2}\right) = \left(\frac{2a}{5}\right)\left(\frac{2}{1}\right) = \frac{4a}{5}$

5. a. $(4)(3) = (2)(x)$

$x = \frac{12}{2} = 6$

b. $n = \frac{13}{15} \times 75 = 65$

c. $s = \frac{36}{12} \times 16 = 48$

d. $(2.5)(w) = (7.5)(5.0)$

$w = \frac{37.5}{2.5} = 15$

6. a. $\sqrt{22} = 4.69$

b. $\sqrt[3]{729} = 9.00$

c. $\sqrt{676} = 26.00$

d. $\sqrt[3]{46.656} = 3.60$

7. a. $\sqrt{16a^2b^4} = 4ab^2$

b. $\sqrt{9t^6} = 3t^3$

8. a. $\sqrt{n^3} = n^{\frac{3}{2}}$

b. $\frac{1}{\sqrt{a}} = \frac{1}{a^{\frac{1}{2}}} = a^{\frac{-1}{2}}$

9. a. $\frac{x^2t}{x^3} = \frac{t}{x} = x^{-1}t$

b. $\sqrt{t^3} = t^{\frac{3}{2}}$

c. $(d^2n)^2 = d^4n^2$

d. $x^2\sqrt{x} = x^{\left(2 + \frac{1}{2}\right)} = x^{\frac{5}{2}}$

10. $\frac{m}{q}\sqrt{\frac{2qv}{m}} = \sqrt{\frac{2qvm^2}{mq^2}}$

$= \sqrt{\frac{2vm}{q}}$

$= (2vm)^{\frac{1}{2}}\, q^{\frac{-1}{2}}$

11. a. 4.56×10^{8}

b. 2.0×10^{-5}

12. a. 0.000000303

b. 97,000,000,000

13. a.

$$(5.2\times10^{-4})(4.0\times10^{8}) = (5.2\times4.0)(10^{-4}\times10^{8})$$
$$= (21)(10^{-4+8}) = 21\times10^{4}$$
$$= 2.1\times10^{5}$$

b.

$$(2.4\times10^{3}) + (8.0\times10^{4}) = (0.24\times10^{4}) + (8.0\times10^{4})$$
$$= (0.24 + 8.0)(10^{4})$$
$$= 8.2\times10^{4}$$

14. a. $2 + 3x - 2 = 17 - 2$

$$\frac{3x}{3} = \frac{15}{3}$$
$$x = 5$$

b. $x - 4 + 4 = 2 + 4 - 3x$

$$x + 3x = 6 - 3x + 3x$$
$$\frac{4x}{4} = \frac{6}{4}$$
$$x = \frac{3}{2}$$

c. $(3)(t - 1) = (3)\left(\frac{x+4}{3}\right)$

$$3t - 3 = x + 4$$
$$x + 4 - 4 = 3t - 3 - 4$$
$$x = 3t - 7$$

d. $(c)(a) = (c)\left(\frac{b+x}{c}\right)$

$$ac - b = b - b + x$$
$$x = ac - b$$

e. $\left(\frac{2x+3}{x}\right)(x) = (6)(x)$

$$2x - 2x + 3 = 6x - 2x$$
$$4x = 3$$
$$x = \frac{3}{4}$$

f. $ax + bx + c - c = d - c$

$$(a + b)(x) = d - c$$
$$x = \left(\frac{d-c}{a+b}\right)$$

15. a. $4x^2 - 19 + 19 = 17 + 19$

$$\frac{4x^2}{4} = \frac{36}{4} = 9$$
$$x = \pm\sqrt{9} = \pm3$$

b. $12 + 9 - 3x^2 = -9 + 9$

$$21 - 3x^2 + 3x^2 = 0 + 3x^2$$
$$\frac{3x^2}{3} = \frac{21}{3}$$
$$x = \pm\sqrt{7} = \pm2.65$$

c. $x = \frac{-b \pm \sqrt{b^2 - 4ac}}{2a}$

$$= \frac{-(-2) \pm \sqrt{(-2)^2 - (4)(1)(-24)}}{2(1)}$$
$$= \frac{2 \pm \sqrt{4 + 96}}{2} = \frac{2 \pm \sqrt{100}}{2}$$
$$= \frac{2 \pm 10}{2} = 1 \pm 5$$

Thus,

$x = -4$ or $x = 6$

d. $x = \frac{-b \pm \sqrt{b^2 - 4ac}}{2a}$

$$= \frac{-(-14) \pm \sqrt{(-14)^2 - (4)(24)(-6)}}{2(24)}$$
$$= \frac{14 \pm \sqrt{196 + 576}}{48} = \frac{14 \pm \sqrt{772}}{48}$$
$$= \frac{14 \pm 27.8}{48}$$

Thus,

$x = -0.3$ or $x = 0.9$

16. $\Delta t = \frac{4.0\times10^2\ \text{m}}{16\ \text{m/s}}$

$= \left(\frac{4.0\times10^2}{16}\right)\left(\frac{\text{m}}{\text{m/s}}\right)$

$= 25\ \text{s}$

17. $v = a\Delta t = (-9.8\ \text{m/s}^2)(5.0\ \text{s} - 0.0\ \text{s})$

$= 49\ \text{m/s}$

18. $\left(\frac{32\ \text{cm}}{1\ \text{s}}\right)\left(\frac{60\ \text{s}}{1\ \text{min}}\right)\left(\frac{60\ \text{min}}{1\ \text{h}}\right)\left(\frac{1\ \text{m}}{100\ \text{cm}}\right)$

$= \left(\frac{32 \times 60 \times 60}{100}\right)\left(\frac{\text{m}}{\text{h}}\right)$

$= 1200\ \text{m/h}$

19. $v = \frac{d}{t} = \frac{100.0\ \text{m}}{9.87\ \text{s}} \times \frac{1\ \text{km}}{1000\ \text{m}} \times \frac{3600\ \text{s}}{1\ \text{h}}$

$= 36.5\ \text{km/h}$

20. **a.** $0.62 = \cos\theta$

$\theta = \cos^{-1}(0.62)$

$= 52°$

b. $0.13 = \cos\theta$

$\theta = \cos^{-1}(0.13)$

$= 83°$

c. $0.53 = \tan\theta$

$\theta = \tan^{-1}(0.53)$

$= 28°$

d. $0.84 = \sin\theta$

$\theta = \sin^{-1}(0.84)$

$= 57°$

21. $\log_3 81 = 4$

$81 = 3^4$

22. $10^{-3} = 0.001$

$\log_{10} 0.001 = -3$

23. $\log x = 3.125$

$10^{\log_{10} x} = 10^{3.125}$

$x \approx 1334$

Math Skill Handbook

Physical Constants			
Quantity	**Symbol**	**Value**	**Approximate Value**
Atomic mass unit	u	$1.660538782 \times 10^{-27}$ kg	1.66×10^{-27} kg
Avogadro's number	N_A	$6.02214179 \times 10^{23}$ mol^{-1}	6.022×10^{23} mol^{-1}
Boltzmann's constant	k	$1.3806504 \times 10^{-23}$ $Pa \cdot m^3/K$	1.38×10^{-23} $Pa \cdot m^3/K$
Coulomb's constant	K	$8.987551788 \times 10^{9}$ $N \cdot m^2/C^2$	9.0×10^{9} $N \cdot m^2/C^2$
Elementary charge	e	$1.60217653 \times 10^{-19}$ C	1.602×10^{-19} C
Gas constant	R	8.314472 $Pa \cdot m^3/mol \cdot K$	8.31 $Pa \cdot m^3/mol \cdot K$
Gravitational constant	G	6.67428×10^{-11} $N \cdot m^2/kg^2$	6.67×10^{-11} $N \cdot m^2/kg^2$
Mass of an electron	m_e	$9.10938215 \times 10^{-31}$ kg	9.11×10^{-31} kg
Mass of a proton	m_p	$1.672621637 \times 10^{-27}$ kg	1.67×10^{-27} kg
Mass of a neutron	m_n	$1.674927211 \times 10^{-27}$ kg	1.67×10^{-27} kg
Planck's constant	h	$6.62606896 \times 10^{-34}$ $J \cdot s$	6.63×10^{-34} $J \cdot s$
Speed of light in a vacuum	c	2.99792458×10^{8} m/s	3.00×10^{8} m/s

SI Prefix		
Prefix	**Symbol**	**Scientific Notation**
femto	f	10^{-15}
pico	p	10^{-12}
nano	n	10^{-9}
micro	μ	10^{-6}
milli	m	10^{-3}
centi	c	10^{-2}
deci	d	10^{-1}
deka	da	10^{1}
hector	h	10^{2}
kilo	k	10^{3}
mega	M	10^{6}
giga	G	10^{9}
tera	T	10^{12}
peta	P	10^{15}

Moments of Inertia for Various Objects			
Object	**Location of Axis**	**Diagram**	**Moment of Inertia**
Thin hoop of radius *r*	through central diameter	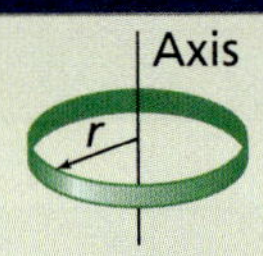	mr^2
Solid, uniform cylinder of radius *r*	through center	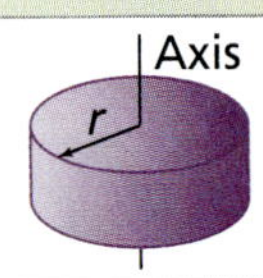	$\frac{1}{2}mr^2$
Uniform sphere of radius *r*	through center	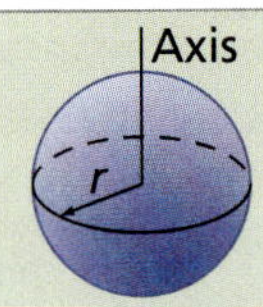	$\frac{2}{5}mr^2$
Long, uniform rod of length *l*	through center	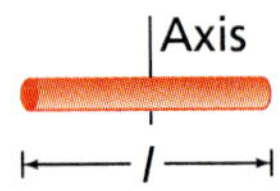	$\frac{1}{12}ml^2$
Long, uniform rod of length *l*	through end	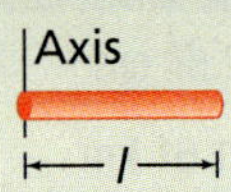	$\frac{1}{3}ml^2$
Thin, rectangular plate of length *l* and width *w*	through center	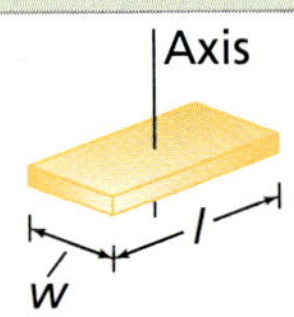	$\frac{1}{12}m(l^2 + w^2)$

Densities of Some Common Substances	
Substance	Density (g/cm^3)
Aluminum	2.70
Cadmium	8.65
Copper	8.92
Germanium	5.32
Gold	19.32
Hydrogen	8.99×10^{-5}
Indium	7.31
Iron	7.87
Lead	11.34
Mercury	13.534
Oxygen	1.429×10^{-3}
Silicon	2.33
Silver	10.5
Water (4°C)	1.000
Zinc	7.14

Melting and Boiling Points		
Substance	Melting Point (°C)	Boiling Point (°C)
Aluminum	660.32	2519
Copper	1084.62	2562
Germanium	938.25	2833
Gold	1064.18	2856
Indium	156.60	2072
Iron	1538	2861
Lead	327.5	1749
Silicon	1414	3265
Silver	961.78	2162
Water	0.000	100.000
Zinc	419.53	907

Reference Tables

Specific Heats			
Material	Specific Heat, C[J/(kg·K)]	Material	Specific Heat, C[J/(kg·K)]
Aluminum	897	Lead	130
Brass	376	Methanol	2450
Carbon	710	Silver	235
Copper	385	Water	4180
Glass	840	Water vapor	2020
Ice	2060	Zinc	388
Iron	450		

Heats of Fusion and Vaporization		
Material	Heat of Fusion, H_f (J/kg)	Heat of Vaporization, H_v (J/kg)
Copper	2.05×10^5	5.07×10^6
Gold	6.30×10^4	1.64×10^6
Iron	2.66×10^5	6.29×10^6
Lead	2.04×10^4	8.64×10^5
Mercury	1.15×10^4	2.72×10^5
Methanol	1.09×10^5	8.78×10^5
Silver	1.04×10^5	2.36×10^6
Water (solid)	3.34×10^5	2.26×10^6

Coefficients of Thermal Expansion at 20°C		
Material	Coefficient of Linear Expansion α (°C^{-1})	Coefficient of Volume Expansion β (°C^{-1})
Solids		
Aluminum	23×10^{-6}	69×10^{-6}
Brass	19×10^{-6}	57×10^{-6}
Concrete	12×10^{-6}	36×10^{-6}
Copper	17×10^{-6}	51×10^{-6}
Glass (soft)	9×10^{-6}	27×10^{-6}
Glass (ovenproof)	3×10^{-6}	9×10^{-6}
Iron, steel	12×10^{-6}	35×10^{-6}
Platinum	9×10^{-6}	27×10^{-6}
Liquids		
Gasoline		950×10^{-6}
Mercury		180×10^{-6}
Methanol		1200×10^{-6}
Water		210×10^{-6}
Gases		
Air (and most other gases)		3400×10^{-6}

Speed of Sound in Various Mediums	
Medium (0°C)	Speed (m/s)
Air (0°C)	331
Air (20°C)	343
Helium (0°C)	972
Hydrogen (27°C)	1310
Water (25°C)	1497
Seawater (25°C)	1533
Rubber	1600
Copper (25°C)	3560
Iron (25°C)	5130
Ovenproof glass	5640
Diamond	12,000

Wavelengths of Visible Light	
Color	Wavelength, λ (nm)
Violet	380–430
Indigo	430–450
Blue	450–500
Cyan	500–520
Green	520–565
Yellow	565–590
Orange	590–625
Red	625–740

Dielectric Constants, k (20°C)	
Vacuum	1.0000
Air (1 atm)	1.00059
Neon (1 atm)	1.00013
Glass	4–7
Quartz	4.3
Fused quartz	3.75
Water	80

Solar System Data								
	Mercury	Venus	Earth	Mars	Jupiter	Saturn	Uranus	Neptune
Mass (kg$\times 10^{24}$)	0.330	4.87	5.97	0.642	1899	569	86.8	102
Average radius (m$\times 10^{6}$)	2.44	6.05	6.38	3.40	71.5	60.3	25.6	24.8
Density (kg/m^3)	5427	5243	5515	3933	1326	687	1270	1638
Albedo	0.068	0.90	0.306	0.250	0.343	0.342	0.300	0.290
Average distance from the Sun (m$\times 10^{9}$)	57.91	108.2	149.6	227.9	778.4	1433.5	2872.5	4498.2
Orbital period (Earth days)	88.0	224.7	365.2	687.0	4332	10,759	30,685	60,189
Orbital inclination (degrees)	7.0	3.4	0.0	1.9	1.3	2.5	0.8	1.8
Orbital eccentricity	0.205	0.007	0.017	0.094	0.049	0.057	0.046	0.011
Rotational period (h)	1407.6	5832.5^R	23.9	24.6	9.9	10.7	17.2^R	16.1
Axial tilt (degrees)	0.03	177.4	23.4	25.2	3.1	26.7	97.8	28.3
Average surface temperature (K)	440	737	288	210	163	133	78	73
Gravitational field strength near the surface (N/kg)	3.7	8.9	9.8	3.7	20.9	10.4	8.4	10.7

R indicates retrograde motion.

The Moon	
Mass	0.073×10^{24} kg
Equatorial radius	1738 km
Mean density	3340 kg/m^3
Albedo	0.11
Average distance from Earth	384×10^{3} km
Orbital period	27.3 Earth days
Synodic period (lunar)	29.53 Earth days
Orbital inclination	5.1°
Orbital eccentricity	0.055
Rotational period	655.7 h
Gravitational field strength near the surface	1.6 N/kg

The Sun	
Mass	1.99×10^{30} kg
Equatorial radius	6.96×10^{8} m
Mean density	1408 kg/m^3
Absolute magnitude	+4.83
Luminosity	3.846×10^{26} J/s
Spectral type	G2 V
Rotational period (equatorial)	609.12 h
Mean energy production	0.1937×10^{-3} J/kg
Average surface temperature	5778 K

Explore **updates to the periodic table.**

Periodic Table

PERIODIC TABLE OF THE ELEMENTS

Key: Element — Hydrogen; Atomic number — 1; Symbol — H; Atomic mass — 1.008; State of matter

Gas · Liquid · Solid · Synthetic

Metal · Metalloid · Nonmetal · Recently observed

Period	1	2	3	4	5	6	7	8	9	10	11	12	13	14	15	16	17	18
1	Hydrogen 1 H 1.008																	Helium 2 He 4.003
2	Lithium 3 Li 6.941	Beryllium 4 Be 9.012											Boron 5 B 10.811	Carbon 6 C 12.011	Nitrogen 7 N 14.007	Oxygen 8 O 15.999	Fluorine 9 F 18.998	Neon 10 Ne 20.180
3	Sodium 11 Na 22.990	Magnesium 12 Mg 24.305											Aluminum 13 Al 26.982	Silicon 14 Si 28.086	Phosphorus 15 P 30.974	Sulfur 16 S 32.066	Chlorine 17 Cl 35.453	Argon 18 Ar 39.948
4	Potassium 19 K 39.098	Calcium 20 Ca 40.078	Scandium 21 Sc 44.956	Titanium 22 Ti 47.867	Vanadium 23 V 50.942	Chromium 24 Cr 51.996	Manganese 25 Mn 54.938	Iron 26 Fe 55.847	Cobalt 27 Co 58.933	Nickel 28 Ni 58.693	Copper 29 Cu 63.546	Zinc 30 Zn 65.39	Gallium 31 Ga 69.723	Germanium 32 Ge 72.61	Arsenic 33 As 74.922	Selenium 34 Se 78.96	Bromine 35 Br 79.904	Krypton 36 Kr 83.80
5	Rubidium 37 Rb 85.468	Strontium 38 Sr 87.62	Yttrium 39 Y 88.906	Zirconium 40 Zr 91.224	Niobium 41 Nb 92.906	Molybdenum 42 Mo 95.94	Technetium 43 Tc (98)	Ruthenium 44 Ru 101.07	Rhodium 45 Rh 102.906	Palladium 46 Pd 106.42	Silver 47 Ag 107.868	Cadmium 48 Cd 112.411	Indium 49 In 114.82	Tin 50 Sn 118.710	Antimony 51 Sb 121.757	Tellurium 52 Te 127.60	Iodine 53 I 126.904	Xenon 54 Xe 131.290
6	Cesium 55 Cs 132.905	Barium 56 Ba 137.327	Lanthanum 57 La 138.905	Hafnium 72 Hf 178.49	Tantalum 73 Ta 180.948	Tungsten 74 W 183.84	Rhenium 75 Re 186.207	Osmium 76 Os 190.23	Iridium 77 Ir 192.217	Platinum 78 Pt 195.08	Gold 79 Au 196.967	Mercury 80 Hg 200.59	Thallium 81 Tl 204.383	Lead 82 Pb 207.2	Bismuth 83 Bi 208.980	Polonium 84 Po 208.982	Astatine 85 At 209.987	Radon 86 Rn 222.018
7	Francium 87 Fr (223)	Radium 88 Ra (226)	Actinium 89 Ac (227)	Rutherfordium 104 Rf (261)	Dubnium 105 Db (262)	Seaborgium 106 Sg (266)	Bohrium 107 Bh (264)	Hassium 108 Hs (277)	Meitnerium 109 Mt (268)	Darmstadtium 110 Ds (281)	Roentgenium 111 Rg (272)	Copernicium 112 Cn (285)	Ununtrium * 113 Uut (284)	Flerovium 114 Fl (289)	Ununpentium * 115 Uup (288)	Livermorium 116 Lv (293)	Ununseptium * 117 Uus (294)	Ununoctium * 118 Uuo (294)

The number in parentheses is the mass number of the longest lived isotope for that element.

* The names and symbols for elements 113, 115, 117 and 118 are temporary. Final names will be approved by IUPAC (International Union of Pure and Applied Chemistry).

Series														
Lanthanide series	Cerium 58 Ce 140.115	Praseodymium 59 Pr 140.908	Neodymium 60 Nd 144.242	Promethium 61 Pm (145)	Samarium 62 Sm 150.36	Europium 63 Eu 151.965	Gadolinium 64 Gd 157.25	Terbium 65 Tb 158.925	Dysprosium 66 Dy 162.50	Holmium 67 Ho 164.930	Erbium 68 Er 167.259	Thulium 69 Tm 168.934	Ytterbium 70 Yb 173.04	Lutetium 71 Lu 174.967
Actinide series	Thorium 90 Th 232.038	Protactinium 91 Pa 231.036	Uranium 92 U 238.029	Neptunium 93 Np (237)	Plutonium 94 Pu (244)	Americium 95 Am (243)	Curium 96 Cm (247)	Berkelium 97 Bk (247)	Californium 98 Cf (251)	Einsteinium 99 Es (252)	Fermium 100 Fm (257)	Mendelevium 101 Md (258)	Nobelium 102 No (259)	Lawrencium 103 Lr (262)

The Elements

Element	Symbol	Atomic Number	Atomic Mass	Element	Symbol	Atomic Number	Atomic Mass
Actinium	Ac	89	(227)	Mendelevium	Md	101	(258)
Aluminum	Al	13	26.982	Mercury	Hg	80	200.59
Americium	Am	95	(243)	Molybdenum	Mo	42	95.96
Antimony	Sb	51	121.760	Neodymium	Nd	60	144.24
Argon	Ar	18	39.948	Neon	Ne	10	20.180
Arsenic	As	33	74.922	Neptunium	Np	93	(237)
Astatine	At	85	(210)	Nickel	Ni	28	58.693
Barium	Ba	56	137.327	Niobium	Nb	41	92.906
Berkelium	Bk	97	(247)	Nitrogen	N	7	14.007
Beryllium	Be	4	9.012	Nobelium	No	102	(259)
Bismuth	Bi	83	208.980	Osmium	Os	76	190.23
Bohrium	Bh	107	(272)	Oxygen	O	8	15.999
Boron	B	5	10.811	Palladium	Pd	46	106.42
Bromine	Br	35	79.904	Phosphorus	P	15	30.974
Cadmium	Cd	48	112.411	Platinum	Pt	78	195.078
Calcium	Ca	20	40.078	Plutonium	Pu	94	(244)
Californium	Cf	98	(251)	Polonium	Po	84	(209)
Carbon	C	6	12.011	Potassium	K	19	39.098
Cerium	Ce	58	140.116	Praseodymium	Pr	59	140.908
Cesium	Cs	55	132.905	Promethium	Pm	61	(145)
Chlorine	Cl	17	35.453	Protactinium	Pa	91	231.036
Chromium	Cr	24	51.996	Radium	Ra	88	(226)
Cobalt	Co	27	58.933	Radon	Rn	86	(222)
Copernicium	Cn	112	(285)	Rhenium	Re	75	186.207
Copper	Cu	29	63.546	Rhodium	Rh	45	102.906
Curium	Cm	96	(247)	Roentgenium	Rg	111	(280)
Darmstadtium	Ds	110	(281)	Rubidium	Rb	37	85.468
Dubnium	Db	105	(262)	Ruthenium	Ru	44	101.07
Dysprosium	Dy	66	162.500	Rutherfordium	Rf	104	(265)
Einsteinium	Es	99	(252)	Samarium	Sm	62	150.36
Erbium	Er	68	167.259	Scandium	Sc	21	44.956
Europium	Eu	63	151.964	Seaborgium	Sg	106	(271)
Fermium	Fm	100	(257)	Selenium	Se	34	78.96
Flerovium	Fl	114	(289)	Silicon	Si	14	28.086
Fluorine	F	9	18.998	Silver	Ag	47	107.868
Francium	Fr	87	(223)	Sodium	Na	11	22.990
Gadolinium	Gd	64	157.25	Strontium	Sr	38	87.62
Gallium	Ga	31	69.723	Sulfur	S	16	32.065
Germanium	Ge	32	72.63	Tantalum	Ta	73	180.948
Gold	Au	79	196.967	Technetium	Tc	43	(98)
Hafnium	Hf	72	178.49	Tellurium	Te	52	127.60
Hassium	Hs	108	(270)	Terbium	Tb	65	158.925
Helium	He	2	4.003	Thallium	Tl	81	204.383
Holmium	Ho	67	164.930	Thorium	Th	90	232.038
Hydrogen	H	1	1.008	Thulium	Tm	69	168.934
Indium	In	49	114.81	Tin	Sn	50	118.710
Iodine	I	53	126.904	Titanium	Ti	22	47.867
Iridium	Ir	77	192.217	Tungsten	W	74	183.84
Iron	Fe	26	55.847	Uranium	U	92	238.029
Krypton	Kr	36	83.798	Vanadium	V	23	50.942
Lanthanum	La	57	138.906	Xenon	Xe	54	131.293
Lawrencium	Lr	103	(262)	Ytterbium	Yb	70	173.04
Lead	Pb	82	207.2	Yttrium	Y	39	88.906
Lithium	Li	3	6.941	Zinc	Zn	30	65.38
Livermorium	Lv	116	(293)	Zirconium	Zr	40	91.224
Lutetium	Lu	71	174.967	Element 113*	Uut	113	(284)
Magnesium	Mg	12	24.305	Element 115*	Uup	115	(288)
Manganese	Mn	25	54.938	Element 117*	Uus	117	(294)
Meitnerium	Mt	109	(276)	Element 118*	Uuo	118	(294)

* Names have not yet been approved by IUPAC.

Safety Symbols	Hazard	Examples	Precaution	Remedy
Disposal	Special disposal procedures need to be followed.	Certain chemicals, living organisms	Do not dispose of these materials in the sink or trash can.	Dispose of wastes as directed by your teacher.
Biological	Organisms or other biological materials that might be harmful to humans	Bacteria, fungi, blood, unpreserved tissues, plant materials	Avoid skin contact with these materials. Wear mask and gloves.	Notify your teacher if you suspect contact with material. Wash hands thoroughly.
Extreme Temperature	Objects that can burn skin by being too cold or too hot	Boiling liquids, hot plates, dry ice, liquid nitrogen	Use proper protection when handling.	Go to your teacher for first aid.
Sharp Object	Use of tools or glassware that can easily puncture or slice skin	Razor blades, pins, scalpels, pointed tools, dissecting probes, broken glass	Practice common-sense behavior and follow guidelines for use of the tool.	Go to your teacher for first aid.
Fume	Possible danger to respiratory tract from fumes	Ammonia, acetone, nail polish remover, heated sulfur, moth balls	Be sure there is good ventilation. Never smell fumes directly. Wear a mask.	Leave foul area and notify your teacher immediately.
Electrical	Possible danger from electrical shock or burn	Improper grounding, liquid spills, short circuits, exposed wires	Double-check setup with teacher. Check condition of wires and apparatus.	Do not attempt to fix electrical problems. Notify your teacher immediately.
Irritant	Substances that can irritate the skin or mucous membranes of the respiratory tract	Pollen, moth balls, steel wool, fiberglass, potassium permanganate	Wear dust mask and gloves. Practice extra care when handling these materials.	Go to your teacher for first aid.
Chemical	Chemicals that can react with and destroy tissue and other materials	Bleaches such as hydrogen peroxide; acids such as sulfuric acid, hydrochloric acid; bases such as ammonia, sodium hydroxide	Wear goggles, gloves, and an apron	Immediately flush the affected area with water and notify your teacher.
Toxic	Substance may be poisonous if touched, inhaled, or swallowed.	Mercury, many metal compounds, iodine, poinsettia plant parts	Follow your teacher's instructions.	Always wash hands thoroughly after use. Go to your teacher for first aid.
Flammable	Flammable chemicals may be ignited by open flame, spark, or exposed heat.	Alcohol, kerosene, potassium permanganate	Avoid open flames and heat when using flammable chemicals.	Notify your teacher immediately. Use fire safety equipment if applicable.
Open Flame	Open flame in use, may cause fire.	Hair, clothing, paper, synthetic materials	Tie back hair and loose clothing. Follow teacher's instruction on lighting and extinguishing flames.	Notify your teacher immediately. Use fire safety equipment if applicable.

Eye Safety

Proper eye protection should be worn at all times by anyone performing or observing science activities.

Clothing Protection

This symbol appears when substances could stain or burn clothing.

Radioactivity

This symbol appears when radioactive materials are used.

Handwashing

After the lab, wash hands with soap and water before removing goggles.